T0204030

Networks of the Future

CHAPMAN & HALL/CRC
COMPUTER and INFORMATION SCIENCE SERIES

Series Editor: Sartaj Sahni

PUBLISHED TITLES

ADVERSARIAL REASONING: COMPUTATIONAL APPROACHES TO READING THE OPPONENT'S MIND
Alexander Kott and William M. McEneaney

COMPUTER-AIDED GRAPHING AND SIMULATION TOOLS FOR AUTOCAD USERS
P. A. Simionescu

COMPUTER SIMULATION: A FOUNDATIONAL APPROACH USING PYTHON
Yahya Esmail Osais

DELAUNAY MESH GENERATION
Siu-Wing Cheng, Tamal Krishna Dey, and Jonathan Richard Shewchuk

DISTRIBUTED SENSOR NETWORKS, SECOND EDITION
S. Sitharama Iyengar and Richard R. Brooks

DISTRIBUTED SYSTEMS: AN ALGORITHMIC APPROACH, SECOND EDITION
Sukumar Ghosh

ENERGY-AWARE MEMORY MANAGEMENT FOR EMBEDDED MULTIMEDIA SYSTEMS:
A COMPUTER-AIDED DESIGN APPROACH
Florin Balasa and Dhiraj K. Pradhan

ENERGY EFFICIENT HARDWARE-SOFTWARE CO-SYNTHESIS USING RECONFIGURABLE HARDWARE
Jingzhao Ou and Viktor K. Prasanna

EVOLUTIONARY MULTI-OBJECTIVE SYSTEM DESIGN: THEORY AND APPLICATIONS
Nadia Nedjah, Luiza De Macedo Mourelle, and Heitor Silverio Lopes

FROM ACTION SYSTEMS TO DISTRIBUTED SYSTEMS: THE REFINEMENT APPROACH
Luigia Petre and Emil Sekerinski

FROM INTERNET OF THINGS TO SMART CITIES: ENABLING TECHNOLOGIES
Hongjian Sun, Chao Wang, and Bashar I. Ahmad

FUNDAMENTALS OF NATURAL COMPUTING: BASIC CONCEPTS, ALGORITHMS, AND APPLICATIONS
Leandro Nunes de Castro

HANDBOOK OF ALGORITHMS FOR WIRELESS NETWORKING AND MOBILE COMPUTING
Azzedine Boukerche

PUBLISHED TITLES CONTINUED

Networks of the Future:
Architectures, Technologies, and Implementations

Edited by

Mahmoud Elkhodr
Western Sydney University, Australia

Qusay F. Hassan
Mansoura University, Egypt

Seyed Shahrestani
Western Sydney University, Australia

CRC Press is an imprint of the
Taylor & Francis Group, an **informa** business

A CHAPMAN & HALL BOOK

CRC Press
Taylor & Francis Group
6000 Broken Sound Parkway NW, Suite 300
Boca Raton, FL 33487-2742

First issued in paperback 2020

© 2018 by Taylor & Francis Group, LLC
CRC Press is an imprint of Taylor & Francis Group, an Informa business

No claim to original U.S. Government works

ISBN 13: 978-0-367-57288-4 (pbk)
ISBN 13: 978-1-4987-8397-2 (hbk)

Visit the Taylor & Francis Web site at
http://www.taylorandfrancis.com

and the CRC Press Web site at
http://www.crcpress.com

Contents

PART I   Cognitive Radio Networks

PART II   5G Technologies and Software-Defined Networks

PART III *Efficient Solutions for Future Heterogenous Networks*

PART IV *Big Data and the Internet of Things*

Contents

Preface

The Internet constitutes the largest heterogeneous network and infrastructure in existence. It is estimated that more than 3.4 billion people accessed the Internet in 2016. The number of mobile subscriptions has already exceeded the world population. The estimated 2.5 exabytes (2.5×10^{18} bytes) of global data exchanged per month in 2014 is forecast to grow at a staggering compound annual growth rate of 57% to reach 24.3 exabytes per month in 2019. This rapid explosion of data can be attributed to several factors, including the advances in wireless technologies and cellular systems, and the widespread adoption of smart devices, fueling the development of the Internet of Things (IoT).

With the ubiquitous diffusion of the IoT, cloud computing, 5G, and other evolved wireless technologies into our daily lives, the world will see the Internet of the future expanding and growing even more rapidly. Recent figures estimate that the number of connected devices to the Internet will rise to 50 billion by 2020. The IoT is a fast-growing heterogeneous network of connected sensors and actuators attached to a wide variety of everyday objects. Mobile and wireless technologies including traditional wireless local access networks (WLANs); low-and ultra-low-power technologies; and short-and long-range technologies will continue to drive the progress of communications and connectivity. The rapid growth of smart devices that connect to each other and to the Internet through cellular and wireless communication technologies forms the future of networking. Pervasive connectivity will use technologies such as 5G systems, cognitive radio (CR), software-defined networks, and cloud computing amongst many others. These technologies facilitate communication among the growing number of connected devices, leading to the generation of huge volumes of data. Processing and analysis of such "Big Data" bring about many opportunities, which, as usual, come with many challenges, such as those relating to efficient power consumption, security, privacy, management, and quality of service. This book is about the technologies, opportunities, and challenges that can drive and shape the networks of the future. We try to provide answers to fundamental and pressing research challenges including architectural shifts, concepts, mitigation solutions and techniques, and key technologies in the areas of networking.

The book consists of 22 chapters written by some established international researchers and experts in their field from various countries. These chapters went through multiple review cycles and were handpicked based on their quality, clarity, and the subjects we believe are of interest to the reader. It is divided into four parts. Part I consists of five chapters. It starts with a discussion on CR technologies as promising solutions for improving spectrum utilization to manage the ever-increasing traffic of wireless networks. This is followed by exploring the advances in CR spectrum sensing techniques and resource management methods.

Part II presents the latest developments and research in the areas of 5G technologies and software-defined networks (SDN). After highlighting some of the challenges that SDN faces, various opportunities and solutions that address them are discussed. It also discusses SDN security solutions for policy enforcement and verification, and explores the application of SDNs in the network intrusion detection context. This part of the book then moves to discuss and present solutions to the most pressing challenges facing the adoption of 5G technologies. In this direction, the new paradigm known as fog computing is examined in the context of 5G networks. A new re-authentication schema for multiradio access handover in 5G networks is also presented. This part then concludes with a chapter that compares and investigates the existing and developing 5G simulators, 5G test beds, projects, and other 5G-based federated platforms.

Part III is focused on efficient solutions for future heterogeneous networks. It consists of a collection of six chapters that discuss self-healing solutions, dealing with network virtualization, QoS in heterogeneous networks, and energy-efficient techniques for passive optical networks and wireless sensor networks.

The final part of this book covers the areas of IoT and Big Data. It consists of five chapters that discuss the latest developments and future perspectives of Big Data and the IoT paradigms. The first three chapters discuss topics such as data anonymization, which is presented as one of the pioneer solutions that can minimize privacy risks associated with Big Data. This part also includes a chapter that advocates employing a data location-aware application scheme to improve the performance of data transfer among clusters. Part IV of this book then ends with two chapters on IoT. The first chapter presents a survey on the current state of IoT security. The second chapter discusses the latest research on beamforming technologies in the IoT.

This book is intended for a broad audience. It is a collection of works that researchers and graduate students may find useful in exploring the latest trends in networking and communications. It can also be used as a resource for self-study by advanced students. The book can also be of value to cross-domain practicing researchers, professionals, and business stakeholders who may be interested in knowing about the future networking landscape.

Acknowledgments

We would like to thank everyone who participated in this project and made this book a reality. In particular, we would like to acknowledge the hard work of authors and their cooperation during the revisions of their chapters.

We would also like to acknowledge the valuable comments of the reviewers which have enabled us to select these chapters out of the so many chapters we received and also improve their quality.

Lastly, we would like to specially thank the editorial team at CRC Press/Taylor & Francis Group, particularly, Randi Cohen for her support throughout the entire process and Todd Perry who greatly managed the production of the book. We also thank Balasubramanian Shanmugam, project manager from DiacriTech, and his team for taking care of the copyediting process of this book.

Mahmoud Elkhodr
Qusay F. Hassan
Seyed Shahrestani

Reviewers

1. **Steven Gordon**, Senior Lecturer, Central Queensland University, Australia
2. **Yufeng Lin**, Associate Proferssor, Central Queensland University, Australia
3. **Ertuğrul Başar**, Assistant Professor, Istanbul Technical University, Turkey
4. **Elias Yaacoub**, Associate Professor, Arab Open University, Lebanon
5. **Chintan M. Bhatt**, Assistant Professor Chandubhai S. Patel Institute of Technology, India
6. **Mehregan Mahdevi**, Director and Lecturer, Victoria University, Australia
7. **Jahan Hassan**, Senior Lecturer, Royal Melbourne Institute of Technology, Australia
8. **Nabil Giweli**, Lecturer, Western Sydney University, Australia
9. **Farnaz Farid**, Lecturer, Western Sydney University, Australia
10. **Mohamed Al Zoobi**, Lecturer, Western Sydney University, Australia
11. **Belal Alsinglawi**, Western Sydney University, Australia
12. **Rachid Hamadi**, Lecturer, Royal Melbourne Institute of Technology, Australia
13. **Ahmed Dawood**, Lecturer, Western Sydney University, Australia
14. **Omar Mubin**, Senior Lecturer, Western Sydney University, Australia

Editors

Dr. Mahmoud Elkhodr completed his PhD degree in information and communication technologies at Western Sydney University (Western), Australia. Mahmoud was awarded the International Postgraduate Research Scholarship (IPRS) and Australian Postgraduate Award (APA) in 2012–2015. He was awarded the High Achieving Graduate Award twice, in 2010 and 2012. Mahmoud has authored several journal articles and book chapters and presented at prestigious conference venues. He is currently editing two books on the future of networking and 5G technologies to be published in 2017. His research interests include the Internet of Things, e-health, human–computer interactions, security, and privacy.

Dr. Qusay F. Hassan is an independent researcher and a technology evangelist with 15 years of professional experience in ICT. He is currently a systems analyst at the United States Agency for International Development in Cairo, Egypt, where he deals with large-scale and complex systems. Dr. Hassan received his BS, MS, and PhD from Mansoura University, Egypt, in computer science and information systems in 2003, 2008, and 2015, respectively. His research interests are varied, including IoT, SOA, high-performance computing, cloud computing, and grid computing. Dr. Hassan has authored and coauthored a number of journal and conference papers as well as book chapters. He recently published a book, *Applications in Next-Generation High Performance Computing* (published in 2016 by IGI Global), and he is currently editing two new books on the Internet of Things to be published in 2017. Dr. Hassan is an IEEE senior member and a member of the editorial board of a number of associations.

Dr. Seyed Shahrestani completed his PhD degree in electrical and information engineering at the University of Sydney, Australia. He joined Western Sydney University (Western), Australia, in 1999, where he is currently a senior lecturer. He is also the head of the Networking, Security, and Cloud Research (NSCR) group at Western. His main teaching and research interests include networking, management and security of networked systems, artificial intelligence applications, health ICT, IoT, and smart environments. He is also highly active in higher degree research training supervision, with successful results.

Contributors

Sharul Kamal Abdul Rahim
Wireless Communication Centre
Universiti Teknologi Malaysia
Johor, Malaysia

Mansaf Alam
Department of Computer Science
Jamia Millia Islamia
New Delhi, India

Mohammed Al-Zobbi
School of Computing, Engineering and
    Mathematics
Western Sydney University
New South Wales, Australia

Ertuğrul Başar
Faculty of Electrical and Electronics
    Engineering
Istanbul Technical University
Maslak, Istanbul

M. Majid Butt
CONNECT Centre
Trinity College Dublin
Dublin, Ireland

Rizwan Aslam Butt
Department of Electrical Engineering
University Technology Malaysia
Johor, Malaysia

Hon Cheung
School of Computing, Engineering and
    Mathematics
Western Sydney University
Sydney, Australia

Acácio F. P. P. Correia
Department of Computer Science
University of Beira Interior
Covilhã, Portugal

Himansu Das
Department of Computer Science &
    Engineering
Kalinga Institute of Industrial Technology
Bhubaneswar, India

Ahmed Dawoud
School of Computing, Engineering and
    Mathematics
Western Sydney University
Sydney, Australia

Alperen Eroğlu
Department of Computer Engineering
Middle East Technical University
Ankara, Turkey

Arman Farhang
CONNECT Centre
Trinity College Dublin
Dublin, Ireland

Farnaz Farid
School of Computing,
    Engineering and Mathematics
Western Sydney University
Sydney, Australia

Giancarlo Fortino
Department of Informatics, Modeling,
    Electronics and Systems
University of Calabria
Rende, Italy

Mário M. Freire
Department of Computer Science
University of Beira Interior
Covilhã, Portugal

Carlo Galiotto
CONNECT Centre
Trinity College Dublin
Dublin, Ireland

Nabil Giweli
School of Computing, Engineering and
    Mathematics
Western Sydney University
Sydney, Australia

Sufi Tabassum Gul
Department of Electrical Engineering
Pakistan Institute of Engineering and Applied
    Sciences (PIEAS)
Islamabad, Pakistan

K. Hemant Kumar Reddy
Department of Computer Science &
    Engineering
National Institute of Science and Technology
Berhampur, India

Sevia Mahdaliza Idrus
Department of Electrical Engineering
University Technology Malaysia
Johor, Malaysia

Pedro R. M. Inácio
Department of Computer Science
University of Beira Interior
Covilhã, Portugal

Toni Janevski
Faculty of Electrical Engineering and
    Information Technologies
Saints Cyril and Methodius University
Skopje, Republic of Macedonia

Suhanya Jayaprakasam
Wireless Systems Laboratory, Engineering
    Building
Hanyang University
Seoul, South Korea

Samiya Khan
Department of Computer Science
Jamia Millia Islamia
New Delhi, India

Adnan Kılıç
Department of Computer Engineering
Middle East Technical University
Ankara, Turkey

Stojan Kitanov
Faculty of Informatics
Mother Teresa University
Skopje, Republic of Macedonia

Amit Kumar
Department of Computer Science and
    Engineering
Indian Institute of Technology (Indian School
    of Mines)
Dhanbad, Jharkhand

Chee Yen Leow
Wireless Communication Centre
Universiti Teknologi Malaysia
Johor, Malaysia

Keqin Li
Department of Computer Science
State University ofNew York
New Paltz, New York

Mir Muhammad Lodro
Electrical Engineering Department
Sukkur IBA University
Sukkur, Sindh, Pakistan

Irene Macaluso
CONNECT Centre
Trinity College Dublin
Dublin, Ireland

Abdul Majid
Department of Computer & Information
    Sciences (DCIS)
Pakistan Institute of Engineering and Applied
    Sciences (PIEAS)
Islamabad, Pakistan

Daniel Malafaia
Department of Electronics,
    Telecommunications and Informatics
    University of Aveiro
Aveiro, Portugal

Nicola Marchetti
CONNECT Centre
Trinity College Dublin
Dublin, Ireland

Jasmina McMenamy
CONNECT Centre
Trinity College Dublin
Dublin, Ireland

Navera Karim Memon
Department of Telecommunication Engineering
Mehran University of Engineering & Technology
Jamshoro, Pakistan

Siti H. Mohammad
Faculty of Electrical Engineering
Universiti Teknologi Malaysia
Skudai, Malaysia

Shahram Mollahasani
Department of Computer Engineering
Middle East Technical University
Ankara, Turkey

Hari Om
Department of Computer Science and
    Engineering
Indian Institute of Technology (Indian School
    of Mines)
Dhanbad, Jharkhand

Asad Ullah Omer
Department of Electrical Engineering
Pakistan Institute of Engineering and Applied
    Sciences (PIEAS)
Islamabad, Pakistan

Ertan Onur
Department of Computer Engineering
Middle East Technical University
Ankara, Turkey

Sandeep Pirbhulal
Shenzhen Institutes of Advanced Technology
Chinese Academy of Sciences
Shenzhen, China

Arnidza Ramli
Faculty of Electrical Engineering
Universiti Teknologi Malaysia
Skudai, Malaysia

Diptendu Sinha Roy
Department of Computer Science &
    Engineering
National Institute of Science and Technology
Berhampur, India

Chun Ruan
School of Computing, Engineering and
    Mathematics
Western Sydney University
New South Wales, Australia

Musa G. Samaila
Department of Computer Science
University of Beira Interior
Covilhã, Portugal

João B. F. Sequeiros
Department of Computer Science
University of Beira Interior
Covilhã, Portugal

Madad Ali Shah
Electrical Engineering Department
Sukkur IBA University
Sukkur, Sindh, Pakistan

Kashish A. Shakil
Department of Computer Science
Jamia Millia Islamia
New Delhi, India

Ali Hassan Sodhro
Electrical Engineering Department
Sukkur IBA University
Sukkur, Sindh, Pakistan

Ana Tomé
Department of Electronics,
    Telecommunications and Informatics
University of Aveiro
Aveiro, Portugal

Fahim A. Umrani
Department of Telecommunication Engineering
Mehran University of Engineering &
    Technology
Jamshoro, Pakistan

José Vieira
Department of Electronics,
    Telecommunications and Informatics
University of Aveiro
Aveiro, Portugal

Ömer Yamaç
Department of Computer Engineering
Middle East Technical University
Ankara, Turkey

Qiang Yang
College of Electrical Engineering
Zhejiang University
Hangzhou, China

Hüsnü Yıldız
Department of Computer Engineering
Middle East Technical University
Ankara, Turkey

Nadiatulhuda Zulkifli
Department of Electrical Engineering
University Technology Malaysia
Johor, Malaysia

Part I

Cognitive Radio Networks

1 Cognitive Radio with Spectrum Sensing for Future Networks

Nabil Giweli, Seyed Shahrestani, and Hon Cheung

CONTENTS

1.1   INTRODUCTION AND BACKGROUND

Traditionally, the radio frequency spectrum (RFS) is divided into frequency bands and regulated in most countries by their governments. Each country has a spectrum management process for allocating the frequency bands to licensed users based on technical and economic aspects. Although each country is independently allocating its RFS, governments regulate their RFSs in compliance with international and regional standards. In general, RFS bands are statically allocated as licensed or unlicensed. Each licensed band is strictly assigned to a licensed user. On the other hand, an unlicensed RFS band can be accessed and shared freely by anyone within some transmission constraints. Therefore, unlicensed bands are in high demand as many wireless technologies and radio devices are designed to work in these free RFS bands. The industrial, scientific, and medical (ISM) bands are well-known examples of unlicensed RFS bands that include the frequency bands 902–928 MHz, 2400–2483.5 MHz, and 5725–5850 MHz. These bands are widely used, especially by wireless devices and systems based on Wi-Fi and Bluetooth communication technologies, under several specific regulations regarding operational requirements, such as power transmission and antenna gain (Cui and Weiss, 2013). As a result of regulation for the RFS bands, many licensed RFS bands are underutilized, in terms of frequency, time, and location, as found by several surveys on RFS occupancy conducted in different regions around the world (Barnes et al., 2013; Palaios et al., 2013; SiXing et al., 2012). In the near future, current RFS regulations may not be able to handle the rapidly growing usage of wireless communication technologies for various applications with highly expected increases in transmission data rate requirements. Therefore, the wireless research community and RFS regulatory organizations face the challenges of achieving high utilization of the overall RFS and need to overcome capacity scarcity in high-demand frequency bands. To overcome these challenges, there will most likely be innovative wireless technologies and revisions of the RFS regulations (Masonta et al., 2013).

From a technical perspective, cognitive radio (CR) technology is a promising solution for achieving significant RFS utilization and to open a new opportunity in the sharing of the unlicensed RFS. The concept of CR was introduced in 1999 by Joseph Mitola (Mitola and Maguire, 1999). Under the CR concept, a radio device, called the CR device, enables an unlicensed user, called the secondary user (SU) or CR user, to use a licensed RFS band without harming the operation of the licensed user, also called the primary user (PU). A licensed band is generally underutilized by the primary user, as mentioned before. This underutilization usually results in the licensed band having possible unoccupied periods, called holes or white spaces (WSs). These unoccupied spectrum holes may be present based on time, frequency, and space. By the use of CR technology, SUs are able to opportunistically share access to these holes of the licensed bands, in addition to unrestricted access to the unlicensed bands. Therefore, more bandwidth is available to the increasing number of mobile and wireless users and the increasing number of internetworking applications.

The remainder of this chapter is organized as follows: the CR concept and its main functions are explained in Section 1.2. Several challenges facing each of the CR functions are discussed in Section 1.3. Different types of sensing methods are briefly defined and evaluated in Section 1.4. Factors for selecting the proper sensing techniques are pointed out in Section 1.5. CR standardization efforts with sensing considerations are briefly reviewed in Section 1.6. Then, sensing function implementation issues in White-Fi networks are investigated in Section 1.7. The conclusion of this chapter along with some suggestions is presented in Section 1.8.

1.2   COGNITIVE RADIO TECHNOLOGY

In this section, CR technology is discussed in more detail, taking into account the challenges and research areas of this technology.

1.2.1 Cognitive Radio Definition

There are several proposed definitions for CR. The Federal Communications Commission (FCC) states a simple definition: "a radio that can change its transmitter parameters based on interaction with the environment in which it operates" (FCC, 2003).

Some other definitions provide more details about the features of the CR system. The following list is a summary of the main proposed features (Haykin, 2005; Lala et al., 2013; Steenkiste et al., 2009):

- *Awareness*: It is aware of its surrounding environment basically by a built-in sensing capability.
- *Intelligence*: It is fully programmable and consists of an intelligent wireless communication system that is capable of learning from and reasoning with the information gathered from its environment.
- *Adaptivity*: It is dynamically reconfiguring its operational parameters, such as transmitting power, carrier frequency, networking protocols, and modulation strategy, so that it can adapt to the variation of RFS conditions and application requirements in a real-time manner.

These features are designed to achieve the following main two objectives of the CR system (Lala et al., 2013):

- *Highly reliable communication*
- *Efficient utilization of RFS*

In the context of the previous two features, CR technology is projected to provide a universal radio platform that will be programmable for general-purpose applications for wireless systems (Steenkiste et al., 2009). This technology will become more feasible based on the progress of improvement in related technologies, noticeably in radio equipment and software, digital signal processing, machine learning, and wireless networking. Most of these technologies can fall under one of the two pillar notions of the CR concept: the software-defined radio (SDR) and dynamic spectrum access (DSA). The SDR concept was proposed by Joseph Mitola in the early 1990s (Mitola, 1995). Under the SDR concept, one hardware platform can be reconfigured on the fly to support a wide range of wireless communication standards, frequency bands, and technologies (Mao et al., 2013). In contrast, DSA is the technology exploiting the SDR capability to allow wireless systems to efficiently access the unused or less-congested portions of the RFS without interfering with the PU's operations (Kumar et al., 2011). The term *dynamic* in DSA refers to the paradigm shift of spectrum management from the static approach to the dynamic approach where a wireless system can opportunistically use different spectrum bands including licensed and unlicensed bands (Qing and Sadler, 2007). The CR system combines SDR and DSA capabilities to dynamically access the optimal spectrum band and reconfigure its operational parameters to effectively adapt to the current operating conditions (Akyildiz et al., 2008).

1.3 COGNITIVE RADIO FUNCTIONS AND CYCLE

To achieve the objectives and features of CR systems, several functions have to be executed in a logical way, called the CR cycle. The first prototype of the CR cycle was illustrated by Mitola (Mitola and Maguire, 1999). According to Mitola's CR cycle model, the main functions of a CR are spectrum sensing, spectrum management/decision, spectrum sharing, and spectrum mobility, as shown in Figure 1.1. Specifically, a CR system requires that a CR used by an SU be able to do the following:

1. Sense the surrounding spectrum to determine spectrum holes/WSs, and detect the PU's presence.

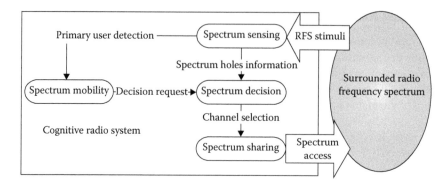

FIGURE 1.1   Basic CR cycle.

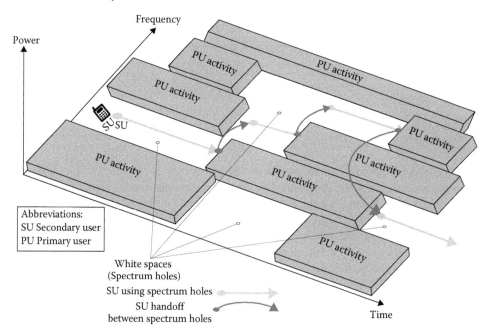

FIGURE 1.2   Handoff in CR. PU, primary user; SU, secondary user.

2. Analyze and decide which spectrum hole is the most suitable to satisfy the user/application requirements.
3. Share the spectrum with other SUs, if there are any, with the best possible fairness.
4. Seamlessly switch to another suitable spectrum hole to avoid harmful interference to the detected PU who is starting to use the spectrum band (Marinho and Monteiro, 2012).

The operation of these functions in a CR device will result in possible sequential spectrum handovers between spectrum holes during communication, as shown in Figure 1.2.

The handover has to be seamless and have little impact on the QoS of the application running on the CR (Lala et al., 2013). In the CR communication scenario, it is an essential requirement that the SUs will not cause any problems or QoS degradation to the PU's communications.

The CR technology relies on information which can be gathered about the surrounding RFS. Detecting the PU's presence (i.e., mainly determining whether the PU is using the spectrum band) is the major activity of a CR. However, even this basic detection requires sophisticated sensing technologies which are able to produce accurate outcomes that avoid false positive detection of the presence of the PU (where the detection gives false indication of the PU's presence) and avoid failing to detect some of the available spectrum holes (Bo et al., 2009). The nature of the electromagnetic signals makes accurate sensing a complicated operation. Specifically, the signal-to-noise ratio (SNR) required for detection, the multipath fading of the PU's signal, and the changing of noise/interference level with time and location are three main factors that may affect sensing accuracy (Molisch et al., 2009; Zeng et al., 2010).

The hidden PU problem is another issue, in which a CR cannot detect the presence of the PU because of signal problems caused by fading and shadowing of the signal while the PU is actually in the same transmission range as the CR (Yucek and Arslan, 2009). Implementing extremely flexible sensing functions is another challenge to cope with as the radio environment's characteristics and the various types of PU systems are changing rapidly (Cabric et al., 2004). Notably, any PU's QoS degradation caused by using CR technology will be an essential obstacle for practical applications of the CR solution.

Alternative solutions, instead of sensing, were proposed so that a CR device can determine the WS availability per location based partially (e.g., geolocation) or totally (e.g., senseless) on a central database service (Gurney et al., 2008; Murty et al., 2012). Although the geolocation database (GDB) approach may help in the more efficient determination of WSs, it raises several system and networking issues (Nekovee, 2010). For example, this approach will not be suitable for an uncentralized CR network. Also, it requires a CR device to constantly provide its location, which consumes power and is an error-prone operation (Murty et al., 2012). Concisely, the sensing operation is required in any CR system as the primary approach of determining the WSs. It is essential to understand the comparison between the leading, well-known spectrum-sensing methods to point out the main factors of selecting a suitable sensing method based on the current CR system operational conditions.

1.4 COGNITIVE RADIO CHALLENGES

Implementing CR functions faces several technical challenges which hinder the practical use of CR technology. In this section, some of the CR challenges are briefly described.

1.4.1 Spectrum Sensing and White Space Determination: The Challenges

The operation of a CR device relies on the information which can be gathered about the surrounding RFS. Hence, a spectrum-sensing function is very important as imperfect spectrum-sensing results increase error rates for both primary and secondary systems (Haddadi et al., 2013).

Detecting the PU's activities (i.e., determining whether the PU is using the spectrum band) is the major task required for successful CR operation. However, even this basic detection requires sophisticated sensing technologies which are able to produce an accurate outcome that avoids the false alarm problem of the presence of the PU (where the detection gives a false indication of the PU's presence) and avoids failing to detect some of the available spectrum holes (Bo et al., 2009). The nature of electromagnetic signals makes accurate sensing a complicated operation. Implementing an extremely flexible sensing function is another challenge to cope with due to the rapid change of the radio environment's characteristics and the various types of PU systems (Cabric et al., 2004). Exchanging the sensing information amongst CR devices on a large network is expected to increase sensing accuracy. How CR radios can cooperate and how much overhead is added by this cooperation are open questions for such improvement in sensing accuracy (Cabric et al. 2004).

1.4.2 The Impact of Sensing Operations on the QoS of Applications

In CR, a CR device is capable of observing some of the RF environment's characteristics through the sensing operation using one of the sensing methods. This sensing should be conducted periodically

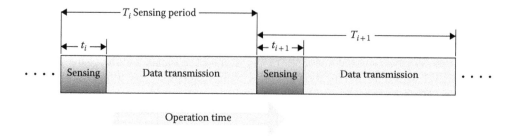

FIGURE 1.3   Simple structure of CR frames based on a sensing operation.

before sending a transmission frame to the wireless channel medium, as shown in Figure 1.3. During the sensing slot time t_i, where $i = 1, 2, 3, etc.$ representing the order of transmission frames, a CR device cannot transmit data because of hardware limitations. After the sensing time, the CR device can transmit data in the same channel or a new, vacant channel. The transmission channel is selected based on the decision made by the CR device in response to the sensing outcome (i.e., the presence or absence of the PU). The transmission starts after the sensing time and continues until the next sensing period, also called the CR frame. The transmission time $(T_i - t_i)$ depends on the sensing time, t_i, and the frame time, T_i, which is based on the design of how frequent the sensing will be conducted. Typically, the sensing operation should be limited and less frequent, as much as possible, without affecting sensing accuracy (Kae Won, 2010). The sensing time, t_i, and the frame time, T_i, can be designed to be fixed for all frames, i.e., all i values, or they could be designed to vary for each frame based on the design goals (Yiyang et al., 2009).

Increasing the sensing time, t_i, and conducting the sensing more frequently (i.e., decreasing T_i) increases the probability of correctly detecting the PU's presence. This, in turn, leads to more protection to the PU from interference by SUs and more utilization of the spectrum by all the users involved. On the other hand, this leads to QoS degradation for SUs, and the degradation can be measured by several parameters such as throughput, delay, and MAC layer process overhead (Xin-Lin et al., 2011). Thus the design of the sensing time and the frequency of sensing should account for the trade-off between protecting the PU's QoS and improving the QoS of SUs.

1.4.3   SPECTRUM MANAGEMENT CHALLENGES

The cognition concept in CR is reflected by the learning and intelligent capabilities that can be performed in the spectrum management function. The spectrum management function is heavily dependent on the information available about the spectrum by WS determination spectrum-sensing techniques, which determine the presence of WSs. As a consequence, all the challenges expected in WS determination will have an impact on the spectrum management decision. The QoS requirements of CR applications have to be considered in the selection criteria of the available spectrum holes (Lala et al., 2013). Spectrum management is the central function in CR architecture (Mitola, 2000). Therefore, it has to deal with complex and multilayer issues related to technical and regulatory policy requirements in real time to achieve spectral efficiency.

1.4.4   SPECTRUM MOBILITY CHALLENGES

In CR, an SU can share its allocated spectrum with a PU in two ways (i.e., overlay and underlay) (Akyildiz et al., 2008). In overlay spectrum sharing, the SU has to vacate the spectrum once the PU appears. In underlay spectrum sharing, the SU attempts to operate under an acceptable noise level for the PU during the PU's transmission. Therefore, the SU has to handoff from one spectrum band to another, whenever it is required, during its operation. Typically, the handoff process is required to be executed with seamless data transmission and QoS assurance (Zhongshan et al., 2013). A similar challenge has been addressed for vertical handover in fourth-generation (4G) heterogeneous wireless networks where several handoff algorithms are proposed (Yan et al., 2010). However, the fluctuating nature of spectrum resources in CR networks gives rise to several unique challenges that were not considered in these algorithms (Lee and Akyildiz, 2012). In CR networks, available spectrum holes may not be contiguous and could be found over a wide frequency range.

1.4.5 SPECTRUM SHARING CHALLENGES

Technically, in CR's basic concept, an SU has the same priority of using an available spectrum hole as any other user. Thus, a hole that has better characteristics is a more extensive target by a high number of competing SUs in the same location and time. There are two possible behaviors of SUs sharing the available spectrum (i.e., cooperative or noncooperative behaviours) (Mohammed et al., 2014). In a noncooperative scenario, the SUs are not willing to cooperate, and, in general, selfish behavior is expected. In such a scenario, it is a challenge for each competing SU to avoid degradation of its own performance caused by other competitors (Xinbing et al., 2010). Such a problem will not only affect the competing SUs, but also will possibly result in an inefficient use of the spectrum (Yuan and Tsang, 2009). In a cooperative scenario, the SUs are willing and capable of cooperating to achieve fair sharing of the available spectrum. However, conducting such cooperative spectrum sharing gives rise to the challenge of how to exchange information and control the signals required for cooperation among different SUs while minimizing the transmission and cooperation overhead (Masonta et al., 2013).

1.4.6 OTHER CR SYSTEM CHALLENGES

In addition to the previously mentioned challenges related to the four CR functions, several other challenges exist that relate to the upper layers of the communication protocol stack. The CR functions are mainly performed at the medium access control (MAC) and physical (PHY) layers. However, the protocols running at the application, transport, and network layers are also affected by CR functions. For instance, the routing protocol has an important role in CR ad hoc networks (CRAHNs), and a routing protocol proposed for typical ad hoc wireless networks has to be modified to adapt the dynamic spectrum behaviour of CRAHNs (Akyildiz et al., 2009; Cesana et al., 2011). Security issues are another important area of research in CR, and a recent survey of these issues is provided in Marinho et al., (2015). Providing solutions to CR challenges must account for real-time operational requirements, system scalability, and resource constraints and complexity (Marinho and Monteiro, 2012). Therefore, provisioning QoS in CR requires the consideration of the various factors and components that make it a complicated and challenging task.

1.5 SENSING METHODS

The main challenge facing sensing techniques is how to improve spectrum-sensing performance by mainly increasing the positive detection probability and decreasing the false detection probability. A sensing technique with higher positive detection probability provides more protection to the PU. An SU with a lower probability of false detection of the presence of the PU has a greater chance of using the available spectrum holes and therefore has a greater chance of achieving a higher throughput on the CR network. The design of a sensing technique is constrained by an acceptable level of false detection (Ying-Chang et al., 2008). In addition, improving sensing performance faces a range of trade-off challenges and various constraints, such as application requirements, hardware capability, complexity, and required infrastructure (Ghasemi and Sousa, 2008). Several sensing techniques have been proposed in literature to improve sensing performance. The proposed sensing methods can be classified mainly into three categories: methods with no prior information required, methods based on prior information, and methods based on SU cooperation.

1.5.1 NO PRIOR INFORMATION REQUIRED (BLIND SENSING)

No prior information about the PU's signal is necessary for the sensing methods under this category. Hence, blind sensing is also used to refer to this type of sensing method. However, prior information about the noise power of the targeted spectrum may be required for better performance. Otherwise, a reasonable estimation of the noise power is alternatively used. Two of the well-known methods under this category are energy detection and covariance-based detection.

1.5.1.1 Energy Detection

Also called radiometry or periodogram, energy detection is the most adopted method for spectrum sensing because of its low implementational complexity and computational overhead (Zeng et al., 2010). In this method, the energy detector is used to detect a narrowband spectrum and then the observed signal energy level is compared with a predefined threshold. The channel is occupied by the PU if the detected signal energy is greater than the threshold or unoccupied if it is lower. Because of this simplicity, this technique requires the shortest sensing duration, t, per frame compared to the other common sensing techniques (Kim and Shin, 2008). Generalizing the use of this method faces several challenges as a consequence of its simplicity. First, selection of the threshold used for detection is an issue when the noise level is unknown or uncertain over time (Tandra and Sahai, 2008). Second, under low a SNR, poor performance is expected, and it is difficult to differentiate between modulated signals, including signals of other SUs, noise, and interference (Zeng et al., 2010). Finally, an energy detector is ineffective to detect spread-spectrum signals (Cabric et al., 2004).

1.5.1.2 Covariance-Based Detection

This method is based on exploiting the covariance of the detected signal and compares it with the covariance of the noise where statistical covariance matrices of signal and noise are usually different (Zeng and Liang, 2007, 2009). The main improvement of this method is the ability to overcome the energy detection shortcoming to distinguish between signal and noise, particularly with low SNR and without any prior information about the PU's signal and channel noise. This goal is achieved at the expense of adding computational overhead as it requires computing the covariance matrix of the received signal samples (Rawat and Yan, 2011). In addition to complexity, other mentioned drawbacks of energy detection are still counted in covariance-based detection.

There is ongoing research to improve the performance of the blind-sensing approach, such as in Christopher Clement et al. (2015) and Li et al. (2012). However, the sensing methods that work with blind information about the PUs' signals have limited performance, particularly for spread spectrum and in situations where other SUs are sharing the same spectrum.

1.5.1.3 Based on Prior Information

Methods belonging to this category rely on partial or full information about the PU's transmission signal to differentiate it from other signals and noise.

1.5.1.4 Cyclostationary Feature Detection

This method is based on distinguishing the PU's signal from noise, interference, and other signals by identifying its cyclostationary features (Enserink and Cochran, 1994; Lundén et al., 2007). These cyclostationary features are associated with the signal modulation type, carrier frequency, and data rate. Hence, the CR node needs sufficient prior information about these unique features of the PU's signal so it can perform a cyclostationary analysis on the detected signal to identify matched features (Yucek and Arslan, 2009). For this method to perform better than energy detection, an adequate number of sample sets in the frequency domain are necessary. As a consequence, better performance accrues more complexity and sensing time at the expense of the available throughput (Yucek and Arslan, 2009).

1.5.1.5 Correlation Detection

Also called waveform-based sensing or coherent sensing, correlation detection involves the expected correlation/coherent between signal sampling being identified to detect the PU's signal based on previous knowledge about its waveform pattern (Yucek and Arslan, 2009). The accuracy of the sensing increases when the length of the known signal pattern of the PU is increased (Tang, 2005). The main drawback of this method is that it needs a large amount of information about the signal patterns of the PUs to achieve high performance, which is not practical for all CR systems.

1.5.1.6 Radio Identification–Based Sensing

This method is based on prior information about the transmission technologies used by the PU. Several features of the received signal are exploited and then classified to determine if the signal demonstrates the PU's signal technology (Yucek and Arslan, 2006). Fundamentally, feature extraction and classification techniques are used in the context of the European Transparent Ubiquitous Terminal (TRUST) project (Farnham et al., 2000). For detecting the signal features, the radio identification method may use one of the known sensing techniques, such as energy detection (Yucek and Arslan, 2009). Although radio identification improves the accuracy of energy detection, the complexity that is added and accuracy achieved will mainly depend on the sensing technique used.

1.5.1.7 Matched Filtering

Under this classification, this method is considered the best as it achieves a higher detection probability in a short detection time compared to other methods which are similarly based on prior information (Shobana et al., 2013a, 2013b). The sensed signal is passed through a filter which will amplify the possible PU's signal and attenuate the noise signals. The filter makes detection of the presence of the PU's signal more accurate (Shobana et al., 2013a). The filter, which is called a matched filter, has to be tuned based on some features of the PU's signal, such as required bandwidth, operating frequency, used modulation, and frame format (Yucek and Arslan, 2009). One of the disadvantages of this method is in implementation where different PU signal types require different dedicated hardware receivers. This disadvantage makes the method impractical to implement and also leads to higher power consumption if the method is implemented based on current hardware technologies. Figure 1.4 shows a comparison between noncooperative sensing methods

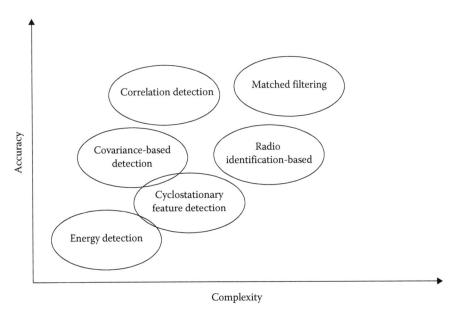

FIGURE 1.4 Estimated comparison between noncooperative sensing methods based on complexity and accuracy.

based on accuracy and complexity metrics. From this comparison, we can realize that none of these sensing methods can represent an optimal sensing with high accuracy and low complexity.

1.5.1.8   Based on SU Cooperation

The main principle of this approach is that SUs share their sensed information of the spectrum, and this individual sensing operation is called local sensing. The use of sensed information from all SUs helps in producing more accurate sensing outcomes than relying only on local sensing. Moreover, when the cooperating SUs are spatially distributed, it helps in overcoming the hidden PU problem and other limitations of local sensing (Chien-Min et al., 2014; Wang et al., 2014). Sensing cooperation can also reduce the local sensing cost (e.g., sensing time duration and energy consumption) while maintaining sensing quality by scheduling the sensing operation among cooperative SUs (Dongyue et al., 2014). The sensing method used by an individual CR user can be based on one of the sensing algorithms for local sensing, such as energy detection and cyclostationary feature detection (Akyildiz et al., 2011). Using the cooperation approach for sensing raises three main questions pointed out in Cabric et al. (2004): *(a) How much can be gained from cooperation? (b) How can cognitive radios cooperate? (c) What is the overhead associated with cooperation?*

In some environments, cooperative sensing may lose its advantages as far as an individual SU is concerned. For instance, increasing the local sensing frequency in individual high-mobility SUs is more efficient in terms of sensing accuracy and overhead than cooperating with other SUs (Cabric et al., 2004). In cooperative sensing, the improvement of sensing is more noticeable when the number of cooperative SUs is increased. However, more SUs being involved in cooperative sensing generally increases the cooperation overhead in terms of the amount of data exchanged and the time required for the exchange (Yulong at al., 2012). Hence, optimal sensing performance should be achieved under the trade-off between the local sensing overhead and the cooperative sensing overhead with a minimized total sensing overhead.

1.6   FACTORS AFFECTING THE SELECTION OF A PROPER SENSING METHOD

From reviewing the main proposed sensing method in the previous section, it is perceived that there is no one sensing method that can be used efficiently in all possible CR operational conditions. Hence, the sensing method that provides optimal performance should be selected based on the given operational conditions. Selecting the proper sensing method for a certain CR operational condition depends on various factors. In this section, the notable factors are listed and discussed in the following subsections.

1.6.1   CR Device Capability

A CR device designed with limited hardware resources and power capacities will not be able to support a wider range of sensing methods. Some methods require sophisticated hardware components and higher power consumption (e.g., the matched-filter method) compared to simple ones such as the energy detection method. An ideal CR device should be able to be reconfigured on the fly to support a broad range of sensing methods. In practice, a CR device's actual capability will limit the range of sensing methods that can be supported.

1.6.2   QoS Required for Applications Running on the CR Device

QoS requirements differ based on the applications running on a CR device. The sensing delay and transmission throughput vary from one sensing method to another within the same conditions. As a result, the sensing operation used on a CR device has a direct impact on the QoS of an application running on the device, mainly in terms of the throughput and delay. As sensing is a repetitive operation, a CR device should be able to select a proper sensing method with the least impact on the QoS of the running application. Other operational requirements must also be taken into account. For example, the PU protection should have higher priority than the QoS requirements of an SU.

1.6.3   A PRIORI INFORMATION

The extent of information available about the characteristics of the PUs and the communication media is a major factor influencing the selection of a proper sensing method. For instance, insufficient information about the PU's signal excludes the use of the matched-filter method. The CR device should be able to change the sensing method based on the information that becomes available about the PU's signal or the SNR of the targeted spectrum by sensing.

1.6.4   LEVEL OF PROTECTION REQUIRED FOR THE PU

The selection of the sensing method must be considered with regard to the degree of protection necessary for the PU. The sensing method may vary depending on available frequency bands and types of services. For instance, analog television (TV) service is more robust against interference than digital TV service (Ghasemi and Sousa, 2008). Hence, a sensing method that provides less protection (i.e., lower PU detection probability) should be used only when the PU is more tolerant of interference, such as in analog TV services.

1.6.5   CR NETWORK MODE AND CAPABILITY

The network mode and capability are important factors to CR systems in deciding between cooperative and noncooperative sensing approaches. In CR networks with infrastructure and centralized topology, a method based on cooperative sensing is more suitable than that based only on local sensing. Hence, the capability of such a CR network depends on how much management ability it can provide for WS determination to its CR devices. Furthermore, the capacity of a CR network relies on how much information the network can gather and provide to its CR devices about the PUs' signals and the ambient spectrum.

1.7   CR STANDARDIZATION EFFORTS WITH SENSING CONSIDERATION

Nowadays, analog TV bands are the most attractive bands for such opportunistic use of the spectrum by SUs. This option is favorable because analog TV bands show high availability of WSs as many TV channels switched to digital TV bands. Moreover, their use by the PUs can be reliably predicted due to their fixed operational schedules, which can be obtained through GDB services (Gurney et al., 2008). Furthermore, the analog TV spectrum is located below 1 GHz and hence offers more desirable propagation characteristics compared to the higher frequency bands. For instance, these lower frequencies experience less attenuation. As such, it is expected that the first proposed CR standard, IEEE 802.22, is related to TV WS bands (Cordeiro et al., 2005). Thus far, CR standardization efforts are mostly for using TV WSs. Three requirements for SU devices have been considered in the well-known proposed standards: the ability to access the GDB, the facility to sense very low power TV signals, and the capability to make intelligent decisions based on processing gathered information (Rahman et al., 2012).

The sensing requirement is the main focus in this chapter. Regarding standardization, how CR devices are accessing and sharing the communication medium is an important aspect regarding sensing implementation. The existing medium access approaches can be classified as time-slotted, random-access, or hybrid protocols (Akyildiz et al., 2009; Cormio and Chowdhury, 2009). In the time-slotted MAC protocol, an SU accesses the spectrum only at a given time slot using time-division multiple access (TDMA) procedures. In a TDMA-based system (e.g., IEEE 802.22) the central controller is responsible for managing the time slots and determining the sensing scheduling strategy (Ren et al., 2012). Wi-Fi 802.11 systems are mostly based on either the random-access protocol, also called the contention protocol, or hybrid protocols. The carrier-sense multiple-access with collision avoidance (CSMA/CA) mechanism is proposed for wireless devices based on various 802.11 standards to share the ISM bands (IEEE, 2007). The concept is that a Wi-Fi device checks the channel occupancy before transmitting over it. Typically, this checking is accomplished using a simple sensing technique (i.e., energy detection). If the measured energy level exceeds the threshold, the channel is in use; otherwise, it is idle. The channel can be utilized only if it is inactive, and the transmission will be deferred for an arbitrary time if the channel is found to be occupied. This mechanism limits the maximum contiguous transmission time of each of the contended users in the same ISM band for fair channel sharing. For CR devices, the SUs have to compete for the same available WSs without interfering with the PUs of these bands. Therefore, the CSMA/CA and spectrum sensing used in Wi-Fi need to be modified to handle the new requirements of operations in WS bands. An IEEE 802.11 protocol with CR capability is often referred to as CR Wi-Fi, White-Fi, Wi-Fi Like, or IEEE 802.11af. IEEE 802.11af is the first-draft standard for CR networks based on Wi-Fi technology to operate in TV WSs (IEEE, 2014). Although the proposed 802.11af amendment includes several modifications and enhancements to the IEEE 802.11 standard to meet legal requirements to operate in TV WSs, there are remaining technical requirements and challenges that have not been addressed (Baykas et al., 2012). The IEEE 802.11af standard contains a procedure for obtaining information about the available TV WSs within the range of operation from the GDB (IEEE, 2015). However, the sensing operation is not addressed yet in the standard as an alternative way to protect the PU under the absence of the GDB. Also, the standard is limited to TV WS bands. For a perfect CR device, a White-Fi device should be able to operate in any available WS bands and not be limited to TV WS bands, as long as the PU is protected. The sensing operation is an essential function in such White-Fi devices. Thus, any future standardization effort to handle this capability should support backward compatibility with legacy IEEE 802.11 standards in addition to the current IEEE 802.11af.

1.8 SPECTRUM SENSING IMPLEMENTATION ISSUES IN WHITE-FI NETWORKS

One of the core features of White-Fi devices with CR capability is their ability to identify the spectrum holes and detect the PU's appearance in the current operational channel. The principal function to achieve that is sensing the surrounding spectrum. Several implementation issues facing this principal function are discussed in this section.

1.8.1 DISTINGUISHING A PU TRANSMISSION FROM SU TRANSMISSIONS

In CR technology, to protect the PU, an SU will transmit on the PU's channel only when it is not used by the PU and will stop transmitting when the PU appears. The SU can use GDB to obtain a list of the available channels within its location. This list will be valid for a specific period for that location. Hence, the SU has to periodically obtain the new updates from the GDB or obtain them immediately when the SU moves to a new location. This approach is proposed for analog TV WS bands where it is possible to gather the operational schedule of the different PUs using these bands and provide them to SUs on one GDB. The database has to be accessed using another communication method, such as via satellite, which both the GDB server and the SU have the ability to access. The GDB approach cannot be used by all possible SUs as not all SUs have access to different types of communication channels. Also, a PU may not have a fixed schedule to use its channel, or a PU may be unwilling to provide this information to the GDB. Therefore, spectrum sensing is the most essential and useable approach for detecting the PU.

The sensing capability could be built into a CR device, and it may be used without depending on the GDB during operation. Several proposed sensing methods make use of information obtained from a central controller or exchange spectrum information between SU peers in a cooperative sensing. Other sensing methods require prior information about the characteristics of the PU's transmission signals. Blind-sensing methods, such as the energy detection method and the covariance-based detection method, can be effectively used to detect the presence of signals from other users on the sensed channels. However, these blind-sensing methods cannot distinguish if the present signal belongs to a PU or another SU. For a more accurate decision about the appearance of a PU's signal, sufficient prior knowledge about the PU's signal is required. The matched-filter sensing method is a well-known sensing method that is based on complete prior knowledge about the PU's signal. To accurately detect the presence of a PU from its signal, the matched-filter sensing method requires a dedicated receiver for each different possible transmission technology used by the PU (Jaiswal et al., 2013). In general, sensing accuracy is a function of the probability of positive detection (P_D) and the probability of false detection (P_{FA}) under the given operational conditions. False detection is sometimes called false-alarm detection. Typically, achieving higher sensing accuracy requires more overhead regarding the sensing duration, sensing frequency, and computation. The trade-off between the sensing accuracy and the various overheads is required for better performance (Yucek and Arslan, 2009). Usually, P_D and P_{FA} are determined under the assumption that all the detected signals are the PU's signals, while in reality, some detected signals could be from other SUs.

1.8.2 DISTINGUISHING WHITE-FI SUs FROM OTHER SUs

Different SUs could exist and operate within the same coverage area. In addition to protecting the PU, for an SU within the same coverage area of the other SUs, its sensing operation has to detect the presence of the other SUs and share the available WSs with the other SUs. On a White-Fi network, two scenarios are possible (i.e., pure White-Fi SUs and heterogeneous SUs).

1.8.2.1 Pure White-Fi SUs

In this scenario, all SUs sharing the same channel are White-Fi users. A White-Fi user who is using the shared channel has to leave the channel if the PU appears; otherwise, the channel can be shared with other White-Fi users using a CSMA/CA mechanism. The main issue for White-Fi users sharing the same channel is how to distinguish between the PU and the other White-Fi users. To distinguish the PU from other White-Fi users, a longer sensing duration and a more complex sensing method need to be used. Moreover, complete knowledge of the PU's signal is required if there is no GDB for that channel. Once the PU detection issue is addressed, the White-Fi users can share the WS bands based on a CSMA/CA mechanism. Typically, the sensing used in CSMA/CA is not able to distinguish a PU's signal from other Wi-Fi user signals. Therefore, White-Fi users can share the WSs by using the existing technologies for Wi-Fi channel access control, but these technologies are not able to distinguish between PUs' signals and other White-Fi users' signals.

1.8.2.2 Heterogeneous SUs

In this scenario, White-Fi users and SUs based on different wireless technologies exist in the same available WSs. Under this scenario, a White-Fi user needs to distinguish between three types of users (i.e., the PU, the other White-Fi users, and the other non-White-Fi SUs). As an example of such a scenario, White-Fi users and IEEE 802.22 users are sharing the same TV WSs. While White-Fi users can back off when the WS channel is occupied by 802.22 transmissions, the 802.22 users do not have the ability to detect White-Fi transmissions using spectrum sensing. This will result in a negative impact on the performance of the SUs and on their ability to share the network in a fair manner, particularly for SUs using the 802.22 standards, as shown in Ghosh et al. (2011), where a GDB is used to determine the TV WSs. When the GDB is not available, the different SUs sharing the same WSs rely solely on sensing to distinguish between the SUs' signals and the PU's signal. The blind sensing methods (e.g., energy detection) cannot distinguish between these different signals. Hence, the performance of secondary networks is expected to be worse in the absence of the

GDB, and only a sensing approach is used. Achieving fair sharing of a WS between heterogeneous CR systems becomes very challenging.

1.8.3 COMPLICATIONS OF OPERATIONAL ADJUSTMENTS

The 802.11-based networks have different operational adjustments that affect their performance. Different operational situations may require different settings of the operational parameter. Hence, for improvement of performance, such settings should be adjusted intelligently and dynamically to adapt to the real-time change of the operational conditions. The size of the data frame, the use of control frames, and the frame classification for prioritization to enhance QoS are the main operational adjustments that affect the sensing operation. The 802.11af standard inherits these possible adjustments from previous non-CR 802.11 standards. Thus, these possible adjustments and their performance should be researched under the new operational requirements of the 802.11af standard with consideration of the compliance of general regulations by big standards organizations. For example, the IEEE workgroup has proposed the packet fragmentation mechanism; when the size of a packet exceeds a given threshold, it is fragmented into smaller-sized frames at the MAC layer. The resulting fragmented MAC frames are transmitted independently and acknowledged separately. As the wireless medium is error prone, the fragmentation mechanism was proposed to reduce the influence of noise errors on the transmitted data and improve reliability (Yazid et al., 2014). In CR networks, it also reduces the transmission period. Consequently, the fragmentation becomes more important in the case of CR operation. On the one hand, fragmentation helps to increase the protection of the PU from an SU transmission. On the other hand, it may require conducting the sensing in a more frequent manner, which adds additional sensing overhead. Another important factor related to these benefits of fragmentation is the sensing operation. The sensing should be performed before any SU transmission to avoid collision with the transmissions from other users sharing the same channel. In CR networks where the SU is assigned a dedicated transmission channel based on frequency or time, the sensing is usually conducted periodically in a fixed time interval sequentially with the data transmission period. In random-access networks, this approach cannot be carried out efficiently because the data transmission period is not fixed, and it depends on the unpredictable channel conditions. Accordingly, the sensing frequency should be a function of the size of the fragmentation. Lower fragmentation size causes smaller transmission frames and more frequent sensing. In general, for the same sensing method, a longer sensing duration helps to reach higher detection probability and less false-alarm probability under the same spectrum conditions. To our knowledge, there is not much research on how to optimize the sensing duration, the fragmentation threshold, and consequently the sensing frequency in random-access networks to improve overall QoS of the running applications. Further investigation should also cover various sensing methods as they produce different sensing accuracy under given conditions. For instance, an optimal sensing duration is found to produce a maximum throughput, as shown in Wong and Chin (2010). Also, the capability of distinguishing the PU's signal from other possible signals must be taken into account.

Instead of fragmenting a larger frame into smaller frames, a frame-aggregation mechanism is used to aggregate the upper layer frames to form larger frames at the MAC layer (e.g., Aggregated MAC Protocol Data Unit [A-MPDU] in IEEE 802.11n). This opposite approach to fragmentation is used when the transmission channel width is large enough for larger frames to be sent during a short time. The fragmentation approach is used for more reliability at the cost of more protocol overhead for legacy 802.11, whereas the aggregation approach is used to reduce the overhead for high-speed IEEE 802.11 networks (Sidelnikov et al., 2006). For Aggregated MAC Service Data Unit (A-MSDU), the aggregated frames belong to the same service category and are sent to the same destination. The maximum A-MSDU length that an 802.11n base station can receive is either 3839 bytes or 7935 bytes (Charfi et al., 2013). A-MPDU upper-layer frames being sent to the same destination are aggregated to form a frame with one of the following lengths: 8191, 16383, 32767, or 65535 bytes (Charfi et al., 2013). Several MSDUs can be aggregated in one larger MPDU and then one or more MPDUs are encapsulated into a single PHY protocol data unit (PPDU) to be sent to the PHY layer (Ginzburg and Kesselman, 2007). More studies are required to find the optimum size of the PPDU based on various factors, such as the sensing duration, the available channel width, and the desired PU protection.

1.8.4 QoS Degradation

The main concern in White-Fi devices is how much QoS degradation will be experienced. Each function of the CR has its overhead that will contribute to degrading the overall QoS. The spectrum-sensing operation is conducted periodically, while other CR operations (i.e., spectrum decision, mobility, sharing, and access) are executed only when the PU is detected. Hence, spectrum sensing has the most effect on QoS, especially for networks with low PU activity where spectrum sensing mostly returns negative results.

The measurement of QoS differs based on the range of QoS metrics that are set as the benchmark of that measurement. Delay, jitter, throughput, and packet loss are the most used metrics (Charfi et al., 2013). Reducing the overhead and hence the effectiveness of the sensing operation is constrained by the required protection for the PU. Therefore, the desired P_D and P_{FA} should be achieved by trading off between sensing accuracy and its impact on the QoS metrics (Yucek and Arslan, 2009).

The IEEE 802.11e standard is proposed to enhance QoS in IEEE 802.11 networks (IEEE, 2005). As applications have different requirements, in 802.11e, the frames belonging to different applications are prioritized with one of the eight user priority (UP) levels. In contrast, previous IEEE 802.11 standards use the distributed coordination function (DCF) mechanism at MAC layer where the best-effort service is provided equally to all traffic streams from different applications to access the medium. In IEEE 802.11e, the hybrid coordination function (HCF) is used for prioritizing traffic streams to enhance QoS on top of the DCF. The HCF function accommodates two medium access methods (i.e., a distributed contention-based channel access mechanism, called enhanced distributed channel access [EDCA], and a centralized polling-based channel access mechanism, called HCF controlled channel access [HCCA]). Based on the UP, the EDCA defines four access categories (AC); voice (AC_VO), video (AC_VI), best effort (AC_BE), and background (AC_BK). These categories are assigned different priorities, ranging from highest to lowest, respectively. The category AC_VO has top priority and is usually given to traffic carrying voice information. It is followed by the AC_VI category for video traffic and then the AC_BE category for data traffic. The category AC_BK has the lowest priority and is usually assigned to unnecessary data traffic. Each AC has a contention window (CW) which has a specified minimum size and maximum size (i.e., CWmin and CWmax). Also, an arbitration interframe space (AIFS) value and a transmit opportunity (TXOP) interval are used to support QoS prioritization (Inan et al., 2007). Instead of using fixed distributed interframe space (DIFS), also called DCF interframe space, the AIFS value is a variable value calculated based on the AC. The AIFS value determines the time at which a node defers access to the channel after a busy period and before starting or resuming the back-off duration. Hence, the time for a station to wait for the channel to become idle before it starts sending data is calculated based on the AC category of the data AC (Bianchi et al., 2005). However, the short interframe space (SIFS) value is used as the shortest interframe space (IFS) value for transmitting high-priority frames, such as data acknowledgment frames and CTS frames. The TXOP is a new concept introduced in IEEE 802.11e to limit the time of transmission for a given station. On one hand, if a frame to be transmitted requires more than the TXOP interval, the frame should be fragmented into smaller frames, each of which can be transmitted within a TXOP interval (Yang, 2004). On the other hand, the use of a TXOP interval limits the number of smaller frames to be aggregated.

Support for the IEEE 802.11e standard becomes mandatory in new IEEE802.11 devices. The IEEE 802.11e standard can handle the coexistence with legacy stations but with poorer QoS (Bianchi et al., 2005). In the case of White-Fi, the sensing duration will impact the effectiveness of the IEEE 802.11e standard on improving QoS in IEEE 802.11 networks. The impact of the sensing operation should be investigated under different settings of the associated parameters, in particular, CW, AIFS, and TXOP. For the frames belonging to the categories AC_VI and AC_VO, the AIFS and CW values are set smaller than those for the frames of the categories, AC_BE and AC_BK, to reduce the delays. The sensing duration should be varied based on the AC without compromising PU protection. Therefore, when AC_VI and AC_VO frames are to be sent, the sensing duration should be as minimized as possible.

1.8.5 SELECTING THE PROPER SENSING METHOD AND PARAMETERS

On the CSMA/CA-based networks, the sensing function plays an important role in their operations. The energy detection method is commonly used for sensing on these networks because of its simplicity and low overhead. However, in a CR environment, energy detection drawbacks, in particular, its inability to recognize the PU's signal among other signals, reduce its effectiveness. The matched-filter sensing method can overcome this drawback and provide higher detection accuracy at the cost of more sensing time, complexity, and power consumption. Therefore, the sensing method should be selected during operation by trading off between the sensing requirements and its implications. The general factors that should be considered for selecting the proper sensing method are discussed in our previous work in Giweli et al. (2015). For instance, the QoS requirements, the available information about the PU's signal, and the required protection for the PU are some of these factors. In White-Fi, more specific factors should be considered, such as fragmentation and RTS thresholds. The proper sensing method for given operating environments and requirements should be selected in real time for the optimal use of the available sensing techniques. Also, the selected sensing method's parameters (e.g., sensing duration and interval) should be adjusted to achieve the best sensing accuracy with the minimum overhead. According to the IEEE standards for sensing in TV WSs, the time required for both sensing and handoff operations should not excessed 2 seconds, while the false-alarm and detection probabilities should be less than 0.1 and 0.9, respectively (Sesham and Sabat, 2015). Different sensing methods can achieve these requirements in different SNR levels of the channel with various sensing times. For example, the energy detection method requires around 15 ms sensing time at a −20 dB SNR, while the cyclostationary detection method needs around 50 ms at the same SNR (Bagwari et al., 2015a; Bagwari and Tomar, 2015b). Therefore, the selection of the proper sensing method should be performed in real time to meet the various requirements optimally within these restrictions.

An SU may run different applications that may require diverse QoS requirements. For instance, real-time applications are more sensitive to delay and jitter than email applications. Therefore, the QoS requirements will depend on the applications running on the CR device and may vary during its operation. For better QoS, the sensing parameters should be adjusted accordingly in real time. This imposes the need of a cross-layer design to handle exchanging information between different layers (Saki et al., 2014). The sensing operation has an impact on the delay, achieved throughput, and packet loss. The main parameters of the sensing operation that affect QoS are sensing duration, sensing frequency, and sensing accuracy (i.e., P_D and P_{FA}). These parameters can be adjusted to mitigate QoS degradation caused by sensing under constraints of protecting the PU and achieving high spectrum utilization.

1.9 CONCLUSION

CR technology offers new, possibly free frequency bands and enhanced spectrum utilization. This chapter provides a critical review of the new technology of CR and its trends. The concept and the motivation behind CR are discussed to evaluate its worthiness. In particular, the need of CR as one of the core technologies for future networks is illustrated. The main functions of CR operation are explained with related challenges. Furthermore, potential CR network architectures and their corresponding standards are investigated to identify the research areas that require more attention. Although CR is considered a promising revolution for wireless technologies, it faces many challenges which have to be addressed. Most of these challenges are related to the sensing operation in CR networks. For instance, conducting accurate sensing that can guarantee a high level of protection to the PUs from SU interference has a significant impact on the QoS. To avoid such an impact, PU protection is mainly based on the GDB approach in the IEEE 802.22 and IEEE 802.11af proposed standards. Though the GDB approach is suitable for TV WS bands, it cannot be generalized for other potential WS bands. Hence, the need for CR technology that provides the required PU protection based on spectrum-sensing approach becomes an essential requirement for the widespread adoption of CR technologies. For White-Fi networks, when and how spectrum sensing should be conducted to prevent interference efficiently with the PU in addition to complying with the contention mechanism under the IEEE 802.11af standard are still open questions. On one hand, the White-Fi device needs sophisticated sensing techniques to detect the PU accurately and distinguish it from other possible SUs. On the other hand, this more accurate sensing requires a longer sensing duration, which causes more QoS degradation. Hence, whenever the GDB is available, or the virtual sensing is effective, the sensing operation should be avoided or reduced because of its overhead and QoS degradation. However, the spectrum-sensing approach becomes essential when the GDB approach is not available. This study pointed out the importance of considering the operational requirements and sensing methods parameters in the design of an efficient sensing strategy. The design should be automated and respond in real time to the changing QoS requirements.

REFERENCES

Akyildiz, I. F., Lee, W.-Y., and Chowdhury, K. R. (2009). CRAHNs: Cognitive radio ad hoc networks. *Ad Hoc Networks, 7*(5), 810–836. doi:10.1016/j.adhoc.2009.01.001

Akyildiz, I. F., Lo, B. F., and Balakrishnan, R. (2011). Cooperative spectrum sensing in cognitive radio networks: A survey. *Physical Communication, 4*(1), 40–62. doi:10.1016/j.phycom.2010.12.003

Akyildiz, I. F., Won-Yeol, L., Vuran, M. C., and Mohanty, S. (2008). A survey on spectrum management in cognitive radio networks. *IEEE Communications Magazine, 46*(4), 40–48. doi:10.1109/MCOM.2008.4481339

Bagwari, A., Kanti, J., and Tomar, G. S. (2015a). Novel spectrum detector for IEEE 802.22 wireless regional area network. *International Journal of Smart Device and Appliance, 3*(2). 9–25

Bagwari, A., and Tomar, G. S. (2015b). Enriched the spectrum sensing performance of estimated SNR based detector in cognitive radio networks. *International Journal of Hybrid Information Technology, 8*(9), 143–156.

Barnes, S. D., Jansen van Vuuren, P. A., and Maharaj, B. T. (2013). Spectrum occupancy investigation: Measurements in South Africa. *Measurement, 46*(9), 3098–3112. doi:10.1016/j.measurement.2013.06.010

Baykas, T., Kasslin, M., Cummings, M., Kang, H., Kwak, J., Paine, R., et al. (2012). Developing a standard for TV white space coexistence: Technical challenges and solution approaches. *IEEE Wireless Communications, 19*(1), 10–22.

Bianchi, G., Tinnirello, I., and Scalia, L. (2005). Understanding 802.11e contention-based prioritization mechanisms and their coexistence with legacy 802.11 stations. *IEEE Network, 19*(4), 28–34.

Bo, Y., Feng, G., Yanyan, S., Chengnian, L., and Xinping, G. (2009). Channel-aware access for cognitive radio networks. *IEEE Transactions on Vehicular Technology, 58*(7), 3726–3737. doi:10.1109/TVT.2009.2014958

Cabric, D., Mishra, S. M., and Brodersen, R. W. (7–10 November 2004). Implementation issues in spectrum sensing for cognitive radios. Paper presented at the *Conference Record of the Thirty-Eighth Asilomar Conference on Signals, Systems and Computers*, 2004, Pacific Grove, CA, USA.

Cesana, M., Cuomo, F., and Ekici, E. (2011). Routing in cognitive radio networks: Challenges and solutions. *Ad Hoc Networks, 9*(3), 228–248. doi:10.1016/j.adhoc.2010.06.009

Charfi, E., Chaari, L., and Kamoun, L. (2013). PHY/MAC enhancements and QoS mechanisms for very high throughput WLANs: A survey. *IEEE Communications Surveys & Tutorials, 15*(4), 1714–1735. doi:10.1109/SURV.2013.013013.00084

Chien-Min, W., Hui-Kai, S., Maw-Lin, L., Yi-Ching, L., and Chih-Pin, L. (2–4 July 2014). Cooperative power and contention control MAC protocol in multichannel cognitive radio ad hoc networks. Paper presented at the *Eighth International Conference on Innovative Mobile and Internet Services in Ubiquitous Computing (IMIS)* (pp. 305–309), 2014, Birmingham, UK.

Christopher Clement, J., Emmanuel, D. S., and Jenkin Winston, J. (2015). Improving sensing and throughput of the cognitive radio network. *Circuits, Systems, and Signal Processing, 34*(1), 249–267. doi:10.1007/s00034-014-9845-y

Cordeiro, C., Challapali, K., Birru, D., and Sai Shankar, N. (8–11 November 2005). IEEE 802.22: The first worldwide wireless standard based on cognitive radios. Paper presented at the *First IEEE International Symposium on New Frontiers in Dynamic Spectrum Access Networks, DySPAN,* 2005, Baltimore, MD, USA, pp. 328–337.

Cormio, C., and Chowdhury, K. R. (2009). A survey on MAC protocols for cognitive radio networks. *Ad Hoc Networks, 7*(7), 1315–1329.

Cui, L., and Weiss, M. B. (2013). Can unlicensed bands be used by unlicensed usage. *Paper for the 41st Annual TPRC, September,* 27–29.

Dongyue, X., Ekici, E., and Vuran, M. C. (2014). Cooperative spectrum sensing in cognitive radio networks using multidimensional correlations. *IEEE Transactions on Wireless Communications, 13*(4), 1832–1843. doi:10.1109/TWC.2014.022714.130351

Enserink, S., and Cochran, D. (1994). A cyclostationary feature detector. Paper presented at the *Conference Record of the Twenty-Eighth Asilomar Conference on Signals, Systems and Computers,* 1994, Pacific Grove, CA., Vol. 2, pp. 806–810.

Farnham, T., Clemo, G., Haines, R., Seidel, E., Benamar, A., Billington, S., et al. (2000). IST-TRUST: A perspective on the reconfiguration of future mobile terminals using software download. Paper presented at the *11th IEEE International Symposium on Personal, Indoor and Mobile Radio Communications, PIMRC,* 2000, London, UK, Vol. 2, pp. 1054–1059.

Federal Communications Commission (FCC), E. (2003). *Docket No 03-222 Notice of proposed rule making and order,* Washington, DC, USA. Retrieved from https://apps.fcc.gov/edocs_public/attachmatch/FCC-03-222A1.pdf

Ghasemi, A., and Sousa, E. S. (2008). Spectrum sensing in cognitive radio networks: Requirements, challenges and design trade-offs. *IEEE Communications Magazine, 46*(4), 32–39. doi:10.1109/MCOM.2008.4481338

Ghosh, C., Roy, S., and Cavalcanti, D. (2011). Coexistence challenges for heterogeneous cognitive wireless networks in TV white spaces. *IEEE Wireless Communications, 18*(4), 22–31. doi:10.1109/MWC.2011.5999761

Ginzburg, B., and Kesselman, A. (2007). Performance analysis of A-MPDU and A-MSDU aggregation in IEEE 802.11n. Paper presented at the *IEEE Sarnoff symposium,* 2007, Princeton, NJ, pp. 1–5.

Giweli, N., Shahrestani, S., and Cheung, H. (December 2015). Spectrum sensing in cognitive radio networks: QoS considerations. Paper presented at the *Seventh International Conference on Network & Communications (NETCOM 2015),* Sydney, Australia.

Gurney, D., Buchwald, G., Ecklund, L., Kuffner, S., and Grosspietsch, J. (2008). Geo-location database techniques for incumbent protection in the TV white space. Paper presented at the *3rd IEEE Symposium on New Frontiers in Dynamic Spectrum Access Networks, DySPAN,* 2008, Chicago, IL, pp. 1–9.

Haddadi, S., Saeedi, H., and Navaie, K. (8–9 May 2013). Channel coding adoption versus increasing sensing time in secondary service to manage the effect of imperfect spectrum sensing in cognitive radio networks. Paper presented at the *2013 Iran Workshop on Communication and Information Theory (IWCIT),* Tehran, Iran, pp. 1–5.

Haykin, S. (2005). Cognitive radio: Brain-empowered wireless communications. *IEEE Journal on Selected Areas in Communications, 23*(2), 201–220.

IEEE. (2005). IEEE Standard for information technology—Local and metropolitan area networks—Specific requirements—Part 11: Wireless LAN medium access control (MAC) and physical layer (PHY) specifications—Amendment 8: Medium access control (MAC) quality of service enhancements. *IEEE Std 802.11e-2005 (Amendment to IEEE Std 802.11, 1999 Edition [Reaff 2003]),* 1–212. doi:10.1109/IEEESTD.2005.97890

IEEE. (2007). IEEE Standard for information technology—Telecommunications and information exchange between systems—Local and metropolitan area networks—Specific requirements—Part 11: Wireless LAN medium access control (MAC) and physical layer (PHY) specifications. *IEEE Std 802.11-2007 (Revision of IEEE Std 802.11-1999),* 1–1076. doi:10.1109/IEEESTD.2007.373646

IEEE. (2014). IEEE Standard for information technology—Telecommunications and information exchange between systems—Local and metropolitan area networks—Specific requirements—Part 11: Wireless

LAN medium access control (MAC) and physical layer (PHY) specifications—Amendment 5: Television white spaces (TVWS) operation. *IEEE Std 802.11af-2013 (Amendment to IEEE Std 802.11-2012, as amended by IEEE Std 802.11ae-2012, IEEE Std 802.11aa-2012, IEEE Std 802.11ad-2012, and IEEE Std 802.11ac-2013)*, 1–198. doi:10.1109/IEEESTD.2014.6744566

IEEE. (2015). ISO/IEC/IEEE International Standard—Information technology—Telecommunications and information exchange between systems—Local and metropolitan area networks—Specific requirements—Part 11: Wireless LAN medium access control (MAC) and physical layer (PHY) specifications—Amendment 5. *ISO/IEC/IEEE 8802-11:2012/Amd.5:2015(E) (Adoption of IEEE Std 802.11af-2014)*, 1–204. doi:10.1109/IEEESTD.2015.7226767

Inan, I., Keceli, F., and Ayanoglu, E. (2007). Modeling the 802.11e enhanced distributed channel access function. Paper presented at the *Global Telecommunications Conference, 2007. GLOBECOM'07. IEEE*. Washington, DC, USA, pp. 2546–2551.

Jaiswal, M., Sharma, A. K., and Singh, V. (2013). A survey on spectrum sensing techniques for cognitive radio. Paper presented at the *Proceedings of the Conference on Advances in Communication and Control Systems, 2013*. Aalborg, DE, pp. 74–79.

Kae Won, C. (2010). Adaptive sensing technique to maximize spectrum utilization in cognitive radio. *IEEE Transactions on Vehicular Technology*, 59(2), 992–998. doi:10.1109/TVT.2009.2036631

Kim, H., and Shin, K. G. (2008). In-band spectrum sensing in cognitive radio networks: Energy detection or feature detection? Paper presented at the *Proceedings of the 14th ACM International Conference on Mobile Computing and Networking*, San Francisco, CA, pp. 14–25.

Kumar, A., Wang, J., Challapali, K., and Shin, K. (2011). Analysis of QoS provisioning in cognitive radio networks: A case study. *Wireless Personal Communications*, 57(1), 53–71. doi:10.1007/s11277-010-0006-8

Lala, N. A., Uddin, M., and Sheikh, N. (2013). Identification and integration of QoS parameters in cognitive radio networks using Fuzzy logic. *International Journal of Emerging Sciences*, 3(3), 279–288.

Lee, W.-Y., and Akyildiz, I. F. (2012). Spectrum-aware mobility management in cognitive radio cellular networks. *IEEE Transactions on Mobile Computing*, 11(4), 529–542.

Li, S., Wang, X., Zhou, X., and Wang, J. (2012). Efficient blind spectrum sensing for cognitive radio networks based on compressed sensing. *EURASIP Journal on Wireless Communications and Networking*, 2012(1), 1–10.

Lundén, J., Koivunen, V., Huttunen, A., and Poor, H. V. (2007). Spectrum sensing in cognitive radios based on multiple cyclic frequencies. Paper presented at the *2nd International Conference on Cognitive Radio Oriented Wireless Networks and Communications, CrownCom, 2007*, Orlando, FL, USA, pp. 37–43.

Mao, S., Huang, Y., Li, Y., Agrawal, P., and Tugnait, J. (2013). Introducing software defined radio into undergraduate wireless engineering curriculum through a hands-on approach. Paper presented at the *Proceedings of the 2013 ASEE Annual Conference*, Atlanta, GA, pp. 1–10.

Marinho, J., Granjal, J., and Monteiro, E. (2015). A survey on security attacks and countermeasures with primary user detection in cognitive radio networks. *EURASIP Journal on Information Security*, 2015(1), 1–14. doi:10.1186/s13635-015-0021-0

Marinho, J., and Monteiro, E. (2012). Cognitive radio: Survey on communication protocols, spectrum decision issues, and future research directions. *Wireless Networks*, 18(2), 147–164. doi:10.1007/s11276-011-0392-1

Masonta, M. T., Mzyece, M., and Ntlatlapa, N. (2013). Spectrum decision in cognitive radio networks: A survey. *IEEE Communications Surveys & Tutorials*, 15(3), 1088–1107. doi:10.1109/surv.2012.111412.00160

Mitola, J. (1995). The software radio architecture. *IEEE Communications Magazine*, 33(5), 26–38.

Mitola, J. (2000). Cognitive radio—An integrated agent architecture for software defined radio. Tekn. Dr. dissertation, Royal Institute of Technology (KTH), Stockholm, Sweden.

Mitola, J., and Maguire Jr, G. Q. (1999). Cognitive radio: Making software radios more personal. *IEEE Personal Communications*, 6(4), 13–18.

Mohammed, O. A., El-Khatib, K., and Martin, M. V. (2014). A survey of cognitive radio management functions. Paper presented at the *COCORA 2014, The Fourth International Conference on Advances in Cognitive Radio*, Nice, France, pp. 6–13.

Molisch, A. F., Greenstein, L. J., and Shafi, M. (2009). Propagation issues for cognitive radio. *Proceedings of the IEEE*, 97(5), 787–804.

Murty, R., Chandra, R., Moscibroda, T., and Bahl, P. (2012). SenseLess: A database-driven white spaces network. *IEEE Transactions on Mobile Computing*, 11(2), 189–203. doi:10.1109/TMC.2011.241

Nekovee, M. (6–9 April 2010). Cognitive radio access to TV white spaces: Spectrum opportunities, commercial applications and remaining technology challenges. Paper presented at the *IEEE Symposium on New Frontiers in Dynamic Spectrum, 2010*, Singapore, pp. 1–10.

Palaios, A., Riihijarvi, J., Mahonen, P., Atanasovski, V., Gavrilovska, L., Van Wesemael, P., et al. (9–13 June 2013). Two days of European spectrum: Preliminary analysis of concurrent spectrum use in seven European sites in GSM and ISM bands. Paper presented at the *IEEE International Conference on Communications (ICC), 2013*, Budapest, Hungary, pp. 2666–2671.

Qing, Z., and Sadler, B. M. (2007). A survey of dynamic spectrum access. *IEEE Signal Processing Magazine, 24*(3), 79–89. doi:10.1109/MSP.2007.361604

Rahman, M. A., Song, C., and Harada, H. (6–9 May 2012). Design aspects of a television white space device prototype. Paper presented at the *IEEE 75th Vehicular Technology Conference (VTC Spring), 2012.* Yokohama, Japan, pp. 1–6.

Rawat, D. B., and Yan, G. (2011). Spectrum sensing methods and dynamic spectrum sharing in cognitive radio networks: A survey. *International Journal of Research and Reviews in Wireless Sensor Networks, 1*(1), 1–13.

Ren, P., Wang, Y., Du, Q., and Xu, J. (2012). A survey on dynamic spectrum access protocols for distributed cognitive wireless networks. *EURASIP Journal of Wireless Communication and Networking, 2012,* 60.

Saki, H., Shojaeifard, A., and Shikh-Bahaei, M. (2014). Cross-layer resource allocation for video streaming over OFDMA cognitive radio networks with imperfect cross-link CSI. Paper presented at the *2014 International Conference on Computing, Networking and Communications, ICNC, 2014,* Honolulu, HI, USA, pp. 98–104.

Sesham, S., and Sabat, S. (2015). Spectrum sensing for cognitive radio networks. In A. K. Mishra and D. Lloyd Johnson (Eds.), *White Space Communication* (pp. 117–151). Switzerland: Springer International Publishing.

Shobana, S., Saravanan, R., and Muthaiah, R. (2013a). Matched filter based spectrum sensing on cognitive radio for OFDM WLANs. *International Journal of Engineering and Technology (IJET), 5*(1), 142–146.

Shobana, S., Saravanan, R., and Muthaiah, R. (2013b). *Optimal Spectrum Sensing Approach on Cognitive Radio Systems.* Tehran, Iran: IEEE.

Sidelnikov, A., Yu, J., and Choi, S. (2006, 24–25 August). *Fragmentation/aggregation scheme for throughput enhancement of IEEE 802.11 n WLAN.* Paper presented at the The 3rd IEEE Asia Pacific Wireless Communications Symposium (APWCS 2006), Daejeon, South Korea.

SiXing, Y., Dawei, C., Qian, Z., Mingyan, L., and ShuFang, L. (2012). Mining spectrum usage data: A large-scale spectrum measurement study. *IEEE Transactions on Mobile Computing, 11*(6), 1033–1046. doi:10.1109/TMC.2011.128

Steenkiste, P., Sicker, D., Minden, G., and Raychaudhuri, D. (2009). Future directions in cognitive radio network research. Paper presented at the *NSF Workshop Report.* (Vol. 4, No. 1, pp. 1–40). Arlington, VA, USA. Retrieved from https://www.cs.cmu.edu/~prs/NSF_CRN_Report_Final.pdf

Tandra, R., and Sahai, A. (2008). SNR walls for signal detection. *IEEE Journal of Selected Topics in Signal Processing, 2*(1), 4–17. doi:10.1109/JSTSP.2007.914879

Tang, H. (2005). Some physical layer issues of wide-band cognitive radio systems. Paper presented at the *First IEEE International Symposium on New Frontiers in Dynamic Spectrum Access Networks, DySPAN, 2005,* Baltimore, MD, USA.

Wang, B., Bai, Z., Xu, Y., Dong, P., and Kwak, K. (2014). Dynamical clustering cooperative spectrum sensing with bandwidth constraints in CR systems. Paper presented at the *16th International Conference on Advanced Communication Technology (ICACT), 2014,* Pyeongchang, South Korea, pp. 244–248.

Wong, D. T. C., and Chin, F. (16–19 May 2010). Sensing-saturated throughput performance in multiple cognitive CSMA/CA networks. Paper presented at the *IEEE 71st Vehicular Technology Conference (VTC 2010-Spring), 2010,* Taipei, Taiwan, pp. 1–5.

Xin-Lin, H., Gang, W., Fei, H., and Kumar, S. (2011). The impact of spectrum sensing frequency and packet-loading scheme on multimedia transmission over cognitive radio networks. *IEEE Transactions on Multimedia, 13*(4), 748–761. doi:10.1109/TMM.2011.2148701

Xinbing, W., Zheng, L., Pengchao, X., Youyun, X., Xinbo, G., and Hsiao-Hwa, C. (2010). Spectrum sharing in cognitive radio networks—An auction-based approach. *IEEE Transactions on Systems, Man, and Cybernetics, Part B: Cybernetics, 40*(3), 587–596. doi:10.1109/TSMCB.2009.2034630

Yan, X., Ahmet Şekercioğlu, Y., and Narayanan, S. (2010). A survey of vertical handover decision algorithms in fourth generation heterogeneous wireless networks. *Computer Networks, 54*(11), 1848–1863. doi:10.1016/j.comnet.2010.02.006

Yang, X. (2004). IEEE 802.11e: QoS provisioning at the MAC layer. *IEEE Wireless Communications, 11*(3), 72–79. doi:10.1109/MWC.2004.1308952

Yazid, M., Bouallouche-Medjkoune, L., Aïssani, D., and Ziane-Khodja, L. (2014). Analytical analysis of applying packet fragmentation mechanism on IEEE 802.11b DCF network in non ideal channel with infinite load conditions. *Wireless Networks*, *20*(5), 917–934.

Ying-Chang, L., Yonghong, Z., Peh, E. C. Y., and Anh Tuan, H. (2008). Sensing-throughput tradeoff for cognitive radio networks. *IEEE Transactions on Wireless Communications*, *7*(4), 1326–1337. doi:10.1109/TWC.2008.060869

Yiyang, P., Ying-Chang, L., Teh, K. C., and Kwok Hung, L. (2009). How much time is needed for wideband spectrum sensing? *IEEE Transactions on Wireless Communications*, *8*(11), 5466–5471. doi:10.1109/TWC.2009.090350

Yuan, W., and Tsang, D. H. K. (19–25 April 2009). Distributed power allocation algorithm for spectrum sharing cognitive radio networks with QoS guarantee. Paper presented at the *IEEE INFOCOM, 2009*, Rio de Janeiro, Brazil, pp. 981–989.

Yucek, T., and Arslan, H. (2006). Spectrum characterization for opportunistic cognitive radio systems. Paper presented at the *Military Communications Conference, 2006. MILCOM 2006*. IEEE.

Yucek, T., and Arslan, H. (2009). A survey of spectrum sensing algorithms for cognitive radio applications. *IEEE Communications Surveys & Tutorials*, *11*(1), 116–130.

Yulong, Z., Yu-Dong, Y., and Baoyu, Z. (2012). Cooperative relay techniques for cognitive radio systems: Spectrum sensing and secondary user transmissions. *IEEE Communications Magazine*, *50*(4), 98–103. doi:10.1109/MCOM.2012.6178840

Zeng, Y., and Liang, Y.-C. (2007). Covariance based signal detections for cognitive radio. Paper presented at the *2nd IEEE International Symposium on New Frontiers in Dynamic Spectrum Access Networks, 2007, DySPAN 2007*, Dublin, Ireland, pp. 202–207.

Zeng, Y., and Liang, Y.-C. (2009). Spectrum-sensing algorithms for cognitive radio based on statistical covariances. *IEEE Transactions on Vehicular Technology*, *58*(4), 1804–1815.

Zeng, Y., Liang, Y.-C., Hoang, A. T., and Zhang, R. (2010). A review on spectrum sensing for cognitive radio: Challenges and solutions. *EURASIP Journal of Advance Signal Process, 2010*, 2–2. doi:10.1155/2010/381465

Zhongshan, Z., Keping, L., and Jianping, W. (2013). Self-organization paradigms and optimization approaches for cognitive radio technologies: A survey. *IEEE Wireless Communications*, *20*(2), 36–42. doi:10.1109/mwc.2013.6507392

2 Cognitive Radio and Spectrum Sensing

Daniel Malafaia, José Vieira, and Ana Tomé

CONTENTS

2.1 INTRODUCTION

The expected explosion of the number of wireless devices over the years to come will increase the demand for more radio frequency spectrum (RFS) to allocate new services. The current spectrum policy is based on a fixed spectrum access in which a slice of the spectrum is assigned to one or more dedicated users. This policy of allocating fixed channels prevents rarely used spectrum channels from being occupied. The proliferation of wireless services in past decades that used the fixed spectrum access created an artificial spectrum scarcity problem. Recent studies on the actual spectrum utilization measurements revealed that a large portion of the licensed spectrum experiences low utilization from the licensed users [1,2]. The only solution is to then use the available RFS more wisely. The cognitive radio (CR) concept was proposed to solve this problem and is based on an adequate capability of the radios to sense the spectrum usage by other transmitters.

A CR is built in a software-defined radio (SDR) platform. In this chapter, SDRs are described in Section 2.2, followed by the description of CR architecture in Section 2.3. Spectrum sensing, the ability to determine the occupation of wide bandwidth of the spectrum, is detailed in Section 2.4. The analysis of spectrum sensing is divided in three areas. Spectrum analysis is detailed in Subsection 2.4.1, where several methods are explained; the first and simpler method is the periodogram, followed by filter banks with a uniform DFT polyphasic efficient implementation, the multitaper method, and cyclic spectrum, and finishing with eigenvalues from the covariance matrix. In Subsection 2.4.2, the theory of signal detection, the binary decision of determining a channel occupation, is explored. In Section 4.3, the analysis of spectrum sensing is concluded with the explanation of the noise model, which is used to determine the threshold of the binary decision. In Section 2.5, the comparative analysis of spectrum sensing is explored using the receiver operating characteristic (ROC) curve. In Section 2.6, the chapter ends by exploring the application of spectrum sensing in the recent trend of MIMO communications with highly directional transmissions using spatial spectrum sensing to allow an even more efficient reuse of spectrum resources.

2.2 SOFTWARE-DEFINED RADIO

The SDR was first introduced by Joseph Mitola as a concept for the future of the radio [3]. The SDR is a generic radio platform with the capability to operate at different bandwidths over numerous frequencies, as well as being flexible in using different modulation schemes and waveform formats. Therefore, the SDR has a very flexible analog front end that, in combination with adjustable digital signal processing techniques, is flexible enough to modulate and demodulate any signal entirely in software regardless of the standard or its characteristics.

The transmitter has the ability to create any baseband waveform with upconversion to any central frequency. This signal is converted to the analog domain with a digital-to-analog converter (DAC) and is then amplified and transmitted. On the other hand, the receiver collects a certain bandwidth of the spectrum, amplifies it, and converts it back to the digital domain using an analog-to-digital converter (ADC). After that conversion, the process of detection and demodulation of the signal is carried out by software. A block diagram of the SDR concept is shown in Figure 2.1.

The concept of SDR has experienced substantial and rapid development as portable and flexible hardware platforms are now commercially accessible and can be easily acquired [4,5]. This massive spread of the technology allows new applications based on this architecture to be implemented and produced. For instance, SDR architecture allowed the development of software-defined radar and software-defined networking platforms [6,7].

2.3 COGNITIVE RADIO

Traditional wireless networks are based on the exclusive allocation of the spectrum, where telecommunication service providers buy reserved frequency bands at auction to ensure quality of service (QoS). Personal local wireless networks normally use a license-free spectrum, called ISM bands, where part of the radio spectrum is used for personal, industrial, scientific, and medical utilization. As these bands are becoming crowded with unlicensed equipment, such as Wi-Fi devices, the need for a new paradigm and radio implementation for this unlicensed spectrum is becoming increasingly important. The expected solution is the CR [8].

The CR, built on a software radio platform, is a context-aware, intelligent radio that is able to autonomously reconfigure itself by learning from and adapting to the communication environment. One of the key features of SDR architecture is its flexibility to use a variety of frequency bands

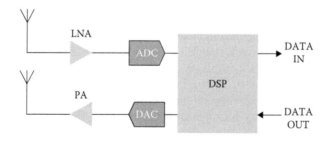

FIGURE 2.1 Block diagram of a software-defined radio.

according to their availability. This flexibility allows for CRs to share parts of the spectrum while avoiding collision with any licensed user or other CRs.

This flexible sharing of the spectrum was initially discussed in the first IEEE DySPAN and named dynamic spectrum access (DSA) [9]. The DSA is also currently called spectrum sharing [10]. It is usually categorized into the following three models:

- **Dynamic exclusive-use model:** An exclusive shared use, where any user can dynamically license, for a given time and location, a number of available frequency bands. Also called licensed shared access (LSA), it allows a most effective use of spectrum resources by permitting temporary allocations of the shared spectrum according to the user's needs [11]. This exclusive shared model is far away from completely eliminating quiet periods in spectrum due to the bursty nature of wireless traffic, but it ensures a predictable QoS for the reserved time.
- **Open-sharing model:** A completely shared use, giving total freedom to the users to manage the sharing of specific slices of the spectrum without intervention of a central regulator entity [12]. This creates the need for the users to constantly sense the allowed spectrum bands and cooperate with the other CRs to detect and use spectrum bands that are not currently being occupied. It is supported by the concept of a periodic listen before talk and limited channel occupancy time.
- **Hierarchical access model:** A model that adopts an access structure where licensed spectrum can be open to secondary users while limiting the interference perceived by primary users, the licensees [13]. The model allows opportunistic access to the licensed spectrum by allowing the secondary users to identify and exploit local and instantaneous spectrum availability in a nonintrusive manner.

The employment of these methods would be expected to alleviate the current spectrum usage while still providing the exclusive spectrum allocation for the primary users. Some specific implementations have already been integrated into the IEEE 802.22 standard for wireless regional networks and the IEEE 802.11af for wireless local networks. Both standards define a hierarchical access model to use licensed frequencies but limit the interference with the primary users, such as analog TV, digital TV, and low-power licensed devices, such as wireless microphones [14–16].

However, the practical implementation of such architecture is surrounded by several challenges both in software and in hardware. In software, new and robust communication protocols for RF devices are needed to facilitate the use of the shared spectrum. In the hardware, there is the necessity to sense the entire viable shared spectrum, which can be widespread and detect which bands are already being used. If the band is not currently under use, the CR should be very flexible and able to change the frequency of the transmission to allocate it into the available band. It should also detect if any user that has the right to a frequency band will start transmitting and should do so in the fastest possible way stop the usage of that section of the spectrum.

These specificities of a DSA create the need for a spectrum-sensing unit that can quickly and accurately identify white spaces (WSs) through a wideband of frequencies. As an example, the current unlicensed frequency bands for RF are incredibly spaced apart in frequency. The bands in the order of GHz are in 2.4 GHz, 5.8 GHz, and 24 GHz [17]. Taking advantage of all these spectrum resources would require an exceedingly wideband CR front end.

2.4 SPECTRUM-SENSING THEORY

Spectrum sensing has been an intense research topic with many different solutions being proposed in the literature to create a reliable method to detect the occupation of a given frequency band, even for unfavorable signal-to-noise ratio (SNR) conditions [18]. To sense a wide span of the RFS and evaluate which frequency bands are being occupied and which are free to use, three tasks need to be ensured.

The first task is spectrum analysis, which is comprised of the frequency analysis of a received RF signal, which involves computing an estimation of its frequency representation. This analysis allows the separation of the acquired signal into frequency bands with uniform or variable bandwidth.

The second task is to determine a noise model, where some of the noise process characteristics are calculated or can even be estimated in real time using the spectrum analysis results in time and/or frequency to estimate the noise model.

The third task is the signal detection, where the goal is to decide channel occupation of the spectrum estimation. The noise model allows the computation of a threshold that is used to make this decision. The most common way to output the occupation is by a hard-binary decision, either a given frequency band is occupied or it is free to use.

A spectrum-sensing unit will then operate as illustrated in Figure 2.2. A wide frequency band of the spectrum is acquired by an ADC. The data is then analyzed by a spectrum estimation algorithm to obtain the most accurate representation of the acquired signal into frequency bands. To achieve real-time spectrum-sensing implementations, the algorithm needs to have low computational costs and be able to detect the small, quiet periods defined in various standards [19]. At the same time, the algorithm needs to have low spectral leakage to have reliable spectrum estimation. This frequency representation is then used to discern if a signal is present in that band.

The noise model is fundamental to the signal detection decision, as it allows to determine if the spectrum analysis data represents a transmitted signal or noise. If the analyzed data does not fit the noise model, then we are in the presence of a signal. The noise model estimation methods can be divided into the following two large groups:

- **Dynamic noise model:** This model is normally used in dynamic threshold energy detection methods [20]. The outputs of the spectrum analysis block are used to calculate the noise floor parameters, such as its probability distribution, variance, and power level. It is expected that the values of the noise parameters change over time due to external effects, such as channel interference from out-of-band emitters, or from effects of the front end

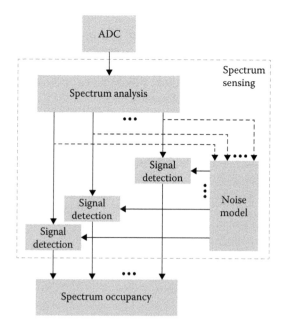

FIGURE 2.2   Spectrum-sensing components block diagram.

itself, such as an automatic gain control system. Due to these effects, a single evaluation of the noise floor is not enough, then requiring a dynamic estimation that is capable of handling energy fluctuations.

- **Fixed noise model:** This model is unaltered during the spectrum sensing and creates a fixed threshold. In 802.11 protocol energy detection, the noise energy is assumed to be always below the hardware sensitivity level plus 20 dBm; anything above that level is assumed to be a signal [21]. For statistical moments-based detection, the noise distribution can be considered a Gaussian distribution, while the signal can be, for instance, modeled by a Raleigh distribution. In that case, the detection can be achieved by estimation of the kurtosis [22]. In eigenvalue-based spectrum sensing, the noise covariance matrix can be assumed to always be a Wishart random matrix, and the noise eigenvalues distribution can be calculated to then allow the application of maximum-minimum detection techniques [23]. In cyclostationary analysis, the noise covariance can be assumed to be chi-square, while the signal can be modeled by a Gaussian distribution, allowing the use of cyclostationary detectors [24].

While some methods assume a model not only for the noise but also for the signal, the signal model is usually not used for signal detection. This is due to the small amount of information that can be used to represent the signal [23]. For instance, it may be assumed that all the detected signals follow a Rayleigh distribution as it is an appropriate distribution model for multipath environments. However, in a direct transmission scenario, with a lack of multipath reflections, the Rayleigh distribution will be a poor model. Other reasons include the large number of modulations that can be used to transmit the RF signal and the wide range of power that can be received. Due to these reasons, the threshold for detection is usually defined based not on the signal, but on the noise model. The model is then used by the signal detection system to calculate a threshold that is used to analyze a frequency span and determine if it is in use while getting the best relationship possible between false and missed detection. The false alarm is usually defined in the communication standard [19].

2.4.1 SPECTRUM ANALYSIS

The chosen method for spectrum analysis can make the difference between the capability to detect weak signals and occluding them under leakage noise. Spectral leakage occurs when the signal energy contributes not only to one frequency band but also spreads to the closest bands. The effects caused by the spectral leakage can be attenuated by using the proper spectrum analysis method. A power spectrum density (PSD) comparative is shown in Figure 2.3.

In this synthetic example, several window functions are used for the same filter bank specifications, and their results are compared to the exact PSD, a sum of three different passband signals. The results show that a filter bank using the Hann function as a prototype filter is able to detect the weakest passband signal that, if using the boxcar function, would be otherwise hidden.

It is then important to know and understand the various methods for spectrum analysis to pick the best ones for a given application. The following are some of the most relevant methods:

- **Periodogram:** This is the most common way of analyzing the spectrum as it has low computational and implementation complexity. It uses a constant number of samples and analzes them using a fast Fourier transform (FFT) that is then squared and the result integrated [25].
- **Filter bank:** A filter bank is an array of bandpass filters that will decompose a given signal into multiple components; each component will then be a frequency sub-band of the original signal. This process shows good performance for spectrum analysis due to its low spectral leakage. Signal decomposition is performed by an analysis filter bank that usually splits the signal using the same prototype filter that is centered in different, but equidistant,

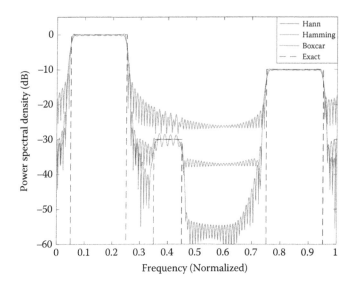

FIGURE 2.3   Example of a signal PSD calculated using a filter bank with three different prototype filters.

central frequencies. This method uses the concept of maximal energy concentration in a given central frequency with a constant bandwidth. A computational efficient implementation is achieved with the polyphasic decomposition of the prototype filter [26].

- **Multitaper:** The multitaper method may be seen as a filter bank that, instead of one, uses multiple orthogonal prototype filters, usually Slepian sequences, to improve the variance and reduce leakage [27].
- **Cyclic spectrum:** This method takes advantage of the cyclostationary behaviour of certain signals to improve the spectrum analysis, making it a good method for signals with periodic autocorrelation behavior, such as digital modulated signals [28]. It has the advantage of the calculated cyclic spectrum features that can be used for the signal modulation recognition [29].
- **Eigenvalues:** Spectrum-sensing literature addresses signal detection with the relation between the eigenvalues of the correlation matrix of the signal [23]. However, the eigenvectors of the matrix can also be considered coefficients of a filter bank. Moreover, the eigenvalues are an estimate of the energy at the output of the corresponding eigenvector filter, as shown in [30]. The uniqueness of the method is that the eigenvector filter bank does not represent equally spaced frequency bands as in Fourier-based methods.

The selection of a method of spectrum analysis should accomplish the best combination of three criteria. It is of critical importance for the method to have the lowest spectral leakage to avoid obtaining a smeared version of the original continuous-time spectrum, which would reduce the ability to detect spectrum holes. Also important is to have the least number of acquired samples needed for a correct representation of the spectrum to allow a fast sense of the spectrum to quickly occupy spectrum opportunities. Finally, the algorithm must not be expensive in terms of computational resources to be deployed in mobile devices and have low power consumption. By achieving these criteria, the optimal system is achieved that, quickly and in real time, can identify spectrum opportunities and avoid possible collision with other concurrent users.

2.4.1.1   Periodogram

The classical method of spectrum analysis is the usage of a periodogram [31]. First used in meteorological phenomena, it allows for a direct spectral estimation of the power spectral density function. However, the method shows some problems—the principle is biased. There are usually two kinds of bias, leakage bias and local bias [32]. Leakage bias is produced by spectral leakage, as a strong peak in a specific frequency will spread into the neighboring bins. Local bias is caused by using a low number of bins that requires that each have a large bandwidth.

A solution for leakage bias is to use a window function before the Fourier transform. For an $x(n)$ discrete data sequence with N points, the windowed spectrogram, $\hat{S}(f)$, can be given by

$$\hat{S}(f) = \left| \sum_{n=0}^{N-1} x(n)w(n)e^{-2\pi ifn} \right|^2 \tag{2.1}$$

where $w(t)$ is the window function defined as a series of weights. This method is also called a single taper, where $w(t)$ is the taper.

2.4.1.2 Uniform DFT Polyphase Filter Banks

To separate the received RF signal into frequency channels, one can use the analysis filter bank. An analysis filter bank is an array of bandpass filters that divide the input signal into various components, each one containing a frequency band of the original input. This process uses a prototype finite impulse response (FIR) filter with the coefficients $h(n) = [h(0), h(1) \dots h(N-1)]$, with N being the number of coefficients. This filter is then translated into frequency to obtain all the outputs of the filter bank. As the number of coefficients can be larger than the number of frequency bins, there is more freedom to design sharper filters using, for example, the window method.

For an efficient implementation, the analysis filter bank is based on the introduction of the DFT after a polyphase decomposition of the input and prototype filter. Assuming an M channel filter bank, this is done by using a decimation by M and decomposing the prototype filter in M phases, $H_k(z)$, $k = 0, \dots, (M-1)$, where $H_0(z)$, $H_1(z)$, $\dots$, $H_M(z)$ represents the polyphase components of the prototype filter.

The analysis filter bank with M channels having transfer functions $G_k(z)$, where $k = 0, \dots, (M-1)$, corresponds to the modulated prototype filter $H(z)$ [33].

$$G_k(z) = H\left(ze^{j\frac{2\pi k}{M}} \right), \quad k = 0, 1, \dots, (M-1) \tag{2.2}$$

This implementation is based on the division of the prototype filter in its various phases, $H_l(z)$, and is given by

$$H_l(z) = \sum_{n=0}^{\frac{N}{M-1}} h(nM+l)z^{-n} \tag{2.3}$$

Assuming that N is a multiple of M, $\dfrac{N}{M-1}$ is the number of coefficients of each phase. The $h(n)$ is the impulse response of the prototype filter, l is the phase, and M is the total number of phases (and number of channels). This practical implementation of this approach will then use a FIR filter. The prototype filter is usually defined by a window function. In spectrum sensing, it is of key importance to avoid the energy of adjacent bands interfering with the frequency band that we want to analyze; this effect is called spectrum leakage. For this purpose, a window with an excellent passband response, rapid decay, and reject band should be used.

In Figure 2.4, three common window functions are evaluated as a low pass with a cutoff frequency of 1/20 of the normalized spectrum using an equal number of coefficients.

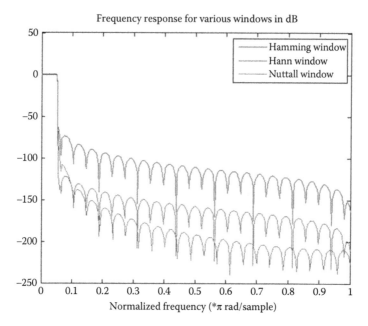

FIGURE 2.4 Evaluation of the frequency response from three common windows used as prototype filters.

It is visible that the best choice of window is the Hann window, as it shows higher attenuation in the rejection band, which will decrease the amount of spectrum leakage.

2.4.1.3 Multitaper

In 1982, Thomson implemented a spectrum-estimation technique that uses more than one taper. In this multitaper spectral analysis method, the data sequence is multiplied not by one but by a sequence of orthogonal tapers that, in this case, are a set of sequences to form N periodograms [34]. The periodograms are then averaged to form the PSD estimation.

The advantage of using this method against the windowed spectrogram is that in the single-taper estimation a large portion of the signal is discarded using the window that in the extremes tends to zero, causing an increase on the variance of the spectrogram.

To estimate the spectrum with this method, the k eigencomponents must first be obtained using

$$\hat{S}_k(f) = \left|Y_k(f)\right|^2 = \left|\sum_{n=0}^{N-1} x(n) w_k(n) e^{-2\pi i f n}\right|^2 \tag{2.4}$$

where $w_k(n)$, $k = 1, 2, ..., K$ are the K orthogonal taper functions. Usually, two types of orthogonal taper functions are used, Slepian sequences or sinusoidal tapers [35].

The most-used taper functions are the Slepian sequences, also called discrete prolate spheroidal sequences (DPSS) [36]. They are defined by K number of sequences with N number of weights and a W bandwidth parameter in the baseband. If modulated in a frequency, W will, of course, be half bandwidth where the energy is concentrated.

So, how should K be chosen? First of all, if the leakage gets worse as K increases, but the variance will also decrease as K rises. For W, if it is increased, the resolution will decrease, but it can be compensated by using more tappers.

The first sequence of Slepian functions have an eigenvalue, λ_0, that is very close to unity, making the first taper the best in terms of spectral leakage for the particular choice of W.

The first $2NW - 1$ sequences also have eigenvalues that are close to unity, which is why $K = 2NW - 1$ is normally used [37]. The remaining sequences have worsening spectral concentration properties, but they can still be used if the objective is to achieve lower bandwidths.

We can then define K as a function of NW. In the literature, NW is sometimes referred to as half-bandwidth resolution and, along with the number of samples, will define our Slepian sequences. If we define $K = 2NW - 1$, then the lower limit of NW by definition will be $NW = 1.5$ as it is the lower amount of half-bandwidth resolution that still has 2 taper sequences. The maximum value of NW is more subjective, with some authors proposing a maximum value of $NW = 16$, as higher values NW will cause the resolution and the overall ideal behavior to be lost and complexity (as more tappers are needed) to increase [38].

Therefore, if a $K \leq 2NW - 1$ is used, where the eigenvalues of each of the sequences is very close to unity, the multitaper spectrum can be given by the direct multitaper estimate.

$$S_{MT}(f) = \frac{1}{K} \sum_{k=0}^{K-1} \hat{S}_k(f)$$

(2.5)

If using a higher number of sequences is intended or even if we only want more precision in our estimation, we do a weighted sum, where the weights are the eigenvalues of each sequence [39].

$$S_{MT}(f) = \frac{\sum_{k=0}^{K-1} \lambda_k^2 \hat{S}_k(f)}{\sum_{k=0}^{K-1} \lambda_k^2}$$

(2.6)

2.4.1.4 Cyclic Spectrum

To identify the presence of a cyclostationary digital signal in a given sampled spectrum, two mathematical operations are usually employed: cyclic autocorrelation and cyclic spectrum. These two methods show unique features when in the presence of a cyclostationary-based signal.

To understand cyclic autocorrelation, we first need to define what correlation is. *Correlation* is a mathematical operation that allows us to measure the degree to which two signals are similar and is the expected value of the inner product of the two signals. For autocorrelation, the procedure is similar, but instead of having two different signals, we use the inner product of the signal with itself at a different time. We know by the definition of a cyclostationary signal that the autocorrelation of a signal, R_x, is periodic. Let us define that period with T_0 and the lag with τ, then

$$R_x(t + T_0, \tau) = R_x(t, \tau)$$

(2.7)

If a periodic correlation exits, then it can be proved that $x(t)$ and its frequency-shifted version, $x(t)e^{j2\pi n/T_0 t}$, are correlated for any $n \in \mathbb{Z}$ [40]. Then we can define the cyclic autocorrelation as

$$R_x^\alpha(\tau) \triangleq E\left\{ x(t+\tau)x^*(t)\bar{e}^{j2\pi\alpha t} \right\}$$

(2.8)

with $\alpha = \dfrac{n}{T_0}$ and * being the signal conjugate. A stochastic process, $x(t)$, will then exhibit cyclostationarity at the cycle frequency α_c if $R_x^{\alpha_c}(t) \neq 0$ [40].

To implement the method in a digital signal processor, the discrete-time cyclic autocorrelation needs to be derived. Using N number of samples, the discrete cyclic autocorrelation is given by

$$R_x^\alpha(\tau) = \frac{1}{N} \sum_{n=0}^{N-1} x(n+\tau)x^*(n)e^{-j2\pi\alpha n}$$

(2.9)

Note that $R_x^\alpha(t)$ can be computed efficiently using the FFT of the product $x(n)x(n+\tau)$ [41].

To derive the cyclic spectrum, the process is similar to the Wiener-Khinchin theorem but applied to cyclic autocorrelation [40]. The cyclic spectrum is then

$$S_x^\alpha(f) = \int R_x^\alpha(\tau)e^{-j2\pi f\tau}d\tau \tag{2.10}$$

with $f = \dfrac{\pm\alpha}{2}$. It can be computed efficiently using the two-dimensional FFT of the cyclic autocorrelation.

2.4.1.5 Eigenvalues of the Covariance Matrix

Let us consider the received data vector $x(n) = [x(0), x(1), ..., x(N-1)]$ with N samples. Its multidimensional variant, with M dimensions (also called the smoothing factor), is obtained by $x_k(n) = [x(k-1+M-1), ..., x(k-1)]$, $k = 1, ..., K$, where $K = N - M + 1$ [42]. These lagged vectors form the columns of the related data matrix X, called a trajectory matrix [43]. The columns of an M-dimension X matrix are given by the x_k vectors.

$$X(n) = \begin{bmatrix} x(M-1) & x(M) & \cdots & x(N-1) \\ x(M-2) & x(M-1) & \cdots & x(N-2) \\ \vdots & \vdots & \ddots & \vdots \\ x(0) & x(1) & \cdots & x(N-M) \end{bmatrix} \tag{2.11}$$

Note that the trajectory matrix has identical entries along its diagonals. Such a matrix is called a Toeplitz matrix. From a segment of N samples of received data $x(n)$ M consecutive output samples can be formed. In practice, the covariance matrix of the input signal is unknown as it is analyzed in only a finite number of samples. Thus, the unstructured classical estimator of $\boldsymbol{R}$, the sample covariance matrix, can be used and is defined as

$$R \triangleq \frac{1}{N}\sum_{n=1}^{N}X[n]*X[n]^H \tag{2.12}$$

where H is the Hermitian [44]. Note that the sample covariance matrix, $\boldsymbol{R}$, is both symmetric and Toeplitz. The eigenvalues of the covariance matrix can then be calculated from the characteristic equation

$$\det(\boldsymbol{R} - \lambda I) = 0 \tag{2.13}$$

where det is the determinant, λ gives us the eigenvalues, and I is the identity matrix. The $\lambda_{max} = \lambda_1 > ... > \lambda_L = \lambda_{min}$ is then the eigenvalue decomposition of the covariance matrix of the input signal.

2.4.2 Signal Detection

With every spectrum analysis measurement, there is the need to decide which frequency bands are occupied or which are free to use. The classic assumption model is to consider that the analyzed signal in a given band can be in one of two states. This binary hypothesis can be formulated in the following way:

$$\begin{aligned} H_0 &: y(n) = w(n) \\ H_1 &: y(n) = s(n) + w(n) \end{aligned} \tag{2.14}$$

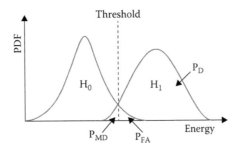

FIGURE 2.5 Probability diagram for the signal detection model.

with $y(n)$ being the band-limited signal, $s(n)$ the received signal coming from a user, and $w(n)$ the noise floor. By analyzing the spectrum, we need to be able to decide what specific frequencies represent noise or signal. This decision is made by assuming a threshold for each frequency band at the output of the spectrum analysis block. Therefore, a binary decision is made by comparing the energy estimates with the threshold, as shown in Figure 2.5.

The most common method for detecting the existence of a signal, in a given SNR, is the energy detection method. For this method, the acquired signal is constantly evaluated against a given threshold. This threshold is either a fixed value or it is adaptive. It is usually defined with the following three parameters in mind:

- **Probability of detection** (P_D), that is, the probability of detecting a signal that is indeed present.
- **Probability of false alarm** (P_{FA}), which is the probability of detecting a signal where only noise exists.
- **Probability of miss detection** (P_{MD}), which is the probability of missing the detection of a signal.

The most popular application of the energy detection method, with an adaptive threshold, is the constant false-alarm rate (CFAR) detector as it avoids the need to know the distribution of the transmitted signal [45]. The CFAR method proposes a dynamic threshold value based on a fixed probability of false alarm. By analyzing a fixed number of samples and using the neighboring ones (the training cells), both previous and future, as a reference for floor noise, we can then define a threshold that follows the CFAR needs [46,47].

This classical application has some faults and cannot be directly applied to maintaining a CFAR in a spectrum-sensing unit. For instance, if a signal with a flat-top frequency response is constantly transmitting the same energy level, then this value could be associated to noise or to a constant-power wideband signal.

To have a true CFAR, we must know the noise floor probability distribution function (PDF). As it is expected that this value does not remain constant in time due to all the variable sources of RF noise that can be near the device, it is important to frequently update the PDF estimation of the noise floor. To do so, we need to constantly evaluate the output from our spectrum analysis algorithms to obtain the noise power mean and variance to then create a threshold that will respect the value of false alarm rate that we want to achieve.

To study the application of signal detection methods, in conjunction with a spectrum analysis algorithm for spectrum sensing, the noise estimation task is then of critical importance.

2.4.3 NOISE MODEL

There are key disadvantages in the fixed noise model estimation. In fixed noise probability distribution–based detection, as previously discussed, the signal can have a similar distribution to the noise, making them indistinguishable from one another. In eigenvalues analysis, the detection is made based on the frequency analysis ratio distribution, making it impossible to differentiate a signal with a flat-top frequency response from the noise floor. Due to this, we will then focus on a dynamic noise model estimation, as it does not suffer from those limitations.

The practical RF noise floor behavior is dynamic both in energy and in frequency response. This is due to internal noise contributions from the used front end and both atmospheric and man-made noise sources. This creates the need for a dynamic noise floor estimation that can be used to create the threshold to classify noise and signal [48].

When the noise floor is not known, its value needs to be estimated; however, this estimation will not give a perfectly accurate value, and estimation error is to be expected. As the noise floor estimation is used to determine the signal detection threshold, this estimation uncertainty will define a minimum value of SNR under which detection is not possible with any desired performance. This limit is called the detection SNR wall and is defined by a certain noise floor uncertainty [49]. As the uncertainty is related to the used number of samples for the estimation, by increasing the observed noise samples to obtain the noise floor estimation, the SNR wall effect can be avoided [50]. Due to this, it is recommended that the dynamic noise floor estimation uses as many samples as possible.

There are two methods to obtain an adaptive estimation for the noise floor. The first method estimates the noise floor for each frequency and is applied if the frequency bands are expected to have a duty cycle of usage. When the channel is not in use, the power value can be used to evaluate the noise floor and get an updated estimation. The second method uses information from frequency bands to estimate the noise floor and is applied if the frequency band is constantly being used, as happens in FM broadcast. An example of an acquisition of the FM broadcast spectrum in a typical urban area is shown in Figure 2.6.

In this case, the unoccupied channels might then be used to estimate the noise floor. This of course requires that the system be precisely equalized to obtain similar noise floor behavior in the entire analyzed frequency spectrum. This method may be unsuitable in presence of noise with non-uniform frequency response (colored noise).

In shared frequency bands, where multiple users are present, it is expected that different sources of signal, depending on the distance to the spectrum unit, will be received at different energy levels. This

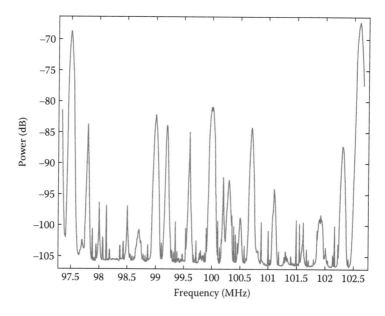

FIGURE 2.6   A power spectrum of part of the FM broadcast spectrum where the broadcast stations and the noise floor in between are visible.

fact occurs in various technologies, such as Wi-Fi and LTE, where the RF bands are shared between all users. When a large amount of users occupy actively the allocated RF band, then only small, sporadic intervals of time exist in which the channel is not occupied. This relatively small period where only the noise floor is present can make the task of estimating its level extremely complex.

It is common that RF protocols have a short period of time where there are no transmissions on the shared medium. For the LTE protocol, in TDD configuration, there is for every radio frame, a 10 ms special frame. This special frame is segmented into three periods; first is the downlink pilot time slot, then a guard period, and finally the uplink pilot time slot. There are no transmissions in the guard period, and the smallest duration for this segment is approximately $71\,\mu s$ [51].

In the 802.11 standard, the distributed coordination function (DCF) determines that a waiting period must exist so that the station can check the medium before transmitting data. This period, where the user needs to sense the channel before using it, is called the DCF interframe space (DIFS). Only after the medium is idle during the DIFS interval is the station allowed to transmit [52]. Even if only one user is using the channel to communicate with the access point, there still is a minimum specified time slot that is not occupied, called the short interframe space (SIFS) interval. The SIFS interval defines the minimum time slot between data fragments and immediate response from the medium user (e.g., data acknowledge) when the medium is not being occupied.

The worst-case scenario is for 802.11n where the spectrum-sensing unit needs to estimate the noise floor using SIFS intervals of $10\,\mu s$ or DIFS intervals of $34\,\mu s$.

One method that can be used to estimate the noise floor that does not require any *a priori* knowledge is to acquire enough samples to allow a precise estimation, based on observed data, of all the acquired unobservable underlying processes of the probability density function. These processes can be divided into the following two groups:

- **Noise floor** (H_0), when no user is present on the allocated RF band and all that is observable is the noise produced by the environment and by the RF front end itself.
- **User occupation** (H_1), when a given user transmits a signal with a given and approximately constant power.

The probability distribution function of an acquired time interval signal will then be the sum of all the processes' probability densities.

If an energy estimator that averages a large number of samples is used by the central limit theorem, the histogram of the analyzed process converges toward a Gaussian distribution [53]. The result in a spectrum-sensing scenario will then be the sum of all the Gaussian components, giving rise to a PDF that follows a Gaussian mixture model (GMM). The lowest mean component will then be the noise floor and the highest value components will be different users transmitting at different powers or at different distances.

2.5 METHOD PERFORMANCE ANALYSIS

A spectrum-sensing system is composed of three blocks: spectrum analysis, signal detection, and Wodel estimation. All of these components are chosen and combined to achieve the best implementation for a given application. Due to this, it is of key importance to establish a performance criterion between spectrum-sensing implementations to be able to compare them. While there are many aspects that could be used to compare the performance of spectrum-sensing methods, there is a method that can analyze the performance over a wide range of conditions. This method is called the ROC curve.

The ROC curve is a standard comparison method in spectrum-sensing systems. The graph is plotted with the probability of false alarm (P_{Fa}) in the horizontal axis and the probability of detection (P_D) in the vertical axis. The curve points are determined by the binary classification of the analyzed data. The data samples are classified as being part of either the noise floor or the signal.

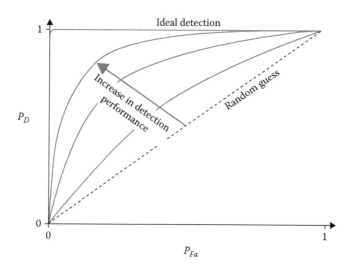

FIGURE 2.7 Performance analysis in the ROC space.

The better the accuracy of this classification, the closer the ROC curve will be to the point of $P_D = 1$ and $P_{Fa} = 0$, as seen in Figure 2.7.

To obtain an empirical ROC curve for a given spectrum-sensing method, a data segment composed by both signal and noise data must be analyzed. In this data, there must be *a priori* knowledge of which point belongs to each state (signal or noise). The data is then classified using a varying threshold. The lowest and highest values that can be attributed to the threshold are given by data segment boundary values. By varying the threshold along these values, it is possible to obtain the empirical classification of the analyzed data. For each threshold, the binary results are then compared to the known attributions of each data point (signal or noise) to obtain the P_{Fa} and P_D. The final ROC curve will then be a two-dimensional plot where each (P_{Fa}, P_D) point is given by a different threshold. Using a changeless threshold vector, this process can be done as many times as needed to obtain a closer approximation to the theoretical ROC curve. This method is independent from any threshold calculation technique.

If the interest is to evaluate different methods of threshold calculation, an identical empirical ROC curve determination can be done. The difference is that in this situation, there is not a varying threshold; the threshold is calculated for each data segment. In this empirical determination, each ROC curve point is not determined by a different threshold but is obtained for each data segment. This is of critical importance for comparing different noise floor estimation techniques and threshold calculation methods, usually for CFAR.

The simplest comparative method for the performance of spectrum-sensing methods is then to compare the ROC curves obtained from each one. Usually the closest curve to the point $P_D = 1$ and $P_{Fa} = 0$ shows the best accuracy for detection with the lowest percentage of false alarms. If the plotted curves are close to each other, visual comparison may not be possible. In those situations, the area under the curve (AUC) parameter is normally used. The AUC is obtained by calculating the integral of the ROC curve. An estimation of it can be obtained by numerical integration of the empirically obtained ROC curve, $f(x)$. Assuming a uniformly spaced grid of N points with a a grid spacing of h, the AUC can be obtained by

$$\hat{AUC} = \frac{h}{2} \sum_{x=0}^{N-1} (f(x_{k+1}) + (f(x_k))$$ (2.15)

There is an expected empirical error due to the effect of the trapezium rule, more evidently by splitting the range from zero to one into a number of subintervals, that approximates the curve and sums the area of the trapezium.

2.6 SPATIAL SPECTRUM SENSING

In the 3rd Generation Partnership Project (3GPP) release 7, MIMO was introduced in mobile networks. Currently, LTE allows up to 8×8 MIMO, and implementation of a greater number of antennas, as massive MIMO arrays, is currently under discussion [54–56]. The usage of multiple antennas allows for beamforming transmissions, thus supporting highly directional communications and improving radio link performance in comparison with omnidirectional communication.

Since 802.11n, Wi-Fi also supports MIMO. While 4×4 implementations were supported initially, the 802.11ac standard supports up to 8×8 [57]. With the expected increase of MIMO usage and beamforming implementations, the determination of the angle of arrival of RF transmissions will be of incredible importance in the future.

This is the basis of spatial spectrum sensing—determining which frequency bands of the RFS are currently under occupation and their direction. For this purpose, each detected transmission is analyzed by their direction of arrival (DoA) to allow a spatial spectrum-sensing analysis.

The DoA describes the direction from which a RF signal is received [58]. In a receiver with an antenna array, the DoA can be calculated by measuring the phase difference between the individual antennas. This is due to the difference in propagation distances from the transmitted signal to the individual receiving antennas. Each antenna will obtain a different phase shift of the same original RF signal. Assuming a two-antenna receiver array in a far-field case, where it is assumed that the wave propagates in parallel through space, we have the scenario shown in Figure 2.8 [59].

The sensed RF signal arrives to the two elements with a path length difference, l, that can be obtained from

$$l = d \sin \theta \qquad (2.16)$$

where d is the distance between the two receiver antennas, and θ is the DoA of the incoming wave. Let α be the phase difference between the two receiving elements that can be expressed by

$$\alpha = 2\pi\lambda l \qquad (2.17)$$

where λ is the wavelength of the incoming wave. By replacing the path length difference, l, the DoA can then be obtained by

$$\theta = \sin^{-1}\left(\frac{\alpha}{2\pi d / \lambda}\right) \qquad (2.18)$$

In a dual-antenna system, the phase difference can be calculated using the inner product of the two antennas' received data. The inner product will then give the phase relationship between the two antennas.

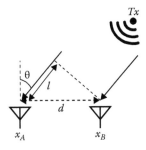

FIGURE 2.8 Direction of arrival scenario with a *Tx* transmission being received by two elements, x_A and x_B.

$$\alpha(n) = \angle x_A(n)x_B^*(n) \tag{2.19}$$

where $x_A(n)$ and $x_B(n)$ are the complex sampled signal from the first and the second antenna, respectively, * indicates the complex conjugate, and $\angle$ denotes the phase angle of the complex values.

2.7 CONCLUSION

In this chapter, the topic of CR, with focus on the spectrum-sensing component of the concept, was explored. The dynamic spectrum access models for CR were explained. The components that form a spectrum-sensing unit were identified and analyzed. Several spectrum analysis methods were described, and the noise model and signal detection components of a spectrum-sensing unit were explored. Furthermore, to evaluate spectrum-sensing methods, the ROC curve for comparative performance analysis was described. The chapter ends with the recent trend of spatial spectrum sensing that will allow an even more efficient reuse of spectrum resources by also using the direction of transmission in spectrum sensing.

REFERENCES

1. M. H. Islam *et al.*, "Spectrum survey in Singapore: Occupancy measurements and analyses," *Proceedings of the 3rd International Conference on Cognitive Radio Oriented Wireless Networks and Communications, CrownCom*, pp. 1–7, 2008.
2. R. Urban, T. Korinek, and P. Pechac, "Broadband spectrum survey measurements for cognitive radio applications," *Radioengineering*, vol. 21, no. 4, pp. 1101–1109, 2012.
3. J. Mitola, "Software radios: Survey, critical evaluation and future directions," *IEEE Aerospace and Electronic Systems Magazine*, vol. 8, no. 4, pp. 25–36, Apr. 1993.
4. S. Galih, M. Hoffmann, and T. Kaiser, "Low Cost Implementation for Synchronization in Distributed Multi Antenna Using USRP/GNU-Radio," *Proceedings of the 1st International Conference on Information Technology, Computer, and Electrical Engineering*, pp. 457–460, 2014.
5. M. B. Sruthi, M. Abirami, A. Manikkoth, R. Gandhiraj, and K. P. Soman, "Low cost digital transceiver design for Software Defined Radio using RTL-SDR," *2013 International Multi-Conference on Automation, Computing, Communication, Control and Compressed Sensing, 2013 iMac4s*, pp. 852–855, 2013.
6. Y. Kwag, J. Jung, I. Woo, and M. Park, "Multi-Band Multi-Mode SDR Radar Platform," *Synthetic Aperture Radar (APSAR)*, pp. 46–49, 2015.
7. A. Banerjee and J. Cho, "Phantom Net: Research Infrastructure for Mobile Networking, Cloud Computing and Software-Defined Networking," *MOBILE PLATFORMS*, vol. 19, no. 2, pp. 28–33, 2015.
8. J. Mitola and G. Q. Maguire, "Cognitive radio: making software radios more personal," *IEEE Personal Communications*, vol. 6, no. August, pp. 13–18, 1999.
9. Q. Zhao and B. M. Sadler, "A Survey of Dynamic Spectrum Access," *Signal Processing Magazine, IEEE*, vol. 24, no. May, pp. 79–89, 2007.
10. T. Irnich, J. Kronander, Y. Selen, and G. Li, "Spectrum sharing scenarios and resulting technical requirements for 5G systems," *IEEE International Symposium on Personal, Indoor and Mobile Radio Communications, PIMRC*, pp. 127–132, 2013.
11. J. Khun-Jush, P. Bender, B. Deschamps, and M. Gundlach, "Licensed shared access as complementary approach to meet spectrum demands: Benefits for next generation cellular systems," *ETSI Workshop on Reconfigurable Radio Systems*, no. December, pp. 1–7, 2012.
12. A. Kliks, O. Holland, A. Basaure, and M. Matinmikko, "Spectrum and License Flexibility for 5G Networks," *IEEE Communications Magazine*, no. July, pp. 42–49, 2015.
13. H. Wang, G. Noh, D. Kim, S. Kim, and D. Hong, "Advanced sensing techniques of energy detection in cognitive radios," *Journal of Communications and Networks*, vol. 12, no. 1, pp. 19–29, Feb. 2010.
14. D. Niyato, E. Hossain, and Z. Han, "Dynamic spectrum access in IEEE 802.22-based cognitive wireless networks: a game theoretic model for competitive spectrum bidding and pricing," *Wireless Communications, IEEE*, vol. 16, no. April, pp. 16–23, 2009.

15. A. B. Flores, R. E. Guerra, E. W. Knightly, P. Ecclesine, and S. Pandey, "IEEE 802.11af: A standard for TV white space spectrum sharing," *IEEE Communications Magazine*, vol. 51, no. October, pp. 92–100, 2013.
16. C. Stevenson, G. Chouinard, S. Shellhammer, and W. Caldwell, "IEEE 802.22: The first cognitive radio wireless regional area network standard," *IEEE Communications Magazine*, vol. 47, no. 1, pp. 130–138, Jan. 2009.
17. ITU-R, "Radio Regulations, Edition of 2012." [Online]. Available: http://www.itu.int/dms_pub/itu-s/oth/02/02/S02020000244501PDFE.pdf.
18. E. Axell, G. Leus, E. G. Larsson, and H. V. Poor, "State-of-the-art and recent advances-Spectrum Sensing for Cognitive Radio," *IEEE Signal Processing Magazine*, vol. 29, no. 3, pp. 101–116, 2012.
19. H. Afzal, M. R. Mufti, M. Nadeem, I. Awan, and U. S. Khan, "The Role of Spectrum Manager in IEEE 802 . 22 standard," *Journal of Basic and Applied Scientific Research*, vol. 4, no. 4, pp. 280–289, 2014.
20. G. Yu, C. Long, M. Xiang, and W. Xi, "A Novel Energy Detection Scheme Based on Dynamic Threshold in Cognitive Radio Systems," *Journal of Computational Information Systems*, vol. 8, no. 6, pp. 2245–2252, 2012.
21. M. S. Gast, *802.11n: A Survival Guide*. "O'Reilly Media, Inc.," 2012.
22. L. Lu, H. Wu, and S. Iyengar, "A novel robust detection algorithm for spectrum sensing," *Selected Areas in Communications...*, vol. 29, no. 2, pp. 1–11, 2011.
23. Y. Zeng and Y. Liang, "Eigenvalue-based spectrum sensing algorithms for cognitive radio," *IEEE Transactions on Communications*, vol. 57, no. 6, pp. 1784–1793, Jun. 2009.
24. S. Maleki, A. Pandharipande, and G. Leus, "Two-stage spectrum sensing for cognitive radios," *2010 IEEE International Conference on Acoustics, Speech and Signal Processing*, pp. 2946–2949, 2010.
25. T. Yucek and H. Arslam, "A Survey of Spectrum Sensing Algorithms for Congnitive Radio Applications," *IEEE Communications Surveys & Tutorials*, vol. 97, no. 1, pp. 805–823, 2009.
26. T. Mehta, N. Kumar, and S. S. Saini, "Comparison of Spectrum Sensing Techniques in Cognitive Radio Networks," *InternatIonal Journal of ElectronIcs and CommunIcatIon Technology*, vol. 7109, 2013.
27. D. D. Ariananda, M. K. Lakshmanan, and H. Nikoo, "A survey on spectrum sensing techniques for Cognitive Radio," *2009 Second International Workshop on Cognitive Radio and Advanced Spectrum Management*, pp. 74–79, 2009.
28. J. Antoni, "Cyclic spectral analysis in practice," *Mechanical Systems and Signal Processing*, vol. 21, no. 2, pp. 597–630, Feb. 2007.
29. S. Li and Y. Wang, "Method of modulation recognition of typical communication satellite signals based on cyclostationary," *2013 ICME International Conference on Complex Medical Engineering*, no. 4, pp. 268–273, May 2013.
30. A. M. Tomé, D. Malafaia, J. Vieira, A. R. Teixeira, and E. Lang, "Singular Spectrum Analysis for Narrowband Signal Enhancement and Detection," *Digital Signal Processing*, pp. 1–18, 2016.
31. A. Schuster, "On the investigation of hidden periodicities with application to a supposed 26 day period of meteorological phenomena," *Journal of Geophysical Research*, vol. 3, no. 1, p. 13, 1898.
32. G. a. Prieto, R. L. Parker, D. J. Thomson, F. L. Vernon, and R. L. Graham, "Reducing the bias of multitaper spectrum estimates," *Geophysical Journal International*, vol. 171, no. 3, pp. 1269–1281, Oct. 2007.
33. A. Mertins, *Signal Analysis: Wavelets, Filter Banks, Time-Frequency Transforms and Applications*. John Wiley & Sons, 1999.
34. D. Thomson, "Spectrum estimation and harmonic analysis," *Proceedings of the IEEE*, vol. 70, 1982.
35. K. S. Riedel and a. Sidorenko, "Minimum bias multiple taper spectral estimation," *IEEE Transactions on Signal Processing*, vol. 43, no. 1, pp. 188–195, 1995.
36. D. Slepian and H. Pollak, "Prolate spheroidal wave functions, Fourier analysis and uncertainty," *Bell System Technical Journal*, 1961.
37. P. Mitra, "Neural Signal Processing: Quantitative Analysis of Neural Activity," *Washington, DC: Society for Neuroscience.*, pp. 1–98, 2008.
38. O. A. Alghamdi, M. A. Abu-rgheff, and M. Z. Ahmed, "MTM Parameters Optimization for 64-FFT Cognitive Radio Spectrum Sensing using Monte Carlo Simulation," *Second International Conference on Emerging Network Intelligence*, no. c, pp. 107–113, 2010.
39. P. S. Reddy and P. Palanisamy, "Multitaper Spectrum Sensing Using Sinusoidal Tapers," *2013 2nd International Conference on Advanced Computing, Networking and Security*, pp. 47–50, Dec. 2013.
40. W. a. Gardner, A. Napolitano, and L. Paura, "Cyclostationarity: Half a century of research," *Signal Processing*, vol. 86, no. 4, pp. 639–697, Apr. 2006.
41. G. . Giannakis, E. V. K. Madisetti, and D. B. Williams, "Cyclostationary Signal Analysis," in *Statistical Signal Processing Section of Digital Signal Processing Handbook*, CRC Press, 1999.

42. M. Hamid, K. Barbé, N. Björsell, and W. Van Moer, "Spectrum sensing through spectrum discrimina-
tor and maximum minimum eigenvalue detector: A comparative study," *IEEE I2MTC - International
Instrumentation and Measurement Technology Conference, Proceedings*, pp. 2252–2256, 2012.

43. N. Golyandina, V. Nekrutkin, and A. A. Zhigljavsky, *Analysis of Time Series Structure : SSA and Related
Techniques*. CRC Press, 2001.

44. W. Pan, H. Wang, and L. Shen, "Covariance matrix based spectrum sensing for OFDM based cognitive
radio," *2012 International Conference on Systems and Informatics (ICSAI2012)*, no. Icsai, pp. 1426–
1430, 2012.

45. S. M. Kay, *Fundamentals of statistical signal processing - Detection Theory*. Prentice Hall, 1998.

46. A. De Maio, A. Farina, and G. Foglia, "Design and experimental validation of Knowledge-based CFAR
detectors," *IEEE National Radar Conference - Proceedings*, no. January, pp. 128–135, 2006.

47. V. Amanipour and A. Olfat, "CFAR detection for multistatic radar," *Signal Processing*, vol. 91, no. 1,
pp. 28–37, Jan. 2011.

48. J. J. LEHTOMÄKI, R. VUOHTONIEMI, K. UMEBAYASHI, and J.-P. MÄKELÄ, "Energy Detection
Based Estimation of Channel Occupancy Rate with Adaptive Noise Estimation," *IEICE Trans. Commun.*,
vol. E95–B, no. 4, pp. 1076–1084, 2012.

49. R. Tandra and A. Sahai, "SNR Walls for Signal Detection," *IEEE Journal of Selected Topics in Signal
Processing*, vol. 2, no. 1, pp. 4–17, Feb. 2008.

50. A. Mariani, A. Giorgetti, and M. Chiani, "SNR Wall for Energy Detection with Noise Power Estimation,"
2011 IEEE International Conference on Communications (ICC 2011), pp. 1–6, 2011.

51. T. Specification, "TS 136 211 - V12.3.0 - LTE; Evolved Universal Terrestrial Radio Access (E-UTRA);
Physical channels and modulation (3GPP TS 36.211 version 12.3.0 Release 12)," vol. 0, 2014.

52. A. Duda, "Understanding the Performance of 802.11 Networks," *2008 IEEE 19th International
Symposium on Personal, Indoor and Mobile Radio Communications*, no. 1, pp. 1–6, 2008.

53. A. Papoulis and S. Pillai, *Probability, random variables, and stochastic processes*. McGraw-Hill, 2002.

54. 3GPP, "Overview of 3GPP Release 7," 2007. [Online]. Available: www.3gpp.org/specifications/
releases/73-release-7.

55. K. Werner, H. Asplund, B. Halvarsson, N. Jaldén, D. V. P. Figueiredo, and A. K. Kathrein, "LTE-A Field
Measurements : 8 x 8 MIMO and Carrier Aggregation," *Vehicular Technology Conference (VTC Spring)*,
2013 IEEE 77th, 2013.

56. S. Mumtaz, K. M. Saidul Huq, and J. Rodriguez, "Direct mobile-to-mobile communication: Paradigm for
5G," *IEEE Wireless Communications*, vol. 21, no. 5, pp. 14–23, 2014.

57. M. S. Gast, *802.11ac: A Survival Guide*. O'Reilly Media, 2013.

58. G. Vinci, F. Barbon, B. Laemmle, R. Weigel, and A. Koelpin, "Wide-Range , Dual Six-Port based
Direction-Of-Arrival Detector," *7th German Microwave Conference (GeMiC)*, pp. 1–4, 2012.

59. M. M. Hyder and K. Mahata, "Direction-of-arrival Estimation Using a Mixed L2,0 Norm Approximation,"
IEEE Transactions on Signal Processing, vol. 58, no. 9, pp. 4646–4655, 2010.

3 Machine Learning Techniques for Wideband Spectrum Sensing in Cognitive Radio Networks

Sufi Tabassum Gul, Asad Ullah Omer, and Abdul Majid

CONTENTS

3.1 INTRODUCTION

Today, wireless communication has become an essential part of our lives. The whole world is accelerating toward the advancement of technology, and the demand for wireless communication increases daily, so to keep up with demand, wireless communication must advance at the same pace. The backbone of wireless communication is the radio spectrum as this communication is being carried out with the help of electromagnetic waves which are differentiated on the basis of frequencies. Radio spectrum is the range of those frequencies that can be electromagnetically radiated (3 Hz to 300 GHz). The whole radio frequency (RF) band can be categorized as a *licensed band* or an *unlicensed band*. A licensed band is issued by some specific governing body within a country or a region. This governing body must ensure that its regional spectrum allocation is within the limits defined by the International Telecommunication Union (ITU) (e.g., in the US, the Federal Communications Commission [FCC] deals with spectrum allocation issues). The regulatory body must ensure that the licensed users are protected from interference which may be due to other systems. Licensed users face this interference when some other users try to access the same licensed band. The licensed users have a band of frequencies (depending upon the demand), a geographical region, and allowable operational parameters. An unlicensed band is free for all, but users must abide by the rules which are finalized by their respective regulatory agencies to avoid interference with other users communicating in the same band.

Increasing demand of this limited radio spectrum is challenging advancement in wireless communication technologies. Numerous multiple-access techniques are deployed at the user end, and various other steps have been taken (e.g., cell splitting, cell sectoring in mobile communication) to increase the utilization of bands to some extent, but the problem still remains to be solved. In an unlicensed band, transmitter, say, a primary transmitter, is transmitting and at the same time in the vicinity of this primary transmitter, some other system's receiver comes in, where both are using the same band, then the second system's receiver will have noisy reception if the level of interference is greater than some predefined value. This is the drawback of direct spectrum access by unlicensed users. Nonetheless, the demand of unlicensed spectrum access is ever increasing, and hence there is need to propose smarter solutions which can fulfill this growing demand [1].

For licensed bands, it has been discovered that they are not always occupied by their users, and hence these bands are underutilized. Different surveys have been performed for spectrum measurements. The authors of [2] demonstrated this underutilization during a spectrum survey performed in Singapore in 2008 when they found multiple *spectrum holes (SHs)* in the licensed band and concluded that certain bands are underutilized. In this chapter, the term *spectrum hole* refers to such frequencies or bands of frequencies that are not active in the communication (i.e., the users are not transmitting at that frequency or a band of frequencies at that particular time, making it available for use).

In a survey performed by Habibul Islam et al. [2], the highest occupancy was found in GSM-900 and the broadcasting bands while a moderate occupancy was detected in GSM-1800, radar and digital cellular bands. Likewise, in some bands, low or no utilization was seen. In another survey carried out in downtown Berkely, the percentage utilization of a 6 GHz wide signal received at a sampling rate of 20GS/s for 50 μs duration was estimated and plotted against power spectral density (PSD) [3]. The authors of [3] discovered that the band from 0~2 GHz was almost 50% occupied while the percentage occupancy of rest of the band was less than 10%.

It is worth mentioning that this situation is not limited to Singapore and Berkeley; in fact, this scenario is prevailing all over the world. Although spectrum usage is different in different countries, a broader picture is clear now. There are SHs in the licensed bands that can be opportunistically used while meeting the *primary user* (PU) constraints. The PU is the registered user of the licensed bands. By opportunistically using the SHs, the spectrum utilization can be significantly improved and this can be made possible by *cognitive radio* (CR), which is the promising solution for this spectrum scarcity issue. It was Joseph Mitola who first introduced the concept of CR in 1999 [4]. The FCC opened a new horizon named IEEE 802.22 when it learned about these SHs and the underutilization of licensed spectrum. This charter defines itself as, "The charter of IEEE 802.22, the Working Group on Wireless Regional Area Networks (WRANs), under the PAR approved by the IEEE-SA Standards Board is to develop a standard for a CR-based PHY/MAC/air-interface for use by license-exempt devices on a non-interfering basis in spectrum that is allocated to the TV Broadcast Service" [5]. The decision for the utilization of a band (i.e., whether the band is free and can be allocated) is made by locating SHs in the gathered information of the spectrum, which can be performed by CR. To understand CR, let us first introduce the *software-defined radio (SDR)*, which is the enabler of CR.

3.1.1 SOFTWARE-DEFINED RADIO

The term *software radio* was coined by Joe Mitola in 1991 to refer to the class of reprogrammable or reconfigurable radios [6]. SDR is an intelligent radio whose hardware is generic and whose parameters can be changed by updating the software; hence, there is no need to change the hardware whenever a change is required [7]. These parameters, in return, can change different modulation, demodulation, and filtering techniques etc. so the hardware remains the same, and by simply updating the software changes, the functionality of the radio can be made adaptable. In other words, the same piece of hardware can perform different functions at different times. In recent times, with tremendous increases in technology development and proliferation, design of multi-standard equipment with several wireless access modes is also gaining popularity. In this context, parametrization techniques can help in the optimal design of multi-standard SDR devices [8,9]. With all these benefits, SDR becomes a potential solution in the scenarios that require an adaptable and reconfigurable radio to incorporate different changes from time to time. The shortcoming of SDR is that it does not have a sensing property. So an extended form of SDR is considered in which the property of *parameter sensing* is added. This extended version of SDR is called CR.

3.1.2 COGNITIVE RADIO

Cognitive radio is a tool whose operational parameters can be changed at runtime. The Federal Communications Commission (FCC) defines CR as, "A radio or system that senses its operational electromagnetic environment and can dynamically and autonomously adjust its radio operating parameters to modify system operation, such as maximize throughput, mitigate interference, facilitate interoperability, access secondary markets" [10]. One of the main capabilities of a CR device is sensing, making it aware of its environment. A CR must quickly and robustly determine the available portions of the radio spectrum. It must have the capability to provide a generic picture of the entire radio spectrum of interest. The remainder of the processing carried out by the CR device depends on faithful initial sensing. It can be safely said that spectrum sensing is the backbone of CR. As a result, CR is highly recommended for efficient utilization of SHs as it can dynamically vary its radio operating parameters and easily handle the randomly allocated SHs. CR can help in the opportunistic use of the radio frequency spectrum by maximizing the utilization of licensed bands while helping to avoid interference caused to PU (licensed user) by *secondary users* (SUs) (unlicensed user). This CR block is basically implemented in the SU that needs to use the SH opportunistically while the PU has no knowledge of this mechanism and does not need to be modified or worried. This is so because the PU always has the highest priority, and as soon as it wants to communicate, it will be given the charge back by SU. A SU can use the band only when it is not occupied by a PU, and as soon as a secondary detects the presence of a PU, it vacates the occupied SH and jumps to a newer one.

Spectrum sensing can be divided broadly into narrowband spectrum sensing (NSS) and wideband spectrum sensing (WSS) [11]. For NSS, many classical techniques have been proposed (e.g., energy detection, matched-filter detection, and cyclostationarity-based detection). However, these techniques cannot be applied directly for WSS. First, the wideband is subdivided into equal, non-overlapping sub-bands and then the classical spectrum-sensing techniques can be applied to these sub-bands. Energy detection is selected in this chapter because of its simplicity. To determine the presence or absence of PU the energy thresholds for each sub-band are to be found simultaneously. The energy thresholds should be optimized considering all the imposed constraints. The ultimate objective is the simultaneous determination of the optimum threshold vector for all sub-bands that allows maximization of the aggregated opportunistic throughput of the CR network while keeping interference to the PU minimal. For optimal detection of wideband sensing, the optimum threshold is needed for correct allocation. To understand this optimum threshold, let us consider the following example. Suppose that M is the minimum magnitude among all frequencies which are in use by the PU, and N is the maximum magnitude among all frequencies which are not in use by the PU (and are available for the SU), then the optimum threshold must be between M and N to ensure perfect allocation without error. If it is not between these two limits, then the wrong allocation will be made, which will result in either a disturbance in the PU's communications or non-allocation of available SH to the SU and consequently inefficient spectrum utilization. Since the optimum threshold is the parameter upon which the decision is to be made, this crucial parameter must be calculated with some efficient and intelligent techniques to avoid any wrong allocation or underutilization of the licensed spectrum. To find the optimal threshold vector, different machine learning techniques (MLTs) can be applied in the CR networks [12,13]. In this chapter, two optimization techniques, namely *genetic algorithms* (GA) and *particle swarm optimization* (PSO), are employed to find the thresholds vector. The simulation results show that PSO outperforms GA. Its execution time is much less, and its implementation is easier than GA.

3.1.3 CHAPTER ORGANIZATION

This chapter is organized as follows. In Section 3.2, some classical estimation techniques are discussed in detail along with their basic block diagrams. Estimation techniques are further divided into parametric and non-parametric estimation techniques. In Section 3.3, the basic system model for WSS is explicated and then the problem formulation of both frameworks, multiband joint detection (MJD) and multiband sensing-time-adaptive joint detection (MSJD), are discussed. A new feature of sensing duration is introduced in the later framework. A brief scenario is also discussed which elaborates the probabilities of *receiver operating characteristics* (ROC). The problem is also formulated under convex optimization. In Section 3.4, the need for optimization using the MLTs in WSS is discussed, and principles of GA and PSO are described. In Section 3.5, simulation results are shown where the uniform detection method is simulated in the beginning, and later, GA along with PSO is implemented, and the results are compared in the form of execution time and net throughput. Simulation results show that PSO takes less time compared to GA, and it is proposed for both frameworks. Finally, Section 3.6 gives conclusions of this work.

3.2 CLASSICAL SPECTRUM-SENSING TECHNIQUES

Spectrum sensing is the process of gathering information about the magnitudes of frequencies. It is an essential task that CR has to perform in a very efficient way. High accuracy of CR in spectrum sensing will result in accurate SH allocation, but the complexity feature also has to be considered.

Different surveys have been studied where various spectrum-sensing techniques are discussed. In [14–16] energy detection, matched-filter detection, and cyclostationarity feature detection for spectrum sensing are discussed. In [17], energy detector and cyclostationary features for cooperative spectrum sensing are presented. These techniques, also called narrowband spectrum-sensing techniques, are employed when the frequency band is sufficiently narrow such that its response can be considered flat. More sophisticated techniques, namely filter bank spectrum sensing, multitaper spectrum sensing, wavelet-based spectrum sensing, and waveform-based sensing, are also addressed by the authors of [15,16]. Some of these techniques along with their merits and demerits are discussed in the following sections.

3.2.1 ENERGY-BASED SENSING

In this sensing technique, detection of PU's presence is based upon the sensed energy. It is a non-coherent detection scheme. If *a priori* knowledge of a primary signal is not available to the receiver, then a non-coherent technique is preferred, and in this case, energy detection is the optimal spectrum-sensing technique [18]. It can be categorized into two types: in the first, the energy is calculated, whereas in the second, the periodogram of the time domain signal is estimated. For optimal detection of the received signal, $O\left(\dfrac{1}{SNR^2}\right)$ samples are required [18]. Energy is measured by integrating the signal, that is, first squaring the signal and then taking its average, as shown in Figure 3.1.

The periodogram is calculated by taking the *Fast Fourier Transform (FFT)* of the signal and then, after squaring the magnitude of the FFT, its average is calculated, as shown in Figure 3.3.

The results are compared with a threshold which tells about the decision to be made as follows:

$$\text{Energy / Periodogram} < \text{threshold} \rightarrow \text{Band / Channel available}$$

$$\text{Energy / Periodogram} < \text{threshold} \rightarrow \text{Band / Channel not available}$$

(3.1)

3.2.2 MATCHED FILTER–BASED SENSING

Matched filter–based sensing is a coherent detection and is applicable only when knowledge of a primary signal is *a priori* to the SU. It is so because CR has to demodulate the received signal, hence different features necessary for correct demodulation are required. If this information is corrupted and becomes non-coherent, then CR could give poor results. A matched filter is a linear filter, and it maximizes the signal-to-noise ratio (SNR). In this technique, a filter whose impulse response, which is modified from a reference signal, is convolved with the received signal. This impulse response is modified by taking a mirrored replica of the reference signal and then shifting it in the time domain. For each PU, it requires a dedicated receiver, which is a major disadvantage of a matched filter. For optimal detection of the received signal, $O\left(\dfrac{1}{SNR}\right)$ samples are required [18]. The block diagram of matched-filter implementation is shown in Figure 3.4.

3.2.3 WAVEFORM-BASED SENSING

As evident from its name, this type of sensing is based on the known signal patterns that are transmitted to achieve synchronization [19]. If such patterns are known at the secondary side, then after the correlation of the received signal with an already known copy of this pattern, sensing can be performed. Patterns can be transmitted at the start or middle of each burst, or they can be sent in a regular pattern.

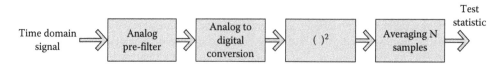

FIGURE 3.1   Energy calculation.

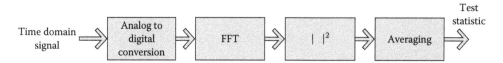

FIGURE 3.2   Periodogram estimation.

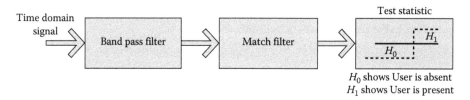

H_0 shows User is absent
H_1 shows User is present

FIGURE 3.3   Matched filter–based sensing.

3.2.4   CYCLOSTATIONARY DETECTION

This type of sensing is particularly suited for an environment where one cannot get rid of noise or interference and wants to detect modulated signals. By exploiting periodicity embedded intentionally in the received signal or in the statistical properties of the received signal, such as mean and autocorrelation, which are periodic in nature, one can faithfully detect the primary users [20]. With the help of this embedded periodicity, the primary signal can be distinguished from noise because stationary noise does not have a periodic behavior. In this sensing method, a particular band is selected from the received time domain signal and then its FFT and correlation are taken, followed by averaging and then feature detection, as shown in Figure 3.4.

3.2.5   FILTER BANK–BASED SPECTRUM SENSING

In this technique, a set of multiple bandpass filters are used, which are called filter banks. These banks are further divided into analysis and synthesis filter banks. In an analysis filter bank, the signal is decomposed into different parts (sub-bands). The number of these sub-bands depends upon the number of bandpass filters through which the signal is passed. In a synthesis filter bank, the signal is recovered by merging sub-bands after passing them through their respective filters. The input is passed through a set of bandpass filters, and their outputs are then squared and averaged, hence finding the spectral power estimate of each sub-band [21].

3.2.6   WAVELET TRANSFORM–BASED ESTIMATION

This technique is widely used in image processing for the detection of edges. In image processing, edges indicate that there is a sudden change in frequency, which is considered a change in the magnitudes of the frequencies. If this technique is applied to the *power spectral density* of a wideband channel, then the edges can be detected (i.e., transitions from/to occupied to/from

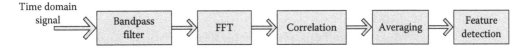

FIGURE 3.4 Cyclostationary feature detection.

vacant band). This can be done by detecting different bands, each with two edges, and then finding the power within those edges, which will give an estimate of the utilization of all bands (occupied/vacant) [22].

Among all the techniques described so far, the energy detection technique has less complexity because it does not require *a priori* knowledge of the PU and complicated receivers as in the case of matched filters. However, if complexity is not an issue, then filter bank techniques are a good choice. Within the energy detection techniques, the figure of merit of Blackman-Tukey is better compared to the conventional periodogram and its other variants. This is mainly because of its low variability along with a considerable resolution [23]. In the next section, WSS is explained.

3.3 WIDEBAND SPECTRUM SENSING (WSS)

Wideband spectrum sensing is very useful for CRs because the wider the band, the higher the number of free spaces or SHs that can be utilized by the SU. However, complex hardware and sophisticated software algorithms will be required to cope with the demand of WSS. Since WSS makes a single binary decision for the entire band, narrowband spectrum-sensing techniques, as discussed previously, cannot be applied directly for WSS because, by doing so, all the individual SH that lie with the wideband cannot be identified. A MJD technique for wideband spectrum sensing was proposed in 2009 [24], in which the authors treated spectrum sensing as an optimization problem and focused upon convex optimization to solve the problem formulation. Soon after, the MJD framework became popular, and much of the research work that was carried out by the CR community was based on this framework. In 2010, the same MJD framework was used for spectrum sensing, but at the optimization end, the author considered GA [1] instead of convex optimization. In 2011, some modifications in MJD were proposed, and it was named MSJD [25,26]. The basic themes of MJD and MSJD are the same, although in MSJD, sensing time is made adaptive and hence gives control between the speed and the quality of sensing. This feature was missing in MJD because sensing time was kept fixed. Let us describe the basic MJD/MSJD framework, the system model, and the optimization process.

3.3.1 BASIC FRAMEWORK

Consider a channel as a narrow sub-band whose state is random; that is, for some time, it is occupied by its PU, and for the remaining time, the PU is not communicating over this sub-band. This partial occupation of bandwidth allows opportunistic access; that is, when the PU is not present, the vacant sub-band can be allotted to some other user (e.g., the SU) for efficient utilization of the channel/sub-band, otherwise the resource will be underutilized. So the status of a sub-band can be either present or absent, making it a binary hypothesis testing problem, as shown in Table 3.1.

Then the detector is applied where a threshold (classifier) is needed, and either of the two decisions, mentioned in Table 3.1, are obtained. $H_{i,k}$ or $D_{i,k}$ shows the hypothesis and detection results of the kth sub-band where $i = 0, 1$. The SU relies upon the decision of the detector. If the detector declares it as vacant (D_0), then the sub-band can be allotted to the SU, whereas if it declares it as occupied (D_1), then PU will keep using the sub-band and will not be interfered with by the SU as SH allocation is not achieved.

TABLE 3.1

Binary Hypothesis and Detector Results

H_0 or D_0	Sub-band vacant
H_1 or D_1	Sub-band occupied

3.3.2 System Model

Let the system be described by following equations:

$$H_{0,k} : X_k = V_k$$

$$H_{1,k} : X_k = H_k S_k + V_k \text{ for } k = 1, 2, ..., Z \tag{3.2}$$

These relations are in the frequency domain where $H_{0,k}$ is referred as the hypothesis for the absent class; that is, the corresponding k^{th} sub-band is free from the PU's communication and can be allotted for opportunistic use. Similarly $H_{1,k}$ is referred to as the hypothesis for the present class; that is, the corresponding k^{th} sub-band is being utilized, so this sub-band is not available. X_k is the received signal, H_k is the channel gain, and S_k is the primary signal transmitted. V_k is the Gaussian noise having power σ_v^2. The signal power for the primary signal transmitted in the kth sub-band is assumed to be 1. The decision is made by the energy detector by computing energy over a specified number of samples. The following equation finds the energy of the kth sub-band:

$$Y_k = \frac{1}{M} \sum_{n=1}^{M} |x(n)|^2 \text{ for } k = 1, 2, ..., Z \tag{3.3}$$

The decision is performed by comparing the calculated energy with the threshold γ_k for the kth sub-band. The sub-band will be declared as free or occupied when its computed energy becomes less or greater than its corresponding threshold.

$$H_0 : Y_K > \gamma_k$$

$$H_1 : Y_K < \gamma_k \text{ for } k = 1, 2, ..., Z \tag{3.4}$$

Here, Y_K, which is the energy of the kth sub-band, acts as a random variable. If there is a large number of samples (e.g., $M > 10$), then the PDF of Y_K is approximated with the help of the central limit theorem, and Y_K comes under normal distribution. The distribution of Y_K under both hypotheses is as follows:

$$Y_k \sim N(\mu_{0,k}, \sigma_{0,k}^2) \text{ under hypothesis } H_{0,k}$$

$$Y_k \sim N(\mu_{1,k}, \sigma_{1,k}^2) \text{ under hypothesis } H_{1,k} \tag{3.5}$$

To know the role of this random variable, Y_K, and the threshold γ_k, and to get an idea about the probability of a false alarm and the probability of detection, which are used in the computation of throughput and interference, consider the following scenario in which a single k^{th} sub-band is discussed. Consider Q samples of Y_K over a single sub-band k, where each Y_K is computed from M

samples ($M > 10$). Half of the samples of Y_K (i.e., $\frac{Q}{2}$) are taken when the sub-band is actually free or when the PU is actually absent, so these $\frac{Q}{2}$ samples belong to an *absent class*. The other half is taken when the sub-band is actually occupied or when the PU is actually present over that sub-band, so these $\frac{Q}{2}$ samples belong to the other class, the *present class*. This is the actual representation before detection, but because of noise, our signal will be randomly affected, and after detection, different results are obtained. At the detection end, the absent and present classes appear as shown in Figure 3.5, where both classes are overlapping, which makes it impossible to detect both classes perfectly at a same time.

This overlapping is because of the noise impairment in the channel. The distribution of the absent class is $N(\mu_{0,k}, \sigma_{0,k}^2)$, while that of present class is $N(\mu_{1,k}, \sigma_{1,k}^2)$, and $\mu_{0,k}$ and $\sigma_{0,k}^2$ are the mean and standard deviation of the absent class, while $\mu_{1,k}$ and $\sigma_{1,k}^2$ are the mean and standard deviation of the present class. The classifier in Figure 3.5 is the threshold which is to be compared with Y_K. The right side ($Y_K > \gamma$) is detected as present and named the *detected present region*, and the left side ($Y_K < \gamma$) is detected as absent and named the *detected absent region*. The intersection of the absent class and the detected absent region, shown as Area 1 in Figure 3.5, is the probability of identifying the SHs, which refers to those events when the sub-band was actually free and the classifier also declared the sub-band as free ($Y_K < \gamma$). The intersection of the present class and the detected present region is the probability of detection, shown by Area 2 in Figure 3.5. This corresponds to those events when the sub-band was occupied and the classifier also declared it as occupied ($Y_K > \gamma$). Area 3 shows the intersection between the present class and the detected absent region and constitutes the probability of *missed detection*. This situation can be considered an event when the sub-band was actually occupied and the classifier misclassified it as free ($Y_K < \gamma$). Because of this misclassification, the SU will consider this sub-band to be a SH and will start its transmission in this sub-band, resulting in interference with the PU that is already transmitting in that sub-band (unbearable condition). Area 4 shows the intersection of the absent class and the detected present region, which is the probability of false alarm. This is an event when a free sub-band is misclassified as occupied ($Y_K > \gamma$) and the SU who is willing to use that sub-band is not allotted this free sub-band, reducing the maximum achievable throughput (a condition which is not desired). All these scenarios are summarized in Table 3.2.

In a nutshell, this γ is a crucial factor in decision making. The correct decision is based on the correct choice of the γ vector for different sub-bands. Optimization is needed to find the optimum threshold vector, for which MLTs can prove very handy and are discussed later.

3.3.3 MULTIBAND JOINT DETECTION (MJD)

According to MJD, a wideband channel is divided into multiple narrow sub-bands where narrowband-sensing techniques can be applied over these independent sub-bands. The design objective is to find the optimal threshold (classifier) vector $\gamma = [\gamma_1, \gamma_2, \dots \gamma_K]$ for K narrow sub-bands, where γ_k shows the threshold for the kth sub-band. The optimal classifier vector is achieved by limiting the aggregate interference, $I(\gamma)$, and maximizing the net throughput, $Q(\gamma)$.

3.3.3.1 Problem Formulation for MJD

The problem formulation of MJD is described in this section is as follows.

3.3.3.1.1 Signal to Noise Ratio (SNR)

SNR for the kth sub-band is defined by the following relation:

$$\Omega_k = \frac{E\{|S_k^2|\}H_k^2}{\sigma_\upsilon^2} \tag{3.6}$$

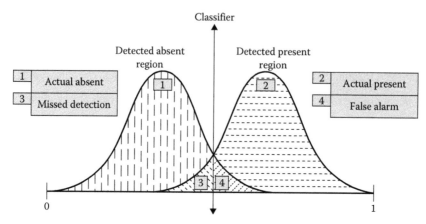

FIGURE 3.5   Classification scenario.

TABLE 3.2
Summary of Classification Scenarios

Regions	Probabilities	Hypotheses	Actual Class	Classifier Status
Area 1	Probability of identifying a SH (P_{sh})	$P(D_{0,k} \mid H_{0,k})$	Absent	$Y_K < \gamma$
Area 2	Probability of detection (P_d)	$P(D_{1,k} \mid H_{1,k})$	Present	$Y_K > \gamma$
Area 3	Probability of missed detection (P_{md})	$P(D_{0,k} \mid H_{1,k})$	Present	$Y_K < \gamma$
Area 4	Probability of false alarm (P_f)	$P(D_{1,k} \mid H_{0,k})$	Absent	$Y_K > \gamma$

In a multiband environment, if the primary signal power in each sub-band is assumed to be 1, and if each sub-band has different channel gain (i.e., H_k) while the noise variance is the same for all, then we can say that each sub-band will have different SNRs.

3.3.3.1.2   Computation of P_f and P_d

Let us assume that $M > 10$, and consider Y_k to be normally distributed. Now consider Figure 3.5 where P_f and P_d can be computed as the tail probability using the following relation:

$$P_k^r(\gamma_k) = Q\left(\frac{\gamma_k - \mu_{0,k}}{\sigma_{0,k}}\right) \tag{3.7}$$

$$P_d^k(\gamma_k) = Q\left(\frac{\gamma_k - \mu_{1,k}}{\sigma_{1,k}}\right) \tag{3.8}$$

Mean and variance of both classes are as follows:

$$\mu_{0,k} = \sigma_v^2$$

$$\mu_{1,k} = \sigma_v^2(\Omega_k + 1)$$

$$\sigma_{0,k} = \frac{\sigma_v^4}{M} \tag{3.9}$$

$$\sigma_{1,k} = \frac{\sigma_v^4(2\Omega_k + 1)}{M}$$

Each sub-band has different statistics for both classes because of different SNRs. For each sub-band, P_f and P_d are to be computed against corresponding thresholds by the same procedure, as discussed earlier, and then the following vectors can be used to proceed toward the net throughput and aggregate interference:

$$P_F(\gamma) = [P_f^1(\gamma_1),\ P_f^2(\gamma_2),\ \dots\ P_f^z(\gamma_z)]^T$$
$$P_D(\gamma) = [P_d^1(\gamma_1),\ P_d^2(\gamma_2),\ \dots\ P_d^z(\gamma_z)]^T$$

(3.10)

3.3.3.1.3   Net Throughput and Aggregate Interference
Vector r_K shows the maximum achievable throughput over the kth sub-band where $r = [r_1, r_2, \dots, r_z]^T$, while c_K shows the cost caused if the transmission is carried out in the kth sub-band where $c = [c_1, c_2, \dots, c_Z]^T$. The remaining equations to be computed are

$$P_{SH}(\gamma) = 1 - P_F(\gamma)$$
$$P_{MD}(\gamma) = 1 - P_D(\gamma)$$

(3.11)

The net throughput can be found as

$$R(\gamma) = r^T[P_{SH}(\gamma)]$$

(3.12)

The aggregate interference can be determined by

$$I(\gamma) = c^T[P_{MD}(\gamma)]$$

(3.13)

3.3.3.1.4   Limitations on Sub-Bands and Threshold Bounds
There are some limitations upon each sub-band. First, its interference must not exceed a predefined level, and second, its utilization must not be less than a predefined level. α_k and β_k are defined as the per-band maximum interference factor and the per-band minimum utilization factor in the kth sub-band, respectively, therefore

$$P_{MD}(\gamma) \le \alpha_k$$
$$P_F(\gamma) \le \beta_k$$

(3.14)

Similarly, for the whole band, we have $\alpha = [\alpha_1, \alpha_2, \dots, \alpha_Z]$ and $\beta = [\beta_1, \beta_2, \dots, \beta_Z]$. The threshold for each sub-band is also bound by its minimum and maximum allowed values, and it must be varied within this range. These minimum and maximum allowed values are formulated by [1] as follows:

$$\gamma_{\min,k} \le \gamma_k \le \gamma_{\max,k}$$

$$\gamma_{\min,k} = \mu_{0,k} + \sigma_{0,k}Q^{-1}(\beta_k)$$

(3.15)

$$\gamma_{\max,k} = \mu_{1,k} + \sigma_{1,k}Q^{-1}(1 - \alpha_k)$$

3.3.3.1.5   Problem Formulation under Convex Optimization
The problem is not convex, but it can be reformulated under convex optimization using the following constraints

$$\text{Max}_\gamma \ R(\gamma)$$

$$\text{s.t. } I(\gamma) \leq \varepsilon$$

$$P_{MD}(\gamma) \leq \alpha \tag{3.16}$$

$$P_F(\gamma) \leq \beta$$

where $0 < \alpha \leq \dfrac{1}{2}$ and $0 < \beta \leq \dfrac{1}{2}$.

3.3.4 MULTIBAND SENSING-TIME-ADAPTIVE JOINT DETECTION

In MSJD, a new feature of periodic sensing is included in which the sensing time is adaptive [25,26].

3.3.4.1 Periodic Sensing

As soon as the SU finds a vacant channel, it starts its transmission and can utilize this vacant channel until the PU reappears and wants the charge back. However, there must be a way to check whether the PU wants to communicate on this channel. The solution lies in periodic sensing; after a fixed time, let us say T seconds, the SU must sense the spectrum again and check its status, and if it is still vacant, then it can still utilize that channel, otherwise the SU has to find some other SH and tune its parameters according to that new SH to continue its communication. The SU must ensure that if it is communicating over a channel, after T seconds, it must stop its communication first and then perform sensing, as both tasks cannot be performed simultaneously. The frame structure considered by [26] is shown in Figure 3.6 in which each frame has a duration of T seconds.

One slot of τ seconds is for sensing the spectrum, where the other slot of $T - \tau$ seconds is used for communication purposes. The number of samples required for detection is estimated by the relation $M = \tau f_s$, where f_s is the sampling frequency of sensing the spectrum, and it is fixed for the whole band. The sensing time depends upon environmental conditions, such as channel degradation and noise impairment. For those sub-bands where SNR is very good, the sensing slot duration can be decreased by taking less averaging samples, causing the time duration for the transmission slot to increase, which will ultimately increase the net throughput. So to get the most suitable sensing slot duration while considering the SNR and the probability of missed detection for getting better throughput results, sensing time also needs to be optimized.

3.3.4.2 Two-Step Optimization

MSJD is a two-step optimization framework which optimizes the thresholds for all sub-bands to get maximum throughput while limiting the interference in the first step and then computes the probability of missed detection, P_{md}, and uses it in the next step of optimization where sensing

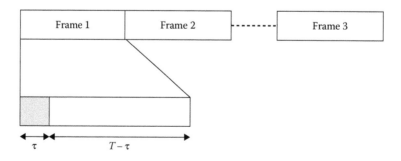

FIGURE 3.6 Frame structure for periodic sensing.

time, τ, is optimized. So we have two separate objective functions with different constraints. The total number of iterations is approximately 2 to 4; while in each iteration, it has to perform both steps. In the first step, the threshold vector is optimized, and the P_{md} computed in this step is used to calculate the new estimated probability of false alarm, $P_{\hat{f}}$, which is ultimately used in the optimization function for sensing time in the second step. The problem formulation of MSJD is explained next.

3.3.4.3 Problem Formulation for MSJD

3.3.4.3.1 Signal to Noise Ratio (SNR)
The received SNR for *the kth* sub-band is as follows:

$$\Omega_k = \frac{E\{|S_k^2|\}H_k^2}{\sigma_v^2} \tag{3.17}$$

3.3.4.3.2 Computation of P_f and P_d
P_f and P_d for the kth sub-band can be computed by the following relation:

$$P_f^k(\gamma_k, \tau_k) = Q\left(\left(\frac{\gamma_k}{\sigma_v^2} - 1\right)\sqrt{\tau f_s}\right)$$

$$P_d^k(\gamma_k, \tau_k) = Q\left(\left(\frac{\gamma_k}{\sigma_v^2} - \Omega_k - 1\right)\sqrt{\frac{\tau f_s}{2\Omega_k + 1}}\right) \tag{3.18}$$

The mean and variance of both classes are as follows:

$$\mu_{0,k} = \sigma_v^2$$
$$\mu_{1,k} = \sigma_v^2(\Omega_k + 1)$$
$$\sigma_{0,k} = \frac{\sigma_v^4}{\tau f_s}$$
$$\sigma_{1,k} = \frac{\sigma_v^4(2\Omega_k + 1)}{\tau f_s} \tag{3.19}$$

The following vectors will be obtained for all sub-bands:

$$P_F(\gamma, \tau) = [P_f^1(\gamma_1, \tau), P_f^2(\gamma_2, \tau), ..., P_f^Z(\gamma_Z, \tau)]^T$$

$$P_D(\gamma, \tau) = [P_d^1(\gamma_1, \tau), P_d^2(\gamma_2, \tau), ..., P_d^Z(\gamma_Z, \tau)]^T \tag{3.20}$$

3.3.4.3.3 Net Throughput and Aggregate Interference
The methodology here is the same as in the case of MJD; r_K and c_K show the maximum achievable throughput and the cost of transmission (if the PU is already present) over the kth sub-band, where $r = [r_1, r_2, ..., r_Z]^T$ and $c = [c_1, c_2, ..., c_Z]^T$. Other equations are

$$P_{SH}(\gamma, \tau) = 1 - P_F(\gamma, \tau)$$

$$P_{MD}(\gamma, \tau) = 1 - P_D(\gamma, \tau) \tag{3.21}$$

The net throughput is given by

$$R(\gamma,\tau) = \frac{T-\tau}{T} r^T [P_{SH}(\gamma,\tau)] \tag{3.22}$$

The aggregate interference is

$$I(\gamma,\tau) = c^T [P_{MD}(\gamma,\tau)] \tag{3.23}$$

The net throughput is modified by the sensing factor. If we keep T constant and increase τ, then more time is spent in sensing the spectrum, while the slot for data transmission is decreased, and because of the longer duration of the sensing slot, CR will have more time and ultimately more opportunities for correct SH detection; hence, there is a trade-off in the net throughput and in the aggregate interference.

3.3.4.3.4 Limitations on Sub-Bands and Threshold Bounds

The maximum allowed interference and minimum utilization that CR must provide are given by the following relations where α_k is defined as the maximum interference factor and β_k as the minimum utilization factor over the kth sub-band:

$$P_{MD}(\gamma,\tau) \leq \alpha_k$$

$$P_F(\gamma,\tau) \leq \beta_k \tag{3.24}$$

Similarly, for the whole band, $\alpha = [\alpha_1, \alpha_2, \ldots, \alpha_Z]$ and $\beta = [\beta_1, \beta_2, \ldots, \beta_Z]$. The threshold for each sub-band is also bound by its minimum- and maximum-allowed values, and it must be varied within this range. These minimum- and maximum-allowed values are formulated as follows:

$$\gamma_{min,k} \leq \gamma_k \leq \gamma_{max,k}$$

$$\gamma_{min,k} = \sigma_\upsilon^2 \left[\frac{Q^{-1}(\beta_k)}{\sqrt{\tau f_s}} + 1 \right] \tag{3.25}$$

$$\gamma_{max,k} = \sigma_\upsilon^2 \left[\frac{\sqrt{2\Omega_k + 1}}{\sqrt{\tau f_s}} Q^{-1}(1-\alpha_k) + \Omega_k + 1 \right]$$

Some other parameters are

$$P_k^f(\tau) = Q\left(\sqrt{2\Omega_k + 1} Q^{-1}(1 - P_m^k) + \sqrt{\tau f_s} \Omega_k \right) \tag{3.26}$$

$$\tau_{min,k} = \frac{1}{\Omega_k^2 f_s} \left[Q^{-1}(\beta_k) - \sqrt{2\Omega_k + 1} Q^{-1}(1 - P_m^k) \right]^2 \tag{3.27}$$

3.3.4.4 Problem Formulation under Convex Optimization

The problem is not convex, but using constraints, it can be reformulated under convex optimization as follows:

$$\text{Max}_\gamma \ R(\gamma,\tau)$$

$$\text{s.t. } I(\gamma,\tau) \leq \varepsilon$$

$$P_{MD}(\gamma,\tau) \leq \alpha_k \tag{3.28}$$

$$P_F(\gamma,\tau) \leq \beta_k$$

$$\tau \leq \tau_{max}$$

where $0 < \alpha \le \dfrac{1}{2}$ and $0 < \beta \le \dfrac{1}{2}$.

The MJSD problem is further divided into two steps. In the first step of optimization, the problem (P1) is formulated as follows:

$$\min_{\gamma} \; R_{mis}(\gamma) = \sum_{k=1}^{K} r_k P_f^k(\gamma_k) \quad \text{(P1)}$$

$$s.t. \; I(\gamma) \le \varepsilon$$

(3.29)

In the second step, the following optimization problem is considered:

$$\min_{\tau} \; R_{miss}(\tau) = \sum_{k=1}^{K} r_k \left(\left(1 - \frac{\tau}{T}\right) P_f^{'k}(\tau) + \frac{\tau}{T} \right) \quad \text{(P2)}$$

$$s.t. \; \tau \ge \text{argmax}\left\{ \tau_{\min}^1, \tau_{\min}^2, ..., \tau_{\min}^K \right\}$$

(3.30)

3.3.4.4.1   Issue with Convex Optimization

In both MJD and MSJD, the problem is reformulated under convex optimization with limitations upon the maximum per-band interference factor and minimum utilization factor; that is

$$0 < \alpha \le \frac{1}{2}$$

$$0 < \beta \le \frac{1}{2}$$

(3.31)

Now the objective function will be convex only if the constraint factors α_k and β_k remain within the previously mentioned limits.

MSJD is better than MJD as it exploits the sensing time parameter, τ, by optimizing it along with the optimization of the threshold vector; hence, it is better strategy which ultimately contributes in the net throughput. The simulation results presented in Section 3.5 establish the superiority of MSJD over MJD. However, it should be noted that due to limitations in the convex optimization domain, there is a loss of generality. However, this optimization problem can be solved without these limitations by employing MLTs. In next section, MLTs are discussed, and selected techniques are implemented to solve the problem formulation of both MJD and MSJD frameworks.

3.4   OPTIMIZATION USING MACHINE LEARNING TECHNIQUES

Machine learning techniques (MLTs) play a significant role in the field of science and engineering. The application of these techniques is found in many domains (e.g., effective web search, speech recognition, self-driving cars, face recognition, character recognition by robots). There are various reasons for using MLTs in today's life [27]. With the help of MLTs, various new structures can be discovered. In a time-dependent environment, if a machine that uses learning techniques detects a change in the environment and adapts it, the constant redesign requirement would be overcome. In some scenarios, it may happen that input and output data are available, but we do not have a good transfer function that describes the relationship between input and desired output. In such cases, machine learning methodologies can be utilized to adapt the internal structure of the transfer function that can give the desired output for a set of inputs. These methods can improve the existing machine design during the job. Machine learning methods have been effectively used in optimization problems. Optimization can be simply described as a sequence of steps through which one can find alternative ways to achieve a specific target and then can select the most suitable way of getting that particular target while remaining within some predefined bounds. Various problems regarding this sphere can be treated as optimization problems. In an optimization problem, we have an objective function, and our target is to get the maximum or minimum value of this objective function. The decision to get the maximum or minimum relies on the problem scenario, while the procedure for finding the maximum or minimum of that particular objective function is called problem optimization. Graphically, the objective function can be realized as a landscape that is irregular in shape containing deep valleys and mountains like structure e.g. a Rastrigin function which is a non-convex function having many local minima [28].

The objective function is expressed in terms of different decision variables, where the combination of these decision variables are varied in a particular manner, and the result of the objective function is changed. By iteratively modifying these decision variables, we get closer to the optimal solution. There are various optimization techniques which actually differ from each other in the manner in which they modify these decision variables. Each technique has its own unique strategy for modifying these decision variables. The bounds, called constraints, depend upon the decision variable, and these constraints are limited by different parameters such as resources, input, or some environmental restrictions. When decision variables are modified (after each iteration), two important results are investigated. The first is whether the value of the objective function gets closer to the optimal solution, and the second is checking whether the constraints are still satisfied. Mathematically, the optimization problem can be described as follows:

$$\text{Optimize } z = f(\upsilon_1, \upsilon_2, \upsilon_3, ..., \upsilon_Q) \tag{3.32}$$

$$\text{Subject to } w(\upsilon_1, \upsilon_2, \upsilon_3, ..., \upsilon_Q)\{\leq \text{ or } = \text{ or } \geq\} \ b_j \ \text{ for } j = 1, 2, ..., J \tag{3.33}$$

where f is the objective function, w is the constraint function, and $\upsilon_1, \upsilon_2, \upsilon_3, ..., \upsilon_Q$ are the decision variables.

The problem formulations of MJD and MSJD are treated as convex optimization problems which apply mathematical limitations, such as per-band utilization and interference factors. These mathematical limitations can be handled easily by MLTs. Machine learning techniques can also solve those systems where convex optimization stops working because of a violation of a convexity region. While implementing these techniques, different issues such as inflections and discontinuity of function can be ignored. These are some of the main reasons for the massive popularity of MLTs for optimization. In addition, these techniques are intelligent and efficient tools that can accomplish the task of finding the global optimum. There are different optimization techniques, such as *Genetic Algorithms* (GA) [27, 29], *Genetic Programming* (GP) [30, 31], *Neural Networks* (NN) [32], *Support Vector Machines* (SVM) [32] and *Particle Swarm Optimization* (PSO) [33] etc. Some of these techniques are explained in the following sections.

3.4.1 Genetic Algorithm (GA)

In the 1960s, I. Rechenberg presented the idea of evolutionary computing, and in the 1970s, John Holland introduced GA in the United States (University of Michigan). GA not only provides an alternative method for problem solving, but in fact, it gives better results compared to traditional searching methods. Although GAs are computationally time expensive, compared to gradient search methods, there are fewer chances that GA gets stuck at local optima. The traditional methods use gradient information, while GA can work without such information [33]. GA can also solve the problem in irregular search space. Solutions for difficult and high-dimensional problems can be discovered rapidly with GA. In those scenarios in which mathematical analysis is not available or traditional methods fail to search or in which complex and large search spaces are to be faced, GA can accomplish the task efficiently. Another advantage of GA is that it can easily handle constraints of various kinds. These attractive features allow GA to be applied to different scenarios such as estimation of parameters, modeling, function optimization, and various other applications of machine learning [34].

3.4.1.1 Basic Theme and Working Principle

GA works on the basis of evolution (i.e., survival of the fittest, or natural selection). It came from the biological concept where the best available chromosomes are paired to form new species, and those chromosomes which become weak are not considered in the production of the next generation. The chromosomes are treated as potential solutions to the problem within the whole search space and represented as strings (variable), where each bit of string is a representation of a gene. A set of strings form a population set and is known as a generation. A generation at the kth point can be represented as

$$G_k = \left\{ \gamma_1^k, \gamma_2^k, \gamma_3^k, ..., \gamma_N^k \right\} \tag{3.34}$$

where N is the population size.

The generation contains the potential solution to the problem, so corresponding to each element of generation, fitness is evaluated from the fitness function, and the score is represented as

$$S_k = \left\{ F\left(\gamma_1^k\right), F\left(\gamma_2^k\right), F\left(\gamma_3^k\right), ..., F\left(\gamma_N^k\right) \right\} \tag{3.35}$$

where $F(\gamma)$ is the fitness function.

Initially, a population set is randomly generated where each element of the set appears to be the solution. Each entity of this set is a variable (usually a binary string) which appeared as a competitor to the other variables. For each variable/solution, fitness is evaluated from the fitness function, and corresponding to their fitness, a new population is created. The new population is formed by applying genetic operators to the solutions of a population set. There are three genetic operators in GA, namely the reproduction operator, crossover operator, and mutation operator [29].

- *Reproduction operator*: Once the fitness function is evaluated with all potential solutions within the population set, the unfit solutions will be discarded, and the solutions with high fitness values will be reproduced, and they will take the place of the discarded solution. So after selecting good strings, the reproduction operator reproduces them to take the place of the discarded one(s).
- *Crossover operator*: After the removal of the unfit solutions from the population set and applying the reproduction operator, the population set is left with all good strings. Now the crossover operator selects good strings (two parent chromosomes at a time) and then chooses a crossover point within the strings, and at the end, it swaps both strings after the crossover point. By doing so, the crossover operator creates two new offspring. The algorithm runs faster when crossover operators are used; otherwise, if only the mutation operator is considered, then the algorithm will be very slow, so it is better to use both.
- *Mutation operator*: This operator does not require comparison of good and bad strings like the reproduction operator, nor does it consider two particular operators at a time; instead, it focuses on one string at a time and randomly selects a bit of that string and then flips it. The mutation operator is considered a background operator whose aim is to ensure that the algorithm does not become stuck in some local optimum.

3.4.1.2   Sequence of Algorithm

The steps of the algorithm are as follows:

1. Create a new population randomly; it spreads the solution set in the whole domain to reach the global optimum while considering the bounds (lower and upper if any) for the elements of the population set.
2. Evaluate the objective function where constraints are also considered; the result is then mapped into the fitness function, where the fitness score is evaluated corresponding to each element. If the constraints formulated in the constraint function are not violated, then the fitness score is simply evaluated and then stored; otherwise, if a violation occurs, then a penalty is applied, and that fitness score will not be considered.
3. Create a new population from the current population with the help of the genetic operators (reproduction, crossover, and mutation) where the weak contestants are dropped, and the strong ones are selected and then modified, forming a new generation.
4. Check the end condition; if it is satisfied, then stop here, and store the best solution; otherwise, go to step 2, and repeat the process until the stopping criteria is met.

3.4.2   PARTICLE SWARM OPTIMIZATION (PSO)

PSO is a very useful tool in function optimization. It falls under the field of evolutionary algorithms and emerged from biological sciences. The theory of PSO was developed by psychologists, biologists, and ornithologists [35]. For modeling those types of objects whose representation in the form of surfaces or polygons was not possible, such as clouds and smoke, Reevs proposed particle systems 1983 [36]. In 1986, Craig Reynolds simulated a flock of birds using particle systems [37]. An ornithologist named Frank Heppner and U. Grenander did more detailed work in the area of bird flock animation in 1990 [38]. In 1995, Kennedy and Eberhart implemented the algorithm of PSO and published their results [39]. Later, in 1998, a new parameter called inertia weight was introduced by Shi and Eberhart, which resulted in better performance and control of the algorithm [40,41].

3.4.2.1   Basic Theme and Working Principle

A swarm is defined as a group of entities in which all members of the group are joined to solve a distributed problem, and they achieve their goal by acting on their respective local environments and communicating with each other during the whole procedure [42].

The basis of swarm intelligence is on the principles of social psychology where each member of the swarm shows social behavior in solving the problem by interacting with each other. PSO is a stochastic optimization tool, and it is in the form of an algorithm based upon swarm intelligence whose aim is function optimization, which it efficiently performs by spreading the particle in the whole search space. Let us say a flock of birds is in search of food. If any bird finds some clue of food, it will communicate its findings to its neighbors and then all the birds will converge to the target. Similarly, in optimizing some function, a set of particles is created that acts as a potential solution to the problem, and each particle is considered a point in N dimensional space. These N dimensions show that there are N unknown parameters which are to be optimized, while M is the total number of particles. A particle at the kth iteration can be represented as $X_k = \{x_1, x_2, ..., x_N\}$, whereas the record of the best position of each particle is maintained in a separate vector $P_k = \{p_1, p_2, ..., p_N\}$, which contains the personal best of each particle. The particle which gives the best result is stored as the gbest particle $\left(P_{gd}^k\right)$, and the score corresponding to this particle is termed as the gbest score. The velocity of the particle that is, the rate of change of its position, is stored as velocity vector

$V_k = \{v_1, v_2, \ldots, v_N\}$. To update velocity and current position vectors, in 1995, Kennedy and Eberhart gave the following relations:

$$V_{kd} = V_{kd} + c_1 * r \text{ and } ()*(p_{kd} - x_{id}) + c_2 * r \text{ and } ()*(p_{gd} - x_{id})$$

$$X_{kd} = X_{kd} + V_{kd} \text{ for } k = 1, 2, \ldots, M \text{ and for } d = 1, 2, \ldots, N$$

(3.36)

where c_1 and c_2 are constants usually set in the range of 2 to 4, and r and $()$ show the random function.

In 1998, Shi and Eberhart modified the equations of PSO by adding an inertia parameter as follows [34,35]:

$$V_{kd} = w + V_{kd} + c_1 * r \text{ and } ()*(p_{kd} - x_{id}) + c_2 * r \text{ and } ()*(p_{gd} - x_{id})$$

$$X_{kd} = X_{kd} + V_{kd} \text{ for } k = 1, 2, \ldots, M \text{ and for } d = 1, 2, \ldots, N$$

(3.37)

There are some other parameters, such as V_{min} and V_{max}, used to limit the velocity of all the particles in each iteration and P_{min} and P_{max}, which are the minimum and maximum bounds for the particles' position. In the modified version, the inertia weight, w, was added, which controls the level of search in the following way. If w is small, then it makes PSO a local search algorithm where it finds the global optimum fast but finds it only when an acceptable solution lies in the initial particle set; otherwise, it will not. If w is large, then PSO will act as a global search algorithm where exploitation of the new areas will be performed. However, by doing so, there is less chance of success as more iterations will be performed to find the global optimum. When w is medium, the number of iterations required to achieve the global optimum will be moderate, and it will have a high chance of success in finding the global optimum.

3.4.2.2 Sequence of Algorithm

The following steps are considered for the implementation of PSO:

1. Set all of the following parameters of PSO with suitable values:

$$w, c_1, c_2, V_{min}, V_{max}, P_{min}, P_{max}$$

2. Create and randomly initialize a set of particle positions, X, such that they cover the whole search space. If the search space is bounded (by P_{min} and P_{max}), then during initialization, the position of the particles is maintained within the bounds P_{min} and P_{max}. The velocity vector, V, is also initialized, but in each iteration, V must remain within the bounds V_{min} and V_{max}.
3. Find the fitness score by evaluating the fitness function corresponding to each particle.
4. Compare the current score with the P vector which stores the previous best score of the particles. If the problem is a maximization problem and the new score of a particle is more than its previous best, then the current particle position and score will be updated; otherwise, no modification will be performed.
5. Compare the current score with the previous best particle in the whole search space. Again, if a maximization problem is considered and the current score is more than the group's previous best score, then update the index in gbest. Update velocity and position.
6. Move to step 2, and repeat the procedure until the stopping criteria is met.

To solve the optimization problem in MJD, PSO is better than GA because PSO has the advantage of easy implementation, and it has to handle less parameters, causing it to take far less time than the execution of GA. This time parameter is very important for CR as CR has to perform in real-time scenarios, so the quicker it performs, the better it is. In the next section, results of simulations carried out in the MATLAB® environment are shown. These results show that PSO's execution time is far less than GA's.

3.5   RESULTS AND DISCUSSION

In this section, results of simulations are presented, including different cases of the MJD and MSJD frameworks implemented with both techniques (i.e., GA and PSO). Implementation of uniform detection of all sub-bands is also shown. For simulations, let us develop an illustrative scenario. Assume that there are eight sub-bands, and each sub-band has its corresponding throughput and associated cost. The ultimate aim is to find a vector of optimum thresholds for all the sub-bands simultaneously. If a sub-band is in use by the PU, and at the same time, if any SU tries to transmit over that band, then a penalty will be faced in terms of the cost factor, which contributes in the aggregate interference. Whereas if the PU is absent and the SU detects it as absent and then communicates over that sub-band, then it adds up in the net throughput, which is the actual purpose of CR (i.e., to increase the overall throughput while limiting the aggregate interference).

Suppose that the set of maximum achievable throughputs over the eight sub-bands (in kbps) is

$$r_k = \{612, 524, 632, 139, 451, 409, 909, 401\}$$

the interference cost vector is

$$c_k = \{1.91, 8.17, 4.23, 3.86, 7.16, 6.05, 0.82, 1.3\}$$

and the path gain for all sub-bands is

$$H_k = \{0.5, 0.3, 0.45, 0.65, 0.25, 0.6, 0.4, 0.7\}$$

Assume that the power of the transmitted signal in each sub-band is 1, and the noise power level, σ_v^2, is also 1. Let the number of samples, M, for MJD be 100, while in the MSJD case, $M = \tau * f_s$ where f_s is 100 samples/sec. The parameter τ is optimized, and its range is 1 to 2. The duration of the time slot is 30 sec. Different values of per-band interference and utilization factors are taken, and simulations are performed. These factors are mentioned in each figure, which show the results obtained using simulations performed in the MATLAB environment. A uniform detection mechanism is considered for all eight sub-bands where a single threshold is applied over all sub-bands, and it is varied from the minimum-allowed value to the maximum-allowed value. Corresponding to each threshold, interference and throughput is computed and stored and then the result is plotted, as shown by the straight line in the lower part of Figure 3.7. Then the MJD framework is considered, and GA is used as the optimization technique. The results of GA are plotted in the upper part of Figure 3.7 for both the non-convex and convex functions.

Similarly, results for the MJD framework with the PSO optimization technique are shown in Figure 3.8. The plots of uniform detection are also shown in Figure 3.8 to make a comparison with the results of PSO. The aggregate throughput achieved using GA and PSO is much higher compared to the uniform threshold.

In the next simulation, the probability of false alarm (P_f) and the probability of missed detection, (P_{md}) are compared for both the GA and the PSO. The main function is executed three times for two cases: one where a non-convex region is considered and the other where a convex region is considered. Results for the third execution for both cases are shown in Figures 3.9 and 3.10, respectively. These results show that the performance of GA and PSO is comparable.

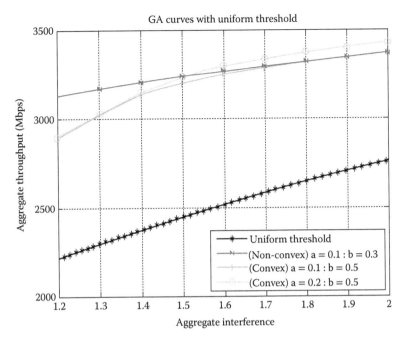

FIGURE 3.7 Plot of MJD curves implemented with GA along with uniform detection.

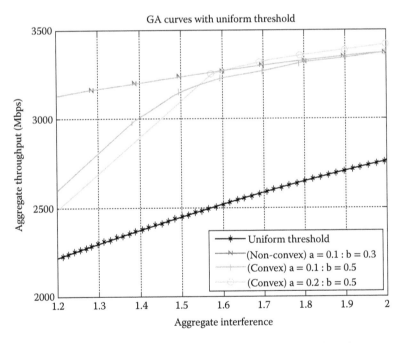

FIGURE 3.8 Plot of MJD curves implemented with PSO along with uniform detection.

A comparison of the execution time and the corresponding throughput using GA and PSO is performed for the three executions, and the results obtained for the convex region case are plotted in Figure 3.11. The results show that although the aggregate throughput for GA and PSO is approximately the same, the execution time of PSO is far less than GA in all three executions.

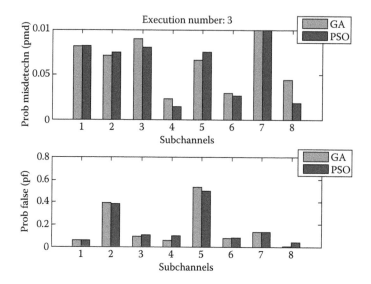

FIGURE 3.9   Execution 3. Comparison of P_f and P_{md} for non-convex function.

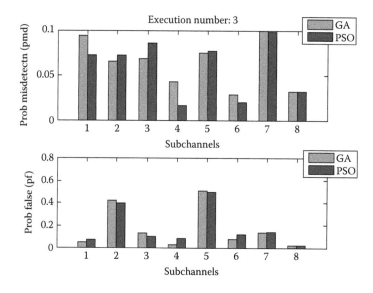

FIGURE 3.10   Execution 3. Comparison of P_f and P_{md} for convex function.

Afterward, the MSJD framework is implemented with PSO, and the results are shown in Figure 3.12. Finally, the uniform detection is compared with both frameworks (i.e., MJD and MSJD), where the optimization technique implemented in both frameworks is PSO, and the results are shown in Figure 3.13. In all the simulations performed, the results of using PSO are superior to GA. As a result, PSO is proposed as the best choice for optimization in MJD and MSJD frameworks.

3.6   CONCLUSIONS

In this chapter, a non-coherent detection technique (i.e., energy-based detection for narrowband spectrum sensing) was selected owing to its very low complexity level. Among the periodogram and its other variant, the Blackman-Tukey method appeared to be a better choice owing to its low variability and hence low figure of merit (favorable condition) compared to the other techniques. Energy detection, which is a narrowband spectrum-sensing technique, cannot be applied directly for

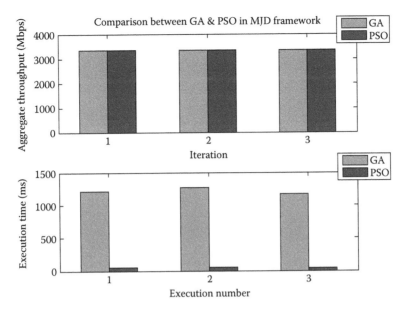

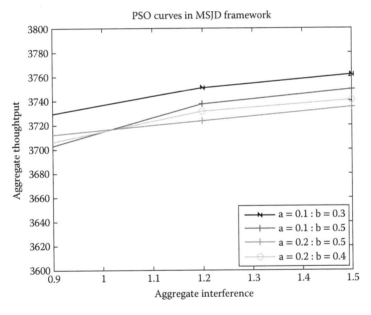

FIGURE 3.11   Execution time and throughput comparison of three executions for convex function.

FIGURE 3.12   Plot of MSJD curves implemented with PSO.

wideband spectrum sensing, and hence there is a need to devise a strategy to use energy detection for WSS. In this context, two frameworks, namely MJD and MSJD, were considered for wideband spectrum sensing. For both frameworks, an illustrative scenario was considered. MSJD incorporates a new feature of sensing time duration that MJD does not support, and hence the problem formulation was modified accordingly. With this new feature, the throughput function became more generalized. The simulation results proved that MSJD works better compared to MJD. By selecting the sensing time adaptively, both the speed and the quality of spectrum sensing could be adjusted. A uniform detection scheme was also discussed and simulated for the problem formulation of the frameworks which gives a nominal throughput, but when optimization came in, the throughput curves were significantly improved. For optimization, GA and PSO were considered. Optimization

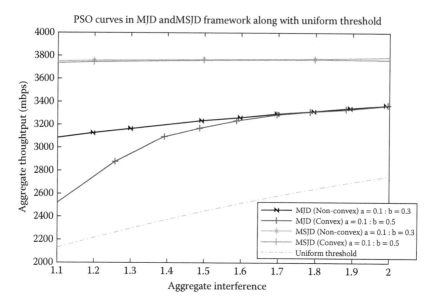

FIGURE 3.13   Comparison of uniform detection with MJD and MSJD frameworks implemented with PSO.

using GA and PSO performed in MJD/MSJD frameworks gives considerably better results compared to uniform detection. The simulation results show that PSO took much less time compared to GA with the MJD/MSJD framework, so PSO turns out to be the better choice for optimization.

TABLE OF ALL ADOPTED NOTATIONS

$H_{0,k}$	Hypothesis for the absent class
$H_{1,k}$	Hypothesis for the present class
X_k	Received signal
H_k	Channel gain
S_k	Primary signal transmitted
V_k	Gaussian noise having power σ_υ^2
Y_K	Energy of the kth sub-band
$\mu_{0,k}$	Mean of absent class
$\sigma_{0,k}^2$	Standard deviation of absent class
$\mu_{1,k}$	Mean of present class
$\sigma_{1,k}^2$	Standard deviation of present class
P_{sh}	Probability of identifying a SH
P_d	Probability of detection
$P_d^k(\gamma_k)$	Probability of detection for k^{th} sub-band
P_{md}	Probability of missed detection
P_f	Probability of false alarm
$P_f^k(\gamma_k)$	Probability of false alarm for k^{th} sub-band
$\gamma = [\gamma_1, \gamma_2, ..., \gamma_K]$	Optimal threshold (classifier) vector
$r = [r_1, r_2, ..., r_Z]^T$	Maximum achievable throughput
$c = [c_1, c_2, ..., c_Z]^T$	Incurred transmission cost

$I(\gamma)$	Aggregate interference
$Q(\gamma)$	Net throughput
Ω_k	SNR for k^{th} sub-band
α_k	Per-band maximum interference factor
β_k	Per-band minimum utilization factor
τ	Spectrum-sensing time
T	Frame duration time
f_s	Sampling frequency
f	Objective function
w	Constraints function
υ_Q	Decision variable
$G(\gamma)$	Generation in GA
$F(\gamma)$	Fitness function in GA
S_k	Score in GA
w	Inertia weight
$P_k = \{p_1, p_2, ..., p_N\}$	Particle best position vector
P_{min}, P_{max}	Minimum and maximum bounds for the position of particles
$V_k = \{v_1, v_2, ..., v_N\}$	Particle velocity vector
V_{min}, V_{max}	Minimum and maximum bounds for the velocity of particles

REFERENCES

1. Sanna M. and Murroni M. Opportunistic wideband spectrum sensing for cognitive radios with genetic optimization. Paper presented at the *IEEE International Conference on Communications (ICC)*, 1–5 May 2010.
2. Md Habibul Islam et al. Spectrum survey in Singapore: Occupancy measurements and analyses. Paper presented at the *3rd International Conference on Cognitive Radio Oriented Wireless Networks and Communications, CrownCom*, 1–7 May 2008.
3. Danijela B. C. and Robert W. B. Cognitive Radios: System Design Perspective. PhD thesis, EECS Department, University of California, Berkeley, December 2007.
4. Mitola J. and Maguire G. Q. Cognitive radio: Making software radios more personal. *IEEE Personal Communications*, 6, 13–18, August 1999.
5. IEEE 802.22. Working Group on Wireless Regional Area Networks. Available at: www.ieee802.org/22/.
6. Mitola J. III. *Software Radio Architecture: Object Oriented Approaches to Wireless Systems Engineering*. Hoboken, NJ: Wiley, 2000.
7. Reed J. H. *Software Radio: A Modern Approach to Radio Engineering*. Upper Saddle River, NJ: Prentice Hall PTR, 2002.
8. Jondral F. *Parameterization: A Technique for SDR Implementation*. London: Wiley, 2002.
9. Gul S. T., Al-Ghouwayel A., Moy C., and Louët Y. A novel design of reconfigurable Fourier transform operator over C and GF(Ft) for future multi-standards SDR equipments. *International Journal of Communication Networks and Distributed Systems*, 4(4), 345–375, 2010.
10. Federal Communications Commission. Notice of proposed rule making and order: Facilitating opportunities for flexible, efficient, and reliable spectrum use employing cognitive radio technologies, February 2005.
11. Sun H. et al. Wideband spectrum sensing for cognitive radio networks: A survey. *IEEE Wireless Communications*, 20(2), 74–81, April 2013.
12. Clancy C. et al. Applications of machine learning to cognitive radio networks. *IEEE Wireless Communications*, 14(4), 47–52, August 2007.
13. Amraoui A. et al. Intelligent wireless communication system using cognitive radio. *International Journal of Distributed and Parallel Systems*, 3(2), 91–104, March 2012.
14. Garhwal A. and Partha P. B. A survey on spectrum sensing techniques in cognitive radio. *Journal of Computer Science and Communication Networks*, 1(2), 196–206, October–November 2011.

15. Subhedar M. and Birajdar G. Spectrum sensing techniques in cognitive radio networks: A survey. *International Journal of Next-Generation Networks*, 3(2), 37–51 June 2011.
16. Yucek T. and Arslan H. A survey of spectrum sensing algorithms for cognitive radio applications. *IEEE Communications Surveys & Tutorials*, 11(1), 116–130, 2009.
17. Ian F. A., Brandon F. L., and Balakrishnan R. Cooperative spectrum sensing in cognitive radio networks: A survey. *Physical Communication*, 4(1), 40–62, 2011.
18. Sahai A., Hoven N., and Tandra R. Some fundamental limits on cognitive radio. Paper presented at the *42nd Allerton Conference on Communication, Control, and Computing*, pp. 1662–1671, 2004.
19. Nasser A. et al. Efficient spectrum sensing approaches based on waveform detection. Paper presented at the *3rd International Conference on e-Technologies and Networks for Development (ICeND)*, 2014.
20. Nafkha A. et al. Cyclostationarity-based versus eigenvalues-based algorithms for spectrum sensing in cognitive radio systems: Experimental evaluation using GNU radio and USRP. Paper presented at the *IEEE 11th International Conference on Wireless and Mobile Computing, Networking and Communications (WiMob)*, 2015.
21. Farhang-Boroujeny B. Signal processing techniques for spectrum sensing and communication in cognitive radios. Paper presented at the *SDR 08 IEEE Technical Conference and Product Exposition*, 2008.
22. Xiaomin L. et al. A feature detector based on compressed sensing and wavelet transform for wideband cognitive radio. Paper presented at the *IEEE 24th International Symposium on Personal Indoor and Mobile Radio Communications (PIMRC)*, 2013.
23. Monson H. H. *Statistical Digital Signal Processing and Modeling*. India: Wiley, 2009.
24. Quan Z. et al. Optimal multiband joint detection for spectrum sensing in cognitive radio networks. *IEEE Transactions on Signal Processing*, 57, 1128–1140, March 2009.
25. Paysarvi-Hoseini P. Design of Optimal Frameworks for Wideband/Multichannel Spectrum Sensing in Cognitive Radio Networks, MS thesis, University of Alberta, June 2011.
26. Paysarvi-Hoseini P. and Norman C. B. Optimal wideband spectrum sensing framework for cognitive radio systems. *IEEE Transactions on Signal Processing*, 59, 1170–1182, March 2011.
27. Siddique N. and Adeli H. *Computational Intelligence Synergies of Fuzzy Logic, Neural Networks and Evolutionary Computing*. UK: John Wiley & Sons, 2013.
28. MATLAB and Optimization Toolbox, Natick, MA: The MathWorks Inc., 2016. Available at: https://www.mathworks.com/products/optimization.html (Accessed August 2016).
29. Melanie M. *An Introduction to Genetic Algorithms*. USA: The MIT Press, 1999.
30. Sumathi S., Ashok Kumar L., and Surekha P. *Computational Intelligence Paradigms for Optimization Problems Using MATLAB®/SIMULINK®*. Boca Raton, FL: CRC Press, Taylor & Francis Group, 2016.
31. Iba H., Kumar Paul T., and Hasegawa Y. *Applied Genetic Programming and Machine Learning*. Boca Raton, FL: CRC Press, Taylor & Francis Group, 2010.
32. Haykin S. *Neural Networks and Learning Machines*, 3rd edition. NJ, USA: Pearson Education, 2009.
33. Nils J. N. *Introduction to Machine Learning: An Early Draft of a Proposed Textbook*, Stanford, CA: Stanford University, December 1996.
34. Matthew B. W. A Genetic Algorithm for Resource-Constrained Scheduling, PhD thesis, Massachusetts Institute of Technology, 1996.
35. Clerc M. *Particle Swarm Optimization*, UK: ISTE Ltd., 2006.
36. Dorigo M., de Oca M. A., and Engelbrecht A. Particle swarm optimization. *Scholarpedia*, 3(11), 1486, 2008.
37. Craig W. R. A distributed behavioral model. Paper presented at the *ACM SIGGRAPH Conference of Computer Graphics*, vol. 21, July 1987.
38. Frank H. H. *A Stochastic Nonlinear Model for Coordinate Bird Flocks*. USA: AAAS Publications, 1990.
39. Eberhart R. and Kennedy J. A new optimizer using particle swarm theory. Paper presented at the *MHS '95 Proceedings of the Sixth International Symposium on Micro Machine and Human Science*, pp. 39–43, October 1995.
40. Shi Y. and Eberhart R. A modified particle swarm optimizer. Paper presented at the *IEEE International Conference, Evolutionary Computation Proceedings, IEEE World Congress on Computational Intelligence*, pp. 69–73, May 1998.
41. Shi Y. and Eberhart R. Parameter selection in particle swarm optimization. Paper presented at the *EP '98 Proceedings of the 7th International Conference on Evolutionary Programming VII* (V. Porto, N. Saravanan, D. Waagen, and A. Eiben, eds.), vol. 1447 of *Lecture Notes in Computer Science*, pp. 591–600, Springer Berlin/Heidelberg, 1998.
42. Jesper N. H. The swarming body in semiotics around the world. Paper presented at the *Proceedings of the Fifth Congress of the International Association for Semiotic Studies*, 1994.

4 Resource Management Techniques in Licensed Shared Access Networks

*M. Majid Butt, Jasmina McMenamy, Arman Farhang,
Irene Macaluso, Carlo Galiotto, and Nicola Marchetti*

CONTENTS

The advent of mobile Internet has led to a phenomenal growth in mobile data traffic over the past few years. This trend is expected to continue considering the envisioned services of the fifth-generation (5G) mobile communication systems that will be required to provide ubiquitous connectivity, support of various verticals, and ten-fold improvements in data rates and latency compared to 4G. Spectrum is therefore at the heart of 5G, and its flexible use and better utilization are two of the key components when addressing its scarcity and fragmented availability. For this, spectrum-sharing paradigms such as licensed shared access (LSA)—a licensing approach designed to enable sharing of spectrum bands with low incumbent activity—become increasingly important. LSA builds on the concept of vertical sharing in which a licensed entity, called an LSA licensee, utilizes spectrum resources unused by the incumbent network(s). LSA rules ensure the protection of the incumbent from harmful interference by the transmissions from the LSA licensees. Moreover, with LSA, the aim is also to provide consistency in quality of service (QoS) for the LSA licensees, typically mobile network operators (MNOs), by enabling exclusive access to spectrum resources not otherwise used by the incumbent.

Starting from the fundamental aspects of LSA, this chapter extends the discussion to the proposed advances in LSA spectrum management framework within the European Advanced Dynamic Spectrum 5G mobile networks Employing Licensed shared access (ADEL) project [1]. This chapter also provides an overview of the literature on spectrum management aspects in LSA and presents distinct resource management algorithms, two of which consider fairness, and the third evaluates an auction-based spectrum allocation. The chapter is organized in the following way: Section 4.1 describes the architecture and central aspects of LSA; Section 4.2 reviews the existing literature on LSA and spectrum management; Section 4.3 addresses two spectrum allocation algorithms based on fairness—one provides strictly fair spectrum allocation among licensee networks, while the second provides fairness on a long-term basis only; Section 4.4 presents an auction-based LSA spectrum allocation, which in addition to spectrum sharing, also considers infrastructure sharing; and Section 5 concludes the chapter.

4.1 LSA FUNDAMENTALS AND ARCHITECTURE

It has been widely accepted that spectrum sharing is an essential requirement to support data traffic growth and ubiquitous and high-bandwidth connectivity. In that respect, the past decade and a half has seen numerous regulatory and standardization initiatives and technological advances that would enable dynamic access to spectrum and its more efficient utilization. Progress in technologies such as cognitive radio (CR)[*] and software-defined radio (SDR)[†] [2] is perceived as key for this paradigm shift. Nevertheless, reaching the vision of a fully developed dynamic access to spectrum requires advances on many fronts, including a well-defined regulatory environment and commercial viability as well as a broad adoption of CR, SDR, and other enabling technologies [3].

In the regulatory domain during the past decade, three main spectrum-sharing models have emerged, namely TV white spaces (WSs) [4–6], a three-tier spectrum access system (SAS)–based sharing model in the 3.5 GHz band in the US [7], and LSA in Europe. This chapter focuses on LSA. Our starting point is the latest status of the work of regulatory and standardization bodies in the field, upon which we propose enhancements in the direction of dynamic LSA.

Since its inception, LSA has been a topic of interest for regulatory and standardization bodies, cellular operators, and the academic community. The concept of LSA stems from the industry initiative, authorized shared access (ASA) [8], to acquire access to additional spectrum for mobile broadband service that would be provided on a shared basis under an exclusive licensing regime. Initially, frequency bands of 2.3 GHz and 3.8 GHz were sought. LSA, as defined by the European Conference of Postal and Telecommunications Administrations (CEPT)[‡] in [9], is a spectrum management tool that enables the sharing of selected frequency bands between the incumbents and licensed users—LSA licensees. The incumbents are the current holders of the right to use the spectrum. The currently designated band in Europe for LSA use is 2.3–2.4 GHz, whose harmonization was completed in 2014 [10]. Typical incumbents include program making and special events (PMSE) applications, telemetry, and other governmental use [11]. The aim with the LSA framework is to protect the incumbents from harmful interference while providing predictable QoS to the LSA licensees through an exclusive use of the LSA-designated spectrum. The use of the band was not restricted to MNOs, although in the first instance, it was envisaged that they would implement the first use cases. It is also foreseen that the LSA licensee and the incumbent will provide different types of services [9]. The access to spectrum by the LSA licensee is determined based on an agreement that specifies the terms of the use of the band, including the requirements on vacating the band upon the incumbent's request. While involvement of a national regulatory authority (NRA) in setting up the LSA agreement between the incumbent and the LSA licensee will vary from country to country, the NRA is responsible for granting the LSA licensee the individual right to use the LSA spectrum.

In relation to the standardization of LSA, the reference design, architecture, and interfaces are specified within the European Telecommunications Standards Institute (ETSI) (www.etsi.org/). Examples of high-level architecture, functional requirements, and deployment scenarios by MNOs, including proposed operational parameters such as transmit power, channel bandwidth, spectrum emission masks, receiver sensitivity, etc., are provided in [12]. In [13], system requirements for the operation of mobile broadband service in the LSA band are presented, including functional and other requirements such as protection of the incumbent, security aspects, and performance. It also specifies exclusion zones

---

[*] CR is defined as a radio aware of its environment, with the ability to learn from it, identify the best spectrum opportunity for efficient and reliable communication, and adjust its operating parameters accordingly.

[†] With SDR, some or all of the radio functionality is realized in software.

[‡] CEPT is a European association that coordinates the activities related to radio spectrum, telecommunication, and postal regulations. In relation to LSA, it ensures technical harmonization between different national administrations.

and restriction zones in relation to the LSA licensees (i.e., MNOs) operations. Exclusion zones are geographical areas where MNOs cannot transmit on the LSA spectrum, whereas in restriction zones, MNOs can transmit albeit under controlled conditions on such parameters as power levels or antennas. The document also requires that, in addition to a scheduled mode of operation, the LSA system must support on-demand operation. That means that in the event of an emergency, an LSA licensee also needs to be able to release spectrum, according to the specified conditions. ETSI system requirements were also reviewed and evaluated from the implementation perspective in [14]. In [15], ETSI reference architecture for LSA is presented. As in [9], it envisages the introduction of two new architectural building blocks: the LSA repository and the LSA controller. The LSA repository contains information on the incumbents' use of spectrum and the requirements on their protection. Its task is to provide the spectrum availability information to the LSA controllers, but it can also receive and store acknowledgment information from an LSA controller. The LSA controller, on the other hand, retrieves information from the LSA repository about the spectrum the incumbent uses and manages the access of the LSA licensee to the available spectrum. The LSA controller may interface one or more LSA repositories and LSA licensees. While [15] does not stipulate in which domain the LSA repository may be located (i.e., whether it is managed by the NRA, the incumbent, or delegated to a third party), it does specify that the LSA controller is within an LSA licensee domain. In this way, it enables the LSA controller to interact with the Operations and Maintenance (O&M) centre system of the LSA licensee to support the reconfiguration of the appropriate transmitters, according to the information from the LSA repository.

The LSA regulation and standardization activities currently focus on long-term sharing arrangements based on fixed-channel plans. With the aim to progress these activities toward a more dynamic approach, the European ADEL project [1] proposes an architecture that supports dynamic LSA configurations, targeting better overall spectrum utilization through the use of advanced radio resource management (RRM) techniques and sensing reasoning. To this end, the basic two-node LSA architecture is complemented with additional modules, as depicted in Figure 4.1, enabling detection of the changes in the radio environment as well as adaptation to these changes that could be caused either by the incumbents or by the LSA licensees. The architecture also allows coordination of access of multiple LSA licensees to the LSA band.

In addition to the LSA repository, which contains information only about the incumbent's spectrum, the ADEL project proposes the use of one or more collaborative spectrum–sensing networks to provide periodic updates about the radio environment. These information sources will be updating a radio environment map (REM), whose role is to reflect the radio environment as accurately as

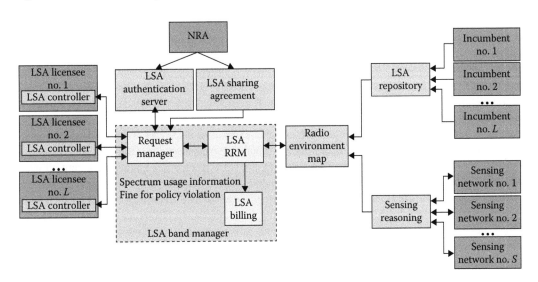

FIGURE 4.1 The LSA system architecture proposed by ADEL [16].

possible. When there is an LSA licensee request for spectrum, the information contained in the REM will be used by the LSA band manager to allocate adequate resources (frequency and power) to that particular LSA licensee. The proposed architecture also includes modules dealing with authentication, storage of the LSA sharing agreement rules, and spectrum usage accounting and billing. These functional modules may be implemented by the same, or by different, physical modules. A detailed description of the functional modules of the proposed LSA system can be found in [17].

The LSA functional architecture discussed here is the one presented in [17] and addresses multiple LSA licensees and multiple incumbent dynamic configurations. It contains a building block responsible for coordinating the access of multiple LSA licensees to the LSA band, thus avoiding the need to have a fixed-band plan, as prescribed by the ETSI standard [15] and by the CORE+ single-licensee trials (http://core.willab.fi). This architecture is also ETSI compliant since each LSA licensee has an LSA controller responsible for translating the spectrum availability information, provided by the LSA band manager, into networking reconfiguration commands.

The LSA band manager contains two sub-blocks: the request manager, which performs priority management according to the LSA spectrum usage rules, and the LSA RRM block, which performs the computation of available resources for assignment to the LSA licensees, based on spectrum usage rules and the information stored in the repository.

4.2   BRIEF LITERATURE OVERVIEW OF LSA SPECTRUM SHARING

Several works have appeared in recent years focusing on different aspects of LSA systems. While some of them are focused on trials, such as [18] and [19], others, which are discussed briefly here, represent research investigations into cellular system performance and advances to LSA. A scheme for offloading macro-cell traffic to a small-cell network using LSA as a basis is proposed in [20]. Based on a game-theoretic approach and taking into account individual utilities of the macro- and small-cell networks, the scheme determines the number of small cells that will be used for offloading using LSA spectrum while increasing energy efficiency. The authors in [21] consider how two parameters of cellular networks, power and antenna tilt, can be optimized to meet the conditions to operate in the LSA band, considering different incumbent services (wireless cameras, video links) and their requirements. The authors use measurements by the MNO's user terminals and additional test points to estimate interference levels caused by MNO's transmissions to determine the feasibility of using the LSA spectrum. Their results show that the best use of LSA spectrum takes place when the locations of the incumbent's users are close to the MNO's users. Outage probability of an LTE system is evaluated in different deployment scenarios, such as in macro and heterogeneous networks with various node densities as in [22], taking into account cumulative interference power in the incumbent region. The authors argue a significant reduction in the size of geographical borders between an LSA licensee and the incumbent when the LSA licensee deploys small cells instead of macro base stations. In [23], the authors provide an interference management scheme, based on a REM, to combat interference caused to the incumbent on the uplink in an LSA system. A distributed antenna system (DAS) architecture in a network virtualization context using fractional frequency reuse is considered in [24]. The paper compares the capacity of cell-edge users between two cases. One is the case when LSA spectrum can be used in combination with single-user multiple-input multiple-output (MIMO) with joint transmission by two remote antennas. In the other scenario, the LSA spectrum is not available, but all users can avail of multi-user MIMO transmission with coordinated beamforming. The paper derives the ratio between the required LSA bandwidth and the cell-edge bandwidth to support a decision on when it is more efficient to use LSA to meet the capacity requirements. In [25], a multicarrier waveform–based, flexible inter-operator spectrum-sharing concept is proposed for 5G communication systems. There, multiple operators obtain access to the shared band, which can be an LSA spectrum band. By adapting waveforms with respect to the out-of-fragment radiation masks, the authors show that the inter-operator interference can be avoided. A one-cell 3GPP LTE system using LSA is studied in [26], in which the authors propose a methodology to model the unreliable operation of an LSA frequency band by employing a multiline queuing system with unreliable servers. Opportunistic beamforming in the LSA band is proposed in [27], in which an LSA licensee coexists with an incumbent in a single-cell scenario. The LSA licensee has instantaneous information on the performance of the incumbent system to protect the incumbents' user's QoS. In [28], cloud RAN and massive MIMO in the context of LSA were analyzed, where the authors evaluate the trade-offs between spectrum and antennas.

The authors of [29] propose a two-tier evolutionary game for dynamic allocation of spectrum resources, enabling the coexistence of incumbents and LSA licensees. The authors present a mechanism for fair decision-making regarding spectrum allocation to LSA licensees, taking into account spectrum demand. An auction-based approach to spectrum sharing in LSA is presented in [30]. There, the authors propose a mechanism to allocate the incumbent's unused spectrum to the access points belonging to a number of LSA licensees. The mechanism, LSA auction (LSAA), combines independent set selection by bidding and a group bid. The goal is a policy aiming for revenue and market regularity. An auction-based approach for spectrum and infrastructure sharing is also proposed in [31]. There, the authors design a hierarchical, combinatorial auction mechanism, based on a Vickrey-Clarke-Groves (VCG) auction, and consider the infrastructure providers and cellular virtual network operators (VNOs). The authors evaluate the allocation with three degrees of freedom (i.e., frequency, power, and antennas) and propose a computationally tractable solution. In [32], an auction mechanism with a mixed graph, which can further quantify and tackle the interference between the LSA licensees, is proposed. Furthermore, to improve the revenue, the merging of bid comparisons is done when grouping nodes in the interference graph.

In the following sections, we consider a more dynamic nature of spectrum access in LSA. As mentioned earlier in Section 4.1, the currently envisaged sharing arrangements between the incumbent and the LSA licensee are foreseen to be maintained in the long term. Here, our aim is to encourage faster allocation (and release) of the spectrum and consider spectrum sharing based on a more immediate MNO's spectrum demands. In that, we discuss and evaluate two distinct approaches to spectrum allocation—one based on fairness and the other based on an auction mechanism.

4.3 FAIR SPECTRUM ALLOCATION SCHEMES

In this section, we assume that there is no formal bidding process involved at the time of spectrum allocation and that the MNOs have agreed a priori on a fair use of shared resources such that every MNO pays the same price and agrees on receiving a fair proportion of the available LSA spectrum. As every MNO is offering the same price for spectrum access, the utility function for the LSA system is to distribute the available spectrum fairly in the "long and short term." We propose spectrum-sharing mechanisms which aim at satisfying spectrum requirements of the allocated MNOs (as much as possible) at a particular spectrum allocation instant and allocating spectrum in a fair manner.

4.3.1 STRICTLY FAIR SCHEME

This scheme aims to provide a fair share of the spectrum to each competing MNO at each *spectrum allocation instant*; that is, available spectrum is distributed among all MNOs (with demand) based on previous allocation history. Each MNO with spectrum demand gets an offer of a nonzero spectrum.

Denoting by $n \in \{1, \cdots, N\}$ the MNO index out of N MNOs, we define the priority index (**PI**), PI_n, for MNO n as

$$PI_n = \lim_{W \to \infty} \frac{\text{BW awarded to MNO } n}{\text{Sum BW allocated by the incumbent}}$$

$$= \lim_{W \to \infty} \frac{\sum_{j=1}^{W} B_n^a(j)}{\sum_{j=1}^{W} B(j)} = \lim_{W \to \infty} \frac{\sum_{j=1}^{W} B_n^a(j)}{\sum_{j=1}^{W} \sum_{n=1}^{N} B_n^a(j)}$$

where $B_n^a(j)$ and $B(j)$ denote the bandwidth awarded to MNO n and the total offered bandwidth by the incumbent at the jth spectrum allocation instant, respectively. Note that we assume that all the offered spectrum is allocated to the MNOs.

Denoting as B, for simplicity, the available spectrum at a single allocation instant, j, the proposed spectrum allocation algorithm operates in the following steps [17]:

1. Initialize the assigned spectrum to every MNO B_n^a with zero in round $i = 1$.
2. In round i, divide the bandwidth, B, in proportion to the PI for each MNO with demand $B_n^d > 0$, that is, the MNO n is allocated spectrum in inverse proportion to its PI such that

$$B_{n,i}^a = B \cdot \left(\frac{1 - PI_n}{\displaystyle\sum_{n=1}^{N} (1 - PI_n)} \right)$$

3. If the spectrum demand for any MNO n is less than $B_{n,i}^a$, the bandwidth $B_{n,i}^a - B_n^d$ becomes the residual bandwidth B_n^r, which is zero otherwise. All the MNOs with $B_{n,i}^a > B_n^d$ do not take further part in the allocation.
4. After completing the allocation procedure in each round i, update the assigned and requested spectrum by $B_n^a = B_n^a + \min\left(B_{n,i}^a, B_n^d\right)$ and $B_n^d = B_n^d - \min(B_{n,i}^a, B_n^d)$, $\forall n$.
5. Set $B = \displaystyle\sum_{n=1}^{N} B_n^r$ for the next round, and go back to Step 2.
6. The process terminates when $B = 0$ or $B_n^d = 0$, $\forall n$.

This algorithm allocates spectrum in a fair fashion to each MNO in every allocation round, since PI depends on spectrum allocation history for every MNO. The drawback of this strictly enforced fairness is that the allocated spectrum to a single MNO may not be sufficient to meet its spectrum demands if the number of MNOs is large, thereby making allocated spectrum less useful.

We use Monte Carlo simulations to evaluate the performance of this algorithm and demonstrate its short- and long-term fairness characteristics. The window size, W, for computing the PI is set to 20 allocation instants to ensure more short-term fairness. The shorter the window size, the more short-term fairness the algorithm will achieve. As the PI computation for each MNO requires bandwidth allocation in the last W instants, the simulations are initialized by having $W - 1$ time slots with zero spectrum allocation and the Wth time slot with allocation depending on a random PI (chosen between 0 and 1) for every MNO. Without loss of generality, in the simulations, $N = 4$, and the incumbent spectrum B is normalized to 100 units. At each spectrum allocation instant, MNOs 1, 2, and 3 choose the demand randomly out of a vector of values [50,100] with uniform probability, while MNO 4 always requires 100 units (resulting in a significantly larger average demand). Ten thousand spectrum allocation instants are simulated to compute the mean spectrum allocation for each MNO.

Figure 4.2 shows the performance of the proposed spectrum allocation algorithm, plotting the spectrum allocation instants 21–220, where the first 20 instants were initialized with zero spectrum allocation and random PI. As all of the MNOs behave symmetrically (including MNO 4), the spectrum allocation statistics are plotted for one MNO only.

The algorithm provides a strictly equal share of available bandwidth from the incumbent to each MNO (25% for $N = 4$) and provides long-term fairness in spite of excessive demand from MNO 4. The short-term allocation for each operator for the first 200 allocation instants is evaluated as well. To study the short-term behavior of the proposed algorithm, let us define the moving average of the allocated spectrum to an MNO in a time slot, t, by

$$\bar{B}(t) = \frac{1}{W} \sum_{j=t-W+1}^{t} B(j)$$

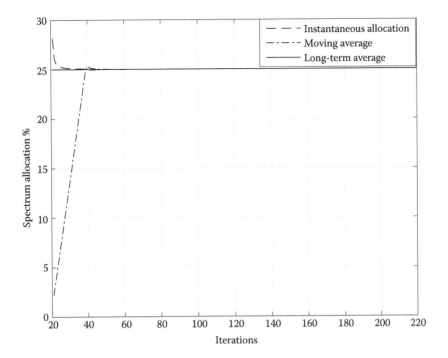

FIGURE 4.2 Performance evaluation for the strictly fair spectrum allocation algorithm.

It is evident that the algorithm allocates spectrum to each MNO in such a way that its moving average (evaluated over W allocation instants) converges to its mean very quickly. After the initialization phase, the algorithm starts dividing the instantly available spectrum equally among the competing MNOs as the *PI*s for all the MNOs converge to the same values. The instantaneous allocation remains constant at 0.25 B for $N = 4$ (strictly fair) if the minimum demand for every MNO is greater than 0.25 B (it is 0.5 B in this example). However, if the minimum possible demand is less than 0.25 B, the instantaneous allocation cannot be 25% all the time, and the moving average slightly diverges from the mean, recovering very quickly in future allocation instances.

4.3.2 LONG-TERM FAIR SCHEME

The proposed spectrum allocation algorithm operates in a proportionally fair manner and assigns spectrum to the operators based on their allocation history in the past, as before. In contrast to the short-term fair algorithm, this algorithm does not aim to provide fair spectrum at every spectrum allocation instant (by providing a nonzero spectrum). However, this algorithm is fair in the long run and aims to meet the spectrum requirements of the MNOs as much as possible at a specific spectrum allocation instant.

Based on *PI* for each operator, apply the following algorithm [33]:

1. Sort the MNOs with respect to *PI* in increasing order and queue them.
2. Offer as much spectrum as possible to the operator at the head of the queue (HOQ) (and with the smallest *PI*) asks for, and remove it from the queue.
3. If the allocated spectrum is less than the MNO's demand, the MNO can refuse to accept the offer.
4. If the MNO accepts the offer, the MNO uses the offered spectrum.
5. If the incumbent spectrum is still available after assignment to the selected HOQ MNO, go back to step 2.
6. Terminate the algorithm either when there are no MNOs with any spectrum demand or when the incumbent-offered spectrum is fully distributed among the MNOs.

The flowchart for the L1 algorithm is shown in Figure 4.3.

We use Monte Carlo simulations to evaluate the performance of the proposed algorithm. The simulation parameters are the same as for the strictly fair algorithm evaluation. Without loss of generality, we assume that an MNO accepts whatever spectrum is offered by the LSA band manager after running the spectrum allocation algorithm.

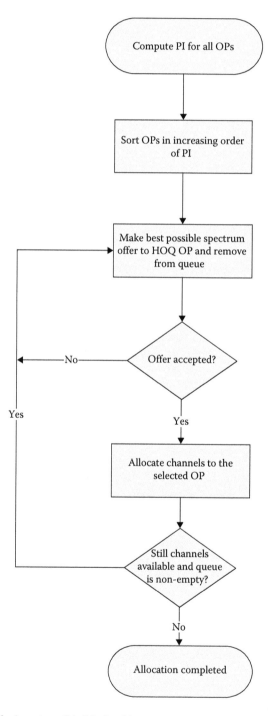

FIGURE 4.3   Flowchart for long-term fair L1 algorithm.

Figure 4.4 shows the mean spectrum allocation to four MNOs. It is clear that the algorithm divides the spectrum among the MNOs uniformly and is fair in the long term (as the strictly fair algorithm was).

Figure 4.5 shows the instantaneous spectrum allocation statistics for the proposed algorithm. As all of the MNOs have symmetrical demand and allocation statistics, we plot statistics for MNO 1 only. The instantaneous allocation for the operator varies between zero and its demand. As apparent from Figure 4.5, when the MNO is allocated full spectrum, it has little chance of accessing the spectrum in the next few allocation instants. Similarly, a long sequence of zero allocation is usually followed by full allocation. This justifies the algorithm's aim to achieve fairness in spectrum

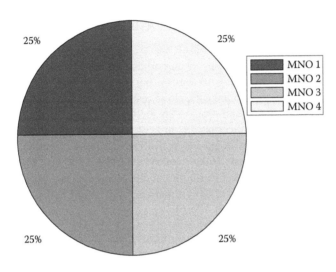

FIGURE 4.4 Spectrum allocation for long-term fair L1 spectrum allocation algorithm.

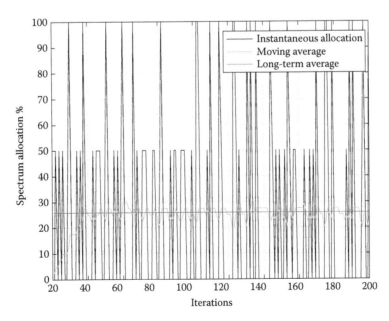

FIGURE 4.5 Performance evaluation for short-term spectrum allocation for MNO 1. The spectrum allocation instants 21–200 are plotted.

allocation for the MNOs in addition to meeting spectrum demands of the MNOs temporarily. It is evident that the moving average of the allocated spectrum for MNO 1 converges to its mean after very few iterations and diverges marginally afterward, which confirms that the algorithm provides reasonably good fairness in a short time span.

It is worth noting that both of the algorithms do not provide any excessive spectrum to MNO 4, which has greater than average demand. On the other hand, if spectrum allocation is provided without taking spectrum allocation history into account, MNO 4 may get additional spectrum during its turn (e.g., on a round robin basis), which will result in an unfair mean spectrum allocation.

4.4   AUCTION-BASED SPECTRUM ALLOCATION

In this section, we introduce an auction-based spectrum management as presented in [34]. We explore the aspects of sharing not only spectrum but also infrastructure by virtual network operators (VNOs), which will be constructing networks using resources from a shared pool, such as base stations, spectrum, core network components, cloud resources, etc. We use the existing LSA spectrum-sharing framework as a basis and propose an auction-based mechanism to allocate resources to the VNOs. As we have seen in Section 4.1, in LSA, the spectrum resources are orthogonally assigned to maintain service quality. Regarding the infrastructure, we consider a cloud-based, massive-MIMO system, in which multiple VNOs can share all the antennas. Antennas are connected to the centralized processing units that reside in the cloud and create a baseband pool. The cloud and fronthaul physical resources are logically separated and shared between VNOs, creating virtual base stations [35]. In this way, each VNO has a virtual slice comprised of infrastructure and radio resources, enabling them to provide distinct services to their users. This section presents the case in which all VNOs offer the same service to their users, evaluated through average user rate. The infrastructure provider may be third party or may be a public network provider. This approach to spectrum and infrastructure management is in line with the radio access network (RAN) sharing scenario proposed by 3GPP in [36], where the participating operators share RAN by utilizing orthogonal portions of the licensed spectrum.

4.4.1   ENHANCED AUCTION-ASSISTED LSA ARCHITECTURE

The proposed enhanced LSA architecture that supports spectrum and infrastructure sharing is depicted in Figure 4.6. It consists of three main building blocks: the LSA architecture with an LSA controller and an LSA repository, an auctioneer, and an infrastructure provider for a given area.

The enhanced LSA architecture, therefore, extends the conventional one and also incorporates cloud RAN, virtualization, and software-defined radio/network concepts [37]. Cloud RAN envisages cloud-based baseband processing, where baseband resources are pooled and shared among different remote radios. Virtualization can be considered a next stage in the evolution of cloud RAN [35], allowing multiple operators to share common infrastructure (baseband, transport, and access) resources as well as spectrum resources. Software radio/networking enhances virtualization, enabling direct programmability of the network. In the context of this chapter, virtualization envisages providing distinct wireless network resources, such as antennas, baseband, fronthaul, and spectrum, to different VNOs. The resources are logically separated, enabling each VNO to manage their resource allocation policy. Here, the spectrum is a public resource, whereas infrastructure may be provided by a third party (e.g. an infrastructure provider) or may be a part of a public (cellular) network. If present, the infrastructure provider is responsible for defining the terms of infrastructure sharing. Should the infrastructure be a part of a public network, the NRA role would need to be extended to set the terms of infrastructure sharing. Furthermore, the new LSA licensees are virtual cellular operators, which now do not own infrastructure. We also envisage that the

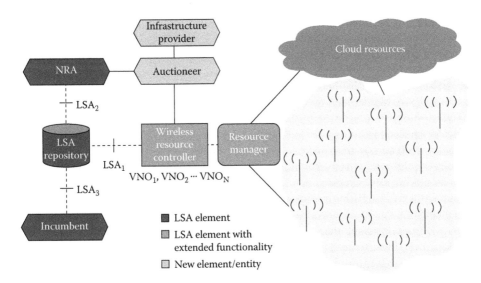

FIGURE 4.6 Proposed enhanced LSA architecture, supported by the auction mechanism. Dashed lines represent the existing LSA interfaces, whereas full lines represent new, required interfaces.

sharing arrangements involve the auction mechanism, where a third party (i.e., the auctioneer) is introduced on behalf of the NRA and infrastructure provider to manage both the spectrum and the infrastructure sharing. The auction mechanism follows the LSA spectrum-sharing rules, where the temporal allocation of spectrum follows the statistics of the incumbent(s) in the band. Concerning the infrastructure sharing, we consider cloud-based, massive-MIMO antennas as a resource that multiple VNOs can share at the same time. Based on the input from the auctioneer and the LSA repository, the wireless resource controller assigns spectrum and infrastructure resources to each VNO (i.e., the appropriate channels, the number of antennas, and the required cloud and fronthaul resources). The wireless resource controller instructs the resource manager to manage the assigned resources. Considering that resources may belong to different entities, the wireless resource controller and resource manager may consist of separate logical units that each control/manage spectrum or infrastructure. It should also be noted that, in general, independent providers may provide different resources (i.e., antennas, cloud, and fronthaul). In this chapter, a single infrastructure provider is responsible for all the resources.

4.4.2 AUCTION PROCEDURE

The auction here is similar to the one in [38], in which a clock auction is performed by a third-party auctioneer for the combined acquisition of spectrum and antennas. In our auction, the bidders (i.e., VNOs) also bid for spectrum and infrastructure resources. Each VNO serves the same number of (its own) users. Furthermore, to comply with the LSA framework and according to the LTE standard, the available spectrum is channelized into blocks of 5 MHz. Thanks to massive MIMO, users of the same VNO can reuse the same spectrum. However, the LSA framework stipulates the orthogonal use of spectrum by VNOs. It should be noted that allocation of resources is valid for the period determined by the type of incumbent and their usage of spectrum. In the case of appearance of an incumbent in a given band, there are a few options as to how the resources can be reassigned, namely (i) the residual spectrum from the current auction can be reassigned to the VNOs that are affected by the appearance of an incumbent, (ii) the auction can be repeated over the updated available spectrum, or (iii) the affected VNOs will be left without LSA spectrum, waiting for the incumbent to evacuate the band.

We consider N VNOs, K users to be served by each VNO, and M distributed antennas. In our model, VNOs lease antennas at a fixed price and acquire access to spectrum via an auction mechanism. It should be noted that when referring to the antenna price, here, we also refer to the required cloud and fronthaul resources. The fixed price associated with the usage of each antenna affects spectrum utilization. Since spectrum and antennas are partially interchangeable resources [38], the demand for spectrum will vary with the cost of antennas. As a case in point, if the cost of antennas is too high, the remaining budget might not be sufficient to acquire the necessary spectral resources for delivering a given rate. In this chapter, we have adopted a clock auction for the assignment of resources to the VNOs. The clock auction operates in two phases, namely, the price discovery (clock) phase and the final assignment phase. The price of spectrum monotonically increases in each round, and VNOs indicate the packages of spectrum and antennas they are willing to buy at a given price. In particular, if the auctioneer detects excess demand for the spectrum after a round of bidding has closed, it increases the posted spectrum price and opens another round of bidding. In our model, in each round, each VNO can XOR two package bids. Each VNO computes the first package bid as the number of antennas and 5 MHz blocks that minimize its cost within its budget constraint while providing its users a minimum rate. The cost is a linear combination of the number of antennas and bandwidth at the prices indicated by the auctioneer. Except for spectrum channelization, this is the same model discussed in [40]. However, since the price of spectrum increases at each round, we also consider a second bidding strategy, which models a less aggressive competitive bidding for the spectrum resources. To calculate the second package bid, each VNO starts from the first set of a number of antennas and spectrum requirements and attempts to minimize the bandwidth requirement by incrementing the number of antennas, provided that the minimum rate requirement is satisfied. Then, each VNO checks if the cost is less than or equal to its available budget and submits a package bid to the auctioneer. This procedure is shown in Figure 4.7, where $a_{ij} \in \{K, K+1, ..., M\}$ and $b_{ij} \in \{5, 10, ..., B\}$ are the number of antennas and spectrum blocks of width 5 MHz that VNO i submits to the auctioneer at bid package j, a_{Max} is the maximum

---

Algorithm 1: THE SECOND BIDDING MECHANISM

Input: First package bid of bidder i, i.e., $\{a_{i1}, b_{i1}\}$
Output: The second package bid of bidder i, i.e.,
$\{a_{i2}, b_{i2}\}$

1 **Initialisation:** $a_{i2} = a_{i2}^{(0)} = a_{i1}$, $b_{i2} = b_{i2}^{(0)} = b_{i1}$, $\ell = 0$

2 $\quad\quad\quad\quad b_{\text{Min}} \leftarrow \text{Rate}(b, a_{\text{Max}}) = \text{Rate}_{\text{Min},i}$

3 **while** $\left(a_{i2}^{(\ell)} \leq a_{\text{Max}} \text{ and } b_{i2}^{(\ell)} \neq b_{\text{Min}}\right)$ **do**

4 $\quad$ $\ell = \ell + 1$

5 $\quad$ $a_{i2}^{(\ell)} = a_{i2}^{(\ell-1)} + 1$

6 $\quad$ $b_{i2}^{(\ell)} \leftarrow \text{Rate}(b, a_{i2}^{(\ell)}) = \text{Rate}_{\text{Min},i}$

7 $\quad$ **if** $\left(c_a a_{i2}^{(\ell)} + c_b b_{i2}^{(\ell)}\right) \leq \beta_i$ **then**

8 $\quad\quad$ $a_{i2} \leftarrow a_{i2}^{(\ell)}$, $b_{i2} \leftarrow b_{i2}^{(\ell)}$

9 **return** $\{a_{i2}, b_{i2}\}$

FIGURE 4.7 Algorithm for the auction-based resource allocation.

number of available antennas, $\text{Rate}_{\text{Min},i}$ is the minimum required rate for the ith VNO, and β_i is the budget for VNO i.

The clock phase of the auction ends when all excess demand is removed from the market. In the ideal situation, both supply and demand completely match. However, this is unlikely to happen in complex multi-item-unit, multi-item-type auctions. Consequently, the approach that is used in this part of the auction may result in the oversupply of spectrum. Namely, this situation will occur if the bidder's private valuation of the minimum required rate is lower than the corresponding cost to acquire spectrum and antennas at the requested price. If this situation arises, the bids are assigned using a revenue-maximizing approach (i.e., using a winner determination algorithm). This algorithm determines which combination of the bids that stood at the last clock price that caused excess demand will maximize the auctioneer's revenue. The winner determination problem can be formulated as follows:

$$\underset{y_{ij}}{\text{Maximize}} \quad \sum_{i=1}^{N}\sum_{j=1}^{2}\left(c_a a_{ij} + c_b b_{ij}\right)y_{ij}$$

$$\text{Subject to} \quad \sum_{i=1}^{N}\sum_{j=1}^{2}\left(y_{ij}b_{ij}\right)\leq B,$$

$$\sum_{j=1}^{2}y_{ij}\leq 1,\ \forall\ i\in\{1,2,\dots,N\}$$

$$y_{ij}\in\{0,1\},\ \forall\ i\in\{1,2,\dots,N\},\ j\in\{1,2\}$$

where $y_{ij}=1$ if package j of bidder i is accepted, otherwise $y_{ij}=0$. c_a and c_b are the costs per antenna and spectrum block, respectively. Finally, B is the total available bandwidth.

4.4.3 NUMERICAL ANALYSIS

The simulated scenario is based on the auctioning strategy explained previously. The scenario includes 15 VNOs competing in a bid to acquire spectrum and infrastructure to meet their requested minimum rate.

The minimum requested rate is the same for all the operators. We consider 10 users per VNO that are randomly distributed in a given area. A total of 64 antennas is available for sharing between the VNOs. The total available spectrum is 50 MHz, where each VNO can acquire spectrum in blocks of 5 MHz, according to the LSA rules. The budget of each VNO is proportional to the rate that it wants to offer to its users—the higher the rate, the greater its budget. The results of the simulated scenario are depicted in Figures 4.8 and 4.9. Figure 4.8 depicts the number of required antennas as a function of the minimum rate and antenna price. Figure 4.9 illustrates two directly related aspects—the required bandwidth and the number of VNOs that can be served. To better understand the trends, the figures should be considered together. Looking along the x-axis in both figures, we can see that for the lowest considered user rate, the same number of antennas and spectrum are required, regardless of the antenna price. In this case, the spectrum is abundant, as VNOs cannot lease less than 5 MHz of spectrum or less than 10 antennas.* This spectrum is, therefore, sufficient to provide the required rate, with the minimum number of antennas. In total,

---

* In C-RAN systems, the number of antennas need to be more than or equal to the number of users to be served by each VNO.

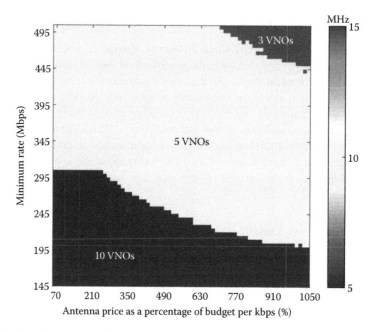

FIGURE 4.8 Number of antennas assigned to each allocated VNO in correspondence to a particular rate requirement (*y*-axis) and antenna cost (*x*-axis). Antenna cost incorporates the required cloud and fronthaul resources.

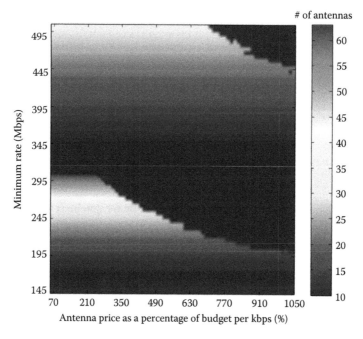

FIGURE 4.9 Number of VNOs in correspondence to a particular rate requirement (*y*-axis) and antenna cost (*x*-axis). Again, antenna cost incorporates the required cloud and fronthaul resources.

10 VNOs are served. As the minimum considered user rate increases (looking along the *y*-axis), the number of antennas increases up to the highest possible number. The VNOs can still serve their users with 5 MHz of spectrum but with an increasing number of antennas. This is the case until the maximum number of antennas is reached and as long as the antenna price is less than a

certain value (approximately 250% of the budget per kbps). For higher antenna prices, it is more cost-effective for VNOs to buy more spectrum than to further increase the number of antennas. The spectrum requirement, therefore, jumps to the region of 10 MHz when 5 VNOs can be served. This trend repeats itself with 10 and 15 MHz of spectrum, serving 5 and 3 VNOs, respectively. It should be noted that this periodicity with the number of antennas is observed only in the case in which discrete spectrum blocks are considered. It is one of the main differences between this and the case in which the continuous spectrum is considered [38].

4.4.4   COMPARISON WITH FIXED SHARING

In this subsection, we compare the results of auction-based sharing with the fixed-based allocation of resources. In that, we consider two approaches—one with the orthogonal and equal allocation of both spectrum and antennas and the other with the orthogonal and equal allocation of spectrum, where all VNOs can utilize all the antennas. Figure 4.10 depicts the number of served VNOs versus the rate requirement for the considered approaches. It should be noted that two different antenna price values are evaluated for the auction-based sharing. The case of fixed sharing with the equal and orthogonal allocation of the spectrum where all the antennas are shared can be considered as a benchmark in terms of system efficiency (i.e., the number of VNOs that can be served with a given minimum rate requirement, but excluding the cost of infrastructure). Namely, as in our study, all VNOs have the same rate requirements, under the assumption of orthogonal spectrum allocation, using all antennas and equally dividing the spectrum among the VNOs is the optimal solution in terms of system efficiency.

The fixed-sharing case with the orthogonal usage of antennas serves the lowest number of VNOs, regardless of the rate. This degradation in the number of VNOs that can be served occurs because virtualization is not exploited (i.e., each VNO uses a smaller number of antennas). As shown in Figure 4.10, the auction-based approach for two different antenna prices under consideration outperforms the fixed-sharing case with the orthogonal utilization of antennas. Furthermore, with a cost of antennas that is less than approximately 250% of the budget, we can always achieve the optimal performance through the auction-based approach.

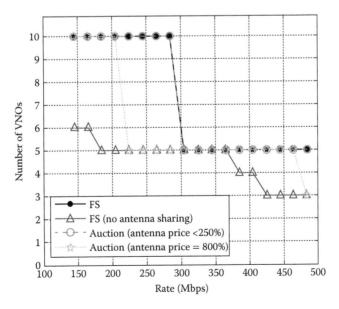

FIGURE 4.10   Fixed sharing versus auction-based sharing.

4.5   CONCLUSIONS

In this chapter, we have described the LSA architecture, followed by the overview of LSA literature and a description of two distinct approaches aimed to promote a more dynamic LSA spectrum allocation. The first approach aims at providing spectrum to multiple MNOs based on fairness. Here, two algorithms are evaluated numerically, and the results show quantitatively that we can guarantee fairness in spectrum allocation regardless of the demand from the MNOs.

We also propose an enhanced, auction-assisted LSA framework, which encompasses not only spectrum but also infrastructure (i.e., cloud-based, massive-MIMO antennas). There, we identify the key architectural aspects required to enhance the LSA framework to avail of this technology. In our numerical evaluation, we observe periodic patterns in the antenna allocations to VNOs when considering a range of minimum rate requirements and antenna prices. Furthermore, we show that the auction-based approach outperforms fixed static sharing with the orthogonal use of spectrum and antennas. Finally, we show that for cases in which the cost of antennas is less than a certain percentage of the budget (per kbps), we can achieve optimal performance in terms of the number of VNOs being served.

It is evident that LSA and other LSA systems will play a key role in dealing with the spectrum scarcity problem in 5G networks and beyond. While LSA has been designed to offer high predictability and certainty for both MNOs and the incumbents, there is progress to be made on the use of licensed shared bands and coexistence aspects. Although the existing MNOs can avail of the existing infrastructure and customer base, they are still cognizant of using shared (licensed) bands. Namely, operational and implemental aspects yet need to be proven for entering and vacating spectrum, management of exclusion and protection zones [39], security [3], and scalability and network-wide deployments that will be supported by the automated operations [40]. Furthermore, the initial costs related to LSA implementation need to be proven to be lower than with the exclusive, dedicated spectrum models. Hence, more advances are needed by industry and academia in relation to business model design and implemental aspects to make LSA a true success.

ACKNOWLEDGMENTS

The project ADEL acknowledges the financial support of the Seventh Framework Programme for Research of the European Commission under grant number 619647. We also acknowledge support from the Science Foundation Ireland under grants No. 13/RC/2077 and No. 10/CE/i853.

REFERENCES

1. European Commission–funded FP-7 Framework Project, *Advanced Dynamic Spectrum 5G mobile networks Employing Licensed shared access (ADEL)*. Available at: www.fp7-adel.eu/.
2. J. Mitola III, *Cognitive Radio. An Integrated Agent Architecture for Software Defined Radio*, Stockholm: Ph.D. dissertation, Royal Institute of Technology (KTH), 2000.
3. S. Bhattarai, J. M. J. Park, B. Gao, K. Bian, and W. Lehr. An overview of dynamic spectrum sharing: Ongoing initiatives, challenges, and a roadmap for future research, *IEEE Transactions on Cognitive Communications and Networking*, 2(2), 11, 2016.
4. FCC, *Unlicensed Use of TV Band and 600 MHz Band Spectrum* [ET Docket No. 14–165; FCC 15–99], 2015.
5. CEPT Report 159, Technical and Operational Requirements for the Possible Operation of Cognitive Radio Systems in the White Spaces of the Frequency Band 470–790 MHz, ECC, 2011.
6. CEPT Report 186, Technical and Operational Requirements for the Operation of White Space Devices under Geo-location Approach, ECC, 2013.
7. FCC, Further Notice of Proposed Rulemaking (FNPRM) in the Matter of the amendment of the Commission's Rules with Regard to Commercial Operations in the 3550–3650 MHz Band [GN Docket No. 12–354], April 2015.

8. Nokia Siemens Networks; Qualcomm, Authorized Shared Access: An evolutionary spectrum authorization scheme for sustainable economic growth and consumer benefit [Input Document FM(11)116 to the 72nd Meeting of the WG FM], May 2011.

9. CEPT, Licensed Shared Access (LSA) ECC Report 205, February 2014.

10. CEPT, ECC Decision (14)02: Harmonised Technical and Regulatory Conditions for the Use of the Band 2300–2400 MHz for Mobile/Fixed Communications Networks MFCN, ECC, June 2014.

11. CEPT Report 56, Technological and Regulatory Options Facilitating Sharing between Wireless Broadband Applications (WBB) and the Relevant Incumbent Services/Applications in the 2.3 GHz Band, ECC, March 2015.

12. ETSI, SR 103 13 Electromagnetic Compatibility and Radio Spectrum Matters (ERM) System Reference Document (SRDoc): Mobile Broadband Services in 2300–2400 MHz Frequency Band under the Licensed Shared Access Regime, Special Report, V1.1.1, July 2013.

13. ETSI, TS 103 154 Reconfigurable Radio Systems (RRS): System Requirements for Operation of Mobile Broadband Systems in the 2 300 MHz–2 400 MHz Band under Licensed Shared Access (LSA), V1.1.1, October 2014.

14. M. Mustonen, M. Matinmikko, M. Palola, T. Rautio, and S. Yrjölä. Analysis of requirements from standardization for licensed shared access (LSA) system implementation, in *IEEE International Symposium on Dynamic Spectrum Access Networks (DySPAN)*, Stockholm: IEEE Conference Publications, September 2015.

15. ETSI, TS 103 235 Reconfigurable Radio Systems (RRS): System Architecture and High Level Procedures for Operation of Licensed Shared Access in the 2300 MHz–2400 MHz Band, December 2015.

16. V. Frascolla, A. J. Morgado, A. Gomes, M. M. Butt, N. Marchetti, K. Voulgaris, and C. B. Papadias. Dynamic licensed shared access: A new architecture and spectrum allocation techniques, in *IEEE Vehicular Technology Conference (VTC-fall)*, Montreal, CA: IEEE Conference Publications, 2016.

17. A. Morgado, A. Gomes, V. Frascolla, K. Ntougias, C. Papadias, et al. Dynamic LSA for 5G networks, in *Proceeding of European Conference Networks and Communications*, Paris: IEEE Conference Publications, June 2015.

18. M. Palola, M. Matinmikko, J. Prokkola, M. Mustonen, M. Heikkilä, et al. Live field trial of licensed shared access (LSA) concept using LTE network in 2.3 GHz band, in *IEEE International Symposium on Dynamic Spectrum Access Networks (DYSPAN)*, McLean, VA: IEEE Conference Publications, 2014.

19. P. Masek, E. Mokrov, A. Pyattaev, K. Zeman, A. Ponomarenko-Timofeev, et al. Experimental evaluation of dynamic licensed shared access operation in live 3GPP LTE system, in *Globecom*, Washington, DC: IEEE Conference Publications, December 2016.

20. M. Hafeez and J. M. H. Elmirghani. Green licensed-shared access, *IEEE Journal on Selected Areas in Communications*, 33(12), 2579–2595, September 2015.

21. E. Pérez, K. J. Friederichs, I. Viering, and J. Diego Naranjo. Optimization of authorised/licensed shared access resources, in *9th International Conference on Cognitive Radio Oriented Wireless Networks and Communications (CROWNCOM)*, Oulu, FL: IEEE Conference Publications, June 2014.

22. T. Wirth, B. Holfeld, D. Wieruch, R. Halfmann, and K.-J. Friederichs. System level performance of cellular networks utilizing ASA/LSA mechanisms, in *1st International Workshop on Cognitive Cellular Systems (CCS)*, Rhine River, DE: IEEE Conference Publications, September 2014.

23. R. C. Dwarakanath, J. D. Naranjo, and A. Ravanshid. Modeling of interference maps for licensed shared access in LTE-advanced networks supporting carrier aggregation, in *IFIP Wireless Days (WD)*, Valencia, SP: IEEE Conference Publications, November 2013.

24. Y. He, E. Dutkiewicz, G. Fang, and M. D. Mueck. Licensed shared access in distributed antenna systems enabling network virtualization, in *1st International Conference on 5G for Ubiquitous Connectivity*, Akaslompolo, FL: IEEE Conference Publications, November 2014.

25. J. Luo, J. Eichinger, Z. Zhao, and E. Schulz. Multi-carrier waveform based flexible inter-operator spectrum sharing for 5G systems, in *IEEE International Symposium on Dynamic Access Networks*, McLean, VA: IEEE Conference Publications, April 2014.

26. V. Y. Borodakiy, K. E. Samouylov, I. A. Gudkova, D. Y. Ostrikova, A. A. Ponomarenko-Timofeev, et al. Modeling unreliable LSA operation in 3GPP LTE cellular networks, in *6th International Congress on Ultra Modern Telecommunications and Control Systems and Workshops (ICUMT)*, St. Petersburg, RU: IEEE Conference Publications, October 2014.

27. N. Taramas, G. C. Alexandropoulos, and C. B. Papadias. Opportunistic beamforming for secondary users in licensed shared access networks, in *6th International Symposium on Communications, Control and Signal Processing (ISCCSP)*, Athens, GR: IEEE Conference Publications, May 2014.

28. I. Gomez-Miguelez, E. Avdic, N. Marchetti, I. Macaluso, and L. Doyle. Cloud-RAN platform for LSA in 5G networks-tradeoff within the infrastructure, in *6th International Symposium on Communications, Control and Signal Processing (ISCCSP)*, Athens, GR: IEEE Conference Publications, 2014.
29. A. Saadat, G. Fang, and W. Ni. A two-tier evolutionary game theoretic approach to dynamic spectrum sharing through licensed shared access, in *IEEE CIT/IUCC/DASC/PICOM*, Liverpool, UK: IEEE Conference Publications, October 2015.
30. H. Wang, E. Dutkiewicz, G. Fang, and M. Mueck. Spectrum sharing based on truthful auction in licensed shared access systems, in *82nd Vehicular Technology Conference (VTC Fall)*, Boston, MA: IEEE Conference Publications, September 2015.
31. K. Zhu and E. Hossain. Virtualization of 5G cellular networks as a hierarchical combinatorial auction, *IEEE Transactions on Mobile Computing*, 2640–2654, October 2016.
32. H. Wang, E. Dutkiewicz, G. Fang, and M. D. Mueck. Framework of joint auction and mixed graph for licensed shared access systems, in *IEEE International Symposium on Dynamic Spectrum Access Networks (DySPAN)*, Stockholm, SE: IEEE Conference Publications, September 2015.
33. M. M. Butt, C. Galiotto, and N. Marchetti. Fair dynamic spectrum access in licensed shared access systems, in *EEE International Symposium on Personal, Indoor and Mobile Radio Communications (PIMRC)*, Valencia, ES: IEEE Conference Publications, September 2016.
34. J. McMenamy, A. Fahrang, N. Marchetti, and I. Macaluso. Enhanced auction-assisted LSA, in *International Symposium on Wireless Communication Systems (ISWCS)*, Poznan, PL: IEEE Conference Publications, September 2016.
35. A. Checko, H. L. Christiansen, Y. Yan, L. Scolari, G. Kardaras, M. S. Berger, and L. Dittmann. Cloud RAN for mobile networks: A technology overview, *IEEE Communications Surveys Tutorials*, 7(1), 405–426, 2015.
36. 3GPP, Study on radio access network (RAN) sharing enhancements, *3GPP TR 22.852 v13.1.0*, September 2014.
37. C. Liang and F. R. Yu. Wireless network virtualization: A survey, some research issues and challenges, *IEEE Communications Surveys Tutorials*, 17(1), 358–380, First quarter 2015.
38. H. Ahmadi, I. Macaluso, I. Gomez-Miguele, L. Doyle, and L. A. DaSilva, Substitutability of spectrum and cloud-based antennas in virtualized wireless networks, *IEEE Wireless Communications Magazine*, 99, 2–8, 2016.
39. M. D. Mueck, V. Frascolla, and B. Badic. Licensed shared access—State-of-the-art and current challenges, in *1st International Workshop on Cognitive Cellular Systems (CCS)*, Rhine River, DE: IEEE Conference Publications, September 2014.
40. S. Yrjölä, P. Ahokangas, and M. Matinmikko. Evaluation of recent spectrum sharing concepts from business model scalability point of view, in *IEEE International Symposium on Dynamic Spectrum Access Networks (DYSPAN)*, Stockholm, SE: IEEE Conference Publications, September 2015.

Part II

5G Technologies and Software-Defined Networks

5 Software-Defined Network Security
Breaks and Obstacles

Ahmed Dawoud, Seyed Shahristani, and Chun Raun

CONTENTS

5.1 INTRODUCTION

The data communication architecture remained constant for decades. As the pace of technologies accelerated, there was a need to adopt a new model to decompose the complexity and inflexibility of the traditional networks. The pillar technologies of software-defined networks (SDN) (e.g., central network control, programmability, and network virtualization) have been under research for decades [1]. OpenFlow introduces the concept of separating control and forward planes and represents novel communication architecture. Despite the significant advantages offered by the new SDN architecture (e.g., flexibility, programmability, and centralization), the model imposes unprecedented security threats [2].

Security is a primary concern in the new model. The SDN controller is a crucial layer in the network. A single point that orchestrates the entire network can be utilized to enhance network security, paradoxically making the centralized architecture more vulnerable to attacks. The controller is an attractive target for the attackers, and the accessibility from the application layer to the forward layer is a severe threat to network resources [1]. This chapter surveys the SDN security concerns as a security enhancer and studies the trade-offs of the SDN model from a security perspective. Additionally, the chapter investigates several SDN security solutions.

The chapter is divided into six sections. The first and the second section introduce the concepts of the SDN (e.g., history, components, and features). The third section discusses SDN as a security enhancer. The fourth section demonstrates the security deficiencies and challenges of the SDN from a security perspective. The fifth section shows the current solutions. The sixth section surveys SDN security applications.

An SDN security article categorizes the current research directions [2]. The study classifies the current research papers into three topics from a security perspective (i.e., analysis, enhancements, and solutions), and it indicates the plane(s) handled by the research. Moreover, the authors demonstrate the security threats in each layer. A comprehensive survey on SDNs by Kreutz et al. inspects security and dependability as a result of design flaws [1]. The study groups the SDN security solutions and applications and highlights the vulnerabilities that risk the controller. This chapter focuses on SDN security enhancements from application perspectives and discusses security flaws of the new major components of the model (i.e., the controller and the southbound API protocol).

5.2 SOFTWARE-DEFINED NETWORKS

Conventional networks are divided into three planes, named management, control, and forwarding layers. The management plane provides services to monitor and configure the network. The control plane generates the data required to establish forwarding tables, which in turn are used by the forwarding plane to direct packets to ingress and egress ports. In traditional network models, both the control and the forward planes are tightly coupled within the single device (e.g., switches and routers). This model is efficient from a performance perspective. However, as the complexity of the networks mounted, a necessity to develop a new architecture emerged. The essence of SDN architecture is the separation of control and forwarding planes. The network devices are merely forwarding devices, and the network control plane is moved to an independent entity named network controller or network operating system. Figure 5.1 shows the three layers of SDNs. The forwarding layer comprises network devices such as switches, routers, and middleboxes, which do not own their logic. The network intelligence resides on the network operating system, called the controller, which abstracts the devices and provides services such as network state and topology information services. Moreover, the controller provides an API to communicate with the application layer, called the northbound API. The southbound API is the communication protocol between the control layer and forward layer devices. A dominated southbound protocol in the SDN model is called OpenFlow [1,3].

The application layer tops the SDN stack. The application introduces programmability, which is a fundamental concept in SDN. Network programmability is the ability to communicate with the network's undelaying devices. Programmability provides opportunities for network innovation with an enormous number of network applications (e.g., network monitoring, traffic engineering, security, and cloud applications).

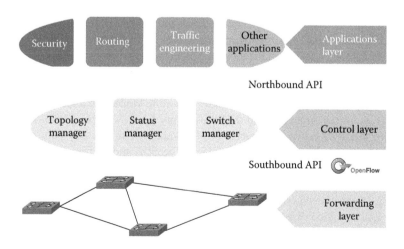

FIGURE 5.1   SDN architecture.

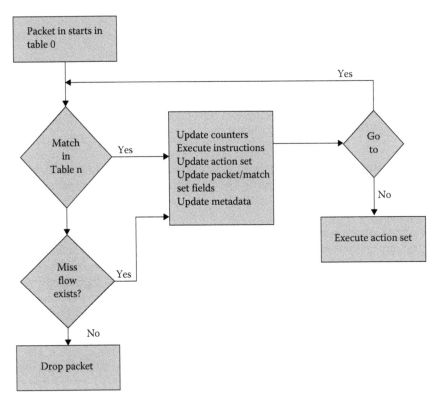

FIGURE 5.2 OpenFlow flowchart.

Centralized control reveals a global view of the network, which facilitates the management and monitoring process. Also, it lessens errors in configuring and deploying network policies. The centralization improves the flexibility; for instance, a pool of devices from various vendors can be deployed and abstracted in the same network.

In OpenFlow protocol specifications, the controller is responsible for modifying the forwarding table in SDN devices. The flowchart in Figure 5.2 depicts the OpenFlow process of incoming packets. On the arrival of a new packet, the switch lookup in the forwarding table searches for a matching forwarding entry. If a record matches the packet fields, a predefined action will be executed. OpenFlow allows a set of the measures being taken, for instance, to DROP, FORWARD, and MODIFY the packet. If no match occurs, the switch forwards the packet to the controller to conduct computations according to the policy issued by the application layer.

5.3 SDN AS A SECURITY ENHANCER

The rigorous and inflexible architecture of the traditional communication networks has hindered their innovation [1]. Multiple attempts to adopt a flexible network model were impelled to separate the control and forward planes. The separation has introduced programmability and centralization features. These features were harnessed to enhance the security of the network. This section probes the SDN applications to enforce and verify network policies and detect and mitigate threats.

5.3.1 POLICY ENFORCEMENT

Network policy is a set of configurations, rules, and constraints that govern network operations (e.g., network access, incident handling, and communications isolation). The architecture imposes the execution of policy enforcement through the network middleboxes. The middleboxes are devices deployed to manipulate the network traffic for specific purposes, for instance, inspection, threat detection, and access control.

Traditionally, two approaches enforce network policy, either by deploying the middleboxes in the network paths between endpoints or by attaching them to middle switches. Both options lack flexibility as they have rigorous deployments [4,5].

SDN architecture allows the following two advantages which were not possible in classic networks:

- **Entire-network coverage:** The network policy is enforced at switching devices by installing flow rules. In conventional networks, middleboxes such as firewalls and intrusion-prevention systems are positioned at specific points in the network, typically at the network entry points (e.g., in a Demilitarized Zone [DMZ]) either on network paths or off network paths. Both deployments are inflexible and incomprehensive. In SDNs, programmable switches are distributed across multiple locations in the network. This architecture avoids a single point of failure and enforces policy inside the network between endpoints.
- **Centralization in SDNs facilitates policy deployment and configuration:** In contrast to existing network middlebox configurations, the network administrators implement and deploy the policy from a single point rather than configuring appliances explicitly.

Historically, OpenFlow protocol was evolved as a successor of the Ethane project. The purpose of Ethane was to define a network policy and enforce it at the switches. The Ethane project was an instantiation of the Secure Architecture for Networked Enterprise (SANE) [6]. The domain controller based on the network policy calculates the flow table entries installed on the switches. The project requires custom switches; hence, network upgrades were expensive. Ethane network integration with current networks did not provide holistic policy enforcement, as there was a probability of the traffic passing through other, non-Ethane custom switches.

SDN features reintroduced a policy enforcement method called active security [7]. The concept includes the following five phases of adaptable network policy:

- Initial configuration of the infrastructure
- Sensing, the controller responsible for collecting data from the network
- Adjusting the configuration, as the controller updates the policy according to the network status
- Forensics, the controller gathers information related to attacks
- Response, the controller initiates a reconnaissance and counterreaction

5.3.2 SECURITY POLICY VERIFICATION

As the complexity of networks escalated, there was a necessity to ensure and verify the attached security policy. The conflict between policies and even the rules in the same policy may lead to network exposure.

FlowGuard is an SDN-based framework to detect firewall policy violations. Upon the update of the network status, FlowGuard will dynamically analyze the path space to detect firewall rule conflicts [8]. FLOVER is another SDN security policy verifier [9]. It is a model-based checking system and was built on a NOX controller to provide formal verification for the security policies. FLOVER transforms flow table rules into a binary tree diagram and applies formal methods to detect rule violations.

5.3.3   INTRUSION DETECTION

Intrusion detection and prevention systems (IDPS) are software or hardware systems dedicated to observing the network for security breaches. The standard IDPS process comprises three stages—collecting data from the network, analyzing the data, and then executing actions in case of threat detection. Substantially, there are three methods to analyze the data for breach detection, named signature based, anomaly detection, and specification based [10]. First, with signature-based detection, a system has a database of predefined violation signatures, and the system matches those signatures against the network activity signatures. Second, with anomaly analysis, the system is concerned with defining abnormal activities. For the system, normal activities are identified in a baseline profile, which the system develops in a learning phase. Third, with with the specification based, also known as stateful protocol analysis, a predefined pattern of protocol behavior is established, a comparison between network activities and the expected behavior defined by protocols, and in the case of a profile violation, an alert is raised. A combination of methods is used to maximize IDPS performance. A study compares various detection methods proposed in [10]. Each method has its pros and cons. A significant weakness in signature based is its inability to detect new attacks, while anomaly detection has a higher false alarms rate. The majority of the commercial implementations use a hybrid approach.

Substantially, a network-based IDPS has a packet- or flow-capturing module [11]. The capturing engine sniffs packets or flows for specific features. Those feature selections rely on the threats targeted by the IDPS.

From the perspective of the SDN, packet and traffic measurements are a current research issue, for example, for traffic engineering, load balancing, monitoring, and security [12–14]. A network central view and programmability provide the necessary assets to develop a robust packet/flow inspection system. SDN consists of three layers, namely the application, the controller, and the forwarding devices. The controller has the capability to communicate with the devices through the southbound protocols (e.g., OpenFlow protocol). OpenFlow allows the API to poll the devices for traffic statistics. Those traffic data are aggregated to the controller, which in turn communicates with the application layer through an OpenFlow interface [15].

The architecture of an anomaly-detection method based on SDN has been proposed in [16]. The framework is a distributed denial-of-service attack detector based on flow inspection. The system has been implemented on a NOX controller. OpenSketch is a significant example of SDN traffic measurement architecture [17]. It is a platform that provides a library to customize measurements to meet specific tasks and sets measurements to detect anomalous behavior. The comprehensive view of the system which lay in the essence of the SDN architecture is a remarkable feature. It reinforces the design of a robust data collection module in intrusion detection systems.

Studies have deployed architecture for an anomaly-detection method based on SDN [16,18]. The concentration on the anomaly detection based on the SDN is supported by the controllability of the traffic. However, there is a necessity to deploy other detection methods in SDNs to exploit the capabilities of SDNs.

Distinctly, SDN architecture can contribute to enhancing detection analysis techniques. Features such as scalability and easiness to configure in the case of anomaly detection can be improved by exploiting the centralized architecture of the SDN. Developing a central analysis module reduces the overhead on the monitored system, which leads to performance improvement.

5.3.4   THREAT RESPONSE

Consistently, the SDN controller has a real-time view of the entire network. Detecting attacks in real time is essential to establishing an active response system. The SDN is a flow-based networking model rather than a destination-based model [1,19]. Traffic control is a key feature in the response module. For example, on the assumption of threat existence, the network middleboxes forward the traffic to virtual appliances or a honeypot for further investigation or forensic processes. Additionally, SDN programmability allows applications, particularity IDPS applications, to communicate with forwarding devices. The flexibility of the architecture facilitates the response mechanisms. For instance, if a section of the network is compromised, the response module isolates infected devices only to mitigate the risks.

5.4 SDN CHALLENGES

Despite the many voices preaching for the promising future of the SDN, various challenges curb the broad adoption of the new model. Contradictorily, the new architecture's primary advantages are the origins of the model's weakness. Performance, scalability, resilience, and security are the main issues to tackle in the context of current research issues in SDNs [1,20].

In contrast to traditional networks, the SDN has performance trade-offs. A tightly coupled data and control plane in a single processing device is performance-oriented architecture. In SDN process flow, the device refers to the controller to perform logical decisions. The delegation of logical processes causes latency and negatively affects the throughput of the devices [20]. The current stream of research concentrates on hardware improvements (e.g., processing chips) [1].

Essential questions about SDN scalability are raised. The network controller is responsible for logically updating forwarding tables on the connected devices pool. In real-world networks, the controller is responsible for processing a huge number of messages sent from the forwarding devices. The question is the matter of how many nodes a controller should support. Increasing the number of installed controllers to expand the network will cause a inconsistency challenge. The consistency is essential in SDN model, as the controller or a set of them should maintain the same view on the network. HyperFlow [21] provides a solution that updates the network state by propagating the events that affect the network state.

A single point of control is equivalent to a single point of failure. Such a configuration is a major threat to network resiliency and fault tolerance. SDN resilience is still an open question [1]. A distributed controller was proposed to improve SDN flexibility [20].

Security threats are critical challenges in traditional networks. Security concerns escalate in SDN networks. The new architecture brings additional challenges that did not exist in traditional networks. Particularly, threats target the control layer [22]. In the following sections, we highlight the security concerns of the controller and the standard southbound protocol OpenFlow.

5.4.1 Controller Security Flaws

Security breaches are common challenges in conventional networks. Security threats are mounting in SDNs as the novel architecture introduces additional features which did not exist in traditional networks. Various studies focus on the challenges related to the SDN architecture. Sezer et al. mentioned three challenges of the SDN model [20]. First, the authors demonstrate the extent to which the performance is negatively affected by increased flexibility in the SDNs. The second drawback is the network's scalability limitations, which bring questions related to controller scalability, and latency associated with the exchange of information between a set of controllers in a distributed controller architecture. The last challenge highlighted relates to security; the authors claim the controller is an attractive target for attackers.

Figure 5.3 depicts seven threat vectors proposed by Kreutz et al. [22]. The following three threats directly relate to the controller itself:

- Attacks on the communications between the controller and the data plane devices
- Attacks on the controller vulnerabilities
- Attacks on the controller originated from untrusted applications

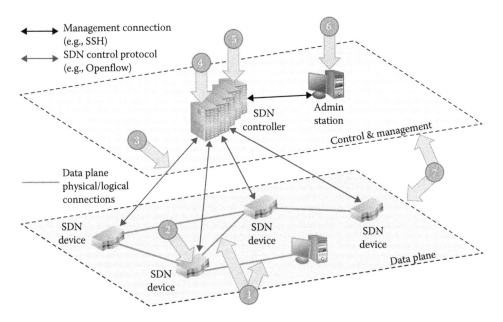

FIGURE 5.3 SDN threat vector. (From Kreutz, D. et al., Towards secure and dependable software-defined networks. In: Proceedings of the 2nd ACM SIGCOMM Workshop on Hot Topics in Software-Defined Networking, 2013, pp. 55–60.)

Security deficiencies are a remarkable ongoing research topic. In an SDN security survey, authors classified the current research on SDNs into three categories—studies to analyze SDN security, studies to enhance SDN security, and research related to the solutions proposed by SDNs [2]. A quick glance at the survey shows that the majority of probes on enhancement are associated with the controller layer.

The OpenFlow protocol security analysis study has revealed several attacks derived from the SDN's unrivaled protocol, for instance, denial-of-service attacks on flow tables and control channels and denial-of-service attacks on the controller itself [23].

Controllers are software systems, which potentially suffer programming vulnerabilities (e.g., buffer, heap, string overflow). Even in a web-terminal managed controller, there is a threat of SQL injection and cross-site scripting (XSS) attacks. A remarkable example of controller vulnerabilities is OpenDaylight programming vulnerabilities. Netconf XML eXternal Entity (XXE) is an OpenDaylight vulnerability exploited to execute an XXE attack on the controller and take control of the entire network.

5.4.2 OpenFlow Protocol Security Flaws

OpenFlow has evolved from SANE/Ethane, which was a Stanford project [1]. It has been a dominant protocol in the SDN southbound layer. Figure 5.4 shows the main components of an OpenFlow switch. The controller communicates with the switch via the control channel to manage one or more flow tables.

The control channel between the controller and the switch is initiated as a Transmission Control Protocol (TCP) connection, with an option for the encryption protocol transport layer security (TLS) to secure the channel. Without an encryption method, the communication between the controller and the forwarding devices is exposed to a man-in-the-middle attack. Kloti et al. have conducted a security analysis for the OpenFlow protocol [23]. The study has deduced that denial-of-service attacks have threatened the flow tables and the communication channels, as

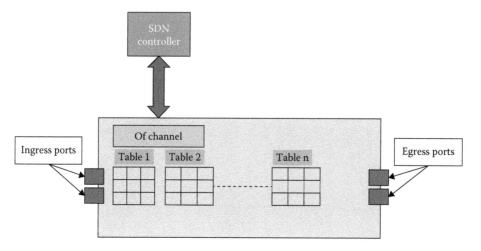

FIGURE 5.4 OpenFlow switch specifications.

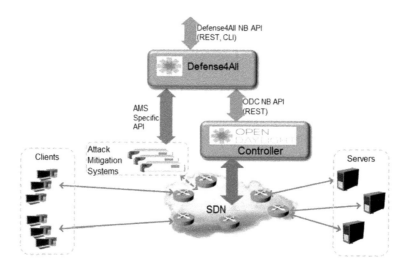

FIGURE 5.5 Defense4All DoS detection application in ODL controller.

the attacker floods those components with OpenFlow rules and requests. Additionally, tampering attacks substantially have targeted the flow tables on the devices by installing rules from untrusted sources.

5.5   SDN SECURITY ENHANCEMENT SOLUTIONS

In the previous section, we addressed various security challenges related to SDN architecture. Remarkably, those threats related to the controller being an attractive target for intruders. For the communications between the controller and the application layer, there is a necessity to authenticate and authorize applications. Conflict in application privileges will propagate to flow rules. FortNox [24] provides role-based authentication and security policy enforcement; it conducts a real-time rule conflict analysis to reveal rule contradictions.

Various intrusion-detection applications have been developed to detect malicious activities in SDNs. Figure 5.5 shows Defense4All; it is an application deployed on an OpenDaylight controller

[25]. Defense4All was developed to detect and mitigate distributed denial-of-service attacks (DDoS). However, the application does not protect the controller itself; rather it deploys a set of rules to protect the network at the network perimeters. The Defense4All application requests the network information from the controller. On detection of malicious activities, the application executes mitigation actions according to its attack response module. A conventional technique to protect the controller is deploying a distributed controller platform [2]. However, significant concerns emerge in a distributed architecture, such as the network performance trade-offs. As multiple controllers exchange information for orchestrated network control, this exchange process results in latency. Another issue relates to data consistency and synchronization on each control point [2,20]. For controller and switching device communication, there is a necessity to cipher the data exchanged over the communication channels, as TCP connections are exposed to various threats. TLS encryptions will provide standard security measures to mitigate man-in-the-middle attacks.

5.6 SDN SECURITY TOOLS

In this section, we survey several SDN security solutions. In Table 5.1, the tools are classified into two categories, as a security enhancer or as a SDN security resolution [2,26]. Through security enhancement tools, the solutions aim to improve network security by utilizing SDN features. SDN

TABLE 5.1
SDN Security Tools Survey

	Solution Domain		Layer			
Security Solution	Security Enhancer	SDN Security Resolution	App	Control	Forward	Description
FRESCO [27]		X		X		Security services composition framework
LiveSec [5]	X		X	X	X	Security policy enforcement
Netfuse [28]		X		X	X	Protection against traffic overload externally (DDoS) or internally
SDN RTBH [35]	X			X	X	DoS mitigation
MAPPER [29]	X		X	X		Policy enforcement
FlowTags [30]	X		X	X	X	Enforce and verify policies
SIMPLE [31]	X		X	X	X	Policy enforcement
OpenSAFE [32]	X		X	X		Verify policies
CloudWatcher [33]	X		X	X		Ensure network packets are inspected
FortNox [24]		X		X	X	Prioritize flow rules to eliminate inconsistency
FLOVER [9]	X		X	X	X	Verify rules
Veriflow [34]	X		X		X	Verify and debug flow rules
OF-RHM [36]	X			X	X	Mutate hosts as a response to threat existence
OrchSec [37]	X		X	X	X	Security application development framework
FlowNAC [38]	X			X	X	Flow-based access control
PermOF [39]		X		X	X	Fine-grained permission and isolation system for SDN apps

security resolutions represent tools developed to improve the security of the SDN itself. Moreover, the table indicates which layers the solution covers.

FRESCO is a security composition framework. It focuses on anomaly detection and mitigation [27]. Netfuse is an example of solutions that address the security flaws in the SDN architecture [28]. It protects the network from denial-of-service attacks. However, there is a significant shortcoming in research related to improving the security of the SDN itself.

The majority of the survey tools focus on using SDNs to enhance security, specifically, policy enforcement solutions. The resilience of the SDN architecture effectively pushes toward the adoption of policy execution and verification applications. MAPPER, FlowTags, SIMPLE, and OpenSAFE are examples of SDN solutions for policy enforcement [29–32]. CloudWatcher [33] controls network flows to guarantee network security devices inspect each flow. Veriflow inspects and verifies flow rules in real time to ensure the flow rules' integrity [34].

5.7   CONCLUSION

SDNs represent a revolutionary change in the data communications model. The separation of control and forward layers accelerates the pace of innovation, increases network resiliency, and decomposes the existing complex and rigorous network paradigm. Distinctly, the SDN architecture impacts the security features. From a positive perspective, SDN characteristics (e.g., centralization and programmability) are exploited to enhance security. Network policy enforcement and verification are significantly flexible in the SDN model. A global view of the system improves threat detection, while network programmability enriches threat countermeasures. However, there are unprecedented threat vectors related to the new paradigm. The majority of the new risks are linked to the controller. Further research is required to enhance control-layer security, as improving the security at this layer will boost the security of the entire paradigm.

REFERENCES

1. Kreutz, D., Ramos, F.M.V., Veríssimo, P.E., Rothenberg, C.E., Azodolmolky S., Uhlig, S. Software-defined networking: A comprehensive survey. In: *Proceedings of the IEEE*, vol. 103, no. 1, pp. 14–76, Jan. 2015. doi: 10.1109/JPROC.2014.2371999.
2. Scott-Hayward, S., O'Callaghan, G., Sezer, S. SDN security: A survey. In: *2013 IEEE SDN for Future Networks and Services (SDN4FNS)*, Trento, Italy, 11–13 November, 2013.
3. McKeown, N. et al. OpenFlow. *ACM SIGCOMM Computer Communication Review*, vol. 38, no. 2, p. 69, 2008.
4. Walfish, M. et al. Middleboxes no longer considered harmful. In: *Proceedings of the Fifth USENIX 467 Conference on Operating Systems Design and Implementation*, 2004, p. 15, 468.
5. Wang, K., Qi, Y., Yang, B., Xue, Y., Li, J. LiveSec: Towards effective security management in large-scale production networks. In: *2012 IEEE International Conference on Distributed Computing Systems Workshops (ICDCSW)*, 2012, pp. 451–460.
6. Casado, M. et al. Ethane: Taking control of the enterprise. In: *Proceedings of the 2007 Conference on Applications, Technologies, Architectures, and Protocols for Computer Communications*, 2007, pp. 1–12.
7. Hand, R., Ton, M., Keller, E. Active security. In: *Proceedings of the 12th ACM Workshop on Hot Topics in Networks*, Nov. 2013, p. 17.
8. Hu, H., Han, W., Ahn, G.-J., Zhao, Z. FlowGuard: Building robust firewalls for software-defined networks. In: *Proceedings of the 3rd Workshop on Hot Topics in Software-Defined Networks*, 2014, pp. 97–102.
9. Son, S., Shin, S., Yegneswaran, V., Porras, P., Gu, G. Model checking invariant security properties in OpenFlow. In: *Proceedings of the IEEE International Conference on Communication*, 2013, pp. 1974–1979.
10. Ghorbani, A.A., Lu, W., Tavallaee, M. *Network Intrusion Detection and Prevention Concepts and Techniques*. New York, NY: Springer, 2010.
11. Mudzingwa, D., Agrawal, R. A study of methodologies used in intrusion detection and prevention systems (IDPS). In: *SoutheastCon, 2012 Proceedings of the IEEE*, Mar. 2012, pp. 1, 6, 15–18.
12. Yassine, A., Rahimi, H., Shirmohammadi, S. Software defined network traffic measurement: Current trends and challenges. *IEEE Instrumentation & Measurement Magazine*, vol. 18, no. 2, pp. 42, 50, Apr. 2015.

13. Agarwal, S., Kodialam, M., Lakshman, T.V. Traffic engineering in software defined networks. In: *INFOCOM, 2013 Proceedings IEEE, 14–19* April 2013, pp. 2211–2219.

14. Ian, F.A., Ahyoung, L., Wang, P., Luo, M., Chou, W. A roadmap for traffic engineering in SDN-OpenFlow networks. *Computer Networks*, vol. 71, pp. 1–30, October 2014.

15. Chowdhury, S.R., Bari, M.F., Ahmed, R., Boutaba, R. PayLess: A low cost network monitoring framework for Software Defined Networks. In: *2014 IEEE Network Operations and Management Symposium (NOMS)*, 5–9 May 2014, pp. 1, 9.

16. Zhang, Y. An adaptive flow counting method for anomaly detection in SDN. In: *Proceedings of the 9th ACM Conference on Emerging Networking Experiments and Technologies*, 2013, pp. 25–30.

17. Yu, M., Jose, L., Miao, V. Software defined traffic measurement with OpenSketch. In: *Proceedings of the 10th USENIX Symposium on Networked Systems Design and Implementation, NSDI*, vol. 13, 2013.

18. Braga, R., Mota, E., Passito, A. Lightweight DDoS flooding attack detection using NOX/OpenFlow. In: *IEEE 35th Conference on Local Computer Networks (LCN)*, 2010, pp. 408–415.

19. Smeliansky, R.L. SDN for network security. In: *2014 First International Science and Technology Conference (Modern Networking Technologies) (MoNeTeC)*, 28–29 Oct. 2014, p. 1, 5.

20. Sezer, S. et al. Are we ready for SDN? Implementation challenges for software-defined networks. *IEEE Communication Magazine*, vol. 51, no. 7, pp. 36–43, Jul. 2013.

21. Tootoonchian, A., Ganjali, Y. HyperFlow: A distributed control plane for OpenFlow. In: *Proceedings of the 2010 Internet Network Management Conference on Research on Enterprise Networking*, 2010, p. 3.

22. Kreutz, D., Ramos, F.M., Verissimo, P. Towards secure and dependable software-defined networks. In: *Proceedings of the 2nd ACM SIGCOMM Workshop on Hot Topics in Software Defined Networks*, 2013, pp. 55–60.

23. Kloti, R., Kotronis, V., Smith, P. 2013. OpenFlow: A security analysis. In: *2013 21st IEEE International Conference on Network Protocols (ICNP), 2013*.

24. Porras, P. et al. A security enforcement kernel for OpenFlow networks. In: *Proceedings of the 1st Workshop on Hot Topics in Software Defined Networks*, 2012, pp. 121–126.

25. ONF Overview. Retrieved on June 17, 2016. Available from www.opennetworking.org/about/onf-overview.

26. Ahmad, I., Namal, S., Ylianttila, M., Gurtov, A. Security in software defined networks: A survey. *IEEE Communications Surveys & Tutorials*, vol. 17, no. 4, pp. 2317–2346, Fourthquarter 2015.

27. Shin, S. et al. FRESCO: Modular composable security services for software-defined networks. *Internet Society NDSS*, Feb. 2013.

28. Wang, Y., Zhang, Y., Singh, V., Lumezanu, C., Jiang, G. NetFuse: Short-circuiting traffic surges in the cloud. In: *Proceedings of the 2013 IEEE International Conference on Communication*, 2013.

29. Sapio, A. et al. MAPPER: A mobile application personal policy enforcement router for enterprise networks. In: *Proceedings of the 3rd European Workshop on Software Defined Networks*, 2014, p. 2.

30. Fayazbakhsh, S., Sekar, V., Yu, M., Mogul, J. FlowTags: Enforcing network-wide policies in the presence of dynamic middlebox actions. In: *Proceedings of the 2nd ACM Workshop on Hot Topics in Software Defined Networks*, 2013.

31. Qazi, Z.A. et al. SIMPLE-fying middlebox policy enforcement using SDN. In: *Proceedings of the ACM SIGCOMM 2013 Conference on SIGCOMM*, Aug. 2013.

32. Ballard, J.R., Rae, I., Akella, A. Extensible and scalable network monitoring using OpenSAFE. In: *Proceedings of the 2010 Internet Network Management Conference on Research on Enterprise Networking*, 2010.

33. Shin, S., Gu, G. CloudWatcher: Network security monitoring using OpenFlow in dynamic cloud networks (or: How to provide security monitoring as a service in clouds?). In: *20th IEEE International Conference on Network Protocols (ICNP)*, 2012, pp. 1–6.

34. Khurshid, A., Zhou, W., Caesar, M., Godfrey, P. VeriFlow: Verifying network-wide invariants in real time. *ACM SIGCOMM Computer Communication Review*, vol. 42, no. 4, pp. 467–472, 2012.

35. Giotis, K., Androulidakis, G., Maglaris, V. Leveraging SDN for efficient anomaly detection and mitigation on legacy networks. In: *Proceedings of the 3rd European Workshop on Software Defined Networks*, 2014, p. 6.

36. Jafarian, J.H., Al-Shaer, E., Duan, Q. OpenFlow random host mutation: Transparent moving target defense using software defined networking. In: *Proceedings of the 1st Workshop on Hot Topics in Software Defined Networks*, 2012, pp. 127–132.

37. Zaalouk, A., Khondoker, R., Marx, R., Bayarou, K. OrchSec: An orchestrator-based architecture for enhancing network-security using network monitoring and SDN control functions. In: *Proceedings of the IEEE Network Operations and Management Symposium*, May 2014.

38. Matias, J., Garay, J., Mendiola, A., Toledo, N., Jacob, E. FlowNAC: Flow-based network access control. In: *Proceedings of the 3rd European Workshop on Software Defined Networks*, 2014, p. 6.

39. Wen, X., Chen, Y., Hu, C., Shi, C., Wang, Y. Towards a secure controller platform for OpenFlow applications. In: *Proceedings of the 2nd ACM SIGCOMM Workshop on Hot Topics in Software Defined Networking*, 2013, pp. 171–172.

6 Fog Computing Mechanisms in 5G Mobile Networks

Stojan Kitanov and Toni Janevski

CONTENTS

6.1   INTRODUCTION

Mobile and wireless networks have made tremendous growth in the past decade. This growth is due to the support of a wide range of applications and services by smart mobile devices such as laptops, smartphones, tablets, and phablets. This resulted in an increased demand for mobile broadband services that demand high data rates, high mobility, low latency, broadband spectrum, and high energy consumption, comparable to the fixed broadband Internet [1].

In this direction, many global research and industrial initiatives have started to work on the building blocks of the next generation of mobile and wireless networks, usually called 5G, or the fifth generation [2–4]. 5G will enable the future Internet-of-Services (IoS) paradigms such as anything as a service (AaaS), where devices, terminals, machines, and smart things and robots will become innovative tools that will produce and use applications, services, and data.

However, 5G will have to support huge mobile traffic volumes, 1000 times larger than those today, in the order of multiples of gigabits per second [5–8]. It will also have to deal with a proliferation of new and complex applications and services, many of which are unknown today. Moreover, mobile devices are resource-constrained devices due to their capabilities and obstacles. These are related to their performance (battery life, memory, storage, bandwidth, processing power), environment (heterogeneity, availability, and scalability), and security (reliability and privacy). In addition, the future Internet will exacerbate the need for improved QoS/QoE, supported by services that are orchestrated on demand and are capable of adapt at runtime, depending on the contextual conditions, to allow reduced latency, high mobility, high scalability, and real-time execution. The emerging applications in the context of the Internet of Everything (IoE), which is a clear evolution of the Internet of Things (IoT), will introduce high mobility, high scalability, real-time, and low latency requirements that raise new challenges on the services being provided to the users [9]. These demands can only be partially fulfilled by existing cloud computing solutions [10].

A new paradigm called fog computing, or briefly fog, has emerged to meet these requirements [11]. Fog extends cloud computing and services to the edge of the network. With its service orchestration mechanisms, it provides data, computing, storage, and application services to end users that can be hosted at the network edge or even end devices such as set-top boxes or access points. The main features of fog are its proximity to end users, its dense geographical distribution, and its support for mobility [12].

By implementing fog computing in 5G networks, end-to-end latency will be significantly reduced to less than a few milliseconds, and it will improve the service quality perceived by mobile users. This high data speed together with the expected pervasive coverage of reliable networks will create opportunities for companies to deploy many new real-time services that cannot be delivered over current mobile and wireless networks [13].

The move from cloud to fog in 5G brings out several key challenges, including the need for supporting the on-demand orchestration and runtime adaptation of resilient and trustworthy fog services. This is essential for the success of the future IoE.

This chapter provides an overview of 5G and fog computing technologies and their research directions. It also proposes and evaluates a hybrid environment service orchestration (HESO) model as a fog computing mechanism for 5G networks in terms of round-trip time latency, throughput, and product latency throughput.

This chapter is organized as follows. Section 6.2 is about 5G networks, their service requirements, and applications and 5G smart user devices. It also explains 5G, the IoE, and the possible existing mechanisms of mobile cloud computing in 5G. Section 6.3 is about fog computing and its features, such as ubiquity of devices, network management, fog connectivity, and fog privacy and security. Section 6.4 explains the possible applications of fog in a 5G network. Section 6.5 proposes the HESO model of fog computing as a possible solution in a 5G network. Section 6.6 performs a quality evaluation of the HESO model in terms of latency, throughput, and product latency throughput. Section 6.7 outlines the open research challenges and issues in fog computing. Finally, Section 6.8 summarizes the chapter.

6.2 5G NETWORKS

The fifth generation of mobile and wireless networks, or 5G, is a name which is used in some research papers and projects to denote the next major phase of mobile telecommunications standards [2–4]. 5G networks will include support of a large number of connected devices and flexible air interfaces and different interworking technologies that are energy efficient and will possess always online capabilities [6]. This transformation will require not only an upgrade of existing systems, but also innovation of new protocols and new access technologies altogether.

The following are three possible migration paths to 5G networks [7–8]:

- A step-by-step evolutionary path focusing on further enhancements of existing technologies
- A revolutionary path using a brand-new, innovative technologies

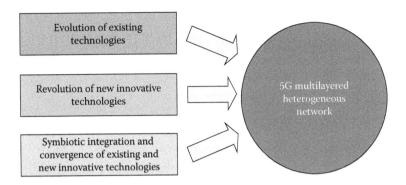

FIGURE 6.1 Technology routes to 5G.

- A symbiotic integration and convergence of existing or new technologies, such as communication, information systems and electronics, multi-radio access technologies, computing techniques, device-to-device communications, bands, links, layers, services, and multiplexing

As shown in Figure 6.1, 5G will be a multilayered, heterogeneous network that will consist of existing 2G, 3G, 4G, and future radio access technologies (RATs). It may also converge many other radio technologies, such as mobile satellite system (MSS), digital video broadcasting (DVB), wireless local access network (WLAN), wireless personal access network (WPAN), and worldwide interoperability for microwave access network (WiMAX), with multi-tiered coverage by macro, pico, femto, relay, and other types of small cells. 5G will support a wide range of applications and services to satisfy the requirements of the information society by the year 2020 and beyond [2–3].

6.2.1 5G Vision and Requirements

The 5G vision and requirements should be defined in multiple dimensions, such as the technology perspective, user perspective, network operator perspective, and traffic and service models [6–8].

From the technology perspective, 5G will be a continuous enhancement and evolution of the present RATs, development of novel RATs, and symbiotic integration and convergence of many technologies to meet the increasing user demand of the future.

From the user's perspective, 5G mobile systems will enhance user experience in many aspects, such as higher demand for data rate and capacity, good performance in terms of pervasive coverage, reliable QoS and battery life of the mobile device, ease of use, affordable price for subscription, safety and reliability, and personalization of the services. 5G should provide user-centric services, where the users can customize the subscription of services and add or remove subscriptions at any time.

From the network operator's point of view, 5G should provide sufficient bandwidth and capacity to support the high data traffic volume at an affordable cost. 5G should provide low-cost, easy deployment and simple, scalable, and flexible operation to decrease CAPEX and OPEX. 5G networks should provide support for backward compatibility with current and legacy networks for investment protection. The future 5G system should support different types of services. The 5G mobile network will be an open service platform to bear all kinds of mobile internet applications, and it will support a more flexible model of operation that will enable both network operators and service providers to generate their own revenue.

The network operators should consider two key traffic models for 5G: high-speed video flow from the server to the subscriber and massive machine-to-machine (M2M), or device-to-device (D2D), communications [5].

TABLE 6.1

QoS Comparison between 4G and 5G Networks

Parameter	4G	5G
Air Link User Plane Latency	10 ms	1 ms
Air Link Control Plane Latency	100 ms	50 ms
Simultaneous Connection Density per Unit km^2	10^5	10^6
Mobility	300 km/hr	500 km/hr
Uplink Cell Spectral Efficiency	1.8 bps/Hz	5 bps/Hz
Downlink Cell Spectral Efficiency	2.6 bps/Hz	10 bps/Hz
Peak Throughput (Downlink) per Connection	100 Mbps to 1 Gbps	10 Gbps to 50 Gbps
Downlink Cell-Edge Data Rate	0.075 bps/Hz/cell	Anywhere 1 Gbps
Uplink Cell-Edge Data Rate	0.05 bps/Hz/cell	Anywhere 0.5 Gbps
Cost Efficiency	10 times	100 times
PDB without Quality Assurance	100 to 300 ms	Undetermined
PDB with Guaranteed Quality	50 to 300 ms	1 ms
PLR for Video Broadcasting	10^{-8} (4k UHD)	10^{-9} (8k UHD)
PLR for M2M Services (without quality assurance)	10^{-3}	10^{-4}
PLR for M2M Services (with guaranteed quality)	10^{-6}	10^{-7}

To fulfill these demands, 5G systems will be required to deliver an order of magnitude of cell capacities and per-user data rate comparable to 4G. It will be a set of telecommunication technologies and services that supports 1000 times more data capacity than today and should provide an ultra-low latency response of less than a few milliseconds. The network should provide a capacity of 50 Gbps per cell and guarantee more than 1 Gbps per user through super-dense networking, regardless of the user's location, including the cell edge. Compared to 4G, the cell spectral efficiency will be increased by 3 to 5 times, and the latency response in the control plane will be reduced to one-half (i.e., to 50 ms). In addition, it will support an ultra-low latency response of 1 ms in the data plane, which is equal to one-tenth of 4G networks [5–8].

To enable the forthcoming IoE [9], 5G should provide anytime, anywhere, anyone, anything (4A) massive and simultaneous connectivity that will accommodate one million different mobile devices per unit square kilometer. It will have flexible and intelligent network architecture with software-based structure that will analyze data in real time and provide intelligent and personalized services [2].

5G will provide reliable, secure operation with more than 99% network availability. There will be a possibility for self-healing reconfiguration and self-optimization. The battery life of mobile devices will be increased by 10 times. Finally, 5G will have low cost for infrastructure and devices and will be 50 to 100 times more efficient in terms of energy usage per bit compared to the legacy systems [6].

The QoS management mechanisms in 5G networks should provide video and VoIP traffic prioritization toward web-search traffic and other applications tolerant to quality [5]. Some of the QoS parameters are packet delay budget (PDB), or maximum packet delay, and packet loss ratio (PLR), which have much lower values to the order of magnitude than 4G.

5G networks will outperform 4G networks. Table 6.1 summarizes the QoS comparison between 4G and 5G networks [5–8].

6.2.2 5G Applications and Services

The service types for all generations of mobile and wireless networks (from 1G to 5G) are given in Figure 6.2. 1G supported only analog voice–type communication. 2G provides digital voice

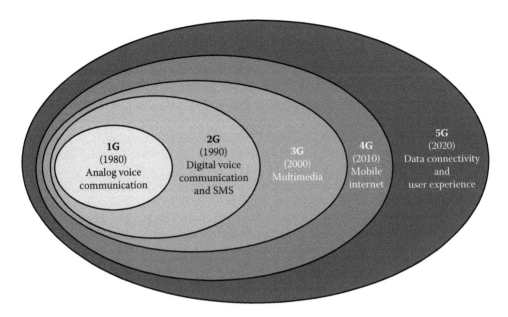

FIGURE 6.2 Service types from 1G to 5G.

communication and short message service (SMS). 3G supports multimedia services as well as the services of 2G. 4G supports the services of 2G and 3G and the mobile Internet. 5G will focus primarily in the following service types: data, connectivity, and user experience. In addition, it will support the services of the previous generations of mobile and wireless networks.

5G will have a user-centric approach, where telecom operators will invest in developing new applications that will provide a ubiquitous, pervasive, seamless, continual, and versatile mobile experience to the end user [3]. The applications will become more personalized and more context aware, recognizing user identity, user location, and user preferences [13].

Therefore, context-rich support services such as context extraction, recommendation, and group privacy should be supported in 5G. Particularly important is the context extraction service that performs data-mining analysis of mobile data combined with other forms of data, such as social networking data and sensor network data, to extract contextual clues relevant to the user. Data-mining services should be able to scale and analyze a large group of people and large quantities of data (big data) to extract collective trends among the population of users in real time. Additionally, recommendation services based on collective group context rather than individual context need to be created and scaled. By using these clues, a layer of recommendation services can be built that creates output which is adjusted to a user, or set of users, with those contextual characteristics [13].

5G can be applied in areas such as health care, education, mobile augmented reality, video game services, disaster relief services, smart cities, smart homes, smart vehicles, robotics, and manufacturing [7–8]. For example, in health care, 5G can be used for such tasks as real-time emergency requests, remote professional prescription of drugs and pills, and remote medical assistance. The connected health-care devices will also send vital signs, such as brain waves, blood pressure, and heartbeat, to an expert system in the hospital in real time to prevent medical emergencies before they occur. In addition, the connected devices can measure the user's athletic performance while he or she exercises or recommend the frequency and the duration of some exercises.

In education, 5G can promote realistic education services with 4D in various fields, online educational services, and multilateral educational services that will result in effective, simultaneous learning. 5G together with the built-in, good-quality cameras of smart phones will allow widespread availability and portability of augmented reality, which is a technology that allows virtual

graphics imagery to exactly overlay physical objects in real time. Some examples include naviga-
tion, public security, real-time text translation, 3D design and visualization, remote instruction, and
shopping assistance.

5G will provide a better user experience for video games using the high-definition hologram (HD
hologram) that will provide huge content representation, implementation of HDTV holograms, and
50-inch hologram display realization. In addition, the recognition sensor will allow vision recognition and situational awareness.

Together with the IoT devices, 5G will support disaster relief services such as location-based
services, high-density user communication, environmental sensors for chemical measurements,
HD-CCTV (closed-circuit television), and emergency services that include recognition (positional
information), estimation of the situation, and the risk and action (service execution).

5G can be used in smart cities, smart homes, and smart vehicles for public transport regulation,
car navigational systems used to find the desired location, vehicle diagnostic services, smart home
environments that will save energy, and to make some recommendations according to the user's
preferences (e.g., a smart refrigerator recommending a recipe requiring ingredients that are already
in the refrigerator).

6.2.3 5G NETWORK ARCHITECTURAL LEVELS

5G networks will be all-IP networks that will provide continued evolution and optimization of the
system concept to provide a competitive edge in terms of both performance and cost [2–4,13]. This
will lessen the burden on the aggregation point, and traffic will move directly from the base station to media gateways. The network architecture in 5G will consist of several levels, as shown in
Figure 6.3.

The innovative service and content providers are located at the top level, and they will accommodate 5G requirements and provide a new ubiquitous and pervasive user experience. New applications and services such as augmented and virtual reality, hologram, mobile ultra-high definition,
and cloud computing will be provided.

One level below is the enabling software-based core platform. At this level, flexible and reliable
flat-IP architecture will converge into different technologies to form a single 5G core, where a range
of intelligent, complex telecommunication network functions can be efficiently implemented. The
5G core will consist of reconfigurable multi-technologies combined in a single core called super
core, nano core, or master core [4]. All network operators will be connected to one single network
core with massive capacity that will eliminate all interconnecting charges and complexities that the
network operator is facing currently, and will reduce the number of network entities in an end-to-
end connection, thus reducing latency significantly. The main functionality in the 5G core will be

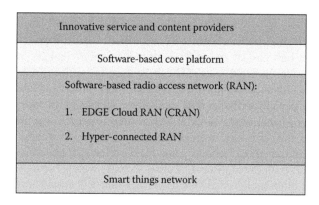

FIGURE 6.3 5G network architectural levels.

the network-as-a-service (NaaS) platform that will allow the configuration of all telecommunication and service functions with virtualized software on a programmable hardware [13].

The third level is the radio access network (RAN) that consists of the EDGE Cloud RAN (EDGE CRAN) service and hyper-connected RAN. EDGE CRAN moves the cloud from the innovative service content provider to the RAN. By placing storage and computing resources at, or close to, the cell site, operators can improve response times, making services feel "snappier" and, uniquely, more responsive to prevailing radio conditions.

Hyper-connected RAN infrastructure acts as a data pipe supporting the massive and ultra-high speed connectivity over ultra-dense networks (UDN) of wireless access with heterogeneous cell arrangement. It will be a multi-layered, heterogeneous network that will consist of existing 2G, 3G, 4G, and future RATs. It may also converge many other radio technologies, such as MSS, DVB, WLAN, WPAN, and WiMAX, with multi-tiered coverage by macro, pico, femto, relay, and other types of small cells. Cognitive radio and software-defined networking together with existing or new modulation and transmission methods can be used for deploying new applications and services that will improve the utilization of the congested radio frequency (RF) spectrum. The RAN infrastructure will be responsible for cellular content caching, radio and resource allocation, mobility, and interference control, among other things.

Finally, the lowest level of the 5G network architecture consists of a smart things network with different smart mobile, IoT, and Web of Things (WoT) end-user devices. This includes various devices from laptops, smartphones, tablets, phablets, and wearable devices to connectivity embedded devices and sensors in such things cars, trucks, and bikes.

The architectural levels in 5G will be defined primarily by the service requirements and the following technologies: nanotechnology, quantum cryptography, and (mobile) cloud computing.

6.2.4   5G SMART MOBILE DEVICES

5G networks will provide ubiquitous and pervasive connectivity with smart capabilities to deal with the communication requirements of smartphones, smart TVs, and smart small devices and to include efficient and effective management of resources. 5G smart mobile devices will be autonomously reconfigurable devices attached to several mobile or wireless networks at the same time and will use the flows of mobile accesses (e.g., 2G, 3G, or 4G) and/or wireless accesses (e.g., Wi-Fi and WiMAX) in a separate and/or an integrated manner (i.e., the multiple connection capability) [3]. Each network access technology will be responsible for handling user mobility, while the terminal will make the final choice among different wireless/mobile networks. An overview of the 5G mobile smartphone device protocol stack layers compared to the OSI layers is given in Figure 6.4.

The physical and data link OSI layers will be based on the open network access layer (i.e., any existing or new mobile and wireless access network, such as LTE/LTE-A, Wi-Fi, and WiMAX) [3].

OSI layers	5G smart mobile device layers
Application layer	Application layer (applications, services, and QoS/QoE management)
Presentation layer	
Session layer	Open transport layer (TCP, RTP, or new protocol version for download)
Transport layer	
Network layer	Network layer
Data link layer	Open network access layer (LTE-A, WiFi, IEEE 802.16m, etc.)
Physical layer	

FIGURE 6.4   5G smartphone device protocol stack layers.

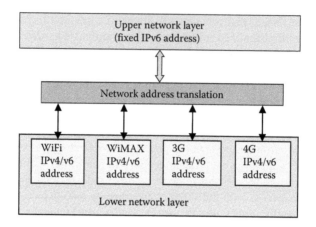

FIGURE 6.5   5G smartphone network layer.

The network layer (see Figure 6.5) will consist of two layers: the upper network layer and the lower network layer [3]. The upper network layer will contain the fixed IPv6 address for the mobile device that will be implemented in the mobile phone by 5G manufacturers. At lower network layers, the Mobile IP (MIP) protocol will be used. Since the mobile device will be simultaneously connected to several mobile and wireless networks at this level, the mobile device will maintain different IP addresses for each of the radio interfaces. Each of these IP addresses will be a care-of address (CoA) for the foreign agent (FA) placed in the mobile phone. This FA will perform the CoA mapping between the fixed IPv6 address and the CoA for the current mobile or wireless network. Thus, the 5G mobile device will maintain a multi-wireless network environment. The network address translation will be performed by the middleware between the upper and lower network layers. In addition, MIP protocol can be used to perform service continuity while the device performs roaming or moves from one wireless access network to another.

At the transport layer, the Open Transport protocol will be used [3]. This will enable the download and install protocol, such as TCP or RTP, or a new transport protocol that is targeted to a specific mobile or wireless network.

The application layer will be used for QoS and QoE management. The QoS and QoE parameters will be stored in a database in the 5G mobile device with the aim to be used by intelligent algorithms running in the mobile terminal as system processes, which at the end shall provide the best wireless connection upon QoS and QoE requirements and personal cost constraints [3].

5G smart devices will contribute significantly to expedite the convergence of information communication technology (ICT) with other industries (e.g., intelligent transport systems [ITS], e-health, and smart grid). In addition, these devices provide status information on an end user (human being),; for example, the health status of patients in the case of e-health application.

However, the increased growth of the traffic and applications does not match the parallel improvement in the processing capabilities of the mobile devices. One possible solution for this obstacle is to enable the mobile devices, whenever possible and convenient, to offload their most energy-consuming tasks to nearby servers [14].

6.2.5   5G AND INTERNET OF EVERYTHING (IoE)

Because of ICT developments, many end-user devices, networks, and services have been acquiring more complicated features and capabilities. At the moment, the Internet is progressively evolving from a network of interconnected computers (Internet of Computers [IoC]) to a network of interconnected objects, IoT, and moreover any-thing communications, called the WoT [15]. The next step in the development of the future internet will be a clear evolution of the IoT into the IoE that will introduce high mobility, high scalability, real-time, and low latency requirements that raise new challenges on the services being provided to the users [9]. Nanotechnology, radio frequency identi-fication (RFID), and sensor network technologies will contribute in this direction.

6.2.5.1 Internet of Things (IoT) and Web of Things (WoT)

The IoT is a system of smart physical objects, or "things," that can be discovered, monitored, con-trolled, or interacted with electronic devices that communicate over various networking interfaces and eventually can be connected to the Internet. The things, or the objects, are interconnected on the basis of interoperable information and communication technologies [15–16]. The objects may have their own IP addresses, can be embedded in complex systems, use sensors to obtain informa-tion from their environment, and/or use actuators to interact with it. Through the exploitation of identification, data capture, processing, and communication capabilities, the IoT makes full use of things to offer services to all kinds of applications while maintaining the required privacy. The IoT will create opportunities for more direct integration between the physical world and computer-based systems, which will result in improved efficiency, accuracy, and economic benefit.

However, the IoT has limited capabilities in the integration of the devices from various manu-facturers into a single application or system. This is because many incompatible IoT protocols exist. This makes the integration of data and services from various devices extremely complex and costly. In addition, no unique and universal application layer protocol for the IoT exists that can work across the many available networking interfaces.

A solution to this problem is to use the standard web application layer protocols, standards, and blueprints for connecting heterogeneous devices. This is called the WoT [15]. The WoT is consid-ered an IoT by integrating smart things not only into the Internet (the network), but into the web (the application layer). It is intended for accessing the physical devices as resources of the web, where the services/applications can be provided through a web-based service environment, or legacy telecom-munications. Thus, any device can be accessed using standard web protocols, and the integration across the systems and applications will be much easier. The services and data offered by objects will be more accessible to a larger pool of web developers, which will enable them to build new scalable and interactive applications, which will be beneficial for everyone.

6.2.5.2 Internet of Everything (IoE)

In the future, everyday objects will interact and communicate with each other. Large amounts of data will circulate to create smart and proactive environments that will significantly enhance the user experience. Smart interacting objects will adapt to the current situation with or without any human involvement. This will lead the IoT and the WoT naturally to evolve into the IoE [9,16–18]. According to Mitchell et al., "IoE brings together people (humans), process (manages the way peo-ple, data, and things work together), data (rich information), and things (inanimate objects and devices) to make networked connections more relevant and valuable than ever before" [18].

People will be able to connect to the Internet in many different ways, and they themselves will become nodes on the Internet, with both static information and a constantly emitting activity system (e.g., health status) [18].

Data will be collected by the smart things, where it will be transformed into useful informa-tion that will be transmitted throughout the Internet to a central source (machines, computers, and people) for further processing, analysis, evaluation, and decision making. The data transformation into information is very important because it will enable us to make faster, more intelligent deci-sions, as well as control our environment more effectively [18].

The smart things, or objects, consist of sensors, consumer devices, and enterprise assets that are connected to the Internet and each other. In the IoE, these smart things will sense more data, become context aware, and provide more experiential information to help people and machines make more relevant and valuable decisions [18].

The process plays an important role in the communication and cooperation among people, data, and things to deliver value in the connected world of the IoE. With the correct process, connections become relevant and add value because the right information is delivered to the right person at the right time in the most appropriate way [18].

In this way, information will turn into actions that will create new capabilities, richer experiences, and unprecedented economic opportunity for businesses, individuals, and countries. For 5G to support the IoE, it must be able to use new emerging technologies creatively over and above the growth of current data sources.

6.2.6 Mobile Cloud Computing in 5G

Cloud computing is a model for enabling ubiquitous, convenient, on-demand network access to a shared pool of configurable computing resources (e.g., networks, servers, storage, applications, and services) that can be rapidly provisioned and released with minimal management effort or service provider interaction [10,18–21].

Mobile cloud computing on the other hand is an integration of cloud computing technology in a mobile environment and provides all the necessary resources to overcome the obstacles of the mobile devices [22–25]. It is an infrastructure used by mobile applications where both data storage and data processing are moved away from the mobile device to powerful, centralized, and high-performance computing platforms located in the clouds.

An overview of a 5G network in a mobile cloud computing environment [13] is given in Figure 6.6. The architecture is proposed on the basis of the architectural levels explained in Subsection 6.2.3.

The top of the architecture consists of innovative content and service providers that provide over-the-top (OTT) applications and services such as augmented and virtual reality, hologram, mobile ultra-high definition, and cloud computing. The software-based single core is enhanced Evolved Packet Core (EPC enhanced), which provides the core functionalities in the network, such as billing and charging, mobility and handover, location and geomanagement, home subscriber server, authorization authentication and accounting (AAA) security, and NaaS, which will allow configuration of all telecommunication and service functions with virtualized software on a programmable hardware. The 5G RAN consists of consists of the EDGE CRAN service and hyper-connected RAN. The EDGE CRAN moves cloud computing functionalities to the RAN. Also, the hyper-connected RAN is a multi-layered heterogeneous network that consists of existing 2G, 3G, 4G, and future RATs that will provide cellular content caching, radio and resource allocation, mobility, and interference control, among other things. Finally, the smart things network consists of smart mobile, IoT, and WoT end-user devices.

Full network function virtualization (NFV) will take place in 5G to satisfy the service requirements. Network virtualization pools the underlying physical resources or logical elements in a network by using the current technologies, such as cognitive and software-defined radios, in the 5G RAN, software-defined networking, and cloud services in 5G core [26].

The NFV functions should cover the control and management of QoS, the service policy, and the prioritization of traffic. Network functions such as signal processing and path computation will be virtualized and offloaded to network management clouds. This will make network operating easier, reduce end-to-end energy consumption, and open the way to new, flexible communication services. The main cloud computing functionality in 5G core will be the NaaS platform, which will allow configuration of all telecommunication and service functions with virtualized software on a programmable hardware [13].

5G will include a mix of a centralized cloud and a distributed cloud using access network and RAN elements [13]. By moving the cloud into the radio access network, called Cloud RAN (CRAN) (i.e., by moving the base station from the cell site into the cloud by placing storage and computing resources at, or close to, the cell site), operators will be able to guarantee the necessary service-level agreement (SLA) and can better support delay-sensitive applications such as virtual desktops and electronic programming guides. They will also improve response times, making services feel "snappier" and, uniquely, more responsive to prevailing radio conditions. This might be useful for

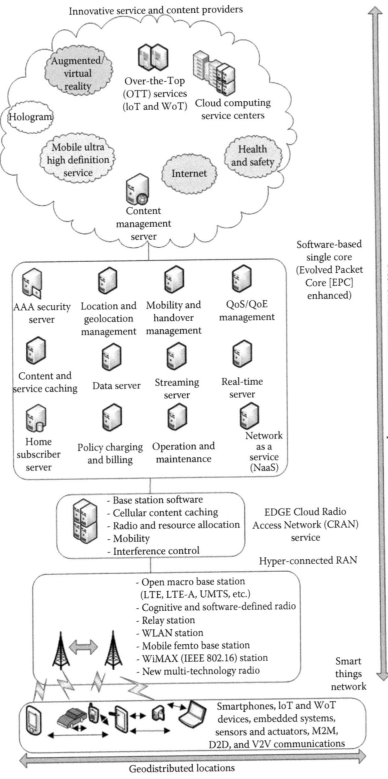

FIGURE 6.6 5G network architecture in a mobile cloud computing environment.

congestion control or rate adaptation for video streams. On the network side, cell-site caching can reduce demand on the backhaul network and potentially play a role in limiting signaling to the core network.

Because 5G RAN will consist of a dense deployment of micro, pico, and femto cells, mobile cloud computing will significantly reduce the latency over the wireless communication channel, as well as the transmit power necessary for computation offloading. Moreover, mobile devices can easily locate a cloud access point. Millimeter wave links, massive MIMO, and multi-cell cooperation can be used to improve the spectral efficiency by reducing the time necessary to transfer the users' offloading requests to the cloud [14].

Mobile cloud computing in a 5G network should support context-rich support services such as a context extraction service, recommendation service, and group privacy service [13,25], where a layer of cloud recommendation services can be built that creates output which is adjusted to a user, or set of users, with those contextual characteristics.

6.3   FOG COMPUTING

The future IoE will become the linkage between extremely complex networked organizations (e.g., telecoms, transportation, financial, health and government services, commodities) that will provide the basic ICT infrastructure that supports the business processes and the activities of society in general [9,18]. Frequently, these processes and activities will be supported by orchestrated cloud services, where a number of services work together to achieve a business objective [10].

Although mobile cloud computing is a promising solution for 5G to cope with the future Internet, it still cannot deal with all future Internet services and applications. This is because the future Internet will exacerbate the need for improved QoS/QoE, supported by services that are orchestrated on demand and that are capable of adapt at runtime, depending on the contextual conditions, to allow reduced latency, high mobility, high scalability, and real-time execution. The emerging wave of the IoT would require seamless mobility support and geodistribution in addition to location awareness and low latency. These demands can be only partially fulfilled by existing cloud computing solutions [10]. Also, existing cloud computing security mechanisms, such as sophisticated access control and encryption, have not been able to prevent unauthorized and illegitimate access to data.

Recently a new paradigm called fog computing, or briefly fog, has emerged to meet these requirements [11]. Fog computing extends cloud computing and services to the edge of the network. Fog will combine the study of mobile communications, micro-clouds, distributed systems, and consumer big data. It is a scenario where a huge number of heterogeneous (wireless and sometimes autonomous) ubiquitous and decentralized devices communicate and potentially cooperate among themselves and with the network to perform storage and processing tasks without the intervention of third parties [12]. These tasks support basic network functions or new services and applications that run in a sandboxed environment. Users leasing part of their devices to host these services get incentives for doing so. The distinguishing fog characteristics are its proximity to end users, its dense geographical distribution, and its support for mobility. Therefore, the fog paradigm is well positioned for real-time big data analytics. Services are hosted at the network edge or even end devices such as set-top boxes or access points [27]. By doing so, fog reduces service latency and improves QoS, resulting in a superior user experience. It supports emerging IoE applications that demand real-time/predictable latency (industrial automation, transportation, networks of sensors and actuators).

The existence of the fog will be enabled by the emerging trends on technology usage patterns on one side and the advances on enabling technologies on the other side. A comparison between fog computing and cloud computing is given in [27], which is summarized in Table 6.2. Fog computing is targeted for mobile users, while cloud computing is targeted for general Internet users. Cloud computing provides global information collected worldwide, while fog computing provides limited, localized information services related to specific deployment locations. Cloud services are located within the Internet, while fog services are located at the edge of the network. There

TABLE 6.2
A Comparison between Cloud and Fog Computing

	Cloud Computing	Fog Computing
Target Type	General Internet users	Mobile users
Service Type	Global information collected worldwide	Limited localized information services related to specific deployment locations
Service Location	Within the Internet	At the edge of the local network
Distance between Client and Server	Multiple hops	Single hop
Number of Server Nodes	Few	Very large
Latency	High	Low
Delay Jitter	High	Very low
Geodistribution	Centralized	Distributed
Security	Undefined	Can be defined
Hardware	Ample and scalable storage, processing, and computing power	Limited storage, processing, and computing power, wireless interface
Deployment	Centralized and maintained by OTT service providers	Distributed in regional areas and maintained by local businesses

is a single hop between smart user devices and the fog computing server, while there are multiple hops between user devices and the cloud computing centers. Fog computing contains a very large number of server nodes, while cloud computing contains only a few server nodes. Latency and the delay jitter are very low in the fog computing environment, while in cloud computing, latency and the delay jitter are very high. Cloud computing centers are centralized, while fog computing nodes are densely geodistributed. Security in cloud computing is undefined, while in fog computing, it can be defined. The hardware in cloud contains ample and scalable storage and very high processing and computing power. On the other hand, the hardware in fog contains limited storage, processing, computing power, and wireless interface. Finally, the cloud computing deployment is centralized and maintained by OTT service providers, while fog computing is distributed in regional areas and maintained by local business, which are close to the end users, and therefore they have more trust in them.

Cloud and fog form a mutually beneficial, interdependent continuum [28]. They are interdependent because coordination among devices in a fog may rely on the cloud. They are also mutually beneficial because certain functions are naturally more advantageous to carry out in fog while others are so in cloud.

6.3.1 Fog Computing Architecture

The fog computing architecture uses one or a collaborative multitude of end-user clients or near-user edge devices to carry out a substantial amount of storage, communication, and management [28]. An overview of fog computing architecture is given in Figure 6.7. It consists of a centralized cloud computing center, an IP/MPLS core network, a RAN network with distributed fog computing intelligence, and a smart things network. Each smart thing device is attached to one of the fog devices in the RAN network. The fog devices could be interconnected to each other, and each of them is linked to the centralized cloud computing center via the IP/MPLS core network.

The RAN network is actually an intermediate fog layer that consists of geodistributed intelligent fog computing servers which are deployed at the edge of networks (e.g., parks, bus terminals, shopping centers). Each fog server is a highly virtualized computing system and is equipped with an onboard, large-volume data storage, computing, and wireless communication facility [27].

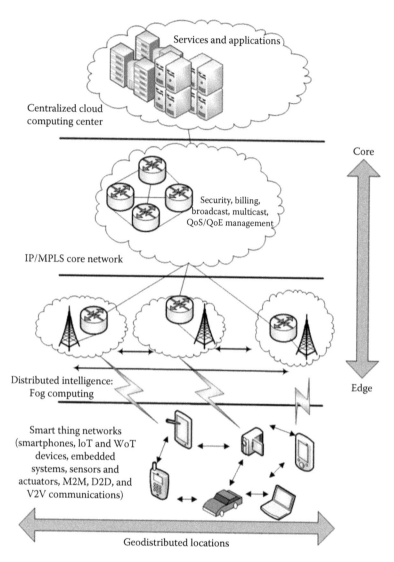

FIGURE 6.7 Fog computing architecture.

The role of fog servers is to bridge the smart mobile device things and the cloud [12]. The geo-distributed, intelligent fog servers directly communicate with the mobile users through single-hop wireless connections using the off-the-shelf wireless interfaces, such as LTE, Wi-Fi, and Bluetooth. They can independently provide predefined service applications to mobile users without assistance from the cloud or the Internet. In addition, the fog servers are connected to the cloud to leverage the rich functions and application tools of the cloud.

Some of the main fog computing features are described in the following subsections.

6.3.2 UBIQUITY OF DEVICES

The main factor that will bring fog into reality is the ubiquity of the devices, whose increase is driven by the user devices and sensors/actuators that enabled the presence and usage of devices everywhere around us for different services and applications. By decreasing the size of the device, the device's portability is increased. However, power consumption is also reduced, which may be crucial in the context of some applications. This can be solved with packaging and power management technologies, such as system-on-chip and system-in-package technologies, 3D microbatteries, and RF-powered computing, that aim to create smaller and more autonomous mobile devices that will run longer at a minimum price [12].

6.3.3 Network Management

The configuration and maintenance of many different types of services running on many heterogeneous devices will only exacerbate the current management problems. Therefore, in the fog computing environment, heterogeneous devices and their running services need to be managed in a more homogeneous manner with the following technologies: NFV; small-edge clouds to host services close to or at the endpoints; and peer to peer (P2P)– and sensor network–like approaches for auto-coordination of applications [12].

NFV is the reaction of telecom operators to their lack of agility and constant need for reliable infrastructures. It is capable of dynamic deployment for on-demand network services and user services. Software-defined networks (SDNs) are one of the main enablers for NFV. For example, the router can be seen as an SDN-enabled virtual infrastructure where NFV and application services are deployed close to the place where they are actually going to be used, which will result in cheaper and more agile operations. However, NFV capabilities still do not reach end-user devices and sensors.

Telecom operators started to deploy clouds in their long-term evolution (LTE) networks closer to the edge (to the user) because the EPC will deliver services close to users (at the edge) more efficiently and will confine traffic there while reducing traffic overload with the help of SDNs. The fog will enable devices to become virtual platforms that can lease some computing/storage capacity for applications to run on them.

The P2P- and sensor network–like approaches exploit the locality and allow endpoints to cooperate to achieve similar results but can scale better and be implemented in a fog. A fog application can be seen as a content distribution network (CDN) where data is exchanged between peers. As a result, the applications and data are no longer required to stay in centralized data centers.

A part of a network and some user devices/sensors can act as mini-clouds in a fog computing environment. The mini-clouds can be implemented by using droplets or small pieces of code that run in a secured manner in devices at the edge with minimum interaction with central coordinating elements, thus reducing the unnecessary and undesired uploads of data to central servers in corporate data centers. The users are able to retain control and ownership of their own data and applications, and the scalability is improved.

6.3.4 Fog Connectivity

The presence of many mobile devices that consume and produce big data at the edge of the network causes a huge bottleneck in the fog [12]. On the physical level, the following technologies can cope with this: LTE-Advanced, Wi-Fi ac, Bluetooth Low Energy, ZigBee, etc.

On the network level, each node must be able to act as a router for its neighbors and must be resilient to node churn (nodes entering and leaving the network) and mobility. Mobile ad hoc networks (MANETs) and wireless mesh networks (WMNs) can provide these functionalities [12]. MANET will enable the formation of densely populated networks without requiring fixed and costly infrastructures to be available beforehand.

WMNs, on the other side, use mesh routers at their cores. Nodes can use the mesh routers to get connectivity or other nodes if no direct link with the routers can be established. Routers will grant access to other networks, such as cellular and WiFi.

On higher levels, some protocols already exist for IoT, such as Message Queue Telemetry Transport (MQTT) and Constrained Application Protocol (CoAP), that provide low resource consumption and resilience to failure [12]. The network and IoT protocols can benefit from data locality, since they no longer need to send all the data around the world all the time, except for potential congestion problems at the edge of the network. Data locality also has a very positive impact on privacy.

6.3.5   FOG PRIVACY AND SECURITY

The greatest concern of fog users is data ownership (i.e., data security and privacy) [29–30]. One method to maintain the privacy is by storing encrypted sensitive data in traditional clouds. The existing cloud computing data protection mechanisms, such as encryption, have failed in preventing data theft attacks, especially those perpetrated by a malicious insider.

However, the value of stolen information can be decreased. This can be achieved through a preventive disinformation attack by using the following additional security features: user behavior profiling, decoys, or a combination of both [30].

User behavior profiling is used to model normal user behavior (i.e., how, when, and how much a user accesses his or her information in the cloud) [30]. Such profiles contain volumetric information (i.e., how many files are accessed and how often). The occurrence of abnormal access to user information in the cloud can be determined by monitoring this normal user behavior, based partially upon the scale and the scope of data being transferred.

Decoys are any bogus information that can be generated on demand. They are used to (1) validate whether data access is authorized when abnormal information access is detected, and (2) confuse the attacker with bogus information [30]. The serving decoys will confuse the malicious attacker into believing that he or she has exfiltrated useful information, but he or she has not. The attacks can be prevented by deploying decoys within the fog by the service customer and within personal online social networking profiles by the individual users.

A combination of decoys and user behavior profiling will provide unprecedented levels of security for the fog and will improve detection accuracy. When access to user information is correctly identified as unauthorized access, the fog security system would deliver unbounded amounts of decoy information to the attacker. Thus the true user data would be protected from unauthorized disclosure. When abnormal access to the fog service is not recognized, decoy information may be returned by the fog and delivered in such a way as to appear completely legitimate and normal. The true owner of the information would identify when decoys are returned by the fog. Hence, the legitimate user could alter the fog responses through a variety of means, such as challenge questions, to inform the fog security system that has inaccurately detected an unauthorized access. At the moment, the existing security mechanisms do not provide this level of security [30].

6.4   FOG COMPUTING FOR 5G NETWORKS

There are four main reasons why fog computing is a suitable solution for 5G [28].

1. **Real-time processing and cyber-physical system control**. Edge data analytics, as well as the actions it enables through control loops, often have stringent time requirements in the order of few milliseconds that can be carried out only at the edge of the network. This is particularly essential for Tactile Internet, which enables virtual reality–type interfaces between humans and devices.
2. **Cognition or awareness of client-centric objectives**. The applications can be enabled by knowing the requirements and the preferences of the clients. This is particularly true when privacy and reliability cannot be trusted in the cloud, or when security is enhanced by shortening the extent over which the communication is carried out.
3. **Increased efficiency by pooling of idle and unused local resources**. The idle and unused gigabytes on many devices, the idle processing power, the sensing ability, and the wireless connectivity within the edge may be pooled within a fog network.

4. **Agility or rapid innovation and affordable scaling**. It is usually much faster and cheaper to experiment with client and edge devices. Rather than waiting for vendors of large boxes inside the network to adopt an innovation, in the fog world, a small team may take advantage of smartphones' application programming interface (API), software development kit (SDK), and proliferation of mobile applications and offer a networking service through its own API.

The following are descriptions of some cases in which fog computing can be applied in 5G [28].

Case 1: *Crowd sensing the states in a 5G base station.* A collection of 5G end-user client devices may be able to infer the states of a 5G base station, such as the number of resource blocks used, by using a combination of passive received signal strength measurement (e.g., RSRQ), active probing (e.g., packet train), application throughput correlation, and historical data mining [31].

Case 2: *OTT network provisioning and content management.* The traditional approach to innovate in the network is to introduce another box inside the network. The fog directly leverages the "things" and phones instead and removes the dependence on boxes in the network altogether. By using end-user client devices for tasks such as universal resource locator (URL) wrapping, content tagging, location tracking, and behavior monitoring, network services can be innovated much faster [28].

Case 3: *Network selection in a heterogeneous environment.* The coexistence of heterogeneous networks (e.g., LTE, femto, Wi-Fi) is a key feature in 5G. Instead of network operator control, 5G will have a user-centric approach where each client can observe its local conditions and make a decision on which network to join [3]. Through randomization and hysteresis, such local actions may emerge globally to converge to a desirable configuration [32].

Case 3: *Borrowing bandwidth from neighbors in M2M or D2D communications.* When multiple devices are next to each other, one device may request the other devices to share their bandwidth by downloading other parts of the same file and transmitting, via Wi-Fi Direct, client to client, etc. [33].

Case 4: *Distributed beam forming.* Fog can also be applied in the physical layer by exploiting multiuser MIMO to improve throughput and reliability when a client can communicate with multiple access points or base stations. For uplink, multiuser beam forming can be used so that the client can send multiple data streams to multiple access points (APs) or base stations (BSs) simultaneously. For downlink, interference nulling can be used to allow the client to decode parallel packets from multiple APs or BSs. These can be done entirely on the client side [34].

As a conclusion, the cloud in a 5G network will be diffused among the client devices, often with mobility too (i.e., the cloud will become fog) [13]. More and more virtual network functionality will be executed in a fog computing environment, and that will provide "mobiquitous" service to the users. This will enable new services paradigms, such as AaaS, where devices, terminals, machines, and smart things and robots will become innovative tools that will produce and use applications, services, and data.

6.5   FOG COMPUTING HYBRID ENVIRONMENT SERVICE ORCHESTRATION MECHANISMS FOR 5G

The 5G network with its cloud computing mechanisms will be unable to deal with future Internet services and applications. This is because the traditional service orchestration approaches that have been applied to cloud services will be unable to adequately handle the forthcoming large-scale and dynamic fog services, since they cannot effectively cope with reduced latency, high mobility, high scalability, and real-time execution [10]. Therefore, it is time to think of a new approach—the so-called 5G network in the fog.

A new HESO is needed that will be capable of ensuring the resilience and trustworthiness of open, large-scale, dynamic services in the fog [13,35]. The HESO orchestrator will be responsible

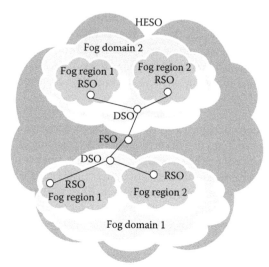

RSO - Regional service orchestrator
DSO - Domain service orchestrator
FSO - Federated service orchestrator
HESO - Hybrid environment service
            orchestrator

FIGURE 6.8   Hybrid environment service orchestrator model for fog computing.

for the composition of service elements available in the fog environment (e.g., sensing, connectivity, storage, processing, platform services, and software services) into more complex fog services (e.g., traffic crowd sensing and trip-planning services) to be offered to the users in the fog environment. The execution of the fog services may involve multiple different components and entities spread across a wide area, increasing complexity in terms of the decision-making process regarding resource allocation to achieve acceptable QoS/QoE levels. To coordinate the execution of the fog services, the orchestration mechanisms need to synchronize and combine the operations of the different service elements to meet the specifications of the composed fog services, including low latency, scalability, and resilience.

The HESO in fog should operate in a loosely coupled mode, resulting in a solution with several levels: regional service orchestrator (RSO), domain service orchestrator (DSO), and federated service orchestrator (FSO), as shown in Figure 6.8.

The RSOs are located at the edges of the fog environment and enable semi-autonomous operation of the different fog regions. This allows the distribution of the load, which provides scalability and much higher proximity to the end users with lower latencies.

The DSOs are responsible for the fog domains and supervise the RSOs below them. This level supports mechanisms that enable intra-domain cooperation between different regions.

The FSO allows fruitful interaction between different fog domains. It is responsible for the management between different fog domains, and, similar to the DSOs, it should be properly adapted to operate in a federated cloud environment. The FSOs will support federation mechanisms to enable cooperation among different fog domains (e.g., belonging to different entities or under the administration of different authorities) and the creation of a multi-domain fog environment able to support service ubiquity.

The HESO model is flexible and scalable and can be implemented in any network technology standard. In particular, its application is important for critical usage cases of IoT devices and the Tactile Internet, which requires 1 ms, end-to-end latency to virtual reality–type interfaces between humans and machines, and big data analytics, which requires real-time processing with a stringent time requirement that can be carried out only in the fog [36].

6.6 HESO MODEL EVALUATION

The performance of the HESO model can be explored in terms of the RTT latency, throughput, and the product latency throughput. The following scenario will be used. There is a region in which a group of N users are located, which are simultaneously covered by several different RANs. Each RAN is connected to several clouds, which can be in the same or different regions with the RANs. The users are assumed to have equally capable smartphone devices and are located at different distances from the RANs. The users can be simultaneously served by the RANs and the clouds.

6.6.1 RTT LATENCY

RTT latency is the time it takes for a single data transaction to occur, meaning the time it takes for the packet of data to travel to and from the source to the destination and back to the source [37,38]. The RTT latency between the user equipment to any cloud via any RAN is equal to

$$RTT = RTT_{RAN} + RTT_{RAN\text{-}CLOUD} \tag{6.1}$$

where RTT_{RAN} represents RTT latency for any RAN network, and $RTT_{RAN\text{-}CLOUD}$ represents the RTT latency between the RAN network and the cloud. The average RTT_{RAN} values for 3G, 4G, and 5G are given in Table 6.3, and the average values of $RTT_{RAN\text{-}CLOUD}$ vary from 50 ms to 500 ms depending on whether the cloud is in the same or different region with the RAN network. For simulation purposes, randomly generated values were used for $RTT_{RAN\text{-}CLOUD}$. Cloud 1 is the closest to the RANs, with the lowest RAN-cloud latency, while cloud 10 is the farthest from the RANs, with the highest RAN-cloud latency.

The simulation results for the RTT latency for different RANs are given in Figure 6.9. RTT latency between the user equipment and any cloud is the lowest for 5G RAN and the highest for 3G RAN, and the user will have the ability to choose the 5G RAN. The RTT latency for any RAN

TABLE 6.3
Round-Trip Time Latency for 3G, 4G, and 5G RAN

Parameter	Network Type		
	3G	4G	5G
RTT_RAN Latency [ms]	70	20	5

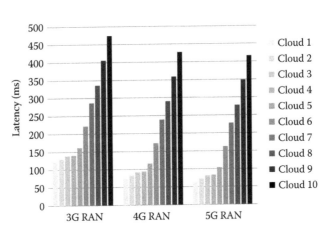

FIGURE 6.9 RTT latency in a 5G network between the user device and the cloud through a particular RAN.

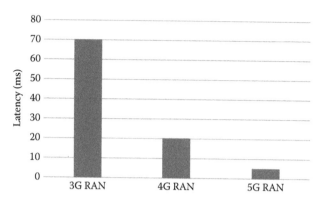

FIGURE 6.10 RTT latency in a 5G network in the fog.

exponentially increases from cloud 1 to cloud 10 because cloud 1 has the lowest latency to all RANs, and cloud 10 has the highest latency to all RANs.

If there is a fog computing node in the RAN (fog RAN) and if the information requested by the user is located in the fog computing node, the RTT latency for any user becomes

$$RTT = RTT_{RAN} \tag{6.2}$$

The RTT latency of a 5G network in the fog environment is shown in Figure 6.10. The RTT latency here is significantly reduced compared to the RTT latency of a 5G network in the cloud, especially for 5G RAN to the order of few milliseconds.

6.6.2 THROUGHPUT

Throughput is the quantity of data that can pass from source to destination in a specific time [36]. The total throughput of any user served by M RANs ($M = 10$) is

$$T = \sum_{i=1}^{M} r_i T_{RAN_i} \tag{6.3}$$

Here, r_i is the weight coefficient equal to 0 or 1 that identifies whether the user will use the flow of a particular RAN in the total throughput. The user throughput for each RAN T_{RAN_i} can be calculated as a ratio of the peak data rate, R, in the RAN and the number of users, N, served by one or all clouds connected to that RAN by

$$T_{RAN_i} = \frac{R}{N} \tag{6.4}$$

The peak data rate for each RAN depends on the distance between the user device and the RAN, where a different modulation coding scheme is used. The modulation coding schemes (i.e., the peak data rate and the distance) are usually defined by the network operator. One such possible configuration for the peak data rate for a 3G, a 4G, and a 5G RAN for a given distance between the mobile device and the RAN is given in Table 6.4.

Each RAN is connected to several clouds, and the user can receive data from a single cloud, several clouds, or all clouds. For simplicity, it is assumed that each user receives data from a particular RAN through all clouds. When all clouds are used, the network overload in the 5G core is significantly reduced, and energy efficiency is significantly improved.

TABLE 6.4

**Peak Data Rate in Mbps for 3G, 4G, and 5G RAN
That Depends on the Distance between the User
and the RAN**

Distance d [m]	Network Type		
	3G	4G	5G
d < 500	672	3900	50000
d < 1000	84	1000	40000
d < 1500	42	600	30000
d < 2000	21	450	20000
d < 2500	14.4	390	15000
d < 3000	7.2	300	10000
d < 3500	3.6	150	5000
d < 4000	1.8	100	2500
d < 4500	0.72	50	1500
d < 5000	0.384	10	1000

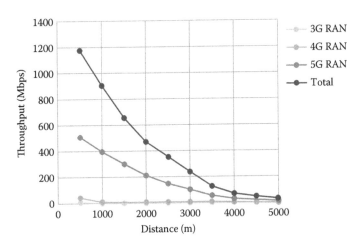

FIGURE 6.11   User throughput in a 5G network in the fog as a function of the distance, for $N = 1000$ users.

The user throughput results, when the user receives data from all clouds, are shown in Figures 6.11 and 6.12. In Figure 6.11, the number of users is fixed to 1000, and the user throughput is represented as a function of the distance. As the distance between the user and the cloud is increased, the user throughput decreases. In Figure 6.12, the distance is less than or equal to 500 m, and the number of users is varied from 100 to 1000. As the number of users increases, the user throughput again decreases. In both figures, 5G offers much higher user throughput than 4G and 3G. This means that a much higher quantity of data can pass through 5G RAN compared to 3G and 4G RANs. Therefore, the smart user device will prefer to choose the 5G RAN.

In 5G, the user equipment will be simultaneously connected to several RANs and several clouds (i.e., it will combine the flows from several RANs and several clouds for a single application or service). This is the total user throughput, which is also shown in Figures 6.11 and 6.12. The user throughput is significantly increased. In addition, the network overload is significantly reduced, and energy efficiency in the 5G network RAN is significantly improved.

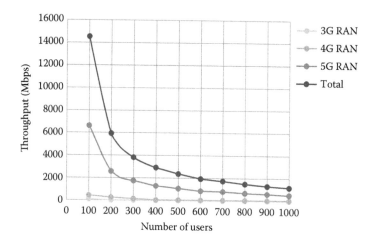

FIGURE 6.12 User throughput in a 5G network in the fog as a function of the number of users, located at a distance of $d < 500$ meters.

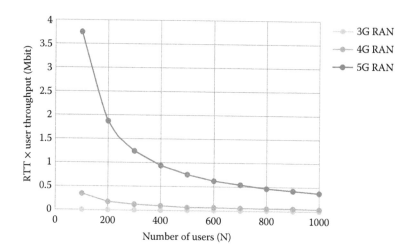

FIGURE 6.13 Product RTT × user throughput for cloud 1.

6.6.3 PRODUCT LATENCY THROUGHPUT

The product *RTT × user throughput* and *RTT × peak data rate* correspond to the bandwidth delay product (i.e., to the maximum amount of data on the network at any given time) that has been transmitted but not yet acknowledged. The simulation results for these products for cloud 1 (closest cloud computing center from the user devices) and cloud 10 (the farthest cloud computing center from the user devices) are given in Figure 6.13 and Figure 6.14, respectively. This product decreases as the number of user devices increases. Here, 5G RAN again outperforms 3G and 4G RANs, and the 5G smartphone device will prefer to choose the 5G RAN.

6.6.4 DISCUSSION OF THE RESULTS

The results clearly demonstrate the benefits of fog computing HESO mechanisms in a 5G network, especially if the flows from different RANs and clouds for a particular application or service requested by the user are combined. This will primarily depend on whether the user requests a

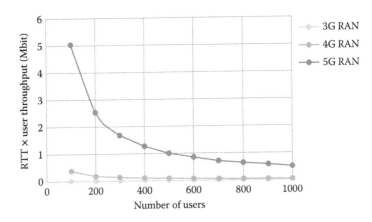

FIGURE 6.14   Product RTT × user throughput for cloud 10.

service that requires high throughput, low latency, optimal product latency throughput, energy efficiency of clouds and RANs, etc.

In particular, the big data analytics that require real-time processing and very often have stringent time requirement can be carried out only in the fog. This is essential for critical usage cases of IoT devices and the Tactile Internet where 1 ms, end-to-end latency is required in the network to provide virtual reality–type interfaces between humans and machines (human–machine interaction and machine–machine interaction) [36].

However, the selection of RAN and cloud flows that will be combined and used for a single service or application by the mobile user will depend on the low latency requirements, high throughput requirements, optimal latency throughput, bandwidth delay, and energy efficiency of the clouds and the RANs. The algorithm for selecting the RAN and cloud flows will be the direction of our future work.

6.7   RESEARCH CHALLENGES IN FOG COMPUTING

Many of the ideas in fog computing have evolved from P2P networks, mobile ad hoc networks (MANETs), and edge computing [28]. Compared to P2P networks, fog is not just about content sharing (or data plane as a whole) but also network measurement, control and configuration, and service definition. Compared to MANET, today there are much more powerful and diverse off-the-shelf edge devices and applications, together with the broadband mobile and wireless networks. Compared to edge computing, in a fog computing environment, edge devices optimize among themselves, and they collectively measure and control the rest of the network.

Fog computing also relocates the operating functions on digital objects of information-centric networks and the virtualized functions in SDNs at the network edge [28]. Because fog computing is a recently emerged paradigm, there are many challenging questions that need to be resolved. The following are some of those questions, which are also applicable for the fog computing environment in 5G:

- **Cloud–fog interface and fog–fog interface [28]:** The cloud will remain useful as the fog arises. The tasks that require real-time processing, end-user objectives, and low-cost leverage of idle resources should go to fog, while the tasks that require massive storage, heavy-duty computation, and wide area connectivity should go to the cloud. Therefore, the cloud–fog interface and the fog–fog interface need to be defined clearly to make the transfer of information much easier.
- **Interactions between smart user device hardware and operating systems (OS) [28]:** Once the operating functions are moved on smart user devices, the interface with their OS and hardware becomes essential, particularly for pooling idle edge resources.

- **Trustworthiness and security [12,28]:** Because of the proximity to end users and locality on the edge, fog computing nodes can act as the first node of access control and encryption, provide contextual integrity and isolation, and enable the control of aggregating privacy-sensitive data before it leaves the edge.
- **Convergence and consistency arising out of local interactions [28]:** Typical concerns of distributed control, divergence/oscillation, and inconsistency of global states become more acute in a massive, under-organized, possibly mobile crowd with diverse capabilities and a virtualized pool of resources shared unpredictably. Use cases in edge analytics and stream mining provide additional challenges on this recurrent challenge in distributed systems.
- A decision of what kind of information to keep between distributed (local) and centralized (global) architectures while maintaining resilience through redundancy [28].
- Exploring the energy efficiency of the fog computing environment through the power consumption of the smart user devices for different types of RANs.
- Algorithm for selecting the most suitable RAN for a particular task to provide the best QoS/QoE.
- Distributed P2P mobile cloud computing among the smart user devices to guarantee a certain QoS/QoE level.

6.8 CHAPTER SUMMARY

This chapter provided an overview of a 5G network and fog computing, their related technologies, and their research directions. It also proposed a HESO model for resilient and trustworthy fog computing services for a 5G network. Then, the HESO was evaluated in terms of RTT latency, user throughput, and product latency throughput. Finally, open research issues for fog computing mechanisms in 5G were outlined.

The results clearly show that 5G will benefit greatly by using the HESO model because its service orchestration mechanisms will effectively cope with the forthcoming services that require reduced latency, high mobility, high scalability, and real-time execution. This is particularly important for critical usage cases of IoT devices and the Tactile Internet, which requires 1 ms end-to-end latency, and big data analytics, which requires real-time processing with stringent time requirement that can be carried out only in the fog.

The cloud in 5G networks will be diffused among client devices, often with mobility too (i.e., the cloud will become the fog). More and more virtual network functionality will be executed in a fog computing environment, and it will provide "mobiquitous" service to the users. This will enable new services paradigms, such as AaaS, where devices, terminals, machines, and smart things and robots will become innovative tools that will produce and use applications, services, and data. This is essential for the success of the future IoE, a clear evolution of the IoT.

5G in the fog will use the benefits of the centralized cloud, CRAN, and fog RAN cloud and the distributed P2P mobile cloud among the devices, which will create opportunities for companies to deploy many new, real-time services that cannot be delivered over current mobile and wireless networks.

REFERENCES

1. T. Janevski. *Internet Technologies for Fixed and Mobile Networks*. Norwood, MA: Artech House, 2015.
2. T. Janevski. *NGN Architectures Protocols and Services* London: Wiley, 2014.
3. T. Janevski. 5G mobile phone concept. *Proceedings of 6th IEEE Consumer Communications and Networking Conference, CCNC 2009*, Las Vegas, Nevada, 1–2, 2009.
4. A. Tudzarov and T. Janevski. Functional architecture for 5G mobile networks. *International Journal of Advanced Science and Technology (IJAST)*, 32: 65–78, 2011.
5. V. Tikhvinskiy and G. Bochechka. Prospects and QoS requirements in 5G networks. *Journal of Telecommunications and Information Technologies*, 1(1): 23–26, 2015.

6. Global mobile Suppliers Association. *The Road to 5G: Drivers, Applications, Requirements and Technical Development*. A GSA (Global mobile Suppliers Association) Executive Report from Ericsson, Huawei and Qualcomm, 2015.
7. SK Telecom Network Technology Research and Development Center 5G White Paper. *SK Telecom's View on 5G Vision, Architecture, Technology and Service and Spectrum*. SK Telecom, October 2014.
8. Datang Mobile Wireless Innovation Center 5G White Paper. *Evolution, Convergence and Innovation*. Beijing, China: Datang Telecom Technology and Industry Group, 2013.
9. B. Brech, J. Jamison, L. Shao, and G. Wightwick. *The Interconnecting of Everything*. IBM Redbook, 2013.
10. S. Zhang, S. Zhang, X. Chen, and X. Huo. Cloud computing research and development trend. *Proceedings of the 2nd IEEE International Conference on Future Networks (ICFN 2010)*, Sanya, Hainan, 93–97, 2010.
11. F. Bonomi, R. Milito, J. Zhu, and S. Addepalli. Fog computing and its role in the Internet of things. *Proceedings of the 1st Edition of the ACM SIGCOMM Workshop on Mobile Cloud Computing (MCC 2012)*, Helsinki, Finland, 13–16, 2012.
12. L. M. Vaquero and L. Rodero-Merino. Finding your way in the fog: Towards a comprehensive definition of fog computing. *ACM SIGCOMM Computer Communication Review Newsletter*, 44(5): 27–32, 2014.
13. S. Kitanov, E. Monteiro, and T. Janevski. 5G and the fog: Survey of related technologies and research directions. *Proceedings of the 18th Mediterranean IEEE Electrotechnical Conference MELECON 2016*, Limassol, Cyprus, 1–6, 2016.
14. S. Barbarossa, S. Sardellitti, and P. Di Lorenzo. Communicating while computing: Distributed mobile cloud computing over 5G heterogeneous networks. *IEEE Signal Processing Magazine International Journal of Advanced Science and Technology (IJAST)*, 1(6): 45–55, 2014.
15. D. Guinard, V. Trifa, F. Mattern, and E. Wilde. From the Internet of things to the web of things: Resource-oriented architecture and best practices. In *Architecting the Internet of Things*, Berlin: Springer-Verlag Berlin Heidelberg, 97–129, 2011.
16. V. L. Kalyani and D. Sharma. IoT: Machine to machine (M2M), device to device (D2D) Internet of everything (IoE) and human to human (H2H): Future of communication. *Journal of Management Engineering and Information Technology (JMEIT)*, 2(6): 17–23, 2015.
17. *Internet of Things vs. Internet of Everything: What's the Difference*. ABI Research White Paper, 2014.
18. S. Mitchell, N. Villa, M. Stewart-Weeks, and A. Lange. *The Internet of Everything for Cities Connecting People, Process, Data, and Things to Improve the 'Livability' of Cities and Communities*. San Jose CA: CISCO, White Paper, 2013.
19. Recommendation ITU-T Y.3501. *Cloud Computing Framework and High-Level requirements*. 2013.
20. Recommendation ITU-T Y.3510. *Cloud Computing Infrastructure Requirements*. 2013.
21. NIST Special Publication 500–291. *NIST Cloud Computing Standards Roadmap*. Version 2, 2013.
22. T. H. Dihn, C. Lee, D. Niyato, and P. Wang. A survey of mobile cloud computing: Architecture, applications, and approaches. *Wireless Communications and Mobile Computing*, Wiley, 13(18): 1587–1611, 2011.
23. S. S. Qureshi, T. Ahmad, K. Rafique, and Shuja-ul-islam. Mobile cloud computing as future for mobile applications: Implementation methods and challenging issues. *Proceedings of the IEEE Conference on Cloud Computing and Intelligence Systems (CCIS)*, Beijing, China, 467–471, 2011.
24. D. Huang et al. Mobile cloud computing. *IEEE COMSOC Multimedia Communications Technical Committee (MMTC) E-Letter*, 6(10): 27–31, 2011.
25. S. Kitanov and T. Janevski. State of the art: Mobile cloud computing. *Proceedings of the Sixth IEEE International Conference on Computational Intelligence, Communication Systems and Networks 2014 (CICSYN 2014)*, Tetovo, Macedonia, 153–158, 2014.
26. J. Marinho and E. Monteiro. Cognitive radio: Survey on communication protocols, spectrum decision issues, and future research directions. *Springer Wireless Networks*, 18(2): 147–164, 2012.
27. H. T. Luan, L. Gao, Z. Li, and L. X. Y. Sun. Fog computing: Focusing on mobile users at the edge. *arXiv:1502.01815[cs.NI]*, 2015.
28. M. Chiang. *Fog Networking: An Overview on Research Opportunities*. White Paper, 2015.
29. I. Stojmenovic and S. Wen. The fog computing paradigm: Scenarios and security issues. *Proceedings of the Federated Conference on Computer Science and Information Systems (FedCSIS)*, ACSIS, 2(5): 1–8, September 2014.
30. S. J. Stolfo, M. B. Salem, and A. D. Keromytis. Fog computing: Mitigating insider data theft attacks in the cloud. *Proceeding of the IEEE Symposium on Security and Privacy Workshops (SPW)*, San Francisco, California, 125–128, 2012.
31. A. Chakraborty, V. Navda, V. N. Padmanabhan, and R. Ramjee. Coordinating cellular background transfers using LoadSense. *Proceedings of ACM Mobicom*, 2013. Miami, Florida, USA, 63–74.

32. E. Aryafar, A. Keshavarz-Haddard, M. Wang, and M. Chiang. RAT selection games in HetNets. *Proceedings of the 32nd IEEE International Conference on Computer Communications IEEE INFOCOM,* Turin, Italy, 998–1006, 2013.

33. Z. Zhang, J. Zhang, and L. Ying. *Multimedia Streaming in Cooperative Mobile Social Networks.* Preprint.

34. Y. Du, E. Aryafar, J. Camp, and M. Chiang. iBeam: Intelligent client-side multi-user beamforming in wireless networks. *Proceedings of the 33rd IEEE International Conference on Computer Communications IEEE INFOCOM,* Toronto, Canada, 817–825, 2014.

35. S. Kitanov and T. Janevski. Hybrid environment service orchestration for fog computing. *Proceedings of the 51st International Scientific Conference on Information, Communication and Energy Systems and Technologies (ICEST 2016),* Ohrid, Macedonia, 2016.

36. G. Fettweis, et al. *The Tactile Internet,* ITU Technology Watch Report, 2014.

37. T. Yeung. *Latency Consideration in LTE Implications to Security Gateway,* Stoke Inc. White Paper, Literature No. 130-0029-001, 2014.

38. Nokia Siemens Networks. *Latency: The impact of latency on application performance,* Nokia Siemens Networks White Paper, 2009.

7 Lightweight Cryptography in 5G Machine-Type Communication

Hüsnü Yıldız, Adnan Kılıç, and Ertan Onur

CONTENTS

7.1 INTRODUCTION TO LIGHTWEIGHT CRYPTOGRAPHY

The number of connected devices will increase with the deployment of the Internet of Things (IoT) and is expected to reach 50 billion by 2020, as shown in Figure 7.1. In the next decade, the amount of transferred data among IoT devices will increase in parallel with the number of devices. A large bandwidth will be required. The present communication protocols will be insufficient to handle massive amounts of data generated by IoT devices and machine-type communication besides human-centric applications. 5G is expected to become the de facto access network for this new paradigm since it will provide low latency, ultra-large capacity, wider coverage, machine-to-machine communication, and green communication. Additionally, IPv6 will substitute for IPv4 in the near

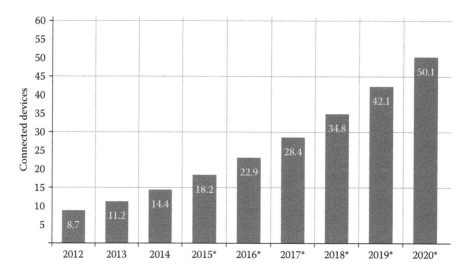

FIGURE 7.1 Prediction of the number of connected devices. (Redrawn from Hotel News Resource. 2015. Internet of Things (IoT): Number of Connected Devices Worldwide from 2012 to 2020 [in Billions]. *Statista— The Statistics Portal.* www.statista.com/statistics/471264/iot-number-of-connected-devices-worldwide, Accessed November 29, 2016.)

future due to the lack of sufficient address space of IPv4. Thanks to IPv6, which has 2^{128} addresses, devices will communicate with each other directly in the network. 5G and IPv6 technologies will pave the way for new security vulnerabilities while having many advantages.

Conventional cryptographic algorithms are employed to provide confidentiality, integrity, availability, and non-repudiation. They are usually designed to satisfy security requirements, and they do not impose any limitation or constraints on energy consumption, computation power, or memory since these algorithms are mostly employed on powerful devices such as servers, desktops, smartphones, and tablets. However, there are many constrained devices that have limited CPU, memory, and power. By 2020, a huge number of constrained devices are expected to be used in smart homes, smart cities, agriculture, healthcare, and transportation to gather information via sensors and to take actions via actuators. While transporting the sensed information to the cloud, new challenges will emerge due to constraints of the employed equipment. Hence, new protocols and security mechanisms have to be developed to deal with these challenges since conventional cryptographic algorithms are not appropriate in the aforementioned tasks. Due to the limited storage, memory, and CPU, there are new solutions such as the Constrained Application Protocol (COAP) (IETF/COAP 2016) as the web transfer protocol, 6LoWPAN (IETF/6LoWPANs 2007) for low-cost Internet connectivity, Message Queuing Telemetry Transport (MQTT) (ISO/MQTT 2016) as the machine-to-machine lightweight messaging protocol for monitoring, Extensible Messaging and Presence Protocol (XMPP) (IETF/XMPP 2011) as the communication protocol for messaging, and Data Distribution Service (DDS) (OMG/DDS 2015) as the data exchange service. However, these protocols and/or services can be eavesdropped by any third party because the existing cryptographic algorithms are not used in constrained devices for establishing secure channels because of the limitations of these devices. Therefore, novel lightweight or ultra-lightweight cryptographic approaches have to be designed. The term *lightweight* does not indicate insecurity, although it does imply energy efficiency. It has to provide a sufficient security level for constrained devices.

The vision of 5G and IoT integration with transportation, smart cities, smart homes, healthcare, agriculture, and other digital systems is illustrated in Figure 7.2. There are many applications envisioned, such as adjusting home temperature, checking unusual situations, measuring the amount of moisture of fields, optimizing traffic density, reducing power consumption, and measuring tension. The data measured by sensors will be transmitted to data centers. To reduce operations on the cloud, there are some cloud-like servers deployed close to the source of the data.

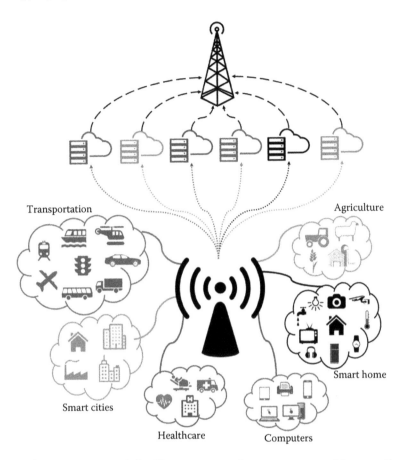

FIGURE 7.2 5G and IoT integration in intelligent transportation systems, smart cities, smart homes, healthcare, agriculture, and other vertical sectors.

To sustain such applications, new devices are being introduced to the market, and the amount of energy consumption has been increasing despite its negative effects on our daily life. One of the well-known effects is global warming since CO_2 emissions are increasing through the production and the consumption of energy. A huge portion of energy is consumed today by factories, as illustrated in Figure 7.3 (U.S. Energy Information Administration 2016). In the near future, constrained devices will also be one of the main energy-consuming domains due to the billions of constrained devices that will take part in our daily life. Moreover, when it is hard to change the batteries of constrained devices frequently, they have to consume their energy sources efficiently. Therefore, the amount of energy consumption should be reduced by employing energy-efficient solutions. Green and secure technology is related to lightweight cryptography since their designs consider the minimum energy consumption with sufficient security principles for constrained devices.

Although the extra energy conservation from using lightweight cryptography per unit might be negligible, the total amount will be considerably large when implemented in a massive network (e.g., IoT network with billions of devices). That is why we present an overall perspective on lightweight cryptography in this chapter. We will present lightweight cryptography standards, block and stream ciphers, hash functions, and possible attacks on lightweight security measures.

This chapter is organized as follows: in Section 7.2, features of constrained devices and IoT security are summarized. In Sections 7.3, 7.4, and 7.5, lightweight block ciphers, lightweight stream ciphers, and lightweight hash algorithms are covered. Some of the algorithms are explained in detail since they provide much greater performance than the others. We also mention lightweight

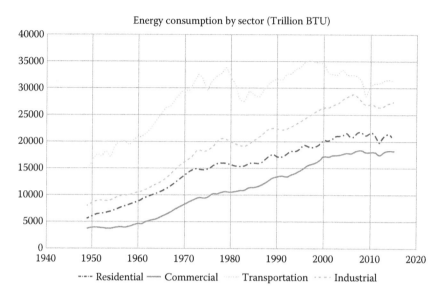

FIGURE 7.3 Energy consumption between 1945 and 2015. (Redrawn from U.S. Energy Information Administration. 2016. U.S. Energy Information Administration Energy Consumption. *U.S. Energy Information Administration*. Accessed August 30, 2016. www.eia.gov/totalenergy/data/monthly/pdf/sec2_2.pdf.)

standards. In addition, key features of the lightweight cryptographic algorithms are listed for comparison. In the last section, we briefly explain attacks.

7.2 CONSTRAINED DEVICES AND IOT SECURITY

One of the main components of IoT is constrained devices. The constraints of these devices can be chip area, energy consumption, program code size, RAM size, communication bandwidth, and execution time (ISO/IEC 2012). They can be used in many areas such as smart cities, smart homes, agriculture, transportation, and healthcare to gather and send information since they are small and portable devices. However, their limitations can be determinative factors, including communication range, application size and its functionalities, data transfer speeds and units, and storage. Therefore, most of the devices are designed to perform only one task. RFC 7228 (IETF 2014) classifies constrained devices according their data and code size, energy limitations, and power usage. Data and code size show how many bytes can be stored in ROM and RAM. The classification of energy limitations refers to the type of power source, which can be event energy limited, period energy limited, lifetime energy limited, or no limitations. Use of power by constrained devices is categorized as follows: always on, normally off, or low power.

The main challenge of IoT originates from its heterogeneity; various devices are assembled into a network, and they interact with each other. Since these devices have different amounts of resources, their protocols should be adapted according to the devices' characteristics. However, conventional protocols do not satisfy the requirements of heterogeneity. From the viewpoint of security, compatible protocols with heterogeneous devices should be standardized for authentication, end-to-end security, and data transfers. On the other hand, communication of heterogeneous devices, which varies from sensors to servers, introduces new threats for IoT. According to the "Security Considerations in the IP-based Internet of Things" draft (Garcia-Morchon et al. 2013), these threats can be listed as cloning things, malicious substitution of things, eavesdropping attacks, man-in-the-middle attacks, firmware replacement attacks, extraction of security parameters, routing attacks, privacy threats, and denial-of-service attacks. These threats can be examined in two groups: bootstrapping phase and operational phase. The bootstrapping phase is the installation of the requirements of the devices (e.g., firmware, key materials) and codifying. The operational phase implies run time.

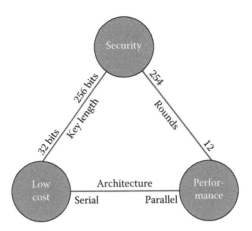

FIGURE 7.4   Design principles of lightweight cryptography. (Redrawn from Poschmann, Axel York. 2009. *Lightweight Cryptography: Cryptographic Engineering for a Pervasive World.* Citeseer.)

The design constraints of lightweight cryptographic algorithms can be categorized as hardware or software. Hardware implementations are compared in accordance with gate equivalents (GEs) and cell libraries that relate to energy consumption. In software implementation, allocated space in RAM and ROM are the key factors for design (NIST 2016). The International Organization for Standardization (ISO) determines lightweight cryptography standards for general purposes (ISO/ IEC 2012), block ciphers (Block Ciphers 2012), stream ciphers (Stream Ciphers 2012), asymmetric techniques (Asymmetric Techniques 2013), and hash functions (Hash-Functions 2016). Lightweight cryptography algorithms are classified as symmetric ciphers, hash functions, and public-key cryptography. Additionally, symmetric ciphers are divided into two groups: stream cipher and block cipher.

As a part of the IoT security, lightweight cryptographic algorithms are designed according to three perspectives: security, performance, and cost. In general, security versus performance, security versus cost, or performance versus cost trade-offs (Figure 7.4) are chosen to design new lightweight cryptographic algorithms since it is hard to balance performance, security, and cost factors simultaneously in the same design due to restrictions imposed by the capabilities of the devices (Poschmann 2009). In fact, cryptographic algorithms form the backbone of security, but security does not consist only of symmetric ciphers, asymmetric ciphers, and hashing algorithms but also of design of the system and its interactions, which determine the overall security.

Public-key cryptographic algorithms have more mathematical operations than other algorithms, which affects power consumption. Indeed, a large portion of energy consumption occurs when wireless communication is used. In Patrick and Schaumont (2016), it is mentioned that public-key cryptographic algorithms consume as much energy as the physical layer of the wireless communication stack. If public-key algorithms are used, the lifetime of the devices will be very short. Therefore, symmetric ciphers and hashing algorithms are usually preferred to provide long battery life.

7.3   LIGHTWEIGHT BLOCK CIPHERS

Lightweight block ciphers concentrate mainly on block size, key size, number of rounds, GE, cell library, and throughput to design an optimal cipher. There have been many block ciphers proposed with different designs, such as fewer rounds, usually supporting a 64-bit block size with different key size options or smaller substitution boxes. In this section, we present some of the algorithms in detail. According to the ISO standard (Block Ciphers 2012), PRESENT (Bogdanov et al. 2007) and CLEFIA (Shirai et al. 2007) are suitable for satisfying the lightweight cryptography requirements. SIMON and SPECK (Ray 2013) ciphers are also presented to cover a broad range of design concerns. We briefly summarize the features of other block ciphers in Table 7.1 (Biryukov and Perrin 2015), which helps us compare block ciphers according to their main characteristics.

TABLE 7.1

Lightweight Block Ciphers

Algorithm	Block Size	Key Sizes	Rounds	Area (GE)	Cell Library (µm)	Structure	Throughput (kb/s, 100 khz)	Power (µW)	References
CLEFIA	128	128	18	4950/5979	0.09	FN	355.6/711.1	NA	Shirai et al. (2007)
		192	22	8536			NA	NA	
		256	26	8482			NA	NA	
DESL	64	64	16	1848	0.18	FN	5.55	0.89	Leander et al. (2007)
DESXL	64	184	16	2168	0.18	FN	44.4	1.6	Leander et al. (2007)
EPCBC	48	96	32	1008	0.18	SPN	12.12	2.21	Yap et al. (2011)
	96			1333				3.63	
HIGHT	64	128	32	3048	0.25	FN	188.2	NA	Hong et al. (2006)
KATAN	32	80	254	802	0.13	LFSR	12.5	0.381	De Canniere et al. (2009)
	48			927			18.8	0.439	
	64			1054			25.1	0.555	
KTANTAN	32	80	254	462	0.13	LFSR	12.5	0.146	De Canniere et al. (2009)
	48			588			18.8	0.234	
	64			688			25.1	0.292	
KLEIN	64	64	12	1220	0.18	SPN	30.9	NA	Gong et al. (2011)
		80	16	1478			23.62	NA	
		96	20	1528			19.1	NA	
LBLOCK	64	80	32	1320	0.18	FN	200	NA	Wu and Zhang (2011)
LED	64	64	32	966	0.18	SPN	5.1	NA	Guo et al. (2011b)
		128	48	1265			3.4	NA	
mCrypton	64	64	12	2420	0.13	SPN	492.3	NA	Lim and Korkishko (2005)
		96		2681			NA	NA	
		128		2949			NA	NA	
MIBS	64	64	32	1396	0.18	FN	200	NA	Izadi et al. (2009)
		80		1530			NA	NA	
Piccolo	64	80	25	1136	0.13	FN	237.04	NA	Shibutani et al. (2011)
		128	31	1197			193.94	NA	

(Continued)

TABLE 7.1 (CONTINUED)
Lightweight Block Ciphers

Algorithm	Block Size	Key Sizes	Rounds	Area (GE)	Cell Library (μm)	Structure	Throughput (kb/s, 100 khz)	Power (μW)	References
PRESENT	64	80	31	1570	0.18	SPN	200	5	Bogdanov et al. (2007)
		128		1886					
PRINTcipher	48	80	48	503	0.18	SPN	100	NA	Knudsen et al. (2010)
	96	160	96	967			100	NA	
Puffin	64	128	32	2577	0.18	SPN	194	NA	Cheng et al. (2008)
Rectangle	64	80	25	1600	0.13	SPN	246	74.31	Zhang et al. (2015)
		128		2064			246	72.15	
SEA	96	96		449	0.13	FN	NA	3.213	Standaert et al. (2006)
SIMON	32	64	32	NA	0.13	FN	NA	NA	Ray (2013)
	48	72/96	36	NA/763			NA	NA	
	64	96/128	42/44	838/1000			NA	NA	
	96	96/144	52/54	984/NA			NA	NA	
	128	128/192/256	68/69/72	1317/NA/NA			NA	NA	
SPECK	32	64	22	NA	0.13	FN	NA	NA	Ray (2013)
	48	72/96	22/23	NA/884			NA	NA	
	64	96/128	26/27	984/1127			NA	NA	
	96	96/144	28/29	1134/NA			NA	NA	
	128	128/192/256	32/33/34	1396/NA/NA			NA	NA	
TWINE	64	80	36	1503	0.09	FN	178	NA	Suzaki et al. (2012)
		128		1866			178	NA	
XTEA	64	128	64	3490	0.13	FN	57.1	19.5	Lu (2009)

Source: Biryukov, Alex, and Leo Perrin. 2015. *Lightweight Cryptography Lounge*. Accessed February 5, 2017. http://cryptolux.org/index.php/Lightweight_Cryptography.
NA = *Not Available.*

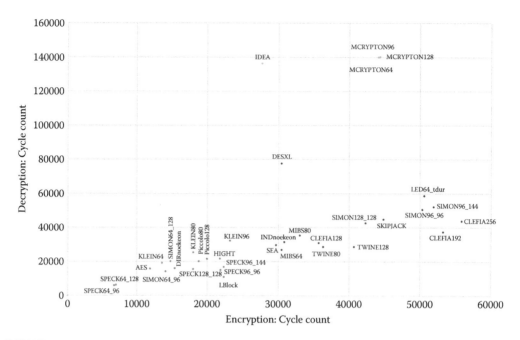

FIGURE 7.5 The number of encryption and decryption cycles of the algorithms on the same sensor.

Lightweight cryptographic algorithms are specially designed considering hardware and software implementations. Their requirements of power consumption, memory, and CPU are optimized considering the selected cell library. To determine these requirements appropriately, algorithms should be tested in their practical environments. Due to the lack of these environments, we classify lightweight block ciphers using their software implementations' results on the same sensor (Cazorla et al. 2013; BLOC Project 2013). The number of encryption and decryption cycles of the algorithms on the same sensor gives an estimation of power consumption, as illustrated in Figure 7.5. This figure shows that we can divide lightweight block ciphers into three groups. From the minimum energy consumption to the maximum energy consumption, they can be divided into three groups. Generally, those ciphers with larger block sizes or key sizes consume a larger amount of energy. Most of these algorithms perform encryption and decryption process with almost the same range of cycle counts, except mCrypton and IDEA. Speck 64-bit block size with 96-bit and 128-bit key sizes has the least power consumption since it is a software-oriented cipher. This situation emphasizes the importance of implementation environments.

7.3.1 PRESENT

PRESENT is one of the lightweight algorithms standardized by the ISO. PRESENT supports the 64-bit block size and 80-bit and 128-bit keys using 1570 GE. It is based on the substitution-permutation network (SPN) (Heys 2010), which uses single 4-to-4-bit S-Box, regular bit permutation for P-Box, and 31 rounds. The encryption and decryption processes need approximately the same physical resources. In addition, encryption keys are processed on the fly. Although it is assumed that PRESENT is secure, it is vulnerable to some types of attacks, such as the related-key rectangle attack on 17 rounds and side-channel attacks (Wu and Zhang 2011; Zhou et al. 2005).

7.3.2 CLEFIA

CLEFIA is another ISO-compatible lightweight block cipher that supports 128-, 192-, and 256-bit key lengths with a 128-bit block size and consists of 18, 22, and 26 rounds, respectively. Its structure is based on a type-2 Feistel network (FN) (Hoang and Rogaway 2010) that consists of 4 data lines and uses 32-bit F-functions in each round. In the F-functions, a diffusion switching mechanism (DSM) (Shirai and Preneel 2004) is used to provide two different diffusion matrices and S-Box is used to prevent cryptographic attacks. A DSM can reduce the number of rounds without compromising the security and accelerate the algorithm.

7.3.3 SIMON and SPECK Family

The SIMON and SPECK algorithms (Ray 2013), designed by NSA, are flexible in terms of different block sizes and key sizes and can be implemented on different platforms (e.g., ASIC, FPGA, microcontroller, and microprocessor). In addition, they provide high throughput on these platforms. Since SIMON and SPECK are very flexible algorithms, they can be used in a wide range of constrained devices. Although each of these algorithms is hardware and software oriented, SIMON has the best performance on hardware, and SPECK has the best performance on software. Therefore, the software-oriented cipher SPECK has fewer rounds and can reuse key scheduling unlike SIMON.

7.4 LIGHTWEIGHT STREAM CIPHERS

Stream ciphers produce keystreams to encrypt a bit or byte stream. Unlike block ciphers, stream ciphers have to be fast and use low memory compared to block ciphers. The ISO recommends two stream ciphers, namely Enocoro (Watanabe et al. 2008) and Trivium (Cannière 2006). We present several stream ciphers and their properties in Table 7.2 (Biryukov and Perrin 2015).

7.4.1 ENOCORO

Enocoro is a stream cipher that uses 80-bit and 128-bit key length with a Panama-like keystream generator (PKSG) (Daemen and Clapp 1998). It uses a 64-bit initialization vector (IV) with both key lengths. The internal state of Enocoro was built in basically by applying two functions, called *rho* and *lambda*. The *rho* function uses linear transformation and S-Box. In contrast to SPN, a substitution-permutation-substitution structure is used to provide stronger encryption. The *lambda* function consists of XOR operations and byte-wise rotation. The algorithm is implemented on ASIC and requires 2.7 K and 4.1 K gate areas for 64-bit and 128-bit key lengths, respectively.

TABLE 7.2
Stream Ciphers and Their Properties

Algorithms	Key Size	IV Size	Cell Library (μm)	Area (GE)	Throughput (kb/s, 100 khz)	Power (μW)	References
Trivium	80	80	0.13	2599	100	5.5	Cannière (2006)
Enocoro	80	64	0.18	2.7 K	28	NA	Watanabe et al.
	128			4.1 K		NA	(2008)
Grain	80	64	0.13	1294	0.1	3.3	Hell et al. (2007)
	128	80		1857	0.1	4.34	
MICKEY 2.0	80	80	0.18	3188	NA	NA	Babbage and Dodd
	128	128		5039	NA	NA	(2006)

Source: Biryukov, Alex, and Leo Perrin. 2015. *Lightweight Cryptography Lounge.* Accessed February 5, 2017. http://cryptolux.org/index.php/Lightweight_Cryptography.

7.4.2 TRIVIUM

Trivium is a synchronous stream cipher with 80-bit key length and 80-bit IV and is designed as a hardware-efficient algorithm as a part of the eSTREAM Project (eSTREAM 2008). Trivium is suitable for constrained devices since the algorithm has a compact design and is fast for applications that require high throughput. In addition, the algorithm uses three feedback registers in the internal state with different key lengths of 84, 93, and 111 bits. For each key and IV, 2^{64} keystream bits are generated.

7.4.3 GRAIN

Grain is another hardware-efficient synchronous stream cipher, which uses 80-bit and 128-bit key lengths and 64-bit and 80-bit IV as inputs, respectively. Its main design consists of three functions—linear feedback shift register (LFSR) (Murase 1992), nonlinear feedback shift register (NFSR) (Fontaine 2005), and an output function. LFSR assures that the period of the keystream is large and provides a good balance for the output. The nonlinearity of the algorithm is related to the existence of the NFSR and the nonlinear output function. The algorithm has an additional option that provides extra speed if the hardware is expandable. Due to some weaknesses, the recommended versions of Grain can be found in Hell et al. (2007) and Ågren et al. (2011).

7.5 LIGHTWEIGHT HASH FUNCTIONS

A hash function takes arbitrary data as input and maps it to a fixed-sized output. It is widely used in numerous fields, such as digital signatures and checksums. Hash functions are expected to be strong against collision, preimage, and second-preimage attacks. There are many design structures for hash functions; however, lightweight hash functions are based on Merkle-Damgård (Charles 1979) and Sponge (Bertoni et al. 2007). ISO 29192-5 (Hash-Functions 2016) recommends three hash functions: PHOTON (Guo et al. 2011a), SPONGENT (Bogdanov et al. 2013), and Lesamnta-LW (Hirose 2010). The main features of hash algorithms are presented in Table 7.3 (Biryukov and Perrin 2015).

7.5.1 PHOTON

PHOTON is a hash function suitable for extremely constrained devices. Its domain extension algorithm uses sponge construction with different internal state and output digest sizes: 100, 144, 196, 256, and 288 bits and 80, 128, 160, 224, and 256 bits, respectively. Although the internal memory used by sponge construction is low, it reduces the speed of the small message hashing. To design a flexible algorithm, researchers modify the squeezing process by using the number of extracted bits on each iteration that affects preimage resistance directly.

7.5.2 SPONGENT

SPONGENT is based on sponge construction that performs a PRESENT-type permutation, which provides robustness. It presents five different hash sizes, 88, 128, 160, 224, and 256, with five different rounds, 45, 70, 90, 120, and 140. Its main design issue is serialization for compact design. SPONGENT focuses on the smallest footprint. Therefore, the SPONGENT algorithm uses a simple round function with a 4- to 4-bit S-Box to minimize the footprint.

7.5.3 LESAMNTA-LW

Lesamnta-LW is a 256-bit hash function with a Merkle-Damgård domain extension algorithm. It provides at least 120-bit security. Unlike other hashing algorithms standardized by the ISO, this algorithm uses a 128-bit key for hashing. Moreover, the algorithm uses a different compression function called LW1, which occupies less memory, since the key size is smaller than the block size.

TABLE 7.3
Lightweight Hash Functions and Their Properties

Algorithms	Size	Round	Internal Size	Cell Library (µm)	Area (GE)	Throughput (kb/s, 100 khz)	Power (µW)	Collision Resistance	Preimage Resistance	Second Preimage	References
PHOTON	80	12	100	0.18	865/1168	2.82/15.15	1.59	64	40	40	Guo et al. (2011a)
	128	12	144	0.18	1122/1708	1.61/10.26	2.29	112	64	64	
	160	12	196	0.18	1396/2177	2.70/20	2.74	124	80	80	
	224	12	256	0.18	1736/2786	1.86/15.69	4.01	192	112	112	
	256	12	288	0.18	2177/4362	3.21/20.51	4.55	224	128	128	
SPONGENT	88	45	88	0.13	738/1127	0.81/17.78	1.57/2.31	80	40	40	Bogdanov et al. (2013)
	128	70	136	0.13	1060/1687	0.34/11.43	2.20/3.58	120	64	64	
	160	90	176	0.13	1329/2190	0.4/17.78	2.85/4.47	144	80	80	
	224	120	240	0.13	1728/2903	0.22/13.33	3.73/5.97	208	112	112	
	256	140	272	0.13	1950/3281	0.17/11.43	4.21/6.62	240	128	128	
Lesamnta-LW	256	NA	384	0.09	8240	125.55	NA	128	256	256	Hirose (2010)
QUARK	136	504	136	0.18	1379/2392	1.47/11.76	2.44/4.07	128	64	64	Aumasson et al. (2013)
	176	704	176	0.18	1702/2819	2.27/18.18	3.10/4.76	160	80	80	
	256	1024	256	0.18	2296/4640	3.13/50.0	4.35/8.39	224	112	112	
ARMADILLO2	80	NA	256	0.18	4030/2923	109/27	NA	40	80	80	Badel et al. (2010)
	128	NA	384	0.18	6025/4353	1000/250	NA	64	128	128	
	160	NA	480	0.18	7492/5406	100/250	NA	80	160	160	
	224	NA	576	0.18	8999/6554	100/25	NA	96	192	192	
	256	NA	768	0.18	11914/8653	100/25	NA	128	256	256	

Source: Biryukov, Alex, and Leo Perrin. 2015. *Lightweight Cryptography Lounge.* Accessed February 5, 2017. http://cryptolux.org/index.php/Lightweight_Cryptography.

7.6 CRYPTOGRAPHIC ATTACKS

7.6.1 CRYPTANALYSIS

Cryptanalysis is a study to discover the hidden or weak aspects of cryptographic systems. There are several types of attacks, which are applicable to different ciphers. Differential and linear cryptanalysis are the most significant ones, and are mostly applicable to private-key block ciphers; they will help us understand the structures of other type of attacks.

7.6.1.1 Differential Cryptanalysis

In differential cryptanalysis (Heys 2010), a cryptographic system is examined to understand whether there exists any input and output difference pairs that occur with higher probability than expected.

Let $P = (P_1 P_2 \ldots P_n)$ be an n-bit plaintext where P_i corresponds to the ith bit of P, and let $C = (C_1 C_2 \ldots C_n)$ be an n-bit ciphertext. $\Delta P := P' \oplus P''$ is called the input difference of two plaintexts, P' and P'', where $\oplus$ denotes the bit-wise exclusive-OR operation, that is $\Delta P_i = P_i' \oplus P_i''$ for every $i \in \{1,\ldots,n\}$. For an ideal block cipher, it is expected that an output difference ΔC (where $\Delta C_i := C_i' \oplus C_i''$ for two ciphertexts, C' and C'', and for every $i \in \{1,\ldots,n\}$) can appear with probability $\dfrac{1}{2^n}$ for a corresponding input difference, ΔP. The aim of differential cryptanalysis is to find a specific pair $(\Delta P, \Delta C)$ such that the output difference, ΔC, of the last round of a cipher appears with a greater (or lesser) probability than $\dfrac{1}{2^n}$ for the given input difference, ΔP. The pair $(\Delta P, \Delta C)$ is called a *differential*. Since it is assumed that the attacker could have different differentials to derive the key, differential cryptanalysis is an example of a chosen plaintext attack, in which the adversary is able to encrypt plaintexts of his choice (Katz and Lindell 2015).

7.6.1.2 Linear Cryptanalysis

In linear cryptanalysis (Heys 2010), a cryptographic system is examined to understand whether there exist linear expressions of plaintext, ciphertext, and round-key bits that occur with a higher probability than expected.

Let $P = (P_1 P_2 \ldots P_n)$ be an input plaintext and $C = (C_1 C_2 \ldots C_n)$ be the output of the second-to-last round of a cipher using SPN structure. For some integer u and v, the aim is to find a linear expression containing bits of plaintext and ciphertext, such as

$$P_{i_1} \oplus P_{i_2} \oplus \ldots \oplus P_{i_u} \oplus C_{j_1} \oplus C_{j_2} \oplus \ldots \oplus C_{j_v} = 0 \text{ (or 1)}$$

which is equal to 0 with a greater probability than $\dfrac{1}{2}$ (or with a lesser probability than $\dfrac{1}{2}$). For an ideal block cipher, the probability should be equal to $\dfrac{1}{2}$. The higher deviation from the probability of $\dfrac{1}{2}$ shows that the block cipher provides poor randomization. Since the attacker knows a set of plaintext and the corresponding ciphertext, linear cryptanalysis is an example of a known plaintext attack in which the adversary knows a set of plaintext and ciphertext pairs. However, the attacker is assumed to be unable to select a random plaintext and know the corresponding ciphertext.

7.6.1.3 Higher-Order Differential Attack

Higher-order differential cryptanalysis (Knudsen 1994; Wang and Yang 2009), which studies a set of differentials over a larger set of texts instead of a single differential, is the generalization of differential cryptanalysis of a single differential over two texts, P and C.

To explain the basics of the attack, assume $f : \mathbb{F}_{2^n} \to \mathbb{F}_{2^n}$ is a function of a block cipher, which maps n-bit plaintext to n-bit ciphertext. The derivative of f at the point $a \in \mathbb{F}_{2^n}$ is defined as

$$\Delta_a f(x) := f(x+a) - f(x) = f(x+a) \oplus f(x)$$

which corresponds to a differential pair, $(a, f(x+a) \oplus f(x))$, for the conventional differential cryptanalysis.

The ith derivative of f at the point $a_1, a_2, \ldots, a_i$ is defined as

$$\Delta_{a_1, a_2, \ldots, a_i}^{(i)} \; f(x) = \Delta_{a_i}(\Delta_{a_1, a_2, \ldots, a_{i-1}}^{(i-1)} \; f(x))$$

It is important to note that using a higher-order differential attack over any five-round Feistel cipher is more efficient than and superior to the attacks that use standard differential cryptanalysis.

7.6.1.4 Impossible Differential Cryptanalysis

Impossible differential cryptanalysis (Li et al. 2011) is a variation of differential cryptanalysis, which is applicable to block ciphers. In differential cryptanalysis, the aim is to find a differential $(\Delta P, \Delta C)$ with a greater (or lesser) probability than $\frac{1}{2^n}$, assuming the plaintext and ciphertext are n-bits long. However, in impossible differential cryptanalysis, the existence of an impossible differential, which has a probability close to zero, becomes important to exploit the vulnerability of a cryptographic algorithm.

7.6.1.5 Boomerang Attack

According to Biham and Shamir (1993), differential cryptanalysis is one of the most powerful attacks known. The higher the deviation from a probability of $\frac{1}{2}$, the less secure the algorithm becomes. Therefore, algorithm designers obtain an upper-bound probability, p, to resist differential cryptanalysis since an attacker should need $\frac{1}{p}$ differentials and considers that the cipher is safe against differential cryptanalysis.

However, impossible differential cryptanalysis shows that having an upper bound, p, for the probability distribution of differentials is not enough since the existence of an impossible differential might allow an attacker to break into the system.

In the boomerang attack (Wagner 1999; Savard 1999), the aim is to exploit any differential-style attack even if the probability distribution does not contain any differential with probability that is too low or high.

The generic attack attempts to generate a quartet structure at an intermediate value halfway through the cipher, as follows:

- The attack assumes four plaintext/ciphertext pairs, $(P, C), (P', C'), (Q, D), (Q', D')$. Let $E(P)$ be the encryption procedure, where P is the plaintext, which can be divided into two procedures, E_o and E_1, such that $E(P) = E_1(E_0(P))$. Note that E_0 represents the first half, and E_1 represents the last half of the cipher.
- Let (Δ, Δ^*) be the differential for E_0 and (∇, ∇^*) be the differential for E_1^{-1}. Choose a plaintext, P, and calculate E_0 and $P' = P \oplus \Delta$.
- Calculate the encryptions of P and P', obtain $C = E(P)$ and $C' = E(P')$, and calculate $D = C \oplus \nabla$ and $D' = C' \oplus \nabla$.
- Compare the results, and calculate the decryptions of D and D' to obtain $Q = E^{-1}(D)$ and $Q' = E^{-1}(D')$.
- Since (Δ, Δ^*) is the differential for encryption, $Q \oplus Q' = \Delta$ should be checked.

7.6.1.6   Meet-in-the-Middle Attack

A meet-in-the-middle (MitM) attack (Biryukov 2011; Kowalczyk 2013; Moore 2010) is a known plaintext attack against ciphers, which rely on multiple encryption procedures in sequence.

After the standardized DES was broken due to a small key size (56-bit keys), designers attempted to increase the security concept by encrypting the data several times using multiple keys. It can be considered that a brute-force attack would require 2^{kxn} operations, where k is the key size and n is the number of encryption procedures.

To show why multiple encryptions might fail, assume the cipher has the following structure:

$$C = E_{l_2} \left(E_{k_1}(P) \right)$$

$$P = D_{k_1} \left(D_{l_2}(C) \right)$$

where E is the encryption procedure and D is the decryption procedure such that $D\big(E(P)\big) = P$ for plaintext P and ciphertext C. The naïve brute-force attack would require $2^{k_1} \times 2^{l_2}$ operations.

In the MitM attack, the attacker knows a set of plaintext/ciphertext pairs and tries to decrease the required number of operations by applying the following steps:

- The attacker forms a set $S = \left\{ E_{k_i}(P) \right\}$ by iterating each possible k_i value.
- For every possible l_j value, the attacker also computes $P' = D_{l_j}(C)$ and checks whether $P' \in S$. A potentially correct key pair (k_i, l_j) is called a candidate key.
- If the number of the candidate's key pair is more than one, the attacker uses additional plaintext/ciphertext pairs in the final brute-force testing phase to find the correct one.

Applying the MitM, the required number of operations becomes $2^{k_1} + 2^{l_2}$ instead of $2^{k_1} \times 2^{l_2}$ despite having $O(2^{k_1})$ or $O(2^{l_2})$. The reason why double DES is also broken is the brute force would require only 2^{57} operations, and triple DES can be favorable in terms of double DES.

7.6.1.7   Related-Key Attack

In the related-key attack (Biryukov and Khovratovich 2009), an attacker tries to exploit similarity between keys and propagates the relation toward the encryption procedure to discover new relations to easily attack or even brute force attack when the key size is small and repetitive. Wired Equivalent Privacy (WEP) is a standard security algorithm in wireless networks and is a famous example of why the related-key concept should be considered carefully because it is based on using the same key per packet in the stream cipher RC4, in which the same key must not be used twice (Maitra and Paul 2008). To prevent this attack, there should not be a simple relation between round keys, and to produce round keys, the designer should prefer cryptographic hash functions or secure key generation methods.

7.7   OPEN RESEARCH ISSUES

In this section, open research issues (Garcia-Morchon et al. 2013; Chan et al. 2004), restrictions, limitations, and security challenges in IoT and lightweight cryptography are mentioned to meet the requirements of the different security aspects through the lifecycle of a "thing" connecting to the Internet.

Physical limitations are the intrinsic issue for an IoT device, which might cause different problems that may block the operations and secure communication between different kinds of devices. Having a restricted memory source causes problems in heterogeneous communication between constrained and powerful devices, while limited CPU and low energy is not always adequate for using public-key cryptography for encrypted, secure communication. Finally, an IoT device should provide resistance against DDoS attacks, which can be a result of physical limitations or undiscovered security vulnerabilities.

Instead of replacing powerful devices, the aim of restricted devices is to cooperate with powerful devices in a heterogeneous environment to solve different communication problems. Therefore, the fragmentation of larger packets of standard communication protocols is crucial for IoT but seriously downgrades the bandwidth and memory usage performance of a thing due to the overhead of extra packages. An excessive number of messages also causes a thing to operate improperly, as if a DoS attack happened.

To have secure communication, using cryptographic algorithms is important in encrypting messages and hence providing confidentiality. Instead of symmetric cryptography, using public-key cryptography may provide a better level of security by using intensive mathematical functions. However, resource-expensive, standard public cryptography algorithms might slow down or completely block the operation of constrained IoT devices. Therefore, new public-key cryptography algorithms for IoT networks should provide as much security as possible compared to standard public-key cryptography algorithms without seriously limiting the operation speed of an IoT device.

According to *USA Today* (Blumenthal and Weise 2016), important companies such as Netflix, Twitter, and Spotify became unreachable on October 21, 2016, as a result of a large-scale DDoS attack caused by constrained home devices. Instead of attacking over a single source, DDoS attacks employ considerably harmless but Internet-connected devices such as DVR cameras and even thermostats that are not configured properly in terms of security considerations and can be easily hacked. Other types of DDoS attacks might target an IoT network consisting of constrained devices to stop operation, and these kinds of attacks are usually hard to notice before the service becomes unavailable. Therefore, a security mechanism should discerningly investigate the source of the packets and find a countermeasure against them.

7.8 CONCLUSION

IoT and machine-type communication will be an integral part of the envisioned 5G mobile networks. The integration of IoT and 5G mobile networks will inevitably cause two problems: energy consumption and security vulnerabilities. On the one hand, battery-driven operation requires energy-efficient operation. On the other hand, energy consumption of IoT will be non-negligible because of large-scale and massive deployments. Although, when compared to the other industries, the energy consumption of machine-type communications might initially seem insignificant, when massive deployment of constrained devices becomes a reality, the total aggregated amount of energy consumed will be huge. Therefore, energy efficiency in all domains, including security of machine-type communications and sensor networks, has to be studied and engineered. In this chapter, we attempted to present lightweight cryptography since conventional algorithms are not appropriate for constrained devices. Lightweight cryptography will play a critical role in the energy efficiency and data security of machine-to-machine and human-to-machine communications in 5G. Therefore, we explained symmetric ciphers and hashing algorithms. Among them, block ciphers may be preferred due to their efficiency. Moreover, they have many alternatives for various constraint environments. Besides stream ciphers and hashing algorithms, public-key algorithms could be used to provide maximum security. However, their power requirements and excess processing load are not appropriate for IoT devices. As a result, lightweight cryptography and its implementations provide solutions for the vulnerabilities of constrained devices and for the right of privacy.

REFERENCES

Ågren, Martin, Martin Hell, Thomas Johansson, and Willi Meier. 2011. A new version of Grain-128 with authentication. In *Symmetric Key Encryption Workshop 2011.*

Aumasson, Jean-Philippe, Luca Henzen, Willi Meier, and María Naya-Plasencia. 2013. Quark: A lightweight hash. *Journal of Cryptology*, 26, 313–339.

Babbage, Steve, and Matthew Dodd. 2006. *The stream cipher MICKEY 2.0.* Accessed August 31, 2016. Available at: www.ecrypt.eu.org/stream/p3ciphers/mickey/mickey_p3.pdf.

Badel, Stéphane et al. 2010. ARMADILLO: A multi-purpose cryptographic primitive dedicated to hardware. In *Proceedings of the 12th International Workshop, Santa Barbara, USA, August 17–20, 2010*, Edited by Stefan Mangard and François-Xavier Standaert, pp. 398–412. Springer.

Bertoni, Guido, Joan Daemen, Michaël Peeters, and Gilles Van Assche. 2007. Sponge functions. In *ECRYPT Hash Workshop*. Citeseer.

Biham, Eli, and Adi Shamir. 1993. *Differential Cryptanalysis of the Data Encryption Standard.* Springer-Verlag.

Biryukov, Alex. 2011. Meet-in-the-middle attack. In *Encyclopedia of Cryptography and Security*, pp. 772–773. Springer.

Biryukov, Alex, and Dmitry Khovratovich. 2009. *Related-key cryptanalysis of the full AES-192 and AES-256.* Springer.

Biryukov, Alex, and Leo Perrin. 2015. *Lightweight cryptography lounge.* Accessed February 5, 2017. Available at: http://cryptolux.org/index.php/Lightweight_Cryptography.

BLOC Project. 2013. *Implementations of lightweight block ciphers on a WSN430 sensor.* Accessed February 5, 2017. Available at: http://bloc.project.citi-lab.fr/library.html.

Blumenthal, Eli, and Elizabeth Weise. 2016. *USA Today.* Accessed February 5, 2017. Available at: www.usa-today.com/story/tech/2016/10/21/cyber-attack-takes-down-east-coast-netflix-spotify-twitter/92507806/.

Bogdanov, Andrey et al. 2007. PRESENT: An ultra-lightweight block cipher. In *Cryptographic Hardware and Embedded Systems, CHES 2007*, Edited by Pascal Paillier and Ingrid Verbauwhede, pp. 450–466. Berlin, Heidelberg: Springer.

Bogdanov, Andrey, Miroslav Knežević, Gregor Leander, Deniz Toz, Kerem Varıcı, and Ingrid Verbauwhede. 2013. SPONGENT: The design space of lightweight cryptographic hashing. *IEEE Transactions on Computers*, 62(10), pp. 2041–2053.

Cannière, Christophe De. 2006. Trivium: A stream cipher construction inspired by block cipher design principles. In *Information Security*, Edited by Sokratis K. Katsikas, Javier López, *Michael Backes, Stefanos Gritzalis, and Bart Preneel*, pp. 171–186. Berlin, Heidelberg: Springer.

Cazorla, Mickaël, Kevin Marquet, and Marine Minier. 2013. Survey and benchmark of lightweight block ciphers for wireless sensor networks. In *Proceedings of the 10th International Conference on Security and Cryptography*, pp. 543–548. New York, NY: SciTePress.

Chan, Haowen, Adrian Perrig, and Dawn Song. 2004. Key distribution techniques for sensor networks. In *Wireless Sensor Networks*, Raghavendra, C.S., Sivalingam, K.M., Znati, T. (eds). Boston, MA: Springer.

Charles, Merkle Ralph. 1979. *Secrecy, Authentication, and Public Key Systems.* Stanford, CA: Citeseer.

Cheng, Huiju, Howard M. Heys, and Cheng Wang. 2008. Puffin: A novel compact block cipher targeted to embedded digital systems. In *DSD '08 11th EUROMICRO Conference on Digital System Design Architectures, Methods and Tools, 2008. IEEE*, pp. Parma, Italy, 383–390.

Daemen, Joan, and Craig Clapp. 1998. Fast hashing and stream encryption with PANAMA. In *International Workshop on Fast Software Encryption*, pp. 60–74. London, UK: Springer.

De Canniere, Christophe, Orr Dunkelman, and Miroslav Knežević. 2009. KATAN and KTANTAN—A family of small and efficient hardware-oriented block ciphers. In *International Workshop on Cryptographic Hardware and Embedded Systems*, pp. 272–288. Springer Berlin Heidelberg.

eSTREAM. 2008. *eSTREAM: The ECRYPT Stream Cipher Project.* Accessed May 12, 2016. Available at: www.ecrypt.eu.org/stream/.

Fontaine, Caroline. 2005. Nonlinear feedback shift register. In *Encyclopedia of Cryptography and Security*, Edited by Henk C. A. van Tilborg, pp. 415–416. Boston, MA: Springer.

Garcia-Morchon, Oscar et al. 2013. *Security considerations in the IP-based internet of things. IETF.* Accessed February 5, 2017. Available at: https://tools.ietf.org/html/draft-garcia-core-security-06.

Gong, Zheng, Svetla Nikova, and Yee Wei Law. 2011. KLEIN: A new family of lightweight block ciphers. In *International Workshop on Radio Frequency Identification: Security and Privacy Issues*, pp. 1–18. Springer Berlin Heidelberg.

Guo, Jian, Thomas Peyrin, and Axel Poschmann. 2011a. The PHOTON family of lightweight hash functions. In *Advances in Cryptology, CRYPTO 2011*, Edited by Phillip Rogaway, pp. 222–239. Springer Berlin Heidelberg.

Guo, Jian, Thomas Peyrin, Axel Poschmann, and Matt Robshaw. 2011b. The LED block cipher. In *International Workshop on Cryptographic Hardware and Embedded Systems*, pp. 326–341. Springer Berlin Heidelberg.

Hash Functions, ISO/IEC 29192-5:2016, Information technology—Security techniques—Lightweight cryptography—Part 5: Hash-Functions, ISO/IEC. 2016. Accessed 8 30, 2016. Available at: www.iso.org/iso/home/store/catalogue_tc/catalogue_detail.htm?csnumber=67173.

Hell, Martin, Thomas Johansson, and Willi Meier. 2007. Grain: A stream cipher for constrained environments. *International Journal of Wireless and Mobile Computing*, 2, 86–93.

Heys, Howard M. 2010. A tutorial on linear and differential cryptanalysis. *Cryptologia*, 26, 189–221.

Hirose, Shoichi et al. 2010. A lightweight 256-bit hash function for hardware and low-end devices: Lesamnta-LW. In *Information Security and Cryptology, ICISC 2010*, Edited by Kyung-Hyune Rhee and DaeHun Nyang, pp. 151–168. Springer Berlin Heidelberg.

Hoang, Viet Tung, and Phillip Rogaway. 2010. On generalized Feistel networks. In *Advances in Cryptology, CRYPTO 2010*, Edited by Phillip Rogaway and Viet Tung Hoang. Springer.

Hong, Deukjo et al. 2006. HIGHT: A new block cipher suitable for low-resource device. In *International Workshop on Cryptographic Hardware and Embedded Systems*, pp. 46–59. Springer Berlin Heidelberg.

Hotel News Resource. 2015. *Internet of Things (IoT): Number of connected devices worldwide from 2012 to 2020 (in billions). Statista—The Statistics Portal.* Accessed November 29, 2016. Available at: www.statista.com/statistics/471264/iot-number-of-connected-devices-worldwide/.

IETF. 2014. Terminology for constrained-node networks. IETF. Accessed February 5, 2017. Available at: https://tools.ietf.org/html/rfc7228.

IETF/COAP. 2016. *The constrained application protocol (CoAP).* Available at: https://tools.ietf.org/html/rfc7252. Accessed November 29, 2016.

IETF/6LoWPANs. 2007. *IPv6 over low-power wireless personal area networks.* Available at: https://datatracker.ietf.org/doc/html/rfc4919. Accessed November 29, 2016.

IETF/XMPP. 2011. *Extensible messaging and presence protocol (XMPP).* Available at: https://tools.ietf.org/html/rfc6120. Accessed November 29, 2016.

ISO/IEC. 2012. *Information technology—Security techniques—Lightweight cryptography—Part 1: General.* Accessed August 30, 2016. Available at: www.iso.org/iso/iso_catalogue/catalogue_tc/catalogue_detail.htm?csnumber=56425.

ISO/IEC 29192-2:2012, Information technology—Security techniques—Lightweight cryptography—Part 2: Block Ciphers, ISO/IEC. 2012. Accessed August 30, 2016. Available at: www.iso.org/iso/home/store/catalogue_tc/catalogue_detail.htm?csnumber=56552.

ISO/IEC 29192-3:2012 Information technology—Security techniques—Lightweight cryptography—Part 3: Stream Ciphers. 2012. Accessed August 30, 2016. Available at: www.iso.org/iso/home/store/catalogue_tc/catalogue_detail.htm?csnumber=56426.

ISO/IEC 29192-4:2013, Information technology—Security techniques—Lightweight cryptography—Part 4: Mechanisms using asymmetric techniques, ISO/IEC. 2013. Accessed August 30, 2016. Available at: www.iso.org/iso/home/store/catalogue_tc/catalogue_detail.htm?csnumber=56427.

ISO/MQTT. 2016. *ISO—Message Queuing Telemetry Transport (MQTT).* Accessed November 29, 2016. Available at: www.iso.org/standard/69466.html.

Izadi, Maryam, Babak Sadeghiyan, Seyed Saeed Sadeghian, and Hossein Arabnezhad Khanooki. 2009. MIBS: A new lightweight block cipher. In *International Conference on Cryptology and Network Security*, pp. 334–348. Springer Berlin Heidelberg.

Katz, Jonathan, and Yehuda Lindell. 2015. *Introduction to Modern Cryptography.* CRC Press.

Knudsen, Lars. 1994. *Truncated and Higher Order Differentials.* Berlin, Heidelberg: Springer.

Knudsen, Lars, Gregor Leander, Axel Poschmann, and Matthew J.B. Robshaw. 2010. PRINTcipher: A block cipher for IC-printing. In *International Workshop on Cryptographic Hardware and Embedded Systems*, pp. 16–32. Berlin, Heidelberg: Springer.

Kowalczyk, Chris. 2013. *Meet-in-the-middle attack.* Accessed February 5, 2017. Available at: www.crypto-it.net/eng/attacks/meet-in-the-middle.html.

Leander, Gregor, Christof Paar, Axel Poschmann, and Kai Schramm. 2007. New lightweight DES variants. In *International Workshop on Fast Software Encryption*, pp. 196–210. Berlin, Heidelberg: Springer.

Li, Ruilin, Bing Sun, and Chao Li. 2011. Impossible differential cryptanalysis of SPN ciphers. *IET Information Security* 5(2), 111–120.

Lim, Chae Hoon, and Tymur Korkishko. 2005. mCrypton—A lightweight block cipher for security of low-cost RFID tags and sensors. In *International Workshop on Information Security Applications*, pp. 243–258. Springer Berlin Heidelberg.

Lu, Jiqiang. 2009. Related-key rectangle attack on 36 rounds of the XTEA block cipher. *International Journal of Information Security* 8(1), 1–11.

Maitra, Subhamoy, and Goutam Paul. 2008. Analysis of RC4 and proposal of additional layers for better security margin. In *International Conference on Cryptology in India*, pp. 27–39. Springer.

Moore, Stephane. 2010. *Meet-in-the-middle attacks*. Accessed February 5, 2017. Available at: http://stephanemoore.com/pdf/meetinthemiddle.pdf.

Murase, Makoto. 1992. Linear feedback shift register. *United States Patent 5,090,035*. February 18.

NIST. 2016. *Report on lightweight cryptography*. Accessed August 30, 2016. Available at: http://csrc.nist.gov/publications/drafts/nistir-8114/nistir_8114_draft.pdf.

OMG/DDS. 2015. *Data distribution service*. Available at: http://www.omg.org/spec/DDS/. Accessed November 29, 2016.

Patrick, Conor, and Patrick Schaumont. 2016. The role of energy in the lightweight cryptographic profile. In *NIST Lightweight Cryptography Workshop, 2016*.

Poschmann, Axel York. 2009. *Lightweight Cryptography: Cryptographic Engineering for a Pervasive World*. Citeseer.

Ray, Beaulieu et al. 2013. *The SIMON and SPECK Families of Lightweight Block Ciphers*. Cryptology ePrint Archive.

Savard, John J. G. 1999. *The Boomerang Attack*. Accessed February 5, 2017. Available at: www.quadibloc.com/crypto/co4512.htm.

Shibutani, Kyoji, Takanori Isobe, Harunaga Hiwatari, Atsushi Mitsuda, Toru Akishita, and Taizo Shirai. 2011. Piccolo: An ultra-lightweight blockcipher. In *International Workshop on Cryptographic Hardware and Embedded Systems*, pp. 342–357. Springer Berlin Heidelberg.

Shirai, Taizo, Shibutani Kyoji, Toru Akishita, Shiho Moriai, and Tetsu Iwata. 2007. The 128-bit blockcipher CLEFIA (Extended Abstract). In *Fast Software Encryption*, Edited by Alex Biryukov, pp. 181–195. Springer Berlin Heidelberg.

Shirai, Taizo, and Bart Preneel. 2004. On Feistel ciphers using optimal diffusion mappings across multiple rounds. In *Advances in Cryptology, ASIACRYPT, 2004*, pp. 1–15. Jeju Island, Korea: Springer.

Standaert, François-Xavier, Gilles Piret, Neil Gershenfeld, and Jean-Jacques Quisquater. 2006. SEA: A scalable encryption algorithm for small embedded applications. In *International Conference on Smart Card Research and Advanced Applications*, pp. 222–236. Springer Berlin Heidelberg.

Suzaki, Tomoyasu, Kazuhiko Minematsu, Sumio Morioka, and Eita Kobayashi. 2012. A lightweight block cipher for multiple platforms. In *International Conference on Selected Areas in Cryptography*, pp. 339–354. Springer Berlin Heidelberg.

U.S. Energy Information Administration. 2016. U.S. Energy Information Administration Energy Consumption. *U.S. Energy Information Administration*. Accessed August 30, 2016. Available at: www.eia.gov/totalenergy/data/monthly/pdf/sec2_2.pdf.

Wagner, David. 1999. The boomerang attack. In *International Workshop on Fast Software Encryption*, pp. 156–170. Berlin, Heidelberg: Springer.

Wang, Yan, and Mohan Yang. 2009. *Higher Order Differential Cryptanalysis on the SHA-3 Cryptographic Hash Algorithm Competition Candidates*.

Watanabe, Dai, Kota Ideguchi, Jun Kitahara, Kenichiro Muto, Hiroki Furuichi, and Toshinobu Kaneko. 2008. Enocoro-80: A hardware oriented stream cipher. In *IEEE 2008 Third International Conference on Availability, Reliability and Security, ARES 08*, Barcelona, Spain: pp. 1294–1300.

Wu, Wenling, and Lei Zhang. 2011. LBlock: A lightweight block cipher. In *International Conference on Applied Cryptography and Network Security*, pp. 327–344. Springer Berlin Heidelberg.

Yap, Huihui, Khoongming Khoo, Axel Poschmann, and Matt Henricksen. 2011. EPCBC-a block cipher suitable for electronic product code encryption. In *International Conference on Cryptology and Network Security*, pp. 76–97. Berlin, Heidelberg: Springer.

Zhang, Wentao, Zhenzhen Bao, Dongdai Lin, Vincent Rijmen, Bohan Yang, and Ingrid Verbauwhede. 2015. RECTANGLE: A bit-slice lightweight block cipher suitable for multiple platforms. *Science China Information Sciences* 58(12), 1–15.

Zhou, YongBin, and DengGuo Feng. 2005. *Side-Channel Attacks: Ten Years after Its Publication and the Impacts on Cryptographic Module Security Testing*. NIST Computer Security Resource Center.

8 Index Modulation

A Promising Technique for 5G and Beyond Wireless Networks

Ertuğrul Başar

CONTENTS

8.1 INTRODUCTION

After more than 20 years of research and development, the achievable data rates of today's cellular wireless communication systems are several thousands of times faster compared to earlier 2G wireless systems. However, unprecedented levels of spectral and energy efficiency are expected from 5G wireless networks to achieve ubiquitous communications between anybody and anything at anytime. [1]. To reach the challenging objectives of 5G wireless networks, the researchers have envisioned novel physical-layer (PHY) concepts such as massive multiple-input multiple-output (MIMO) systems and non-orthogonal multi-carrier communications schemes such as generalized frequency-division multiplexing (GFDM) and filter bank multi-carrier (FBMC) modulation. However, the wireless community is still working relentlessly to come up with new and more effective PHY solutions for 5G wireless networks. There has been a growing interest in index modulation (IM) techniques during the past few years. IM is a novel digital modulation scheme that is shown to achieve high spectral and energy efficiency by considering the indices of the building blocks of the considered communication systems to transmit information bits in addition to the ordinary modulation schemes. Two interesting as well as promising forms of the IM concept are *spatial modulation (SM)* and *orthogonal frequency-division multiplexing with IM (OFDM-IM)* schemes, where the corresponding index-modulated building blocks are the transmit antennas of a MIMO system and the subcarriers of an OFDM system, respectively.

SM techniques have attracted tremendous attention during the past few years after the inspiring works of Mesleh et al. [2] and Jeganathan et al. [3], which introduced new directions for MIMO communications. Despite having strong and well-established competitors, such as vertical

Bell Labs layered space-time (V-BLAST) and space-time coding (STC) systems [4], SM schemes have been regarded as possible candidates for next-generation small-/large-scale and single-/multi-user MIMO systems. Meanwhile, several researchers have explored the potential of the IM concept for the subcarriers of OFDM systems in the past three years after its widespread introduction [5], and it has been shown that the OFDM-IM scheme offers attractive advantages over classical OFDM, which is an integral part of many current wireless communications standards and is also being considered as a strong waveform candidate for 5G wireless networks.

In this chapter, we present the basic principles of these two promising IM schemes, SM and OFDM-IM, and review some of the recent, interesting as well as promising achievements in IM technologies. Furthermore, we discuss the possible implementation scenarios of IM techniques for next-generation wireless networks and outline possible future research directions. Particularly, we investigate the recently proposed generalized, enhanced, and quadrature SM schemes and the application of SM techniques for massive multi-user MIMO (MU-MIMO) and relaying networks. Additionally, we review the recent advances in OFDM-IM technologies, such as generalized, MIMO, and dual-mode OFDM-IM schemes, and provide possible implementation scenarios.

The remainder of this chapter is organized as follows. In Section 8.2, we describe the SM concept and discuss its advantages and disadvantages for next-generation wireless networks. In Section 8.3, we review the most recent as well as interesting advances in SM technologies. In Section 8.4, we present the OFDM-IM scheme and discuss its advantages over classical OFDM. Finally, in Section 8.5, we review the most recent developments in OFDM-IM technologies. Section 8.6 concludes the chapter.

8.2 INDEX MODULATION FOR TRANSMIT ANTENNAS: SPATIAL MODULATION

SM is a novel way of transmitting information by means of the indices of the transmit antennas of a MIMO system in addition to the conventional M-ary signal constellations. In contrast to conventional MIMO schemes that rely on either spatial multiplexing to boost the data rate or spatial diversity to improve the error performance, the multiple transmit antennas of a MIMO system are used for a different purpose in an SM scheme. More specifically, there are two information-carrying units in SM: indices of transmit antennas and M-ary constellation symbols. For each signaling interval, a total of

$$\log_2(n_T) + \log_2(M) \tag{8.1}$$

bits enter the transmitter of an SM system, as seen in Figure 8.1, where n_T and n_R denote the number of transmit and receive antennas, respectively, and M is the size of the considered signal constellation, such as M-ary phase-shift keying (M-PSK) or M-ary quadrature amplitude modulation (M-QAM). The $\log_2(M)$ bits of the incoming bit sequence are used to modulate the phase and/or amplitude of a carrier signal traditionally, while the remaining $\log_2(n_T)$ bits of the incoming bit sequence are reserved for the selection of the index (I) of the active transmit antenna that performs the transmission of the corresponding modulated signal (s). Consequently, the transmission vector of SM, with dimensions $n_T \times 1$, becomes

$$\begin{bmatrix} 0 & \cdots & 0 & s & 0 & \cdots & 0 \end{bmatrix}^{\mathrm{T}} \tag{8.2}$$

whose I^{th} entry is nonzero only, where $[\cdot]^{\mathrm{T}}$ stands for the transposition of a vector. The sparse structure of the SM transmission vector given in Equation 8.2 not only reduces the detection complexity of the maximum likelihood (ML) detector but also allows the implementation of compressed sensing–based low/near-optimal detection algorithms for SM systems.

The receiver of the SM scheme has two major tasks to accomplish: detection of the active transmit antenna for the demodulation of the index selecting bits and detection of the data symbol transmitted over the activated transmit antenna for the demodulation of the bits mapped to the M-ary signal constellation.

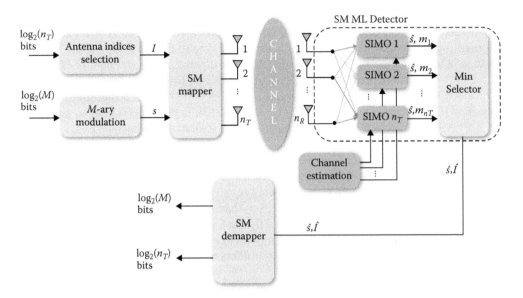

FIGURE 8.1 Block diagram of the SM transceiver for an $n_T \times n_R$ MIMO system. s (or $\hat{s}$) and I (or $\hat{I}$) $\in \{1, 2, \ldots, n_T\}$ denote the selected (or estimated) M-ary constellation symbol and transmit antenna index, respectively, and $m_n, n = 1, 2, \ldots, n_T$ is the minimum decision metric provided by the nth SIMO ML detector.

Unfortunately, the optimum ML detector of SM has to make a joint search over all transmit antennas and constellation symbols to perform these two tasks [6]. In other words, the ML detector of the SM scheme independently implements a classical single-input, multiple-output (SIMO) ML detector for all transmit antennas to find the activated transmit antenna by comparing the corresponding minimum-decision metrics, $(m_1, m_2, \ldots, m_{n_T})$. On the other hand, the primitive suboptimal detector of SM deals with the aforementioned two tasks one by one; that is, first, it determines the activated transmit antenna, and second, it finds the data symbol transmitted over this antenna [7,8]. Therefore, the size of the search space becomes $n_T \times M$ and $n_T + M$ for the ML and suboptimal detectors, respectively. Although the suboptimal detector can obtain a significant complexity reduction, its error performance is considerably worse than the ML detector, which makes its implementation problematic for critical applications.

SM systems provide attractive advantages over classical MIMO systems, which are extensively covered in the literature [9–11]. The main advantages of SM over classical MIMO systems can be summarized as follows:

- *Simple transceiver design*: Since only a single transmit antenna is activated, a single radio frequency (RF) chain can handle the transmission for the SM scheme. Meanwhile, inter-antenna synchronization (IAS) and inter-channel interference (ICI) are completely eliminated, and the decoding complexity of the receiver, in terms of the total number of real multiplications performed, grows linearly with the constellation size and the number of transmit antennas.
- *Operation with flexible MIMO systems*: SM does not restrict the number of receive antennas as the V-BLAST scheme does, which requires $n_R > n_T$ to operate with minimum mean square error (MMSE) and zero-forcing (ZF) detectors.
- *High spectral efficiency*: Due to the use of antenna indices as an additional source of information, the spectral efficiency of SM is higher than that of single-input, single-output (SISO) and orthogonal STC systems.
- *High energy efficiency*: The power consumed by the SM transmitter is independent from the number of transmit antennas while information can be still transferred via these antennas. Therefore, SM appears to be a green and energy-efficient MIMO technology.

As an example, the SM scheme achieves $200(n_T - 1) / (2n_T + 1)\%$ reduction in ML detection complexity (in terms of the total number of real multiplications) compared to V-BLAST for an $n_T \times n_R$ MIMO system operating at a fixed spectral efficiency. This significant reduction is achieved by the activation of a single transmit antenna in SM. Additionally, the sparse structure of SM transmission vectors allows the implementation of several near-/sub-optimal, low-complexity detection methods for SM systems such as matched filter–based detection [12] and compressed sensing–based detection [13]. In terms of the energy efficiency in Mbits/J, improvements of up to 46% compared to V-BLAST are reported for different types of base stations (BSs) equipped with multiple antennas [14].

While the SM scheme has the aforementioned appealing advantages, it also has some disadvantages. The spectral efficiency of SM increases logarithmically with n_T, while the spectral efficiency of V-BLAST increases linearly with n_T. Therefore, a higher number of transmit antennas are required for SM to reach the same spectral efficiency as that of V-BLAST. The channel coefficients of different transmit antennas must be sufficiently different for an SM scheme to operate effectively. In other words, SM requires rich scattering environments to ensure better error performance. Since SM transfers the information using only the spatial domain, plain SM cannot provide transmit diversity as STC systems do.

Considering the advantages and disadvantages of SM systems mentioned previously, we may conclude that the SM scheme provides an interesting trade-off among encoding/decoding complexity, spectral efficiency, and error performance. As a result, SM technologies are regarded as possible candidates for spectrum- and energy-efficient next-generation wireless communications systems [1].

8.3   RECENT ADVANCES IN SM

The first studies on the SM concept date back to the beginning of this century, in which the researchers used different terminologies. However, after the inspiring works of Mesleh et al. [2] and Jeganathan et al. [3], numerous papers on SM have been published, in which the experts focus on generalized spectrum- and energy-efficient SM systems [15–17], low-complexity detector types [8,12,13,18–20], block/trellis coded SM systems with transmit/time diversity [21–23], link adaptation methods such as adaptive modulation [24], transmit antenna selection [25] and precoding [26], performance analysis for different types of fading channels [27,28] and channel estimation errors [29], information theoretical analyses [30], differential SM schemes with non-coherent detection [31], cooperative SM systems [32–37], and so on. Interested readers are referred to previous survey papers on SM [10,11,38] for a comprehensive overview of these studies.

In this section, we review some of the recent as well as promising advances in SM technologies, such as generalized, enhanced, and quadrature SM systems; massive MU-MIMO systems with SM; cooperative SM schemes; and spectrum sharing–based SM schemes, which have the potential to provide efficient solutions toward 5G and beyond wireless networks.

8.3.1   GENERALIZED, ENHANCED, AND QUADRATURE SM SCHEMES

As mentioned previously, the major disadvantage of SM is its lower spectral efficiency compared to the classical V-BLAST scheme for the same number of transmit antennas. Although a considerable number of information bits can still be transmitted by the indices of active transmit antennas, for higher-order modulations and MIMO systems, SM suffers a significant loss in spectral efficiency with respect to V-BLAST due to its inactive transmit antennas.

One of the first attempts to not only increase the spectral efficiency of SM but also ease the constraint on the number of transmit antennas, which has to be an integer power of two for classical SM, has been made by the generalized SM (GSM) scheme [15], where the number of active transmit antennas is no longer fixed to unity. In the GSM scheme, the same data symbol is transmitted over the selected multiple active transmit antennas. Let us denote the number of active transmit antennas by n_A where $n_A < n_T$. Then, for the GSM scheme, $\left\lfloor \log_2 \binom{n_T}{n_A} \right\rfloor$ information bits can be conveyed in

each signaling interval in addition to the $\log_2(M)$ bits transmitted by the M-ary data symbols, where $\lfloor . \rfloor$ is the floor operation, and $\binom{\cdot}{\cdot}$ stands for the binomial coefficient. Since $\log_2(n_T) \leq \left\lfloor \log_2 \binom{n_T}{n_A} \right\rfloor$

for $n_T = 2^n (n = 1, 2, \ldots)$, the spatial domain can be used in a more effective way by the GSM scheme. As an example, for $n_T = 8$, only three bits can be transmitted by the antenna indices in SM, while this can be doubled by GSM for $n_A = 4$. Later, the concept of GSM was extended to multiple-active spatial modulation (MA-SM, also called multi-stream SM) by transmitting different data symbols from the selected active transmit antennas to further boost the spectral efficiency [17]. Therefore, the spectral efficiency of the MA-SM scheme can be calculated as

$$\left\lfloor \log_2 \binom{n_T}{n_A} \right\rfloor + n_A \log_2(M) \tag{8.3}$$

bits per channel use (bpcu), which is considerably higher than that of SM. It should be noted that MA-SM provides an intermediate solution between two extreme schemes: SM and V-BLAST, which are the special cases of MA-SM for $n_A = 1$ and $n_A = n_T$, respectively.

As a strong alternative to SM, GSM techniques have attracted considerable attention in the past few years. It has been shown that compared to V-BLAST, GSM can achieve better throughput and/or error performance. Furthermore, percentage savings in the required number of transmit RF chains have been reported [39]. A closed-form expression has been derived for the capacity of GSM, and the error performance of GSM has been analyzed for correlated and uncorrelated Rayleigh and Rician fading channels [40]. Ordered-block MMSE [41], compressed sensing [42], and reactive tabu search-based [43] low-complexity detectors of GSM, which provide near-ML error performance, have been proposed.

Enhanced SM (ESM) is a recently proposed and promising form of SM [44]. In the ESM scheme, the number of active transmit antennas can vary for each signaling interval, and the information is conveyed not only by the indices of active transmit antennas but also by the selected signal constellations used in transmission. In other words, the ESM scheme considers multiple signal constellations, and the information is transmitted by the combination of active transmit antennas and signal constellations. As an example, for two transmit antennas and four bpcu transmission, the ESM scheme transmits two bits by the joint selection of active transmit antennas and signal constellations, where one quadrature PSK (QPSK) and two binary PSK (BPSK) signal constellations (one ordinary and one rotated) can be used. For two-bit sequences $\{0,0\}$, $\{0,1\}$, $\{1,0\}$, and $\{1,1\}$, the ESM scheme uses the following transmission vectors, respectively: $\begin{bmatrix} S_4 & 0 \end{bmatrix}^T$, $\begin{bmatrix} 0 & S_4 \end{bmatrix}^T$, $\begin{bmatrix} S_2 & S_2 \end{bmatrix}^T$, and $\begin{bmatrix} S_2 e^{j\theta} & S_2 e^{j\theta} \end{bmatrix}^T$, where $S_m, m = 2, 4$ denotes the M-PSK constellation, and $\theta = \pi/2$ is a rotation angle used to obtain a third signal constellation in addition to classical BPSK and QPSK signal constellations. It is interesting to observe that the first two transmission vectors of the ESM scheme correspond to the classical SM using QPSK with a single activated transmit antenna, where the first and second transmit antenna are used for the transmission of a QPSK symbol, respectively. On the other hand, the third and fourth transmission vectors correspond to the simultaneous transmission of two symbols selected from BPSK and modified BPSK constellations, respectively. The reason behind reducing the constellation size from four to two can be explained by the fact that the same number of information bits (two bits for this case) must be carried with M-ary constellations independent from the number of active transmit antennas. Examples of the generalization of the

ESM scheme for different numbers of transmit antennas and signal constellations are available in the literature [44].

Quadrature SM (QSM) is a modified version of classical SM, which is proposed to improve the spectral efficiency while maintaining the advantages of SM, such as operation with a single RF chain and ICI-free transmission [45]. In the QSM scheme, the real and imaginary parts of the complex M-ary data symbols are separately transmitted using the SM principle. For a MIMO system with n_T transmit antennas, the spectral efficiency of QSM becomes $2\log_2(n_T) + \log_2(M)$ bpcu by simultaneously applying the SM principle for in-phase and quadrature components of the complex data symbols. As an example, for $n_T = 2$ and $M = 4$, in addition to the two bits mapped to the QPSK constellation, an extra two bits can be transmitted in the spatial domain by using one of the four transmission vector: $[s_R + js_I \ \ 0]^T$, $[s_R \ \ js_I]^T$, $[js_I \ \ s_R]^T$, and $[0 \ \ s_R + js_I]^T$ for input bit sequences $\{0,0\}$, $\{0,1\}$, $\{1,0\}$, and $\{1,1\}$, respectively, where s_R and s_I denote the real and imaginary parts of $s = s_R + js_I \in S_4$, respectively. It is interesting to note that the first and second elements of these two-bit sequences indicate the transmission position of the real and imaginary parts of s, respectively. Even if the number of active transmit antennas can be one or two for the QSM scheme, a single RF chain is sufficient at the transmitter since only two carriers (cosine and sine) generated by a single RF chain are used during transmission.

In Table 8.1, transmission vectors of SM, ESM, and QSM schemes are given for a 4-bpcu transmission and two transmit antennas, where we considered natural bit mapping for ease of presentation. We observe from Table 8.1 that both ESM and QSM schemes convey more bits by the spatial domain compared to conventional SM, which leads to not only improved spectral efficiency but also higher energy efficiency.

TABLE 8.1

Transmission Vectors ($\mathbf{x}^T$) of SM, ESM, and QSM Schemes for 4-bpcu and Two Transmit Antennas ($n_T = 2$)

Bits	SM	ESM	QSM	Bits	SM	ESM	QSM
0000	$[1 \ \ 0]$	$\left[\frac{1+j}{\sqrt{2}} \ \ 0\right]$	$\left[\frac{1+j}{\sqrt{2}} \ \ 0\right]$	1000	$[0 \ \ 1]$	$\left[\frac{1}{\sqrt{2}} \ \ \frac{1}{\sqrt{2}}\right]$	$\left[\frac{j}{\sqrt{2}} \ \ \frac{1}{\sqrt{2}}\right]$
0001	$\left[\frac{1+j}{\sqrt{2}} \ \ 0\right]$	$\left[\frac{-1+j}{\sqrt{2}} \ \ 0\right]$	$\left[\frac{-1+j}{\sqrt{2}} \ \ 0\right]$	1001	$\left[0 \ \ \frac{1+j}{\sqrt{2}}\right]$	$\left[\frac{1}{\sqrt{2}} \ \ \frac{-1}{\sqrt{2}}\right]$	$\left[\frac{j}{\sqrt{2}} \ \ \frac{-1}{\sqrt{2}}\right]$
0010	$[j \ \ 0]$	$\left[\frac{-1-j}{\sqrt{2}} \ \ 0\right]$	$\left[\frac{-1-j}{\sqrt{2}} \ \ 0\right]$	1010	$[0 \ \ j]$	$\left[\frac{-1}{\sqrt{2}} \ \ \frac{1}{\sqrt{2}}\right]$	$\left[\frac{-j}{\sqrt{2}} \ \ \frac{-1}{\sqrt{2}}\right]$
0011	$\left[\frac{-1+j}{\sqrt{2}} \ \ 0\right]$	$\left[\frac{1-j}{\sqrt{2}} \ \ 0\right]$	$\left[\frac{1-j}{\sqrt{2}} \ \ 0\right]$	1011	$\left[0 \ \ \frac{-1+j}{\sqrt{2}}\right]$	$\left[\frac{-1}{\sqrt{2}} \ \ \frac{-1}{\sqrt{2}}\right]$	$\left[\frac{-j}{\sqrt{2}} \ \ \frac{1}{\sqrt{2}}\right]$
0100	$[-1 \ \ 0]$	$\left[0 \ \ \frac{1+j}{\sqrt{2}}\right]$	$\left[\frac{1}{\sqrt{2}} \ \ \frac{j}{\sqrt{2}}\right]$	1100	$[0 \ \ -1]$	$\left[\frac{j}{\sqrt{2}} \ \ \frac{j}{\sqrt{2}}\right]$	$\left[0 \ \ \frac{1+j}{\sqrt{2}}\right]$
0101	$\left[\frac{-1-j}{\sqrt{2}} \ \ 0\right]$	$\left[0 \ \ \frac{-1+j}{\sqrt{2}}\right]$	$\left[\frac{-1}{\sqrt{2}} \ \ \frac{j}{\sqrt{2}}\right]$	1101	$\left[0 \ \ \frac{-1-j}{\sqrt{2}}\right]$	$\left[\frac{j}{\sqrt{2}} \ \ \frac{-j}{\sqrt{2}}\right]$	$\left[0 \ \ \frac{-1+j}{\sqrt{2}}\right]$
0110	$[-j \ \ 0]$	$\left[0 \ \ \frac{-1-j}{\sqrt{2}}\right]$	$\left[\frac{-1}{\sqrt{2}} \ \ \frac{-j}{\sqrt{2}}\right]$	1110	$[0 \ \ -j]$	$\left[\frac{-j}{\sqrt{2}} \ \ \frac{j}{\sqrt{2}}\right]$	$\left[0 \ \ \frac{-1-j}{\sqrt{2}}\right]$
0111	$\left[\frac{1-j}{\sqrt{2}} \ \ 0\right]$	$\left[0 \ \ \frac{1-j}{\sqrt{2}}\right]$	$\left[\frac{1}{\sqrt{2}} \ \ \frac{-j}{\sqrt{2}}\right]$	1111	$\left[0 \ \ \frac{1-j}{\sqrt{2}}\right]$	$\left[\frac{-j}{\sqrt{2}} \ \ \frac{-j}{\sqrt{2}}\right]$	$\left[0 \ \ \frac{1-j}{\sqrt{2}}\right]$

Note: The most significant bit is the one transmitted by the spatial domain for SM, and the second most significant bit is the additional bit transmitted by the spatial domain for ESM and QSM.

In Figure 8.2, we compare the minimum squared Euclidean distance between the transmission vector (d_{min}), which is an important design parameter for quasi-static Rayleigh fading channels to optimize the error performance of SIMO, SM, ESM, and QSM schemes. In all considered configurations, we normalized the average total transmitted energy to unity to make fair comparisons. It is interesting to note that ESM and QSM schemes achieve the same d_{min} value for 4- and 6-bpcu transmissions. However, as seen from Figure 8.2, QSM suffers a worse minimum Euclidean distance, and as a result worse error performance, compared to the ESM scheme for higher spectral efficiency values, while the ESM scheme requires a more complex and higher-cost transmitter with two RF chains. Finally, the results of Figure 8.2 also prove that the relative d_{min} advantage of IM schemes over classical SIMO scheme increases with increasing spectral efficiency; that is, IM techniques are preferable for higher spectral efficiency values.

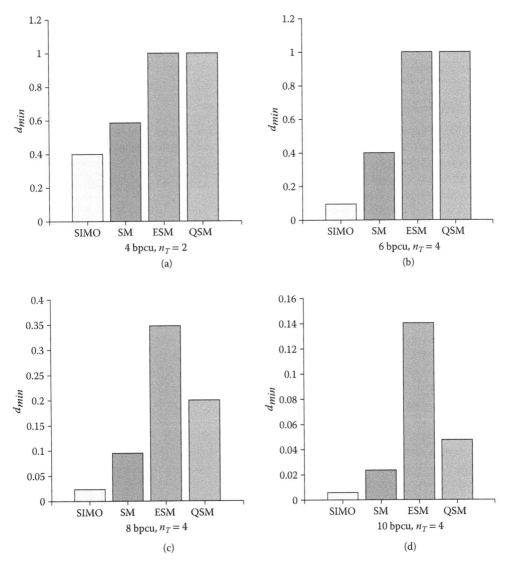

FIGURE 8.2 Minimum squared Euclidean distance (d_{min}) comparison of SIMO, SM, ESM, and QSM schemes for different configurations. (a) 4-bpcu, $n_T = 2$. SIMO:16-QAM, SM:8-PSK, ESM:QPSK/BPSK, QSM:QPSK. (b) 6-bpcu, $n_T = 4$. SIMO:64-QAM, SM:16-QAM, ESM:QPSK/BPSK, QSM:QPSK. (c) 8-bpcu, $n_T = 4$. SIMO:256-QAM, SM:64-QAM, ESM:16-QAM/QPSK, QSM:16-QAM. (d) 10-bpcu, $n_T = 4$. SIMO:1024-QAM, SM:256-QAM, ESM:64-QAM/8-QAM, QSM:64-QAM.

In the past two years, ESM and QSM schemes have attracted significant attention from the community, and several follow-up studies have been performed by researchers. The inventors of ESM have proposed the enhanced spatial multiplexing (E-SMX) scheme, which is based on the multiple signal constellations concept of ESM, to improve the performance of classical V-BLAST [46]. MA-SM and ESM concepts have recently been combined to obtain better error performance with the design of new signal constellations [47]. Moreover, the error performance of ESM has been investigated under channel estimation errors for uncorrelated and correlated Rayleigh and Rician fading channels, and it has been shown that ESM exhibits improved tolerance to channel estimation errors [48]. In the meantime, researchers have explored the error performance of QSM for different types of fading channels [49,50] and cooperative networks [51,52], under channel estimation errors [53] and co-channel interference [54]. Conventional QSM has been extended to the receiver side by the generalized precoding aided QSM scheme [55]. It has been recently shown that near-ML, compressed sensing–based detectors can provide significant reduction in ML detection complexity of the QSM scheme [56,57]. Furthermore, the novel dual IM concept of QSM has also triggered research activities on the design of high-rate SM systems [58].

8.3.2 SM-Based Massive Multiuser MIMO Systems

Massive MIMO systems, in which the BSs are equipped with tens to hundreds of antennas, have been regarded as one of the potential key technologies for 5G wireless networks due to their attractive advantages such as very high spectral and energy efficiency. While the initial studies on MIMO systems generally focused on point-to-point links, where two users communicate with each other, practical MU-MIMO systems are gaining more attention. MU-MIMO systems can exploit the multiple antennas of a MIMO system to support multiple users concurrently.

Within this perspective, the extension of MIMO systems into massive scale provides unique as well as promising opportunities for SM systems. For massive MIMO setups, it becomes possible to transmit a significant number of information bits by the spatial domain even if the number of available RF chains is very limited due to the space and cost limitations of the mobile terminals. Although the spectral efficiency of SM systems becomes considerably lower compared with that of traditional methods, such as V-BLAST for massive MIMO systems, the use of the IM concept for the transmit antennas of a massive MIMO system can provide effective implementation solutions thanks to the inherently available advantages of SM systems. Furthermore, SM is well suited to unbalanced massive MIMO configurations, in which the number of receive antennas is less than the number of transmit antennas [59], and V-BLAST-based systems cannot operate with linear detection methods, such as ZF and MMSE detection. As seen in Figure 8.3, SM techniques can be considered for both uplink and downlink transmissions in massive MU-MIMO systems.

In Figure 8.3a, we consider a massive MU-MIMO system, where K users employ SM techniques for their uplink transmissions. Additional information bits can be transmitted using SM without increasing the system complexity compared to user terminals with single antennas employing ordinary modulations. To further boost the spectral efficiency of the mobile users, GSM, ESM, and QSM techniques can be considered at the users instead of SM. At the BS, the optimal (ML) detector can be used at the expense of exponentially increasing decoding complexity (with respect to K) due to inter-user interference. However, the detection complexity of this detector can be unfeasible in practical scenarios with several users. Consequently, low-complexity, sub-optimal detection methods can be implemented as well by sacrificing the optimum error performance. On the other hand, SM techniques along with precoding methods can be considered at the BS for the downlink transmission, as shown in Figure 8.3b. To support a high number of users, the massive antennas of the BS can be split into subgroups of fewer antennas where SM techniques can be employed for each user [60]. To perform an interference-free transmission for the specific case of two users, the data of

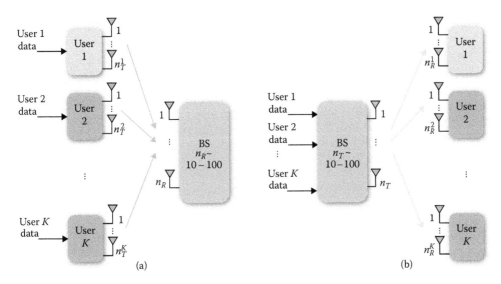

FIGURE 8.3 Massive MU-MIMO systems with SM. (a) An uplink transmission scenario where User k has n_T^k transmit antennas available for SM, and the BS has $n_R \sim 10-100$ receive antennas. (b) A downlink transmission scenario where User k has n_R^k receive antennas, and the BS has $n_T \sim 10-100$ transmit antennas available for SM.

User 1 can be mapped into the antenna indices while the data of User 2 can be conveyed with M-ary signal constellations [61]. The implementation of SM variants discussed in Section 8.3.1 can also be considered at the BS to transmit the data of different users.

8.3.3 Cooperative SM Systems

Cooperative communication, which allows the transmission of a user's data not only by its own antenna but also by the active or passive nodes available in the network, has been one of the hot topics in the wireless communications field in the past decade. Initially, cooperative communication systems have been proposed to create virtual MIMO systems for the mobile terminals due to the problems, such as cost and hardware, associated with the employment of multiple antennas in mobile terminals. However, due to recent technological advances, multiple antennas can be employed at mobile terminals, and cooperative communication systems can efficiently provide additional diversity gains and high data rates by improving coverage. Consequently, relaying technologies have been incorporated into the Long Term Evolution–Advanced (LTE-A) standard for increasing coverage, data rate, and cell-edge performance [62].

Considering the attractive solutions provided by SM techniques and cooperative communications systems, the combination of these two technologies naturally arises as a potential candidate for future wireless networks. Fortunately, SM-based cooperative communications systems can provide new implementation scenarios, additional diversity gains, and higher data rates without increasing the cost and complexity of the mobile and relay terminals due to the recent technological advances. It has been shown by several studies that SM techniques can be efficiently implemented for decode-and-forward (DF) and amplify-and-forward (AF) relaying-based cooperative networks, dual- and multi-hop relay systems, distributed cooperation, and network coding systems that allow bi-directional communications [32–37].

In Figure 8.4, four different cooperative SM transmission scenarios are considered, where S, R, and D respectively stand for the source, relay, and destination node. In Figure 8.4a, a dual-hop network is given, where the communications between S and D are accomplished over an intermediate R. In this dual-hop system, SM techniques can be implemented at S and/or R with DF- or AF-based

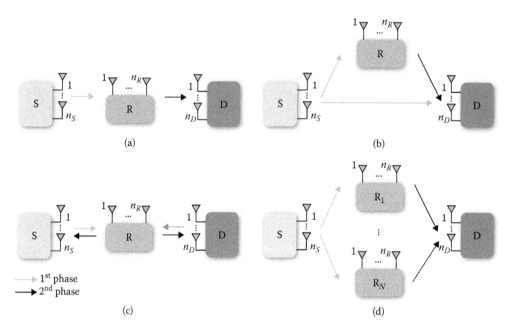

FIGURE 8.4 An overview of cooperative SM systems where n_S, n_R, and n_D denote the number of antennas for source (S), relay (R), and destination (D) nodes, respectively. (a) Dual-hop SM. (b) Cooperative SM. (c) Network-coded SM. (d) Multi-relay and distributed SM.

relaying techniques. The scenario in Figure 8.4a is generally observed in practical networks, where S and D cannot communicate directly due to distance or obstacles; as a result, DF-based, dual-hop relaying has also been incorporated to LTE-A standards. In this relaying scenario, the energy and spectral efficiency of S can be improved by the use of IM techniques compared to the single-antenna case, while multiple RF chains are required at R and D for signal reception. However, considering the uplink transmission from S to D, this would not be a major design problem. In Figure 8.4b, a direct link from S to D is also considered, and R can improve the quality of service of the transmission between S and D by employing different relaying methods, such as incremental and selective relaying.

We consider the bi-directional (two-way) communications of S and D that are accomplished via R in Figure 8.4c. Without network coding, the overall transmission between S and D requires four transmission phases (from S to R, R to D, D to R, and R to S), which considerably reduce the spectral efficiency of the overall system. On the other hand, by using physical-layer network coding (PLNC), the two-way communications between S and D can be performed at two phases, where in the first transmission phase, S and D simultaneously transmit their signals to R with/without SM techniques. In the second transmission phase, R combines the signals received from S and D and then forwards this combined signal to S and D. The use of SM provides some opportunities for R, such as transmitting one user's data with antenna indices and the other's data with constellation symbols.

Finally, a distributed cooperation scenario with N relay nodes, $(R_1, \ldots, R_N)$, is considered in Figure 8.4d. In the first transmission phase, S can use SM techniques to transfer its data to the relays. In the second transmission phase, one or more relays cooperate by forming a virtual SM/ SSK system, and the indices of the activated relays can be considered as an additional way to convey information. This allows the relays to cooperate even if they have single antennas ($n_R = 1$) since their own indices carry information. Furthermore, to improve the spectral efficiency, opportunistic relay selection can be considered for the network topology of Figure 8.4d, where only the selected best relay takes part in the transmission. For all the different cooperation scenarios described previously, the use of GSM/ESM/QSM techniques at S and/or R is also possible to improve the spectral/energy efficiency of the overall system.

8.3.4   Spectrum Sharing–Based SM Systems

Cognitive radio (CR) networks are capable of overcoming with the scarcity and inefficient use of the wireless spectrum by utilizing spectrum sharing. A typical CR network consists of two types of users: licensed and unlicensed users, which are also called the primary users (PUs) and the secondary users (SUs), respectively. The secondary users (or cognitive radio users) are intelligent devices, which can sense the available spectrum as well as recognize the nearby environment to adjust their transmission parameters since these type users are allowed to use the same frequency band along with PUs under the condition of improving or, at least, not degrading the performance of PUs [63]. Since both PUs and SUs use the available spectrum concurrently in CR networks, one of the major problems of these networks becomes the mutual interference generated by the users. Fortunately, SM techniques can be exploited effectively to overcome the interference problems of conventional CR networks.

SM techniques have been implemented for both underlay- and overlay-type CR networks in recent years [61,64–70]. In underlay networks, SUs can use the licensed spectrum band under an interference constraint to PUs. On the other hand, SUs assist the communications of PUs through cooperation to improve the performance of the primary network in overlay networks. Most of the spectrum sharing–based SM studies in the literature consider underlay networks, in which the secondary transmitters consider SM/SSK/QSM techniques in their transmission under an interference constraint [64–66,69,70]. In these studies, the authors investigated the error performance of IM-based CR networks in the presence of partial/full channel state information at the secondary transmitters and perfect/imperfect channel estimation at the secondary receivers. It has been shown that SM and its variants can provide efficient implementation scenarios for underlay networks.

The integration of SM into overlay networks has also been performed in some recent studies [61,67,68]. To mitigate the interference between PUs and SUs, the unique transmission properties of SM are considered. More specifically, the secondary transmitter exploits SM and considers the antenna indices to transmit its own data bits; on the other hand, it uses ordinary M-ary modulation to transmit the primary transmitter's information with the purpose of supporting the communications of the primary network. It has been shown by computer simulations that SM-based systems can achieve better bit error rate (BER) performance compared to conventional cooperative spectrum-sharing systems using superposition coding.

8.4   INDEX MODULATION FOR OFDM SUBCARRIERS: OFDM WITH INDEX MODULATION

Although the concept of IM is generally remembered by the SM scheme, it is also possible to implement IM techniques for communication systems different from MIMO systems. As an example, one can efficiently implement IM techniques for the massive subcarriers of an OFDM system. OFDM-IM is a novel multi-carrier transmission scheme that has been inspired by the IM concept of SM [5]. Similar to the bit mapping of SM, the incoming bit stream is split into subcarrier index selection and M-ary constellation bits in the OFDM-IM scheme. Considering the index selection bits, only a subset of available subcarriers are activated, while the remaining inactive subcarriers are set to zero and not used in data transmission. However, the active subcarriers are modulated as in classical OFDM according to M-ary constellation bits. In other words, the OFDM-IM scheme conveys information not only by the data symbols, as in classical OFDM, but also by the indices of the active subcarriers that are used for the transmission of the corresponding M-ary data symbols.

For an OFDM system consisting of N_F available subcarriers, one can directly determine the indices of the active subcarriers, similar to SM-based schemes. However, considering the massive structure of OFDM frames, IM techniques can be implemented in a more flexible way for OFDM-IM schemes compared to SM-based schemes. On the other hand, keeping in mind the practical values of N_F, such as 128, 256, 512, 1024, or 2048 as in the LTE-A standard, if the subcarrier IM is directly applied to the overall OFDM frame, there could be trillions of possible active subcarrier combinations. For instance, to select the indices of 128 active subcarriers out of $N_F = 256$ available subcarriers, one should consider 5.77×10^{75} possible combinations of active subcarriers, which would make the selection of active subcarriers an impossible task. For this reason, the single and massive OFDM-IM block should be divided into G smaller and more manageable OFDM-IM sub-blocks for the implementation of OFDM-IM. In this divide-and-conquer approach, each sub-block contains N subcarriers to perform IM, where $N_F = G \times N$. For each sub-block, we select K out of N available subcarriers as active according to

$$p_1 = \left\lfloor \log_2 \binom{N}{K} \right\rfloor \tag{8.4}$$

index selection bits, where typical N values could be 2, 4, 8, 16, and 32 with $1 \leq K < N$. It should be noted that classical OFDM becomes the special case of OFDM-IM with $K = N$, that is, when all subcarriers are activated, where a total of $N \log_2 M$ bits can be transmitted per frame.

The block diagrams of the OFDM-IM scheme's transmitter and receiver structures are shown in Figures 8.5a and 8.5b, respectively. As seen from Figure 8.5a, for the transmission of each OFDM-IM frame, a total of

$$m = pG = \left(\left\lfloor \log_2 \binom{N}{K} \right\rfloor + K \log_2 M \right) G \tag{8.5}$$

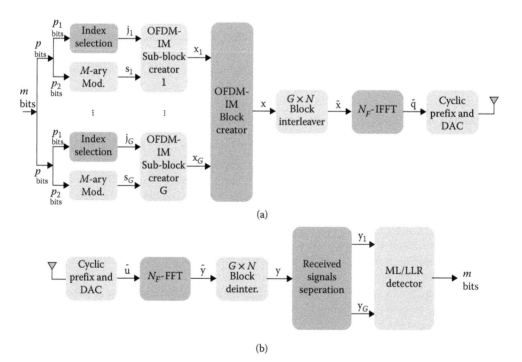

FIGURE 8.5 OFDM-IM system at a glance. (a) Transmitter structure. (b) Receiver structure.

bits enter the transmitter, where $p = p_1 + p_2$ and $p_2 = K \log_2 M$. In Figure 8.5a, $\mathbf{j}_g$ and $\mathbf{s}_g$ denote the vectors of the selected indices and M-ary data symbols with dimensions $K \times 1$, respectively. The operation of the OFDM-IM transmitter can be summarized as follows. First, the OFDM-IM sub-block creator obtains the OFDM-IM sub-blocks $\mathbf{x}_g$, $g = 1, \ldots, G$, with dimensions $N \times 1$ by considering $\mathbf{j}_g$ and $\mathbf{s}_g$. Afterward, the OFDM-IM block creator obtains the main OFDM-IM frame $\mathbf{x}$ with dimensions $N_F \times 1$ by concatenating these G OFDM-IM sub-blocks. After this point, $G \times N$ block interleaving is performed to ensure that the subcarriers of a sub-block undergo uncorrelated wireless fading channels to improve the error performance of the detector. At the last step, inverse fast Fourier transform (IFFT), cyclic prefix (CP) insertion, and digital-to-analog (DAC) conversion procedures are performed for the transmission of the signals through the wireless channel as in classical OFDM systems.

The selection of active subcarriers appears to be a challenging problem for OFDM-IM systems. For this purpose, two different index selection procedures are proposed for OFDM-IM depending on the size of the sub-blocks: reference look-up tables and combinatorial number theory method for lower and higher sub-block sizes, respectively. Examples of these two methods are provided in Figure 8.6. In the first example, two out of four subcarriers are selected as active by considering a reference look-up table with size four. In this case, two bits can determine the indices of the active subcarriers. In the second example, to select the indices of 16 active subcarriers out of 32 total subcarriers, the index selector processes 29 bits. First, these 29 bits are converted into a decimal number. Then, this decimal number is given to the combinatorial algorithm, which provides the required number of active indices.

The task of the OFDM-IM receiver is to determine the indices of the active subcarriers as well as the corresponding data symbols carried by these active subcarriers in conjunction with the index selection procedure used at the transmitter. After applying inverse operations (analog-to-digital (ADC) conversion, CP removal, FFT, and block deinterleaving), first, the received signals are separated since the detection of different sub-blocks can be carried out independently. Unfortunately, the optimal detection of OFDM-IM cannot be accomplished at the subcarrier level as in classical OFDM due to the index information, and the receiver must process the OFDM-IM sub-blocks for detection. The optimum but high-complexity ML detector performs a joint search by considering all possible subcarrier activation combinations and data symbols. On the other hand, a low-complexity, log-likelihood ratio (LLR), calculation-based, near-optimal detector handles each subcarrier independently and determines the indices of the active subcarriers first, and then it detects the corresponding data symbols. This detector calculates a probabilistic measure (LLR) on the active status of a given subcarrier by considering the two scenarios: an active subcarrier (carrying an M-ary constellation symbol) or an inactive one (that is set to zero). This detector is classified as near optimal since it does not consider the set of all legitimate subcarrier activation combinations.

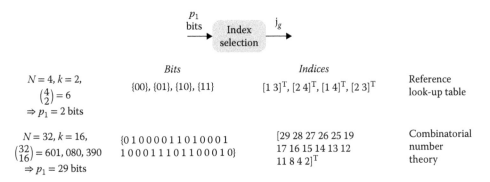

FIGURE 8.6 Two different index selection procedures for OFDM-IM.

It has been shown in the literature that OFDM-IM not only offers attractive advantages over classical OFDM but also provides an interesting trade-off between error performance and spectral efficiency with its flexible system design. The major difference between classical OFDM and OFDM-IM schemes is the adjustable number of active subcarriers of the latter. In other words, the number of active subcarriers of OFDM-IM can be adjusted accordingly to reach the desired spectral efficiency and/or error performance. Furthermore, OFDM-IM can provide a better BER performance than classical OFDM for low-to-mid spectral efficiency values with a comparable decoding complexity using the near-optimal LLR detector. This BER improvement can be attributed to the fact that the information bits carried by IM have lower error probability compared to ordinary M-ary constellation bits. Finally, it has been proved that OFDM-IM provides better performance than classical OFDM in terms of ergodic achievable rate [63].

As a result, we conclude that due to its appealing advantages over OFDM and more flexible system design, OFDM-IM can be considered as a possible candidate for emerging high-speed wireless communications systems. Furthermore, OFDM-IM has the potential to be well suited to machine-to-machine (M2M) communications systems of the next-generation wireless networks that require low power consumption.

8.5   RECENT ADVANCES IN OFDM-IM

The subcarrier IM concept for OFDM has attracted significant attention from the wireless community in recent times since its widespread introduction in 2012–2013 [5,71,72]. OFDM-IM techniques have been investigated in some up-to-date studies that deal with capacity analysis [73] and error performance [74]; the selection problem of the optimal number of active subcarriers [75,76]; subcarrier-level block interleaving for improved error performance [77]; generalization [78], enhancement [79–81], and low-complexity detection of OFDM-IM [82]; combination of OFDM-IM with coordinate interleaved orthogonal designs [83] and MIMO systems [84,85]; and its adaptation to different wireless environments [86,87]. In this section, we focus on three recently proposed and promising forms of OFDM-IM: generalized OFDM-IM, MIMO-OFDM-IM, and dual-mode OFDM-IM systems.

8.5.1   GENERALIZED OFDM-IM SCHEMES

Two generalized OFDM-IM structures have been proposed by modifying the original OFDM-IM scheme to obtain an improved spectral efficiency [78]. In the first scheme, which is called the OFDM-GIM-I scheme, the number of active subcarriers is no longer fixed, and it is also determined according to the information bits. As an example case of $N = 4, K = 2$ with QPSK modulation $(M = 4)$, according to Equation 8.5, $\left\lfloor \log_2 \binom{4}{2} \right\rfloor + 2\log_2(4) = 6$ bits can be transmitted per OFDM-IM sub-block; that is, a total of $4 \times 4^2 = 64$ sub-block realizations can be obtained. On the other hand, the OFDM-GIM-IM scheme considers all activation patterns $(K \in \{0, 1, 2, 3, 4\})$, which means that the number of active subcarriers can take values from zero (all subcarriers are inactive, $K = 0$) to four (all subcarriers are active, $K = 4$), as well as all possible values of M-ary data symbols. A total of

$$\sum_{K=0}^{N} \binom{N}{K} M^K = \binom{4}{0}4^0 + \binom{4}{1}4^1 + \binom{4}{2}4^2 + \binom{4}{3}4^3 + \binom{4}{4}4^4 = 625 \qquad (8.6)$$

possible sub-block realizations can be obtained for which $\lfloor \log_2(625) \rfloor = 9$ bits can be transmitted per sub-block. Consequently, compared to OFDM-IM, OFDM-GIM-I is capable of transmitting more bits per sub-block.

The second generalized OFDM-IM scheme, which is called the OFDM-GIM-II scheme, improves the spectral efficiency further by applying IM independently for the in-phase and quadrature components of the complex data symbols analogous to QSM. In other words, a subcarrier can be active for one component, while being inactive simultaneously for the other component. For the case of $N = 8, K = 4$ with QPSK modulation ($M = 4$), according to Equation 8.5, $\left\lfloor \log_2 \binom{8}{4} \right\rfloor + 4 \log_2(4) = 14$ bits can be transmitted per OFDM-IM sub-block. On the other hand, the OFDM-GIM-II scheme allows the transmission of $\left\lfloor \log_2 \left(\binom{8}{4} (\sqrt{4})^4 \times \binom{8}{4} (\sqrt{4})^4 \right) \right\rfloor = 20$ bits per sub-block, which is 30% higher than that of OFDM-IM.

8.5.2 FROM SISO-OFDM-IM TO MIMO-OFDM-IM

In the first studies on OFDM-IM, the researchers generally investigated SISO configurations and performed comparisons with classical the SISO-OFDM scheme. However, OFDM is generally implemented along with MIMO systems in current wireless communications standards to support high data rate applications, which require increased spectral efficiency. For this reason, MIMO transmission and OFDM-IM principles are combined to further boost the spectral and energy efficiency of the plain OFDM-IM scheme [84,85]. In the MIMO-OFDM-IM scheme, a V-BLAST-type transmission strategy is adopted to obtain increased spectral efficiency. More specifically, the transmitter of the MIMO-OFDM-IM scheme is obtained by the parallel concatenation of multiple SISO-OFDM-IM transmitters (Figure 8.5a). At the receiver of the MIMO-OFDM-IM scheme, the simultaneously transmitted OFDM-IM frames interfere with each other due to the V-BLAST-type parallel transmission; therefore, these frames are separated and demodulated using a novel low-complexity MMSE detection and LLR calculation-based detector. This detector performs sequential MMSE filtering to perform the detection of OFDM-IM sub-blocks at each branch of the transmitter and considers the statistics of the MMSE filtered received signals to improve the error performance. It has been demonstrated via extensive computer simulations that MIMO-OFDM-IM can be a strong alternative to classical MIMO-OFDM due to its improved BER performance and flexible system design. It should be noted that unlike other waveforms, such as GFDM and FMBC, OFDM-IM is a more MIMO-friendly transmission technique and provides improvements in BER performance over classical MIMO-OFDM.

In Figure 8.7, we present the uncoded BER performance of MIMO-OFDM-IM ($N = 4, K = 2$, $M = 2$) and classical V-BLAST-type MIMO-OFDM schemes ($M = 2$) for three MIMO configurations: 2×2, 4×4, and 8×8. In all cases, we obtain the same spectral efficiency values for both schemes to perform fair comparisons. As seen in Figure 8.7, considerable improvements in the required signal-to-noise ratio (SNR) are obtained to reach a target BER value by the MIMO-OFDM-IM scheme compared to classical MIMO-OFDM.

Some other studies that combine OFDM-IM and MIMO transmission principles have also been performed recently. Generalized space-frequency index modulation (GSFIM) [88] combines the OFDM-IM concept with the GSM principle by exploiting both spatial and frequency (subcarrier) domains for IM. It has been shown that the GSFIM scheme can also provide improvements over MIMO-OFDM in terms of achievable data rate and BER performance with ML detection for lower constellations such as BPSK and QPSK. However, the design of low-complexity detector types is an open research problem for the GSFIM scheme. More recently, a space-frequency coded index modulation (SFC-IM) scheme has been proposed to obtain diversity gains for MIMO-OFDM-IM [89]. Even more recently, low-complexity and near-optimal detection algorithms, based on the sequential Monte Carlo theory, have been proposed for emerging MIMO-OFDM-IM schemes [90].

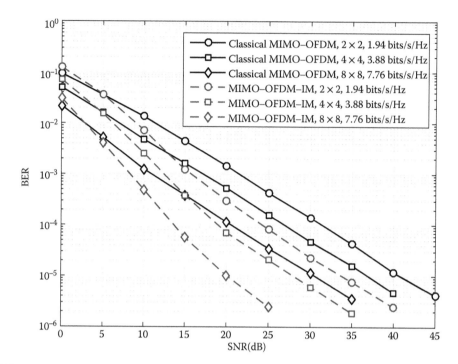

FIGURE 8.7 Uncoded BER performance of MIMO-OFDM-IM and classical MIMO-OFDM schemes for three $n_T \times n_R$ MIMO configurations: 2×2, 4×4, and 8×8. OFDM system parameters: $M = 2$ (BPSK), $N = 4, K = 2$, $N_F = 512$, CP length $= 16$, frequency-selective Rayleigh fading channel with 10 taps, uniform power delay profile, successive MMSE detection. The 3% reduction in spectral efficiency compared to the single-carrier case ($n_T \log_2 M$) is due to CP.

8.5.3 DUAL-MODE OFDM-IM SCHEME

One of the main limitations of the plain OFDM-IM scheme is its limited spectral efficiency due to the inactive subcarriers, which do not carry information for IM purposes. As a result, the BER advantage of OFDM-IM over classical OFDM diminishes with increasing spectral efficiency values. This can be understood by clearly examining Equation 8.5, which shows that the percentage of IM bits decreases by increasing modulation orders. As an example, to achieve the same spectral efficiency as that of classical OFDM, one can set $K = N - 1$ and $N = M$, for which the percentage of IM bits compared to the total number of bits becomes $\frac{100}{M}\%$, and this limits the inherent advantages of OFDM-IM. To transmit a maximum number of bits with IM, one can select $K = N/2$; however, in this case, the spectral efficiency of OFDM-IM cannot compete with that of classical OFDM for the same modulation order in most cases.

The dual-mode OFDM-IM (DM-OFDM) scheme provides a clever solution to overcome the spectral efficiency limitation of OFDM-IM by activating all subcarriers while still exploiting IM [81]. In the DM-OFDM scheme, all subcarriers are modulated, and the index information is carried by the signal constellations assigned to subcarrier groups. Two distinguishable signal constellations, a primary and a secondary constellation, are determined to transmit the data symbols from the active and inactive subcarriers of the OFDM-IM scheme, respectively. In other words, OFDM-IM becomes the special case of DM-OFDM if the secondary constellation contains a single element that is zero. Denoting the sizes of the primary and secondary constellations with M_1 and M_2, respectively, for each DM-OFDM block,

$$\tilde{m} = \tilde{p}G = \left(\left\lfloor \log_2 \binom{N}{K} \right\rfloor + K \log_2 M_1 + (N-K) \log_2 M_2 \right) G \qquad (8.7)$$

bits can be transmitted, where $\tilde{p}$ is the number of bits per DM-OFDM sub-block, G is the number of DM-OFDM sub-blocks, N is the number of subcarriers in a sub-block similar to OFDM-IM with $N = N_F / G$, and K is the number of subcarriers modulated by considering the primary constellation. It should be noted that by letting $M_2 = 1$ in Equation 8.7, that is, by not modulating the second group of subcarriers, the number of bits transmitted in a DM-OFDM block becomes the same as that of OFDM-IM given in Equation 8.5. It has been shown by computer simulations that the DM-OFDM scheme can achieve better BER performance than other OFDM-IM variants by using a near-optimal LLR calculation-based detector. More recently, a generalized DM-OFDM scheme has been proposed [91]. In this scheme, the number of subcarriers modulated by the primary and secondary constellations also changes according to the information bits to further improve the spectral efficiency with marginal performance loss.

8.6 CONCLUSIONS AND FUTURE WORK

IM appears to be a promising digital modulation concept for next-generation wireless communications systems since IM techniques can offer low complexity as well as spectrum- and energy-efficient solutions for emerging single-/multi-carrier, massive single-/multi-user MIMO, cooperative communications, and spectrum-sharing systems. In this chapter, we have reviewed the basic principles, advantages and disadvantages, the most recent as well as promising developments, and possible implementation scenarios of SM and OFDM-IM systems, which are two highly popular forms of the IM concept. In Table 8.2, the pros and cons of the reviewed major IM schemes in terms of the spectral efficiency, ML detection complexity, and error performance are provided. We conclude from Table 8.2 that IM schemes can offer interesting trade-offs among the error performance, complexity, and spectral efficiency; consequently, they can be considered as possible candidates for

TABLE 8.2

Pros and Cons of Several Index Modulation Schemes

	Scheme	Spectral Efficiency	ML Detection Complexity	Error Performance
Single-carrier communications systems	SIMO	Low	Low	Low
	SM	Moderate	Low[a]	Moderate
	GSM	Moderate	Moderate[a]	Moderate
	MA-SM	High	Moderate[a]	Moderate
	ESM	High	Low	High
	QSM	High	Low	High
	V-BLAST	High	High[a]	Moderate
Multi-carrier communications systems	OFDM	Low	Low	Low
	OFDM-IM	Low	Moderate[a]	Moderate
	OFDM-GIM-I	Moderate	High[a]	Moderate
	OFDM-GIM-II	Moderate	High[a]	Moderate
	MIMO-OFDM-IM	High	High[a]	High
	GSFIM	High	High	Moderate
	V-BLAST-OFDM	High	Moderate[a]	Moderate

[a] Lower complexity near-/sub-optimal detection is also possible.

5G and beyond wireless communication networks. However, interesting and challenging research problems remain to be investigated to further improve the efficiency of IM-based schemes. These research challenges can be summarized as follows:

- The design of novel generalized/enhanced IM schemes with higher spectral and/or energy efficiency, lower transceiver complexity, and better error performance
- The integration of IM techniques (such as SM, GSM, ESM, QSM, and OFDM-IM) into massive MU-MIMO systems to be employed in 5G and beyond wireless networks and the design of novel uplink/downlink transmission protocols
- The adaption of IM techniques to cooperative communications systems (such as dual-/multi-hop, network-coded, multi-relay, and distributive networks) and spectrum-sharing systems
- The investigation of the potential of IM techniques via practical implementation scenarios
- Exploration of new digital communications schemes for the application of IM techniques

ACKNOWLEDGMENT

Our studies on cooperative spatial modulation systems are supported by the Scientific and Technological Research Council of Turkey (TUBITAK) under grant number 114E607.

This chapter has been extended from a previous magazine paper [92]. © [2016] IEEE. Reprinted, with permission, from Basar, Ertugrul. Index modulation techniques for 5G wireless networks. *IEEE Communications Magazine 54*, no. 7 (2016): 168–75.

REFERENCES

1. Wang, Cheng-Xiang, Fourat Haider, Xiqi Gao et al. Cellular architecture and key technologies for 5G wireless communication networks. *IEEE Communications Magazine* 52, no. 2 (2014): 122–30.
2. Mesleh, Raed Y., Harald Haas, Sinan Sinanovic, Chang Wook Ahn, and Sangboh Yun. Spatial modulation. *IEEE Transactions on Vehicular Technology* 57, no. 4 (2008): 2228–41.
3. Jeganathan, Jeyadeepan, Ali Ghrayeb, Leszek Szczecinski, and Andres Ceron. Space shift keying modulation for MIMO channels. *IEEE Transactions on Wireless Communications* 8, no. 7 (2009): 3692–703.
4. Jafarkhani, Hamid. *Space-Time Coding: Theory and Practice*. Cambridge: Cambridge University Press, 2005.
5. Basar, Ertugrul, Umit Aygolu, Erdal Panayirci, and Harold Vincent Poor. Orthogonal frequency division multiplexing with index modulation. *IEEE Transactions on Signal Processing* 61, no. 22 (2013): 5536–49.
6. Jeganathan, Jeyadeepan, Ali Ghrayeb, and Leszek Szczecinski. Spatial modulation: Optimal detection and performance analysis. *IEEE Communications Letters* 12, no. 8 (2008): 545–47.
7. Mesleh, Read, Harald Haas, Chang Wook Ahn, and Sangboh Yun. Spatial modulation—A new low complexity spectral efficiency enhancing technique. *First International Conference on Communications and Networking in China*, Beijing, China, 2006.
8. Naidoo, Nigel R, Hongjun Xu, and Tahmid Al-Mumit Quazi. Spatial modulation: Optimal detector asymptotic performance and multiple-stage detection. *IET Communications* 5, no. 10 (2011): 1368–76.
9. Di Renzo, Marco, Harald Haas, and Peter Grant. Spatial modulation for multiple-antenna wireless systems: A survey. *IEEE Communications Magazine* 49, no. 12 (2011): 182–91.
10. Di Renzo, Marco Di, Harald Haas, Ali Ghrayeb, Shinya Sugiura, and Lajos Hanzo. Spatial modulation for generalized MIMO: Challenges, opportunities, and implementation. *Proceedings of the IEEE* 102, no. 1 (2014): 56–103.
11. Yang, Ping, Marco Di Renzo, Yue Xiao, Shaoqian Li, and Lajos Hanzo. Design guidelines for spatial modulation. *IEEE Communications Surveys and Tutorials* 17, no. 1 (2015): 6–26.
12. Tang, Qian, Yue Xiao, Ping Yang, Qiaoling Yu, and Shaoqian Li. A new low-complexity near-ML detection algorithm for spatial modulation. *IEEE Wireless Communications Letters* 2, no. 1 (2013): 90–3.
13. Yu, Chia-Mu, Sung-Hsien Hsieh, Han-Wen Liang, Chun-Shien Lu, Wei-Ho Chung, Sy-Yen Kuo, and Soo-Chang Pei. Compressed sensing detector design for space shift keying in MIMO systems. *IEEE Communications Letters* 16, no. 10 (2012): 1556–9.

14. Stavridis, Athanasios, Sinan Sinanovic, Marco Di Renzo, and Harald Haas. Energy evaluation of spatial modulation at a multi-antenna base station. *2013 IEEE 78th Vehicular Technology Conference (VTC Fall)*, 2013.

15. Younis, Abdelhamid, Nikola Serafimovski, Raed Mesleh, and Harald Haas. Generalised spatial modulation. *2010 Conference Record of the Forty Fourth Asilomar Conference on Signals, Systems and Computers*, Las Vegas, NV, USA, 2010.

16. Jeganathan, Jeyadeepan, Ali Ghrayeb, and Leszek Szczecinski. Generalized space shift keying modulation for MIMO channels. *IEEE 19th International Symposium on Personal, Indoor and Mobile Radio Communications*, Cannes, France, 2008.

17. Wang, Jintao, Shuyun Jia, and Jian Song. Generalised spatial modulation system with multiple active transmit antennas and low complexity detection scheme. *IEEE Transactions on Wireless Communications* 11, no. 4 (2012): 1605–15.

18. Younis, Abdelhamid, Sinan Sinanovic, Marco Di Renzo, Read Mesleh, and Harald Haas. Generalised sphere decoding for spatial modulation. *IEEE Transactions on Communications* 61, no. 7 (2013): 2805–15.

19. Rajashekar, Rakshith, K.V.S. Hari, and Lajos Hanzo. Reduced-complexity ML detection and capacity-optimized training for spatial modulation systems. *IEEE Transactions on Communications* 62, no. 1 (2014): 112–25.

20. Li, Cong, Yuzhen Huang, Marco Di Renzo, Jinlong Wang, and Yunpeng Cheng. Low-complexity ML detection for spatial modulation MIMO with APSK constellation. *IEEE Transactions on Vehicular Technology* 64, no. 9 (2015): 4315–21.

21. Basar, Ertugrul, Umit Aygolu, Erdal Panayirci, and H. Vincent Poor. Space-time block coded spatial modulation. *IEEE Transactions on Communications* 59, no. 3 (2011): 823–32.

22. Sugiura, Shinya, Sheng Chen, and Lajos Hanzo. Coherent and differential space-time shift keying: A dispersion matrix approach. *IEEE Transactions on Communications* 58, no. 11 (2010): 3219–30.

23. Basar, Ertugrul, Umit Aygolu, Erdal Panayirci, and H. Vincent Poor. New trellis code design for spatial modulation. *IEEE Transactions on Wireless Communications* 10, no. 8 (2011): 2670–80.

24. Yang, Ping, Yue Xiao, Lei Li, Qian Tang, Yi Yu, and Shaoqian Li. Link adaptation for spatial modulation with limited feedback. *IEEE Transactions on Vehicular Technology* 61, no. 8 (2012): 3808–13.

25. Rajashekar, Rakshith, K. V. S. Hari, and Lajos Hanzo. Antenna selection in spatial modulation systems. *IEEE Communications Letters* 17, no. 3 (2013): 521–4.

26. Yang, Ping, Yong Liang Guan, Yue Xiao, Marco Di Renzo, Shaoqian Li, and Lajos Hanzo. Transmit precoded spatial modulation: Maximizing the minimum Euclidean distance versus minimizing the bit error ratio. *IEEE Transactions on Wireless Communications* 15, no. 3 (2016): 2054–68.

27. Di Renzo, Marco, and Harald Haas. A general framework for performance analysis of space shift keying (SSK) modulation for MISO correlated Nakagami-*m* fading channels. *IEEE Transactions on Communications* 58, no. 9 (2010): 2590–603.

28. Mesleh, Raed, Osamah S. Badarneh, Abdelhamid Younis, and Harald Haas. Performance analysis of spatial modulation and space-shift keying with imperfect channel estimation over generalized $\eta - \mu$ fading channels. *IEEE Transactions on Vehicular Technology* 64, no. 1 (2015): 88–96.

29. Basar, Ertugrul, Umit Aygolu, Erdal Panayirci, and H. Vincent Poor. Performance of spatial modulation in the presence of channel estimation errors. *IEEE Communications Letters* 16, no. 2 (2012): 176–9.

30. An, Zhecheng, Jun Wang, Jintao Wang, Su Huang, and Jian Song. Mutual information analysis on spatial modulation multiple antenna system. *IEEE Transactions on Communications* 63, no. 3 (2015): 826–43.

31. Bian, Yuyang, Xiang Cheng, Miaowen Wen, Liuqing Yang, H. Vincent Poor, and Bingli Jiao. Differential spatial modulation. *IEEE Transactions on Vehicular Technology* 64, no. 7 (2014): 3262–8.

32. Mesleh, Raed, Salama Ikki, and Mohammed Alwakeel. Performance analysis of space shift keying with amplify and forward relaying. *IEEE Communications Letters* 15, no. 12 (2011): 1350–2.

33. Mesleh, Raed, Salama S. Ikki, El-Hadi M. Aggoune, and Ali Mansour. Performance analysis of space shift keying (SSK) modulation with multiple cooperative relays. *EURASIP Journal on Advances in Signal Processing*, 2012, no. 1 (2012): 1–10.

34. Mesleh, Raed, and Salama S. Ikki. Performance analysis of spatial modulation with multiple decode and forward relays. *IEEE Wireless Communications Letters* 2, no. 4 (2013): 423–6.

35. Wen, Miaowen, Xiang Cheng, H. Vincent Poor, and Bingli Jiao. Use of SSK modulation in two-way amplify-and-forward relaying. *IEEE Transactions on Vehicular Technology* 63, no. 3 (2014): 1498–504.

36. Som, Pritam, and Ananthanarayanan Chockalingam. Performance analysis of space-shift keying in decode-and-forward multihop MIMO networks. *IEEE Transactions on Vehicular Technology* 64, no. 1 (2015): 132–46.

37. Altin, Gokhan, Ertugrul Basar, Umit Aygolu, and Mehmet Ertugrul Celebi. Performance analysis of cooperative spatial modulation with multiple-antennas at relay. *4th International Black Sea Conference on Communications and Networking*, Varna, Bulgaria, 2016, pp. 1–5.

38. Yang, Ping, Yue Xiao, Yong Liang Guan et al. Single-carrier SM-MIMO: A promising design for broadband large-scale antenna systems. *IEEE Communications Surveys & Tutorials* 18, no. 3 (2016): 1687–716.

39. Datta, Tanumay, and Ananthanarayanan Chockalingam. On generalized spatial modulation. *2013 IEEE Wireless Communications and Networking Conference (WCNC)*, Shanghai, China, 2013.

40. Younis, Abdelhamid, Dushyantha A. Basnayaka, and Harald Haas. Performance analysis for generalised spatial modulation. *20th European Wireless Conference*, Barcelona, Spain, 2014.

41. Xiao, Yue, Zongfei Yang, Lilin Dan, Ping Yang, Lu Yin, and Wei Xiang. Low-complexity signal detection for generalized spatial modulation. *IEEE Communications Letters* 18, no. 3 (2014): 403–6.

42. Liu, Wenlong, Nan Wang, Minglu Jin, and Hongjun Xu. Denoising detection for the generalized spatial modulation system using sparse property. *IEEE Communications Letters* 18, no. 1 (2014): 22–5.

43. Narasimhan, Theagarajan Lakshmi, Patchava Raviteja, and Ananthanarayanan Chockalingam. Generalized spatial modulation for large-scale MIMO systems: Analysis and detection. *48th Asilomar Conference on Signals, Systems and Computers*, Pacific Grove, CA, USA, 2014.

44. Cheng, Chien-Chun, Hikmet Sari, Serdar Sezginer, and Yu Ted Su. Enhanced spatial modulation with multiple signal constellations. *IEEE Transactions on Communications* 63, no. 6 (2015): 2237–48.

45. Mesleh, Raed, Salama S. Ikki, and Hadi M. Aggoune. Quadrature spatial modulation. *IEEE Transactions on Vehicular Technology* 64, no. 6 (2015): 2738–42.

46. Cheng, Chien-Chun, Hikmet Sari, Serdar Sezginer, and Yu Ted Su. Enhanced spatial multiplexing—A novel approach to MIMO signal design. *2016 IEEE International Conference on Communications (ICC)*, Kuala Lumpur, Malaysia, 2016.

47. Cheng, Chien-Chun, Hikmet Sari, Serdar Sezinger, and Yu Ted Su. New signal designs for enhanced spatial modulation. *IEEE Transactions on Wireless Communications* 15, no. 11 (2016): 7766–7777.

48. Carosino, Michael, and James A. Ritcey. Performance of MIMO enhanced spatial modulation under imperfect channel information. *2015 49th Asilomar Conference on Signals, Systems and Computers*, 2015, Pacific Grove, CA, USA pp. 1415–9.

49. Younis, Abdelhamid, Raed Mesleh, and Harald Haas. Quadrature spatial modulation performance over Nakagami–m fading channels. *IEEE Transactions on Vehicular Technology* 65, no. 12 (2016): 10227–10231.

50. Alwakeel, Mohammed M. Quadrature spatial modulation performance analysis over Rician fading channels. *Journal of Communications* 11, no. 3 (2016): 249–54.

51. Afana, Ali, Raed Mesleh, Salama Ikki, and Ibrahem E. Atawi. Performance of quadrature spatial modulation in amplify-and-forward cooperative relaying. *IEEE Communications Letters* 20, no. 2 (2016): 240–3.

52. Afana, Ali, Salama Ikki, Raed Mesleh, and Ibrahem Atawi. Spectral efficient quadrature spatial modulation cooperative AF spectrum-sharing systems. *IEEE Transactions on Vehicular Technology* 66, no. 3 (2017): 2857–2861.

53. Mesleh, Raed, and Salama S. Ikki. On the impact of imperfect channel knowledge on the performance of quadrature spatial modulation. *2015 IEEE Wireless Communications and Networking Conference (WCNC 2015)*, New Orleans, LA, USA, 2015.

54. Mesleh, Raed, Salama S. Ikki, and Osamah S. Badarneh. Impact of cochannel interference on the performance of quadrature spatial modulation MIMO systems. *IEEE Communications Letters* 20, no. 10 (2016): 1927–30.

55. Li, Jun, Miaowen Wen, Xiang Cheng, Yier Yan, Sangseob Song, and Moon Ho Lee. Generalised pre-coding aided quadrature spatial modulation. *IEEE Transactions on Vehicular Technology* 66, no. 2 (2017): 1881–1886.

56. Yigit, Zehra, and Ertugrul Basar. Low-complexity detection of quadrature spatial modulation. *Electronics Letters* 52, no. 20 (2016): 1729–31.

57. Xiao, Lixia, Ping Yang, Shiwen Fan, Shaoqian Li, Lijun Song, and Yue Xiao. Low-complexity signal detection for large-scale quadrature spatial modulation systems. *IEEE Communications Letters* 20, no. 11 (2016): 2173–2176.

58. Yigit, Zehra, and Ertugrul Basar. Double spatial modulation: A high-rate index modulation scheme for MIMO systems. *2016 International Symposium on Wireless Communication Systems (ISWCS)*, Poznan, Poland, 2016.

59. Basnayaka, Dushyantha A., Marco Di Renzo, and Harald Haas. Massive but few active MIMO. *IEEE Transactions on Vehicular Technology* 65, no. 9 (2016): 6861–77.

60. Narayanan, Sandeep, Marium Jalal Chaudhry, Athanasios Stavridis, Marco Di Renzo, Fabio Graziosi, and Harald Haas. Multi-user spatial modulation MIMO. *2014 IEEE Wireless Communications and Networking Conference (WCNC)*, Istanbul, Turkey 2014.

61. Ustunbas, Seda, Ertugrul Basar, and Umit Aygolu. Performance analysis of cooperative spectrum sharing for cognitive radio networks using spatial modulation at secondary users. *2016 IEEE 83rd Vehicular Technology Conference (VTC Spring)*, Nanjing, China 2016.

62. 3rd General Partnership Project: Technical Specification Group Radio Access Network: Further Advancements for EUTRA Physical Layer Aspects (Release 9). *Tech. Rep. 36.814 (V9.0.0)*, 2010.

63. Biglieri, Ezio, Andrea Goldsmith, Larry J. Greenstein, Narayan B. Mandayan, and H. Vincent Poor. *Principles of Cognitive Radio.* New York: Cambridge University Press, 2012.

64. Al-Qahtani, Fawaz S., Yuzhen Huang, Marco Di Renzo, Salama Ikki, and Hussein Alnuweiri. Space shift keying MIMO system under spectrum sharing environments in Rayleigh fading. *IEEE Communications Letters* 18, no. 9 (2014): 1503–6.

65. Afana, Ali, Telex M. N. Ngatched, and Octavia A. Dobre. Spatial modulation in MIMO limited-feedback spectrum-sharing systems with mutual interference and channel estimation errors. *IEEE Communications Letters* 19, no. 10 (2015): 1754–7.

66. Bouida, Zied, Ali Ghrayeb, and Khalid A. Qaraqe. Adaptive spatial modulation for spectrum sharing systems with limited feedback. *IEEE Transactions on Communications* 63, no. 6 (2015): 2001–14.

67. Alizadeh, Ardalan, Hamid Reza Bahrami, and Mehdi Maleki. Performance analysis of spatial modulation in overlay cognitive radio communications. *IEEE Transactions on Communications* 64, no. 8 (2016): 3220–32.

68. Babaei, Mohammadreza, Ertugrul Basar, and Umit Aygolu. A cooperative spectrum sharing protocol using STBC-SM at secondary user. *2016 24th Telecommunications Forum (TELFOR)*, Belgrade, Serbia, 2016.

69. Afana, Ali, Salama Ikki, Raed Mesleh, and Ibrahim Atawi. Spectral efficient quadrature spatial modulation cooperative AF spectrum-sharing systems. *IEEE Transactions on Vehicular Technology* 66, no. 3 (2017): 2857–2861.

70. Afana, Ali, Islam Abu Mahady, and Salama Ikki. Quadrature spatial modulation in MIMO cognitive radio systems with imperfect channel estimation and limited feedback. *IEEE Transactions on Communications*, 65, no. 3 (2017): 981–991.

71. Basar, Ertugrul, Umit Aygolu, Erdal Panayirci, and H. Vincent Poor. Orthogonal frequency division multiplexing with index modulation. *2012 IEEE Global Communications Conference (GLOBECOM)*, Anaheim, CA, USA, 2012.

72. Basar, Ertugrul, Umit Aygolu, and Erdal Panayirci. Orthogonal frequency division multiplexing with index modulation in the presence of high mobility. *2013 First International Black Sea Conference on Communications and Networking (BlackSeaCom)*, Batumi, Georgia, 2013.

73. Wen, Miaowen, Xiang Cheng, Meng Ma, Bingli Jiao, and H. Vincent Poor. On the achievable rate of OFDM with index modulation. *IEEE Transactions on Signal Processing* 64, no. 8 (2016): 1919–32.

74. Ko, Youngwook. A tight upper bound on bit error rate of joint OFDM and multi-carrier index keying. *IEEE Communications Letters* 18, no. 10 (2014): 1763–6.

75. Wen, Miaowen, Xiang Cheng, and Liuqing Yang. Optimizing the energy efficiency of OFDM with index modulation. *2014 IEEE International Conference on Communication Systems*, Macau, China, 2014.

76. Li, Wenfang, Hui Zhao, Chengcheng Zhang, Long Zhao, and Renyuan Wang. Generalized selecting subcarrier modulation scheme in OFDM system. *2014 IEEE International Conference on Communications Workshops (ICC)*, Sydney, Australia, 2014.

77. Xiao, Yue, Shunshun Wang, Lilin Dan, Xia Lei, Ping Yang, and Wei Xiang. OFDM with interleaved subcarrier-index modulation. *IEEE Communications Letters* 18, no. 8 (2014): 1447–50.

78. Fan, Rui, Ya Jun Yu, and Yong Liang Guan. Generalization of orthogonal frequency division multiplexing with index modulation. *IEEE Transactions on Wireless Communications* 14, no. 10 (2015): 5350–9.

79. Fan, Rui, Ya Jun Yu, and Yong Liang Guan. Improved orthogonal frequency division multiplexing with generalised index modulation. *IET Communications* 10, no. 8 (2016): 969–74.

80. Wen, Miaowen, Yuekai Zhang, Jun Li, Ertugrul Basar, and Fangjiong Chen. Equiprobable subcarrier activation method for OFDM with index modulation. *IEEE Communications Letters* 20 no. 12 (2017): 2386–2389.

81. Hanzo, Lajos, Sheng Chen, Qi Wang, Zhaocheng Wang, and Tianqi Mao. Dual-mode index modulation aided OFDM. *IEEE Access*, 5 (2017): 50–60.

82. Zheng, Beixiong, Fangjiong Chen, Miaowen Wen, Fei Ji, Hua Yu, and Yun Li. Low-complexity ML detector and performance analysis for OFDM with in-phase/quadrature index modulation. *IEEE Communications Letters* 19, no. 11: 1893–6.

83. Basar, Ertugrul. OFDM with index modulation using coordinate interleaving. *IEEE Wireless Communications Letters* 4, no. 4 (2015): 381–4.

84. Basar, Ertugrul. Multiple-input multiple-output OFDM with index modulation. *IEEE Signal Processing Letters* 22, no. 12 (2015): 2259–63.

85. Basar, Ertugrul. On multiple-input multiple-output OFDM with index modulation for next generation wireless networks. *IEEE Transactions on Signal Processing* 64, no. 15 (2016): 3868–78.

86. Cheng, Xiang, Miaowen Wen, Liuqing Yang, and Yuke Li. Index modulated OFDM with interleaved grouping for V2X communications. *17th International IEEE Conference on Intelligent Transportation Systems (ITSC)*, Qingdao, China, 2014.

87. Wen, Miaowen, Xiang Cheng, Liuqing Yang, Yuke Li, Xilin Cheng, and Fei Ji. Index modulated OFDM for underwater acoustic communications. *IEEE Communications Magazine* 54, no. 5 (2016): 132–7.

88. Datta, Tanumay, Harsha S. Eshwaraiah, and Ananthanarayanan Chockalingam. Generalized space-and-frequency index modulation. *IEEE Transactions on Vehicular Technology* 65, no. 7 (2016): 4911–24.

89. Wang, Lei, Zhigang Chen, Zhengwei Gong, and Ming Wu. Space-frequency coded index modulation with linear-complexity maximum likelihood receiver in The MIMO-OFDM system. *IEEE Signal Processing Letters* 23, no. 10 (2016): 1439–43.

90. Zheng, Beixiong, Miaowen Wen, Ertugrul Basar, and Fangjiong Chen, Multiple-input multiple-output OFDM with index modulation: Low-complexity detector design. *IEEE Transactions on Signal Processing* 65, no. 11 (2017): 2758–2772.

91. Mao, Tianqi, Qi Wang, and Zhaocheng Wang. Generalized dual-mode index modulation aided OFDM. *IEEE Communications Letters* 21 no. 4 (2017): 761–764.

92. Basar, Ertugrul. Index modulation techniques for 5G wireless networks. *IEEE Communications Magazine* 54, no. 7 (2016): 168–75.

9 Seamless and Secure Communication for 5G Subscribers in 5G-WLAN Heterogeneous Networks

Amit Kumar and Hari Om

CONTENTS

9.1 INTRODUCTION

One of the key options in future heterogeneous wireless networks [1] is to constantly offer excellent data services and network connectivity to potable clients through completely different wireless networks, where dissimilar interworking scenarios come into view during client roaming and handover. Today, the cellular operators and internet service providers (ISPs) want to expand wireless access networks (WANs) with various technologies to fulfill the expanding portability and bandwidth necessities. Handover [2] among these advancements needs to be seamless to clients to permit an improved and consistent on-the-move roaming occurrence. In the perspective of mutual complementary features of wide-area coverage at excessive movability and excessive data rate, the idea of integrating more than one wireless access technology into a heterogeneous wireless system is one of the most promising visions of the wireless next-generation networks (WNGNs) [3]. The latter migrates toward the interworking of various wireless technologies, such as 3GPP Long-Term Evolution/Service Architecture Evolution (LTE/SAE) [4], 5G (fifth generation) [5], and wireless local area network (WLAN) [6].

The service environment and communication network of 5G are going to be much richer and more complicated than that of the present-day scenarios. It is perceived that the network communications with 5G will be able to link the vast spectrum of things as indicated by a variety of application-specific necessities: public, equipment, processes, content, computing centers, facts, statistics, goods, and information in adaptable, genuinely cellular, and effective way. The future will include linked sensors, smart meters, vehicles, and smart home devices that will make our present involvement of smart phone and tablet connectivity a thing of the past. The tremendous growth of clients in cellular networks will extend cellular services toward the Internet of things (IoT) [7], which will pressure the promotion of the fifth generation of mobile technologies. The present 4G systems such as 3GPP LTE are able to provide IP services; however, the present cellular networks are largely utilized for speech and call services. The costly data and short transfer speed services are the primary offenders. 5G [8] systems will enhance the data speed up to 10 Gb/s or greater; however, it is impossible that a client will be provided the high data speed capacity due to the very high setup expense. IEEE 802.11 [9] WLANs, the first tool for accessing broadband Internet all over the world, are becoming more popular day by day. These technologies are generally utilized and well suited for small-office and home environments. A failing in the present WLAN standard is the absence of sufficient security procedures and design beyond the fundamental radio access. Combining two systems is expected to yield a product that has their best features intact and their shortcomings removed by the integrated networks. An essential challenge is to resolve the security drawbacks of the interworking networks. This chapter provides a comprehensive discussion on the key enabling technologies for 5G-WLAN heterogeneous networks. The notations used in this chapter are provided in Table 9.1.

TABLE 9.1
List of Notations

Symbol	Description
UE	User equipment
E/D	Symmetric key encryption/decryption
IMSI	International mobile subscriber identity
ID	Identity of user
PW	Password
P_BBU	Public key of BBU
D_BBU	Private key of BBU
SK	Session key
TMSI	Temporary mobile subscriber identity
h(.)	One-way hash function
USIM	Universal subscriber identity module
[ ]*	Asymmetric key encryption/decryption using *
‖	Concatenation operation
xor	Exclusive or operation
s	Secret key of BBU
P_MBS	Public key of MBS
D_MBS	Private key of MBS

The remainder of this chapter is organized as follows. In Sections 9.1 and 9.2, we introduce the requirements of seamless roaming and the requirements of secure communication. Section 9.2 discusses the existing wireless technologies and 5G-WLAN heterogeneous network architecture, and Section 9.3 discusses the related works on handover in heterogeneous wireless networks. Section 9.4 introduces new reauthentication schemes for multi-radio access handoff so as to secure the handover of the user equipment of the 5G network into the WLAN network. Section 9.5 discusses the performance evaluation of the proposed reauthentication schemes for 5G-WLAN heterogeneous architecture with an authentication scheme of a 3GPP LTE–WLAN heterogeneous network in terms of security features, computational cost, and energy cost. Finally, in Section 9.6, the chapter is concluded.

9.1.1 REQUIREMENTS OF SEAMLESS ROAMING

From the viewpoint of portable clients/users, user equipment generally does not recognize at each instance which wireless system is the best one to access for their Internet services due to the accessibility of various nearby WANs belonging to various organizational areas. A network architecture which can give clients access to proper networks in an efficient manner is highly in demand. In the event that a mobile client needs to move from one network to another with continuous services, a mobile system needs to adapt to network changes to keep up the services seamlessly. Service quality is a main right that a client should be given with the service fulfilling his prerequisite, while the network cost and workload should not be an excessive amount. Moreover, since clients are provided with various services, such as video-on-demand and voice-over-IP services, it is expected that a client will achieve similar quality for all the dissimilar services. For this, the network architecture needs to be designed to include a quality mechanism that ensures the clients are served according to their individual QoS necessities.

From the viewpoint of a network administrator, the network infrastructure needs to help a client in recognizing the most efficient network at any time due the accessibility of different wireless technologies. This choice of networks must be done dynamically to consider various hypothetical and reasonable issues (e.g., the system capacity in terms of data transfer speed upper bound and the number of simultaneous clients supported). Handover management [10] is a mechanism to handle portability of clients roaming in the influence of different technologies or networks. When portable clients move from one wireless network to different wireless networks, the mobility mechanisms need to make sure that these sessions are not lost and that the data after roaming can be sent to the clients irrespective of their possible location after a move to another network. Such a mechanism provides rules for clients as well as networks to follow regarding the mechanisms and procedures for consistent services while roaming.

9.1.2 REQUIREMENTS OF SECURE COMMUNICATION

Securing [11] the client's hidden information is an essential purpose of a protection solution and the security architecture for future heterogeneous wireless networks such as the interworking of 5G and WLAN. Security prerequisites should be achieved wherever and whenever the web services are provided. A network operator is required to provide security solutions that detect all kinds of possible attacks, avoid the attacks, and protect from the attacks where the disclosure of Internet protocol–based networks are in large numbers of cellular terminals. The security of futuristic heterogeneous wireless networks can generally be classified into two groups that provide access network security at WANs and terminal security at cellular phone terminals, respectively. Hence, secure communication is a joint task among every component the architecture of heterogeneous networks of the future, which include mobile clients, Wi-Fi networks, cellular networks, regulatory bodies, and cellular device manufacturers. The security prerequisites should first be considered in the early phase of drafting the specifications and guidelines for any business wireless technology. We introduce the architecture of 5G WLAN in the next section.

9.2   FIFTH GENERATION-WLAN (5G-WLAN) ARCHITECTURE

In this section, we discuss 802.11 WLAN, 5G network technologies, and their internetworking.

9.2.1   802.11 WLAN

A WLAN is a wireless computer network that connects at least two gadgets utilizing high-frequency radio waves within a limited region (e.g., school, home, office building, and computer laboratory). This gives a client the capability to move around in a local coverage area while connected to the WLAN and can provide a link to the Internet. Most advanced WLANs are dependent on IEEE 802.11 guidelines, advertised under the Wi-Fi [12] brand name. A WLAN is, in all simplicity, a cable replacement technology that provides typical client devices, such as mobiles, PDA with means to move freely inside the coverage area while maintaining connectivity to the wireless network. The IEEE 802.11 is a collection of physical layer specifications and media access control for imposing WLAN communications' different frequency bands, including 2.4, 3.6, 5.8, and 60 GHz and 900 MHz. The specifications are standardized by the Institute of Electrical and Electronics Engineers' WLAN Standards Committee. The base edition of the WLAN standard was launched in 1997 [12] and has had subsequent amendments [13]. Since then, a progression of IEEE 802.11 guidelines has been proposed to enhance the efficiency of WLAN. The following is a list of a few of those standards.

- **IEEE 802.11a:** This standard uses the same frame format and data link layer protocol as the initial standard; however, the physical layer uses an OFDM-based air interface. It can provide a data rate of up to 54 Mbit/s and works in a 5-GHz band.
- **IEEE 802.11b:** This standard uses a media access method similar to the one defined in the initial standard. It has a maximum data rate of 11 Mbit/s and works in a 2.4-GHz band.
- **IEEE 802.11g:** It uses an OFDM-based communication scheme similar to 802.11a. It can provide an average throughput of about 22 Mbit/s and a data rate of up to 54 Mbit/s. It works in a 2.4-GHz band, which is the same as that of 802.11b.
- **IEEE 802.11n:** It works on a 2.4-GHz band and optionally on a 5-GHz band. It provides data speeds ranging from 54 Mbit/s to 600 Mbit/s.
- **IEEE 802.11ac:** It provides high-throughput Wi-Fi on a 5-GHz band. The specification has a single-link throughput of no less than 500 Mbit/s and multi-station throughput of no less than 1 Gbps.
- **IEEE 802.11ad:** It provides another physical layer for WLANs to work within a 60-GHz millimeter-wave spectrum. This frequency band has radically dissimilar propagation attributes from 2.4-GHz and 5-GHz bands, where the Wi-Fi networks work. It provides a highest data speed of up to 7 Gbit/s.
- **IEEE 802.11ah:** 802.11ah operates on 900-MHz Wi-Fi, which is perfect for low power utilization and long-range information communication. It is also called "the low-power Wi-Fi." It can be used for various objectives, such as extended-range hotspots, outdoor WLANs for cellular traffic offloading, and huge-scale sensor systems. It guarantees a range of up to 1 km with a minimum of 100 kbps throughput and a maximum of 40 Mbps of throughput. It is also capable of penetrating walls and different barriers, which is impressively superior to any past Wi-Fi standards.
- **IEEE 802.11af:** 802.11af utilizes unused television spectrum frequencies to transmit data. It is also called White-Fi. It can be utilized for low-power, wide-area range. It operates on frequencies between 54 MHz and 790 MHz.

A wireless LAN has three basic modes [14] of operation: infrastructure mode, ad hoc network mode, and mixed-network mode.

- **Infrastructure mode:** In this configuration, the wireless network consists of a wireless access point and numerous user equipments. In this mode, any kind of mobile client can communicate with every other type of mobile entity with the assistance of access points.
- **Ad hoc network mode:** In this mode, every mobile client or station is connected directly to different clients without the help of access points.
- **Mixed-network mode:** It is a type of network mode that is created by mixing ad hoc and infrastructure networks.

9.2.2 5G NETWORKS

In the past decade or so, several telecommunication industries started to redesign 4G to 5G wireless systems, which give a lot of broad communication solutions with good system performance and much higher data rates in terms of low packet-delivery latencies, low packet-loss ratios, and high reception ratios. According to the Federal Communications Commission (FCC), 5G technologies [15] will provide a lower Internet connection speed of about 100 Mbps to a user moving at a high speed and a higher Internet connection speed of up to 10 Gbps to support better real-time applications and multimedia applications. The second generation, or 2G, began in 1991 as a collection of standards that administered wireless telephone technology, without much worry for information transmission or the mobile web. The third generation (3G) concentrated on applications in voice communication, mobile Internet, mobile TV, and video calls. The fourth generation (4G) was intended to better support IP telephony (voice over IP), cloud computing, video conferencing, and online gaming and video streaming.

From primary cell phones to 4G LTE, the broadcast communication industry has changed a lot in a couple of years. We have jumped four G's in about as long as it took for Snoop Dogg to become Snoop Lion. Presently, the telecommunication sector is ready to enter the fifth generation, which guarantees 100 to 1000 times the speed of 4G LTE. It simply means a full-length movie can be downloaded in a couple of seconds. 5G is a term used to explain the forthcoming fifth generation of mobile network innovation, and its speed can be up to 1000 times faster than 4G. 5G technology is estimated to provide fresh frequency bands (much wider than the previous bands) along with broader spectral data transfer capacity per frequency channel. It will use millimeter waves (mmWave) for high frequencies greater than 24 GHz to increase throughput and speed. It is expected to allow more efficient data exchange between various gadgets. 5G is in its very early stages, and the networking regulatory bodies have not settled on a standard. It is possible that public demonstrations will occur by the year 2018 due to the fact that South Korea is going to show off its 5G technology at the 2018 Winter Olympics in Pyeongchang and plans to commercialize it by December 2020. The Japanese government has stated its aim to show off its 5G capability for handy smart phone use at the Tokyo Summer Olympics in 2020.

9.2.3 5G-WLAN INTERWORKING ARCHITECTURE

The integrating architecture [5,16,17] represents the interconnection between different authentication servers and new entities within the heterogeneous networks. A detailed outlook of 5G-WLAN's integrating structure is shown in Figure 9.1. Interworking architectures between 5G and WLAN networks are based on a tightly coupled architecture.

In a tightly coupled architecture, non-cellular WLANs are linked to the core network of 5G cellular networks as entry networks to offer Internet connectivity to its users. The interworking architecture is similar to a tree structure, where the root node baseband (BBU) is linked to different core network macro base stations (MBSs) using fiber. A leaf node is the user equipment (UE), and the other intermediate nodes are access points (APs), small-cell base stations (SBSs), and an access gateway (AG). The BBU retrieves the user information from the mobile client and validates the authentication credential provided by the mobile client. The information is exchanged

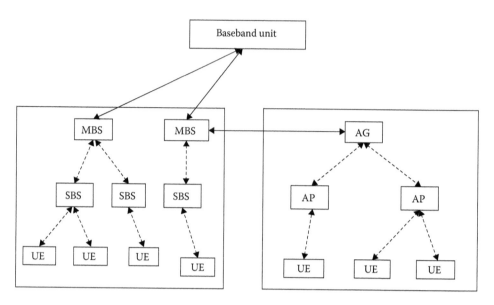

FIGURE 9.1   A basic architecture of a 5G-WLAN integrated network.

from the UE nodes to the SBSs and UEs nodes to the APs using mmWaves. The information is also exchanged from the SBS nodes to the MBS nodes and the AP nodes to the AG nodes using mmWaves. The information is exchanged from the AG nodes to the MBS nodes using fiber cable. In this architecture, a node, called a MSB, interconnects 5G and WLAN. The fundamental responsibilities of MSBs are signaling conversion of mobility/user records and message transferring between two heterogeneous networks. A MSB in a WLAN works as an access router and performs as a general packet radio service (GPRS) support node (GSN) in 5G networks. When UEs roam between 5G and WLAN, the MSB takes the responsibility of GSN functionality in the network to which the mobile client relocated. The wireless network is shown in dotted lines and the wired traffic in solid lines in Figure 9.1.

There are five main entities in a 5G-WLAN; AP, AG, SBS, MBS, and BBU controller, as shown in Figure 9.2. We have shown the network's signaling messages in solid lines and the user's traffic in dotted lines. A cellular terminal normally operates with more than one network interface, including 5G radio, a wireless card, and a universal subscriber identity module (USIM) card. A UE connects to a 5G network using a USIM card inside the coverage of a 5G network and to WLAN using a Wi-Fi card outside of its coverage. The UE informs the BBU via an SBS about its association and constructs a protected tunnel between the UE and the BBU when going into the coverage area of a 5G network.

Several integrated architectures have been discussed for 5G-WLAN heterogeneous networks, where the BBU remains in the top tier and the SBSs work under the administration of the BBU in a lower tier. A macrocell includes all small cells of diverse types, such as microcell (WLAN) and small cell (femtocell), and both tiers have a common frequency band. The small cell improves the service and coverage of a macrocell. There are two types of users in these heterogeneous networks: indoor and outdoor clients. To separate them, an MBS holds large antenna arrays with antenna elements dispersed around the MBS and linked to additional MBSs. The SBSs with large antenna arrays are deployed in every building to communicate with the MBS. All user equipment inside a building has an association to other user equipment either through an AP or an SBS. An SBS or an AP is located in a small area to permit communication among UEs, while a MBS is placed outside the zone to communicate with multiple SBSs and APs. Furthermore, 5G clients may go away from one small cell (SBS) and join a micro cell (AP) more frequently, which can introduce excessive handover latency in an integrated network. Due to micro- and macro-cell deployment in 5G-WLAN networks, there is a need to carry out frequent mutual authentications to stop impersonation and

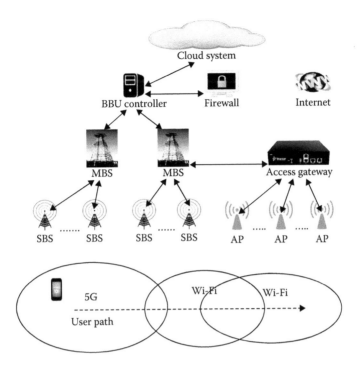

FIGURE 9.2   A tree architecture of a 5G-WLAN network.

MitM attacks. For that reason, efficient, robust, and faster privacy protection and handover authentication schemes must to be developed for 5G-WLAN integrated networks.

9.3   ROAMING MANAGEMENT IN 5G-WLAN INTERWORKING

Handoff and roaming [2,18] provide a facility to a UE associated with a base station to pass to another base station or AP without detaching its connection. Handoff management in heterogeneous networks connected with current cellular networks has many challenges (e.g., security, enhanced routing, lowest latency). Accessing multiple networks, network densification, zero latency, and very high movability make handoff management in 5G-WLAN networks harder. A heterogeneous network consists of UEs, SBSs, MBSs, a BBU controller, and APs. To enable handover between dissimilar wireless networks, various network devices and schemes are required because each network has its own structure in a 5G-WLAN heterogeneous network, rendering repeated establishments of trust connections and validations during movability necessary. The following two forms of handoffs [19] are provided in 5G-WLAN networks:

1. *Single radio access handoff (SRA handoff):* This is a handoff procedure in a network with similar network technology. It has two subtypes.
   - Intra-macrocell handoff: This refers to the handover of UEs from one SBS to other SBSs that operate under a single MBS.
   - Inter-macrocell handoff: This refers to the handover of UEs from one MBS to other MBSs. It might likewise involve the handover of UEs from one SBS to other SBSs that operate under different MBSs. However, if the roaming between small cells of two dissimilar MBSs is not done accurately, the inter-macrocell handoff will be unsuccessful.
2. *Multi–radio access handoff (MRA handoff):* This refers to the handover of UEs from one radio access technology to another radio access technology. It might likewise handoff between SBSs and APs.

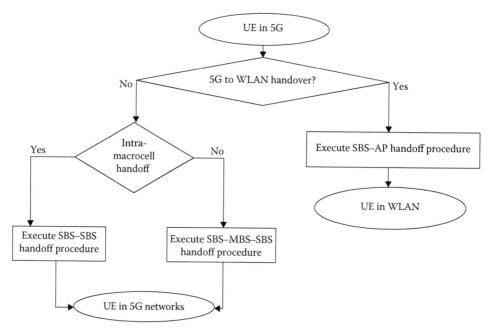

FIGURE 9.3 Handoff procedure during 5G to WLAN handover.

Figure 9.3 explains the handover procedure during handover from 5G to WLAN. Once the UE gets the required data to settle on the handover decision, it can perform one of the three following types of handover procedures:

1. **SBS–SBS handoff:** This is a handoff procedure for intra-macrocell handoff.
2. **SBS–MBS–SBS handoff:** This is a handoff procedure for inter-macrocell handoff.
3. **SBS–AP handoff:** This is a handoff procedure for multi-radio access handoff. The SBS–APhandover procedure is executed when a 5G client is entering a WLAN.

The handling of handovers in a seamless and efficient manner is one of the strongest challenges in heterogeneous networks. These handover actions need be enhanced to guarantee non-perceptible interruption to current communication as a client moves from one network type to another. Unsuccessful handover is typically caused by its long latency. For seamless mobility in 5G and WLAN networks, low handover delay is required. The most important factor of handover delay is the reauthentication procedure required to guarantee the secure passing and confirmation of the network. A fast and frequent handover of UEs over SBSs or SBS–AP is required for a secure, efficient, and robust roaming process for transferring the information. Handover authentication of multi-RAHs (MRAHs) is more difficult because every radio access technology has its own security parameters, and a fixed procedure is used to provide efficient security. Obviously, there is a need to provide overlapped security solutions across various types of radio access technologies. The other challenges (e.g., availability of networks, authorization and access control of UEs, confidentiality of data transmission, integrity of data communication, auditing and accounting of user information, low computation cost, and low energy cost) need to be addressed to create a secure and seamless 5G-WLAN heterogeneous network. The existing privacy and security solutions to heterogeneous networks are not efficient; so, these schemes are not capable of holding large connections simultaneously.

We may imagine a feasible solution for developing a handover reauthentication scheme that must consider end-to-end security, secure data transmission, secure and private storage, threat-resistant

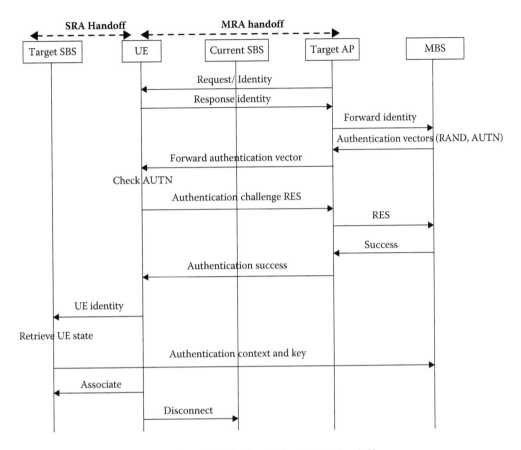

FIGURE 9.4 Authentication procedures for SRA handoff and MRA handoff.

UEs, a legitimate network, and resource access. Figure 9.4 provides an outline of a handover re-authentication methodology between dissimilar networks and inside one network. It can be seen from Figure 9.3 that mutual authentication throughout handover between a client and another WLAN network is acknowledged by matching a particular hashing value. Every time the involved vector [4] includes AUTH, an authentication token sent by the MBS, RAND, an arbitrary number known by the MBS, and a pairwise key. For movability inside a similar network (i.e., SRA handoff), the present serving SBS informs the target SBS of possible handover so that the latter can find the client a cryptographic value from the MBS. We explore the existing handover authentication methodologies and recognize the difficulties in 5G-WLAN heterogeneous networks. The Third Generation Partnership Project (3GPP) [20] has specified a handover message flow and a specific key hierarchy for various mobility circumstances.

Our (designed) session key for diverse handover methodology for various situations, such as 5G-WLAN handover, will enhance handover complexity. The BBU is generally located far away from the user, which requires a delay of up to hundreds of milliseconds due to the frequent transmission of messages among SBSs, UEs, APs, and the BBU for client authentication. Such large delay cannot be accepted for seamless handover. The authors [21,22] have discussed the simplified handover authentication schemes for LTE networks, which involve only UEs and APs for authentication using public-key cryptography. These schemes provide key agreement and mutual authentication with other networks through a three-way handshake without contacting any other entity, such as the BBU. The computational cost, energy cost, and handover delay of a handover authentication procedure are increased due to the overhead for transferring many cryptographic messages (e.g., a digital signature through a wireless interface). Due to this reason, transferring a digital signature is secured but not significant for seamless 5G-WLAN wireless communications. Because of these difficulties, significant and efficient handover authentication mechanisms and secure data transmission are critical for securing 5G-WLAN heterogeneous networks.

9.4 PROPOSED REAUTHENTICATION SCHEME FOR MRAH

In this section, we propose a MRAH authentication scheme for 5G-WLAN to permit frequent roaming between an SBS and an AP. Since the target cell is an AP of WLANs and the source cell is an SBS of 5G networks, the proposed reauthentication scheme utilizes the USIM to share private data for handover authentication between a client and an MBS. We assume that both the target cell APs and source-cell base station SBS are connected through the same MBS. Furthermore, a mobile client can use mobile office through his or her USIM. Our proposed scheme consists of four phases: setup phase, user registration phase, login and authentication phase, and reauthentication phase, which are presented in the following subsections.

9.4.1 SETUP PHASE

To initialize the system, the BBU picks up two large prime numbers, p and q, and calculates $n = p \times q$. The BBU keeps p and q as secret keys and distributes n as a public key. The BBU also takes an integer D_BBU and a prime number P_BBU such that $P\_BBU \times D\_BBU \equiv 1 \mod (p-1)(q-1)$, and a cryptographic one-way hash function, h, characterized as $h(.): \{0, 1\}^* \rightarrow \{0, 1\}^l$, where $\{0, 1\}^l$ is a binary string of fixed length l, D_BBU is the private key, and P_BBU is the public key of the BBU.

9.4.2 USER REGISTRATION PHASE

The registration phase is performed when a UE registers with the BBU or an authentication server. The UE and the BBU execute the following steps:

Step 1: UE selects an identity (ID), a password (PW), IMSI, and a random nonce (a) and computes $UID = h(ID)$, $HPW = h(ID|PW\|h(a))$, $TMSI = h(IMSI)\ XOR\ h(a)$. He sends UID, HPW, and TMSI to BBU for registration through a protected channel.

Step 2: After getting the registration information, BBU calculates the following to share secret data for authentication between UE and BBU:

$$A = h(UID \| HPW \| TMSI)$$

$$B = h(HPW)\ x\ or\ h(TMSI \| s)$$

Step 3: BBU stores [P_BBU, A, B, h (·)] in USIM and hands it over to UE through a protected channel.

Step 4: After getting USIM, UE stores TMSI in his USIM; thus, USIM contains [P_BBU, A, B, TMSI, h (·)].

9.4.3 LOGIN AND AUTHENTICATION PHASE

The login and authentication phase transfers a login request message to the corresponding MBS when the UE wants to get some resource from the BBU. The BBU accomplishes mutual authentication with the UE by checking the login request message of the UE and then accomplishes secure communication by calculating the session key, SK, between the MBS and the UE. Figure 9.5 demonstrates the processes of the login and authentication phases.

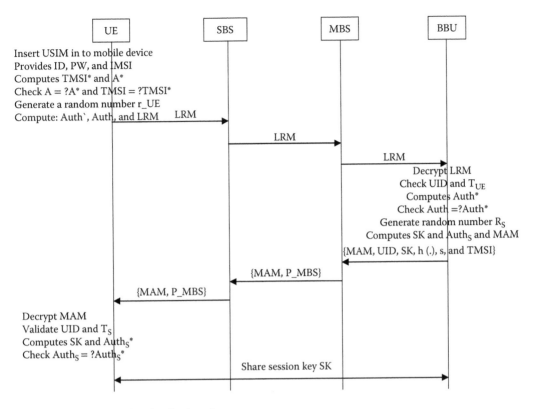

FIGURE 9.5   Login and authentication phases.

Step 1: UE inserts his USIM into a smart cell phone and provides IMSI, PW, ID, and a. After getting it, USIM calculates TMSI* = h(IMSI) XOR $h(a)$ and compares it with the stored TMSI. If it is equal, USIM trusts the nonce, a, and assumes that IMSI has originated from a similar client; otherwise, it terminates the connection. The USIM computes A* = h(UID‖HPW‖TMSI*) using a, ID, and PW and checks whether A* is equal to A. If they are not equal, USIM discards the login request. The USIM creates an arbitrary integer number, r_{UE}, and encrypts {UID, Auth, r_{UE}, T_{UE}} using the public key of BBU and sends the login request message LRM = [UID, Auth, r_{UE}, T_{UE}]$_{P\_BBU}$ to MBS through an insecure channel after computing the following:

$$Auth` = B \text{ x or } h(h(ID|PW \| h(a)))$$

$$Auth = h(Auth`\|r_{UE}\|T_{UE})$$

Step 2: MBS forwards LRM to BBU.

Step 3: After getting the login request message, BBU decrypts LRM utilizing its own private key, D_BBU, and verifies UID and T_{UE}. If this verification does not hold, BBU rejects the login request of UE; otherwise, BBU calculates Auth* = $h(h(TMSI\|s)\|r_{UE}\|T_{UE})$ using s, TMSI, r_{UE}, T_{UE} and checks whether Auth* = Auth. If they are equal, BBU accepts the login request of UE, creates an arbitrary R_s, and calculates a shared session key, SK = $h(r_{UE}\|R_s\|s\|TMSI)$, and Auth$_s$ = HPW‖h(SK‖T_s), where T_s is the current timestamp of BBU, and Auth$_s$ is the information used when BBU authenticates UE. The BBU computes the authentication message MAM = $[UID, Auth_S, R_s, T_S]_{r\_UE}$, where BBU encrypts {UID, $Auth_S$, R_S,T_S} using $r_{\_UE}$. BBU sends MAM, UID, SK, P_MBS, h (.), s, and TMSI to the corresponding MBS.

Step 4: MBS sends the message MAM, P_MBS to UE for mutual authentication.

Step 5: After getting the mutual authentication message, UE decrypts MAM using $r_{\_UE}$ and approves UID and T_S. If this verification does not hold, UE terminates the present session; otherwise, UE calculates SK = $h(r_{\_UE}\|R_s\|s\|TMSI)$, $Auth_S^* = HPW\|h(SK\|T_S)$, and checks whether $Auth_S^* = Auth_S$. If they are not equal, the process is terminated; otherwise, UE effectively authenticates BBU. After completing mutual authentication, UE and BBU can communicate securely in the future using the session key SK.

9.4.4 Reauthentication Phase

When a UE accesses a BBU using a SBS, an MRA handover event takes place for moving to another access point. For that reason, the UE executes one more authentication to the BBU. The reauthentication phase performs the handover authentication with the BBU using the existing authentication data so that the UE does not require the execution of the full authentication phase once more in case of MRA handover. Figure 9.6 demonstrates the reauthentication phase in the proposed MRAH reauthentication scheme.

Step 1. When the handover happens, USIM creates an arbitrary number, r^*, for authentication with BBU. The USIM encrypts {UID, r^*, SK, T_{UE}} using the public key of MBS and transfers $[UID, r^*, SK, T_{UE}]_{P\_MBS}$ to AP for the reauthentication request, where SK is the session key shared between UE and BBU in the previous session, and T_{UE} is the current timestamp of UE.

Step 2. AP forwards $[UID, r^*, SK, T_{UE}]_{P\_MBS}$ to MBS.

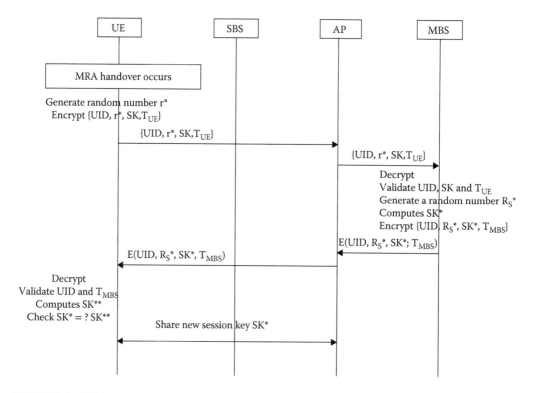

FIGURE 9.6 MRA reauthentication phase.

Step 3. After getting the reauthentication request message, MBS decrypts [UID, $r*$, SK, $T_{UE}]_{P\_MBS}$ using its own particular private key and verifies UID, T_{UE}, and SK. If this verification does not hold, MBS rejects the reauthentication request of UE; otherwise, MBS accepts the reauthentication request of UE, generates a random R_S*, and calculates $SK^* = h(r* \|R_s^*\|s\|TMSI)$ for the following session.

Step 4. The MBS encrypts $\{UID, R_S*, SK^*, T_{MBS}\}$ using $r*$ and transfers the mutual reauthentication message $E(UID, R_S^*, SK^*, T_{MBS})$ to AP, where T_{MBS} is the current timestamp of MBS.

Step 5: AP forwards the mutual reauthentication message $E(UID, R_S^*, SK^*; T_{MBS})$ to UE for reauthentication.

Step 6. After getting the mutual reauthentication message for handover authentication, UE decrypts $E(UID, R_S^*, SK^*; T_{MBS})$ using $r*$ and verifies UID and T_{MBS}. If this verification does not hold, UE terminates the session; otherwise, UE calculates $SK^{**} = h(r*\|R_s^*\|s\|TMSI)$ and checks whether SK^{**} is identical to SK^*. If they are identical, UE effectively reauthenticates BBU, and UE and BBU can communicate securely in the future using the session key SK^*.

9.5 PERFORMANCE EVALUATION

This section presents the performance evaluation of the proposed reauthentication scheme for the 5G-WLAN heterogeneous architecture and compares it with an authentication scheme of a 3GPP LTE–WLAN heterogeneous network in terms of security features, computational cost, and energy cost. In comparison to Elbouabidi et al.'s scheme (2014) of 3GPP LTE–WLAN, it does not just provide mutual authentication and resist many attacks, but it also utilizes the hash algorithm that takes much less computational time and hence is more favorable for small, portable smart phones with limited computing power.

9.5.1 SECURITY ANALYSIS

In this section, we evaluate the security features offered by our proposed scheme.

9.5.1.1 Mobile Device Attack

Regardless of the possibility that an attacker gets or steals the cell phone, he cannot masquerade as a UE. If an attacker gets or steals the cell phone of a UE that is connected to a heterogeneous network, he cannot get the ID and PW of the UE from the cell phone. Additionally, if the attacker impersonates the UE to obtain the secret data P_BBU, A, B, h(.), and TMSI stored in the USIM, he cannot masquerade as a genuine UE without knowing HPW, s, ID, and UID.

9.5.1.2 Password-Guessing Attack

In the proposed MRAH scheme, if an attacker determines the PW of a UE from the transmitted authentication messages over a public channel, he cannot compute a new password HPW of the client without knowing $h(a)$, ID, and the transmitted authentication message. Even if the attacker gets the private data kept in the USIM, he cannot guess the password of the UE because A and B are computed on the basis of HPW.

9.5.1.3 Impersonation Attack

An attacker cannot act as a genuine UE, since he cannot calculate the reauthentication or login request message of the UE. If an attacker tries to compute the login request message, LRM = [UID, Auth, r_{UE}, $T_{UE}]_{P\_BBU}$, or the reauthentication message, [UID, $r*$, SK, $T_{UE}]_{P\_MBS}$, of the USIM, he cannot calculate LRM or the reauthentication message, [UID, $r*$, SK, $T_{UE}]_{P\_MBS}$, because UID, s, a, and SK are unknown to him.

9.5.1.4 Replay Attack

Regardless of the possibility that an attacker replays the message that gets the reauthentication message or the login request message transmitted between the UE and the MBS over a public channel, he cannot masquerade as an authorized client. Suppose the attacker replays an earlier login request message, LRM = [UID, Auth, r_UE, $T_{UE}]_{P\_BBU}$, or reauthentication message, [UID, $r*$, SK, $T_{UE}]_{P\_MBS}$, to impersonate a client; he cannot succeed in a replay attack because the timestamp, T_{UE}, fails to confirm. Furthermore, even if the attacker replays to create a legitimate timestamp, T^*_{UE}, and calculates a new login request message, LRM = [UID, Auth*, r_UE^*, $T_{UE}^*]_{P\_BBU}$, or reauthentication message, [UID, $r**$, SK, $T_{UE}^*]_{P\_MBS}$, he cannot impersonate a legitimate UE because UID, s, $h(a)$, and SK are unknown to him.

9.5.1.5 Eavesdropping Attack

If an attacker monitors the messages transmitted between the UE and the BBU over a public channel, he cannot get the actual message content in the third phase of our proposed scheme. Suppose he gets message LRM = [UID, Auth, r_UE, $T_{UE}]_{P\_BBU}$ and the mutual authentication message MAM = [UID, $Auth_s$, R_s, $T_s]_{r\_UE}$ circulated in the third phase of the proposed scheme; he cannot find the content of the message without knowledge of r_UE and SK. Also, when the attacker gets the reauthentication message [UID, $r*$, SK, $T_{UE}]_{P\_MBS}$ and the mutual reauthentication message E(UID, R_s^*, SK*; T_{MBS}) circulated in the MRAH authentication phase of the proposed scheme, he cannot get the actual content of the message.

9.5.1.6 Mutual Authentication

The proposed scheme offers mutual authentication in both the login and the authentication phase and the MRAH authentication phase. The BBU will authenticate the UE using the *Auth* parameter in the login request message, LRM, and the UE will authenticate the BBU using the *Auth$_S$* parameter in the mutual authentication message, MAM, of the login and authentication phases. Furthermore, the MBS will authenticate the UE using SK and UID in the reauthentication message [UID, $r*$, SK, $T_{UE}]_{P\_MBS}$, and the UE will authenticate the MBS using UID and SK* in the mutual reauthentication message [UID, R_s^*, SK*, $T_{MBS}]_r^*$ of the MRAH authentication phase.

9.5.2 Computational Time

The computational time represents the required running time for the base station and the UE when they execute an authentication scheme. We have computed the running of the fundamental cryptographic algorithms using the CryptoPP library [23] tested on an Intel Core 2 with a CPU frequency of 1.83 GHz under Windows Vista (32-bit mode), as shown in Table 9.2. In Table 9.3, we have

---

TABLE 9.2

Time Costs of Fundamental Cryptographic Algorithms (in ms)

T_{AES}	T_H	T_{RSA}	T_{MIC}
2.56×10^2 Cycles	1.82×10^2 cycles	0.16×10^6 Cycles	0.3522×10^2 Cycles

T_{AES}: Execution time of a 128-bit AES-CBC encryption/decryption
T_H: Execution time of generating a 128-bit message digest using a SHA-1 algorithm
T_{RSA}: Execution time of a 1024-bit RSA encryption/decryption
T_{MIC}: Execution time of VMAC message authentication code

TABLE 9.3

Computational Time

Schemes	Computational Time of MRA Handover
Our proposed scheme	0.0175 msec
Elbouabidi et al.'s scheme	0.066 msec

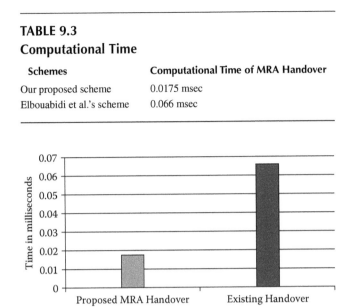

FIGURE 9.7   Computational time of reauthentication scheme.

presented the computational cost comparison of the proposed reauthentication scheme of the MRA handover with the schemes of 3GPP LTE–WLAN handover. As evident from this table, our proposed reauthentication scheme provides a 75% reduction in computational cost as compared to that of Elbouabidi et al.'s scheme [4]. Figure 9.7 demonstrates a bar graph diagram of the computational cost comparison of the existing and proposed MRAH schemes.

9.5.3   ENERGY OVERHEAD

This section provides the energy cost evaluation that analyzes the energy cost utilized by Elbouabidi et al.'s scheme [2] and the proposed reauthentication scheme. The energy cost has been evaluated as the energy consumed in cryptographic and communication operations. The notations of the energy expense of security and communication activities are shown by the following variables.

E_M: Energy expense of sending or receiving a message
E_{MAC}: Energy expense of computing or examining a MAC value using a key
E_{VMAC}: Energy expense of computing or verifying a MAC using a VMAC hashing function
E_{L-H}: Energy cost of key calculation
$E_{ENC-SYM}$: Energy expense of data enciphering or deciphering using symmetric key cryptography
$E_{ENC-ASYM}$: Energy expense of data enciphering or deciphering using asymmetric key cryptography

Elbouabidi et al.'s scheme involves the device in the following energy expending activities: (i) sending and receiving messages, (ii) generation of 2*LK, 2*LID, 2*HK, and 2*HID, (iii) generation and verification of two signatures, (iv) encryption/decryption of six messages by a public key, and (v) encryption/decryption of two messages by a private key. Thus, the overall energy cost of Elbouabidi et al.'s scheme [4] is computed as

$$E_{Elbouabidi} = 6 \times E_{ENC-ASYM} + 2 \times E_{ENC-SYM} + 4 \times E_{MAC} + 8 \times E_{L-H} + 7 \times E_M \qquad (9.1)$$

The proposed reauthentication scheme involves the device in the following energy expending activities: (i) encryption/decryption of one message by an asymmetric key, (ii) generation and verification of two hash values by VMAC, (iii) sending and receiving messages, and (iv) encryption/decryption of one message by a symmetric key. Thus, the overall energy cost is given by

$$E_{Proposed} = 2 \times E_{ENC\text{-}ASYM} + 2 \times E_{ENC\text{-}SYM} + 2 \times E_{VMAC} + 4E_M \qquad (9.2)$$

From Equations 9.1 and 9.2 it can be observed that the proposed scheme of MRAH diminishes the energy consumption as compared to Elbouabidi et al.'s scheme [4] of 3GPP LTE–WLAN heterogeneous networks.

9.6 CONCLUSIONS

This chapter discusses the interworking architecture of 5G-WLAN heterogeneous architecture and roaming management of heterogeneous networks. In future 5G-WLAN heterogeneous networks, deployment of SBSs will bring difficulties in security and privacy issues. We have assessed the current methods for handover and tried to present difficulties on privacy protection and authentication in a heterogeneous network. This chapter also presents a solution for securing a 5G-WLAN heterogeneous system to fulfill the security prerequisites. We have explored the existing handover authentication techniques of the present heterogeneous networks. We found that it is not efficient for the latest heterogeneous networks to adapt the risks from mutually 5G cellular networks and WLAN networks in a sequent manner. In addressing these challenges, we have discussed a USIM-based handover authentication scheme that not only reduces energy consumption, but also reduces computational cost. The proposed MRAH scheme is efficient regarding security functionalities and accomplishes efficient session-key verification, mutual authentication properties, and early incorrect input recognition in the login phase. Regardless of a few difficulties, the future looks bright for 5G-WLAN interworking, and new security provisions will be beneficial for cellular operators and clients. The emerging 5G-WLAN heterogeneous communications will have difficulties with the existing security provisioning methods, particularly with respect to latency and complexity. Broader security and performance analysis will be the theme of our future work. In the future, we would try to present a comprehensive analysis concentrated on the mechanisms of roaming management, handover management, data transmission, or forwarding.

REFERENCES

1. Topside E. Mathonsi, Okuthe P. Kogeda. Handoff delay reduction model for heterogeneous wireless networks. IST-Africa Week Conference, 2016, pp. 1–7, Durban, South Africa.
2. Xiaoyu Duan, Xianbin Wang. Authentication handover and privacy protection in 5G HetNets using software-defined networking. *IEEE Communications Magazine*, 53, no. 4 (2015): 28–35.
3. Mamta Agiwal, Abhishek Roy, Navrati Saxena. Next generation 5G wireless networks: A comprehensive survey. *IEEE Communications Surveys & Tutorials*, 18, no. 3 (2016): 1617–1655.
4. Imen Elbouabidi, Faouzi Zarai, Mohammad S. Obaidat, Lotfi Kamoun. An efficient design and validation technique for secure handover between 3GPP LTE and WLANs systems. *Journal of Systems and Software*, 91 (2014): 163–173.
5. Nisha Panwar, Shantanu Sharmaa, Awadhesh Kumar Singh. A survey on 5G: The next generation of mobile communication. *Physical Communication*, 18 (2016): 64–84.
6. IEEE 802.11r, Wireless LAN Medium Access Control (MAC) and Physical Layer (PHY) Specifications: Fast BSS Transition, IEEE Standard, 2008[online]. Available at:http://ieeexplore.ieee.org/document/4573292/.
7. Mohammad Farash, Muhamed Turkanovi. An efficient user authentication and key agreement scheme for heterogeneous wireless sensor network tailored for the Internet of things environment. *Ad Hoc Networks*, 36 (2016): 152–176.
8. Ian F. Akyildiz. 5G roadmap: 10 key enabling technologies. *Computer Network*, 106 (2016): 17–48.

9. Chun-I Fan, Yi-Hui Lin, Ruei-Hau Hsu. Complete EAP method: User efficient and forward secure authentication protocol for IEEE 802.11 wireless LANs. *IEEE Transaction on Parallel and Distributed System*, 24 (2013): 672–680.

10. Yun Li, Bin Cao, Chonggang Wang. Handover schemes in heterogeneous LTE networks: Challenges and opportunities. *IEEE Wireless Communications*, 23, no. 2 (2016): 112–117.

11. Kashif Faheem, Khalid Rafique. Securing 4G/5G wireless networks. *Computer Fraud and Society* 5 (2015) : 8–12.

12. https://en.wikipedia.org/wiki/IEEE_802.11. Accessed on 25 August 2016.

13. IEEE 802.11, The Working Group Setting the Standards for Wireless LANs. Retrieved 25 August 2016.

14. Iti Saha Misra. *Wireless Communications and Networks: 3G and Beyond*. New Delhi: McGraw Hill Education (India) Pvt Ltd, pp. 16–17, 2013.

15. Jiann-Ching Guey, Pei-Kai Liao, Yih-Shen Chen, et al. On 5G radio access architecture and technology [Industry Perspectives]. *IEEE Wireless Communications*, 22, no. 5 (2015): 2–5.

16. Cheng-Xiang Wang, Fourat Haider, Xiqi Gao, et al. Cellular architecture and key technologies for 5G wireless communication networks. *IEEE Communications Magazine*, 52, no. 2 (2014): 122–130.

17. Naga Bhushan, Junyi Li, Durga Malladi, et al. Network densification: The dominant theme for wireless evolution into 5G. *IEEE Communications Magazine*, 52, no. 2 (2014): 82–89.

18. Borja Bordel Sanchez, Alvaro Sanchez-Picot, Diego Sanchez De. Using 5G technologies in the Internet of Things: Handovers, problems and challenges. *9th International Conference on Innovative Mobile and Internet Services in Ubiquitous Computing (IMIS)*, 2015, pp. 1–6.

19. Maissa Boujelben, Sonia Ben Rejeb, Sami Tabbane. A novel green handover self-optimization algorithm for LTE-A/5G HetNets. *IEEE Conference*, 2016, pp. 1–6, Dubrovnik, Croatia.

20. 3rd Generation Partnership Project, 3GPP TS 33.401 V11.2.0 (2011–12), 3GPP System Architecture Evolution (SAE); Security Architecture (Release 11) [online]. Available at:https://portal.3gpp.org/desktopmodules/Specifications/SpecificationDetails.aspx?specificationId=2296.

21. Jaeduck Choi, Souhwan Jung. A handover authentication using credentials based on chameleon hashing. *IEEE Communications Letters*, 14, no. 1 (2010): 54–56.

22. Jin Cao, Hui Li, Manode Ma, et al. A simple and robust handover authentication between HeNB and eNB in LTE networks. *Computer Networks*, 56, no. 8 (2012): 2119–2131.

23. Cryptographic Algorithms. https://www.cryptopp.com/benchmarks.html((accessed August 15, 2016).

10 Simulators, Testbeds, and Prototypes of 5G Mobile Networking Architectures

Shahram Mollahasan, Alperen Eroğlu,
Ömer Yamaç, and Ertan Onur

CONTENTS

10.1 INTRODUCTION

The recent analyses claim that each person will have tens, if not hundreds, of Internet-connected devices by 2020, and the total worldwide number of devices will rise to 50 billion (Evans, 2011). The devices will range from RFID devices, tablets, and smart phones to augmented- or virtual-reality equipment. The traffic and other quality-of-service (QoS) requirements of these devices will be diverse. 5G systems are expected to be flexible to satisfy a varied set of user and system requirements, such as low latency, high capacity, and high reliability. A huge research and development

185

effort is invested in the challenges of 5G networks. The products of the 5G research activities have to be validated on simulators and testbeds that provide a realistic environment.

Simulators provide an experimentation environment dismissing the need for hardware, and they provide users an insight into the operational performance of the proposed solutions, whereas testbeds provide a more realistic environment in which to conduct practical experiments. To make the 5G systems a reality, 5G testbeds have been deployed all over the world, mostly in a federated fashion.

In this chapter, we present an overview of distinct approaches and reasons to build 5G testbeds and experimentation environments considering various design and research concerns. Testbed requirements and the existent solutions are discussed. Nowadays, we have a huge opportunity to access a wide variety of 5G testbeds that we can employ based on our needs and conditions. The existent and developing 5G simulators, 5G testbeds, projects, and federated platforms, such as FIRE and GENI, are investigated and compared in this chapter. At the end of this chapter, the reader will obtain a comprehensive overview of the existing 5G experimentation testbeds and simulators and the main learning points while designing a testbed for 5G. In addition, this chapter aims at providing readers a guideline to determine the key requirements and a means for choosing an adequate experimentation approach.

The remainder of this chapter is organized as follows. Section 10.2 and Section 10.3 present the widely utilized simulators and emulators. We compare the simulators in terms of learning difficulty, development language, open-source support, graphical user interface support, emulation support, type of simulator, and the other key features, such as flexibility, capacity, frequency spectrum, and any other performance parameters. In Section 10.4, we compare and investigate the existing 5G testbeds in terms of some important topics and requirements, such as mobility support, heterogeneity, coverage, size, protocols, applications, and frequency spectrum. Moreover, we provide an architectural perspective with taxonomy of components, such as the backbone (wired, wireless), hierarchy (tiers), and management system (scripts, interconnections, etc.), and address some constraints, such as capacity, and challenges, such as coverage, energy efficiency, and network reliability.

10.2 5G SIMULATORS AND EMULATORS

There are many different 5G simulation projects in the literature (e.g., Redana et al., 2015; Keysight Technologies, 2016). In this section, we discuss some of the well-known 5G simulator projects such as the 5G NOvel Radio Multiservice adaptive network Architecture (NORMA) project's simulator, SystemVue, New York University's (NYU) wireless 5G simulator, electromagnetic (EM) propagation software, the 5G millimeter-wave (mmWave) module for the ns-3 simulator, OpenAirInterface, and the cloud radio-access network (C-RAN) simulator. At the end, these simulators are compared with each other from a different perspective.

As shown in Figure 10.1, there are three main types of approaches to developing simulators for mobile networks, which can be broadly categorized as link level, system level, and protocol level. The link-level simulators can be used to simulate the wireless channels, and they are generally used for single-cell scenarios because of the high complexity in link-level simulations of multicell models (Zhang et al., 2011). In this type of simulator, the impact of interference by other cells has to be ignored. The impact of noise and fading can be considered in simulations. The system-level simulators

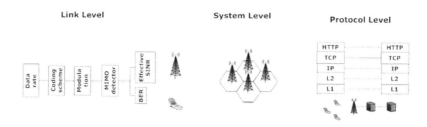

FIGURE 10.1   Various types of simulators.

can be used to analyze coverage, network deployment, and system functions such as fast power control and adaptive scheduling (Cosovic et al., 2009). They are also suitable for simulating spatial processing techniques in depth. Simulation time in system-level simulators is on the order of milliseconds. The protocol-level simulators are used to analyze the performance of multicell and multiuser in layer 3 and above (Cosovic et al., 2009). Various protocols can be modeled in this type of simulator. Due to the long response time of higher layers, simulation time is on the order of seconds up to minutes.

Using simulation tools in comparison with the existent testbeds has some advantages, such as developers do not need to build the system to analyze it. The cost of building an experiment needs to be considered. By dismissing the specific hardware needs, the costs of the experimentation can be reduced when simulators are employed. If the researchers can obtain reliable and credible results by simulations, they may discover unexpected phenomena besides studying the system and acquiring insights. An equally important point is that interpreting and evaluating the real model is not always as easy as the simulated version. However, notice that only insights will be gained by simulations due to the abstraction of many stochastic phenomena that may occur in practice.

In the following, we present the 5G simulators and simulation platforms and discuss their features. After presenting a broad overview of the existent simulators, we compare them.

10.2.1   5G NORMA PROJECT

NORMA (Redana et al., 2015) is a 5G system-level simulator produced by NOvel MObile Radio (NOMOR™) as a European Horizon 2020 (H2020) project. NOMOR developed a simulator based on its real-time network simulator (RealNeS) platform. In this simulator, developers are capable of deploying different cells, such as macro cell, pico cell, and femto cell, that can be employed in heterogeneous networks. Base stations in NORMA can be designed in a flexible fashion, and researchers may study adaptive networks. It is also possible to use different types of transceivers, such as multiple-input multiple-output (MIMO), MIMO minimum mean-squared error (MMSE), and even zero forcing (ZF), in the simulations. NORMA also increases the accuracy of channel models by providing fading, path loss, and user mobility. Different types of parameters for designing antennae such as beamforming and three-dimensional antenna patterns are implemented. Thanks to the parallel computing architecture of NORMA, processing of the collected data can be done with very high speed.

10.2.2   SYSTEMVUE™ ESL SOFTWARE

The simulator SystemVue (Keysight Technologies, 2016), introduced by Keysight Technologies, focusses mainly on electronic system-level (ESL) design to develop the physical layer of a wireless communication system. The SystemVue platform can be used as a general-purpose environment for digital and analog systems. After its 2015.01 release, it also supports 5G systems, and its reference libraries can be directly integrated into MATLAB®. In SystemVue, baseband designers can virtualize radio frequency (RF). It supports a huge number of communications-oriented mathematical functions implemented in a multi-threaded fashion. Besides the advanced scheduler with multi-rate that allows complex topologies to be implemented in SystemVue, developers can also design equipment that can interact through TCP/IP dataflow. Finally, SystemVue facilitates the integration of model-based designs to real-world systems to accelerate the validation procedure. The simulation environment of SystemVue, shown in Figure 10.2, depicts its capability for designing and simulating systems at the same time.

Keysight Technologies has also introduced phased-array beamforming as a software module for SystemVue that can be used as a dynamic and accurate system-level simulator for analyzing active electronically scanned array (AESA) platforms, expanding range, reducing interference, and decreasing power consumption by enabling developers and architects to work on beamforming algorithms in 5G networks. By using this software, one can predict performance degradation of 5G networks in digital and hybrid beamforming architectures. By using AESA systems, architectures can access up to 256 elements in 5G applications. System developers can analyze the performance of

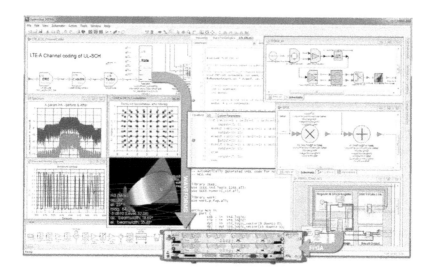

FIGURE 10.2 Demonstration of the SystemVue environment. (From Keysight Technologies. [December, 2016]. SystemVue Electronic System-Level [ESL] Design Software. [Online]. www.keysight.com/en/pc-1297131/systemvue-electronic-system-level-esl-design-software?cc=TR&lc=eng.)

baseband models that can reduce expenses and complexity thanks to the integration of SystemVue, MATLAB, and tools with RF and system design groups. This software is an add-on module for SystemVue and facilitates designers simply and quickly for simulating adaptive beamforming algorithms and multifunction arrays.

10.2.3 NYU WIRELESS 5G MILLIMETER-WAVE STATISTICAL CHANNEL MODEL

One of the first open-source simulators that allows researchers to use three-dimensional statistical spatial channel model data is NYU's wireless simulator. This simulator is based on real experiments that were carried out between 2011 and 2014 over mmWaves. Researchers can analyze the behavior and potentials of the mmWave spectrum that are obtained from outdoor scenarios. The database was populated in 4 years, and statistics were collected in New York City and Austin, Texas. Researchers may employ frequencies ranging from 2 to 73 GHz in this simulator (Samimi and Rappaport, 2016). This simulator facilitates researchers to model multipath propagation. Moreover, researchers can calculate the spread of delay that may occur during transmission due to multiple copies of a signal at the receiver side with different delays and amplitudes. The received power level and angle of arrival (AoA) of signals over urban channels are provided, since in the future, a smart antenna will be able to calculate the optimal angle of arrival and the angle of departure to improve the signal power (Samimi et al., 2013). Furthermore, the statistical channel model and the simulation codes are also provided in MATLAB.

10.2.4 WIRELESS EM PROPAGATION SOFTWARE

Remcom has been recently working on an innovative MIMO simulator using Wireless InSite™ (Remcom, 2016), which is an electromagnetic (EM) propagation software employed for analyzing site-specific wireless channels and radio propagation systems. By using this software, researchers and developers may predict accurate and efficient EM propagation and model communication channels in different environments, such as urban, indoor, rural, and even a combination of the environments. Remcom provides the capability of 5G channel models with 3D structures and MIMO. The user interface of Wireless InSite during the implementation of the outdoor massive MIMO scenario of Rosslyn is shown in Figure 10.3.

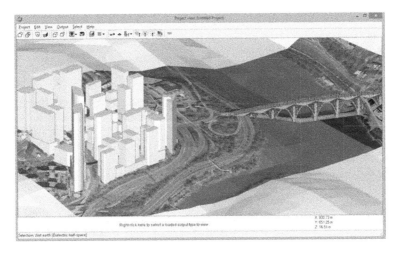

FIGURE 10.3   The user interface of Wireless InSite during implementation of the outdoor massive MIMO scenario of Rosslyn. (From Remcom. [2016]. 5G and MIMO using Wireless InSite. [Online]. www.remcom. com/5g-mimo.)

10.2.5   5G mmWave Module for NS-3 Simulator

NS-3 is an open-source network simulator. A 5G mmWave module has been recently developed for NS-3 (Mezzavilla et al., 2015). These modules model 5G channels, the physical layer, and the MAC layer of mmWave systems. The C++ language is used to develop this module based on the NS-3 Lena module. The main features of this simulator can be briefly summarized as follows:

- It includes a customizable time-division duplex (TDD) implementation.
- Small-scale and large-scale channels can be modeled in addition to MIMO techniques mmWave propagation loss.
- It is equipped with an error detection to prevent redundant retransmission.
- It has a redesigned interference model, and a channel adaptation model based on a feedback loop to provide reliable communication over variable channels.

This simulator is publicly available for researchers. It can be customized to analyze a wide variety of scenarios to develop mmWave devices and to model medium access control and the physical layer for 5G networks.

10.2.6   OpenAirInterface (5G Software Alliance for Democratising Wireless Innovation)

The OpenAirInterface Software Alliance (OSA) is a not-for-profit consortium. It develops open-source software and hardware for evolved packet core (EPC) and Evolved Universal Terrestrial Radio Access (E-UTRAN) networks of the 3rd Generation Partnership Project (3GPP) (Marina et al., 2014). It introduces a software framework for current and future networks to validate and analyze the behavior of wireless access technologies with real-life data. Cloud radio-access networks (C-RANs), software networks, and massive MIMO can be studied using the OSA software framework.

In this platform, the protocol stacks for user equipment (UE) and E-UTRAN Node B (ENodeB) are implemented, conforming to the 3GPP standards. Each node has its own IP address that allows it to communicate with either an existent data stream or a traffic generator. Moreover, this emulator is capable of implementing 3GPP channel models by considering shadowing, path loss, and small-scale fading. As shown in Figure 10.4, this platform focuses on the large-scale experimentation and can be

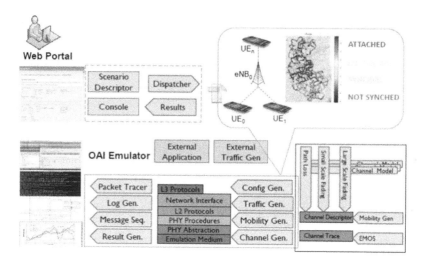

FIGURE 10.4 OpenAirInterface experimentation opportunities. (From Mahesh K. Marina et al. *SIGCOMM Computer Communication Review*, 44, 33–38, 2014.)

used for integration, performance evaluation, and testing realistic systems. It is also possible to run the full protocol stack in a controlled fashion with a wide variety of real-life test cases.

10.2.7 CLOUD RADIO-ACCESS NETWORK SIMULATOR

The advantages of C-RAN over conventional radio-access networks (RAN) are the centralized management of resources and interference, dynamic selection of radio-access technologies to increase system throughput, and decreasing energy consumption. The C-RAN system-level simulator presented in(Mohsen and Hassan (2015)) is capable of centralized user scheduling and global carrier aggregation (CA) per antenna and can perform edge-user joint transmission (JT). This simulator is supported by a realistic channel model inspired by a simplified version of the TU Vienna's LTE system-level simulator (Rupp et al., 2016). It can be used for modeling both homogeneous and heterogeneous networks with a wide variety of cell sizes. Moreover, this simulator can be categorized as a multicell-based, system-level simulator that can construct manageable, concentric hexagonal–shaped tiles. This simulator can deploy micro and macro remote radio heads (RRH) to any position in a mobile network. By combining different types of RRH in a network, heterogeneous networks (HetNet) can be modeled. Furthermore, network types can be managed by inserting cloud-config parameters into this simulator.

10.3 COMPARISON OF SIMULATORS

We compare these tools by considering the following set of parameters:

- Learning curve: represents the complexity of the simulator
- Development language: the language or the programming environment researchers need to learn to use the simulator
- Whether the simulator is open source
- Whether the simulator has built-in support for graphical user interface (GUI)
- The type of the simulator: system, link, or protocol level

We also present the main features of the simulators. We briefly summarize the features of various 5G simulators in Table 10.1. While Wireless InSite and SystemVue are capable of analyzing

TABLE 10.1
Comparison of 5G Simulators

Features	Learning Difficulty	Language	Open Source	GUI	Emulator	Type	Key Features	Reference
NORMA	Medium	RealNes	No	Yes	Yes	System Level	Flexible base stations; variety of transceivers, beamforming, 3D antenna	Redana et al. (2015)
SystemVue	Easy	ESL/MATLAB®	No	Yes	No	System/Link Level	Designing of PHY layer, integrated with MATLAB, general-purpose environment, special module for beamforming	Keysight Technologies (2016)
NYU	Medium	MATLAB	Yes	Yes	No	System Level	Generate realistic spatial and temporal wideband channel impulse responses, calculating AoA, 3D statistical spatial channel model	Samimi and Rappaport (2016)
Wireless InSite	Medium	InSite	No	Yes	No	System Level	Specifically designed for ray tracing, high-fidelity EM solvers, accurate prediction of EM propagation	Remcom (2016)
5G Module	Medium	NS3	Yes	No	No	System Level	NS3 based simulator, fully customizable, modeling of PHY and MAC layers of 5G channels	Mezzavill et al. (2015)
OpenAirInterface	Easy	RTAI	Yes	Yes	Yes	System Level/Link Level	Validate with real data, real channel convolution with PHY signal in real time, testing pre-deployment system, protocol validation	Marina et al. (2014)
C-RAN Simulator	Easy	MATLAB	No	Yes	Yes	System Level	Centralized user scheduling, dynamic selection of RAT, globally CA per antenna, edge-user JT, realistic channel, deploy micro and macro RRH	Mohsen and Hassan (2015)

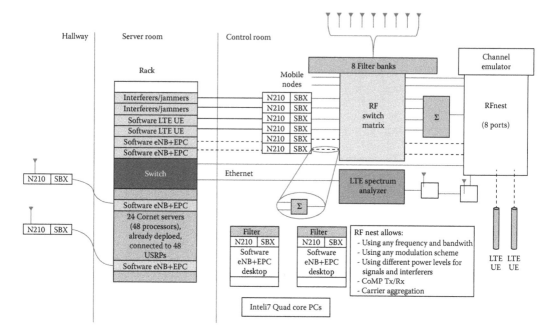

FIGURE 10.5 LTE-CORNET architecture. (From Virginia Tech. [December, 2016]. LTE-Enhanced Cognitive Radio Testbed. [Online]. www.cornet.wireless.vt.edu/lte.html.)

propagation models and communication over a physical channel (PHY), the C-RAN and NYU simulators can be used to analyze network coverage and adaptive scheduling. Some of these simulators work based only on analytic data. Some of them simulate models by using a database that is obtained through real-life experimentation. Some of them are emulators that can practically analyze models by using real hardware and infrastructures.

10.4 5G EXPERIMENTATION TESTBEDS AND PROTOTYPES

We present several existent testbeds, the major federations of testbeds, and their prototypes in this section. The presentation order is random, and the list is not exhaustive. At the end of this section, we present a comparison of the presented testbeds.

10.4.1 LTE-CORNET (LTE-ENHANCED COGNITIVE RADIO TESTBED)

Virginia Tech's Cognitive Radio Network (CORNET) is a large-scale testbed with 48 remotely accessible software radio nodes (Virginia Tech, 2016). The testbed is used in education and research.

Figure 10.5 shows the LTE-CORNET architecture. CORNET provides wide-range experimental tools with an FCC license agreement for several frequency bands. The CORNET nodes support dynamic spectrum access (DSA) and cognitive radio (CR) by providing open-source software and flexible hardware (Virginia Tech, 2016).

10.4.2 NITOS

NITOS provides a heterogeneous experimentation environment consisting of outdoor and indoor networks (NITOS, 2016). It provides an open-source platform using a control and management framework (OMF). A NITOS scheduler together with OMF provides reserving of slices such as frequency spectrum and nodes of the testbed to conduct their experiments to the users. Figure 10.6 shows the testbed architecture of NITOS at the University of Thessaly (NITOS, 2016).

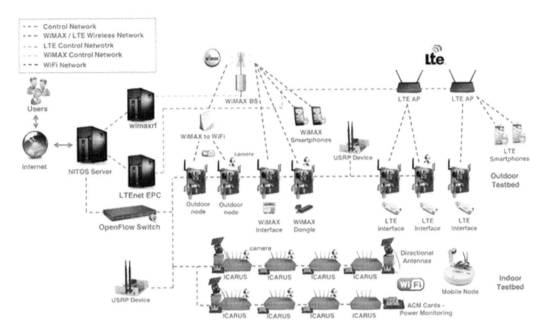

FIGURE 10.6 NITOS testbed architecture at the University of Thessaly. (From NITOS. [August, 2016]. FED4FIRE. [Online]. www.fed4fire.eu/nitos/.)

NITOS has powerful nodes equipped with various wireless interfaces such as Wi-Fi, Bluetooth, and ZigBee (NITOS, 2016). Some of the nodes are mobile. The software-defined radio (SDR) testbed is composed of Universal Software Radio Peripheral (USRP™) devices (NITOS, 2016). USRP instruments are integrated into the NITOS wireless nodes. SDR is a radio communication system and a combination of software and hardware technologies in which some physical layer processing functions are employed via a firmware or modifiable software on programmable processing systems, including digital signal processors (DSP), field programmable gate arrays (FPGA), general-purpose processors (GPP), programmable system on chip (SoC), and the other programmable processors (Wireless Innovation, 2016). USRPs are feasible transceivers that convert a personal computer to a capable wireless system (National Instruments, 2016). USRPs let the researchers program a number of physical layer features supporting dedicated cross-layer or PHY-layer research (National Instruments, 2016). NITOS's testbed also contains multiple OpenFlow switches connected to the NITOS nodes that facilitate experimentation on switching and networking protocols (NITOS, 2016). An OpenFlow switch splits the control and data planes, which makes OpenFlow different from the conventional switches (Braun and Menth, 2014). NITOS's testbed provides reproducibility of experiments and evaluation applications and protocols in the real world (NITOS, 2016).

10.4.3 PHANTOMNET

PhantomNet makes the analysis of current- and next-generation mobile networking technologies possible via its tools and infrastructure (PhantomNet, 2016). The users can conduct end-to-end experiments with various resources, such as mobile handsets, EPC services (OpenEPC, OpenLTE, and OpenAirInterface), hardware access points (SDR-based eNodeBs and IP access), mobile UE, bare-metal nodes, virtual nodes, and other resources taken from the main Emulab infrastructure. PhantomNet hardware resources are connected to a programmatically regulated attenuator matrix to enable managed RAN experimentation. PhantomNet provides scripts to help researchers conduct their mobility experiments.

Figure 10.7 demonstrates the workflow of PhantomNet (PhantomNet, 2016). The experimenters can make an EPC setup with access points and emulated endpoints. A user also can combine

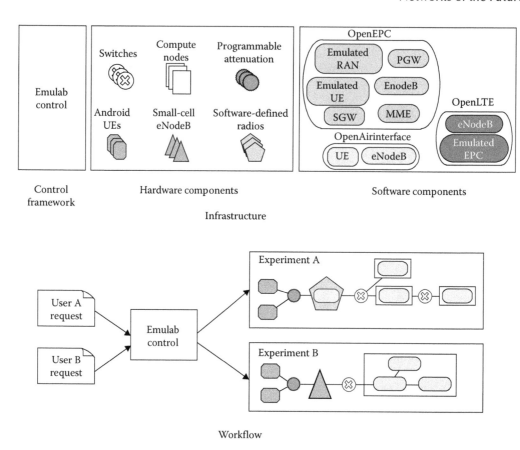

FIGURE 10.7 Workflow of PhantomNet. (From PhantomNet. [December, 2016]. PhantomNet. [Online]. http://phantomnet.org/.)

EPC core software with eNodeBs and UE. The RAN is suited for experiments with software implementations such as OpenAirInterface and SDRs. PhantomNet uses the OpenEPC software suite, which includes authentication, authorization, and accounting (AAA), home subscriber server (HSS), policy and charging resource function (PCRF), serving gateway (SGW), mobility management entity (MME), packet data network gateway (PGW), and emulated eNodeB and UE. PhantomNet uses open-source EPC software, OpenLTE, and OpenAirInterface. This testbed also supports traditional mobility platforms (3GPP). PhantomNet provides configuration scripts and directives to help users while conducting their experiments (PhantomNet, 2016).

10.4.4 5G Massive MIMO Testbed

The 5G massive MIMO testbed is a large-scale antenna system called hyper MIMO, very large MIMO, and full-dimension MIMO (5G Massive MIMO Testbed, 2016). 5G networks must enlarge the capacity and solve the current network problems, such as coverage, reliability, latency, and energy efficiency. Massive MIMO solves these problems by increasing the number of antennae at the base station. This solution is different from the existent technologies since the testbed uses thousands of antennae managed adaptively and coherently. The energy is directed to mobile users in a determined area using precoding methods and hence conserved. The current LTE or LTE-Advanced data networks need a pilot overhead by considering the number of antennae at the base station (Instruments, 5G Massive MIMO Testbed, 2016). Massive MIMO may use TDD between uplink and downlink to handle the pilot overhead considering the channel reciprocity. The channel state information coming from uplink pilots can be used in the precoder of the downlink. Additional challenges exist regarding the scalability of the interfaces, data buses, and the distributed synchronization between a wide range of independent RF transceivers (5G Massive MIMO Testbed, 2016). The 5G massive MIMO testbed uses LabVIEW system design and SDRs in a modular and scalable fashion. Thanks to employing flexible hardware, redeployment can be done for the other configurations, such as a multicell coordinated network or distributed nodes in an ad-hoc network. Moreover, 5G massive MIMO uses a large number of low-power components and decreases the latency. The prototyping system has 20 MHz of real-time bandwidth, including from 2 to 128 antennas, and can be used with a diverse set of UE (5G Massive MIMO Testbed, 2016). The 5G massive MIMO testbed has the following important attributes:

- It has flexible SDRs that receive and send RF signals.
- It provides frequency synchronization and accurate time between the radio heads.
- It provides a lossless and high-throughput bus to manage the large amounts of data (5G Massive MIMO Testbed, 2016).

The 5G massive MIMO testbed employs USRP RIO SDRs and a cabled PCI Express switch box to get multiple USRP RIO connections into a single stream, which simplifies the system in terms of the usage of a large number of channels (5G Massive MIMO Testbed, 2016). 5G massive MIMO uses reconfigurable FPGA processing modules to provide high performance and real-time signal processing on MAC and PHY layers. This testbed has been developed by Nutaq, Lund University, and National Instruments. The prototype of the system uses a system-design software called LabVIEW with the PXI platforms to provide low-latency and high-throughput applications and USRP RIO platforms (5G Massive MIMO Testbed, 2016).

10.4.5   5G PLAYGROUND

5G Playground is an environment that is developed for researchers from anywhere to use and build a 5G testbed cooperatively (FOKUS, 2016). In 5G Playground, it is possible to make an innovative prototype product. Researchers are able to calibrate and analyze new prototypes with other products. Researchers and developers, by use of 5G Playground, not only can design proof of concepts, but also they can use the testbed for evaluating different network parameters related to security, performance, and reliability. Due to the characteristics of 5G heterogeneous networks, we will face a wide variety of requirements in 5G. For addressing these requirements, we can use 5G Playground's customizable functions for localized and multimedia services, providing low delay and high reliability for wireless industrial communication, high availability and security for critical environments, and the possibility of using a satellite for backhauling.

5G Playground is equipped with considerably customizable toolkits that are designed based on available components. Some of these toolkits are the following: Open5GCore is developed for providing scalable and highly customizable scenarios for core networks; OpenSDNCore can be used by features related to backhaul; Open Baton addresses network function virtualization orchestration; and Open5GMTC analyzes connectivity of a huge number of devices (FOKUS, 2016).

10.4.6   METIS PROJECT

The METIS project (Danish et al., 2015) includes more than 140 technical components, such as air-interface technologies, new waveforms, multiple access and MAC schemes, multi-antenna and massive MIMO technologies, multi-hop communications, interference management, resource allocation schemes, mobility management, robustness enhancements, context-aware approaches, device-to-device (D2D) communication, dynamic reconfiguration enablers, and spectrum management technologies. A wide variety of testbeds is implemented in this project for D2D communication, massive machine-type communications, and designing new waveforms.

10.4.7 WINS Testbed

Using USRP B210 and X310 SDR platforms, the WINS testbed provides an experimentation environment covering a large spectrum between 24 MHz and 6 GHz (WINS Lab, 2016). Testbed users can manipulate signal coverage of these SDRs by changing the power level. In addition, these boards facilitate MIMO capacity. Sensor networks are also integrated to implement machine-to-machine (M2M) communication and scenarios related to the IoT.

10.4.8 Federations of Testbeds

A federated testbed is a model in which each autonomous testbed has its own unique properties. However, all testbeds cooperate in a federative fashion to share their resources under a common framework (Ricci et al., 2012). FIRE, GENI, and CREW are three of the federated testbeds for future networks.

10.4.8.1 FIRE

The Future Future Internet Research and Experimentation (FIRE) federation is a foundation that is financially supported by the European Commission, and it was initiated in 2010 (FIRE, 2016). FIRE is a European Open Lab, which can be used for prospective development and research. Due to a huge amount of European participation in FIRE, cutting-edge test facilities are provided for researchers. FIRE focuses on future networks and infrastructures such as smart cities, 5G networks, and eHealth. Moreover, it provides a unique, multidisciplinary environment of networks and infrastructures. This is vital for validating innovative ideas for future networks rapidly by decreasing cost.

The latest generation of FIRE is FIRE+ based on the Horizon 2020 vision. FIRE+ focuses on the next framework program (FIRE, 2016). The FIRE+ project can be divided into the following five key technology areas in which testbeds can be identified and enabled:

- Federation: A group of computing platforms and testbeds that all have an agreement over standards and operations. Moreover, testbeds can be reachable through a web portal.
- Data management: Two FIRE+ projects focus on geospatial mapping and online education.
- The IoT: Eight FIRE+ projects work in a wide variety of IoT fields, such as underwater M2M and wireless sensor networks (WSN).
- Smart cities: Four FIRE+ projects mainly focus on smart city services, mobility back-end services, and consumer buying services.
- Networking: Ten FIRE+ projects work in the field of optical networks, LTE, SDN, and 5G mobile networks.

In the networking compartment, FIRE introduced FIRE LTE testbeds for open experimentation (FLEX) that are designed to improve current FIRE facilities for 4G and beyond technologies, such as 5G mobile networks. This experimentation environment is equipped with highly customizable and adaptable devices, such as macro, pico, and small cells. It is also capable of mobility framework emulation. Some of the key features of FLEX can be summarized as follows:

- FLEX has five operating and online testbeds.
- FLEX is integrated with FIRE, which means that it can use FIRE equipment and tools.
- FLEX can not only be used as a simulation tool, but it is also capable of emulating mobility.
- In its first open call, six new partners have joined FLEX.
- Nine state-of-the-art experiments were done over FLEX in the second open call.

10.4.8.2 GENI

The Global Environment for Networking Innovation (GENI), sponsored by the U.S. National Science Foundation (NSF), is a distributed virtual laboratory for the deployment, development, and validation of concepts in network science, security, and services. GENI is solving widespread issues about ossification of the Internet (Berman et al., 2014). GENI allows users to conduct various kinds of experiments, including protocol design and evaluation, content management, social network integration, clean-slate networking, and in-network service deployment (Berman et al., 2014). In recent years, GENI has been giving a chance to analyze the potential of its technologies, such as software-defined networking (SDN) and GENI racks, in support of university campus network management and applications. The key components of GENI are GENI racks, SDNs, and worldwide interoperability for microwave access (WIMAX). GENI racks include virtualized computation and storage resources. SDNs provide programmable and virtualized network resources. WiMAX includes virtualized cellular wireless communication (Berman et al., 2014).

There are some key experimenter concepts such as sliceability (being able to provide logical infrastructures to tenants of a system) and deep programmability. The concept of sliceability is borrowed from the PlanetLab testbed by GENI (Berman et al., 2014). Sliceability provides virtualization, namely, commonly shared access to resources such as link, storage, node's processor to GENI, by retaining some degree of isolation for concurrent experiments (Berman et al., 2014; Peterson et al., 2003). Slice refers to a set of virtualized resources that are reserved for an experiment, and a slice consists of independently conducted resources, which are called slivers (Peterson et al., 2003). GENI slicing includes computation and network resources (Berman et al., 2014). Deep programmability is the key concept that allows a user to change the behavior of routing, storage, forwarding components, and computation not only near the network edge but also deep inside the network. GENI provides end users who are not associated with a team the ability to opt into the services supported by an experimental slice of GENI. GENI supports a flexible and suited environment for the users in terms of network science, security, services, and the other corresponding fields (Berman et al., 2014).

10.4.8.3 CREW

The purpose of Cognitive Radio Experimentation World (CREW) is to construct an open federated test platform (CREW, 2016). CREW is a federation of five existing testbeds. The first one is an industrial, scientific, and medical (ISM) test environment at iMinds (Wilabt, 2016) embodying IEEE 802.15.4, IEEE 802.15.1, IEEE 802.11, and Universal Software Radio Peripheral software radios. The second one is a licensed cognitive radio testbed consisting of TV bands at TCD, which is subject to reconfigurable radio platform and the USRP (IRIS, 2016). The third one is a WSN testbed at the Technical University of Berlin (TUB), embodying some sensor nodes produced by different companies, spectrum analyzers, USRP software radios, and FPGA platforms (TWIST, 2016). The fourth one is a test environment related to LTE cellular at the Technical University of Dresden (TUD), combining a set of LTE-equivalent base stations, SDRs, and mobile UE (EASYC, 2016). The last one is an outdoor heterogeneous ISM/TVWS (TV white spaces) testbed (VSN, 2016). The CREW federated platform provides a framework for benchmarking and controlled and reproducible experiments. The CREW platform makes experimental validation of cognitive networking and cognitive radio concepts, such as radio sensing, horizontal resource sharing among the heterogeneous environments in the ISM band, cooperation in licensed bands for the heterogeneous networks, and robust cognitive sensor networks to ensure non-interference.

10.4.9 REQUIREMENTS AND COMPARISONS OF 5G TESTBEDS

Presently, we have an opportunity to access a wide variety of 5G testbeds that satisfy many important 5G requirements. A 5G testbed should be flexible, reconfigurable, heterogeneous, complete, and able to give reproducible results based on open-source solutions that are site agnostic and topology agnostic (NetWorld2020, 2014). It should be flexible and contain many distinct options such as separate PHY layers, frequency bands, and components across the network stack. It should be easily reconfigurable to help researchers reshape the testbed depending on the requirements of their solutions. Furthermore, the testbed should manage heterogeneity, such as sensor, platform, device, and protocol heterogeneity. Heterogeneity in a 5G testbed will be useful to facilitate an experimenter in selecting the desired components and in adapting the environment to the experiment (Horneber and Hergenröder, 2014). Another important requirement is that a 5G testbed should be complete to grant the involvement of all components of the 5G systems, from the legacy mobile operators to the virtual operators and from the end user to the M2M applications and the Internet of Things. Reproducible results and outcomes must be achievable by a 5G testbed to assure scientific quality, fair testing, and comparison of various technologies. Moreover, if a 5G testbed provides open-source solutions, then it will enhance its capabilities and share its potential competences with the scientific community. A 5G testbed should be site agnostic as much as possible so that the researchers can easily access and test their technologies and solutions in different situations without requiring physical presence at a site. A 5G testbed should be topology agnostic to include all of the wireless topologies, from small cells to macro cells, and solutions with cellular and satellite technologies (NetWorld2020, 2014).

Several important concepts can be used as the comparison criteria of the testbeds (Horneber and Hergenröder, 2014). Some of these parameters used in Table 10.2 are mobility, coverage and size, support for open source, frequency, heterogeneity, and support for some technologies and applications, such as SDR and MIMO. Mobility is one of the most important concepts for wireless networks, especially for 5G networks (Horneber and Hergenröder, 2014; NetWorld2020, 2014). The number of

TABLE 10.2
Comparisons of the Existing Testbeds

Features	Mobile	Coverage/ Size	Open Source	Heterogeneous	Key Features	Reference
LTE-CORNET	Yes	Large scale	Yes	Yes	Flexible platform, FCC experimental license, modular setup	Virginia Tech (2016)
PhantomNet	Yes	Large scale	Yes	Yes	SDR, EPC software (OpenLTE, OpenEPC, and OpenAirInterface), virtual nodes, emulated end points, off-the-shelf eNodeBs, Emulab infrastructure	PhantomNet (2016)
NITOS	Yes	Large scale	Yes	Yes	SDR, OpenFlow switches, 2.4 GHz and 5 GHz antennas, USRP devices, OMF, NITOS scheduler, remotely accessible, indoor and outdoor testbeds, PHY-layer or cross-layer implementations, slicing, MIMO	NITOS (2016)
5G Massive MIMO Testbed	Yes	Large scale	No	Yes	50 MHz to 6 GHz RF frequencies, 4–28 antennas, SDR, USRP devices, FPGA platforms, TDD, MIMO, large number of antennas at BTS	5G Massive MIMO Testbed (2016)
5G Playground	-	-	No	Yes	Reconfigurable, commercially available components	FOKUS (2016)

nodes and wireless devices and the size of the deployment area are other important attributes that make a testbed small- or large-scale (NetWorld2020, 2014). A 5G testbed will enhance its capabilities and share its potential competences with the scientific community by providing open-source solutions and using open-source software (NetWorld2020, 2014). Device, interface, protocol, sensor, and platform heterogeneities have to be considered covering a broad range of heterogeneous applications such as intelligent transportation, robotics, and entertainment (NetWorld2020, 2014; Horneber and Hergenröder, 2014).

10.5   CONCLUSIONS

In this chapter, the existent 5G solutions such as federated platforms, simulators, projects, and testbeds are reviewed. The requirements of 5G testbeds, namely flexibility, heterogeneity, completeness, reproducibility, open-source support, and site-agnostic and topology-agnostic approaches, are discussed. The existing 5G simulators are analyzed and compared by considering the difficulty of learning them, their development language, open-source support, graphical user interface support, emulation support, type of simulator, and the other key features such as flexibility, capacity, and frequency spectrum. The existent federations, projects, and testbeds are presented including their hardware (e.g., USRP devices, FPGA platforms, ENodeBs, and OpenFlow switches) and software attributes (e.g., open-source software, SDR, and MIMO applications, OpenLTE, OpenAirInterface). The testbeds are compared based on mobility, coverage and size of the testbed, support of open-source software, heterogeneity, protocols, applications, and supported frequency spectrum. While 5G testbeds, simulators, and federations try to satisfy the requirements of future networks, they are subject to significant challenges, such as green operation, minimal latency, coherency, and management of heterogeneity.

ACKNOWLEDGMENT

This work is partially supported by TÜBiTAK Project, No: 215E127.

REFERENCES

CREW. (December, 2016). Cognitive Radio Experimentation World. [Online]. www.crew-project.eu.
Danish Aziz, Katsutoshi Kusume, Olav Queseth, Hugo Tullberg, Mikael Fallgren, Malte Schellmann, Mikko Uusitalo, and Michał Maternia. (2015). METIS Project. [Online]. https://www.metis2020.com/wp-content/uploads/deliverables/METIS_D8.4_v1.pdf
Dave Evans. (April, 2011). *The Internet of Things: How the Next Evolution of the Internet Is Changing Everything,* Cisco Internet Business Solutions Group (IBSG).
EASYC. (December, 2016). Cognitive Radio Experimentation World, EASYC. [Online]. www.crew-project. eu/easyc.
FIRE. (August, 2016). FIRE–Future Internet Research and Experimentation [Online]. www.ict-fire.eu/fire/.
Fraunhofer FOKUS. (December, 2016). 5G Playground [Online]. www.fokus.fraunhofer.de/go/en/fokus_testbeds/5g_playground.
Ivan Cosovic, Gunther Auer, and David Falconer. (2009). Link level procedures. In *Radio Technologies and Concepts for IMT-Advanced,* Martin Döttling, Werner Mohr, and Afif Osseiran (Eds.). UK: Wiley.
IRIS. (December, 2016). Cognitive Radio Experimentation World, IRIS [Online]. www.crew-project.eu/iris.
Jens Horneber and Anton Hergenröder. (2014). A survey on testbeds and experimentation environments for wireless sensor networks. *IEEE Communication Surveys & Tutorials, vol. 16,* no. 4, 1820–1838.
Keysight Technologies. (December, 2016). SystemVue Electronic System-Level (ESL) Design Software. [Online].www.keysight.com/en/pc-1297131/systemvue-electronic-system-level-esl-design-software?cc=TR&lc=eng.
Larry Peterson, Tom Anderson, David Culler, and Timothy Roscoe. (January, 2003). A blueprint for introducing disruptive technology into the Internet. *Computer Communication Review,* vol. 33, no. 1, pp. 59–64.

Mahesh K. Marina, Saravana Manickam, Navid Nikaein, Alex Dawson, Raymond Knopp, and Christian Bonnet. (October, 2014). OpenAirInterface: A flexible platform for 5G research. *SIGCOMM Computer Communication Review,* vol. 44, no. 5, pp. 33–38.

Mark Berman, Jeffrey S. Chase, Lawrence Landweber, Akihiro Nakao, Max Ott, Dipankar Raychaudhuri, Robert Ricci, and Ivan Seskar. (2014). GENI: A federated testbed for innovative network. *Computer Networks,* vol. 61, pp. 5–23.

Markus Rupp, Stefen Schwarz, and Martin Taranetz. (2016) *The Vienna LTE-Advanced Simulators: Up and Downlink, Link and System Level Simulation, Signals and Communication Technology.* Singapore: Springer.

Mathew Samimi, Kevin Wang, Yaniv Azar, George N. Wong, Rimma Mayzus, Hang Zhao, Jocelyn K. Schulz, Shu Sun, Felix Gutierrez, and Theodore S. Rappaport. (2013). 28 GHz Angle of Arrival and Angle of Departure Analysis for Outdoor Cellular Communications Using Steerable Beam Antennas in New York City. In *IEEE Vehicular Technology Conference (VTC Spring),* Dresden, Germany, pp. 1–6.

Mathew K. Samimi and Theodore S. Rappaport. (July, 2016). 3-D Millimeter-Wave Statistical Channel Model for 5G Wireless System Design. *IEEE Transactions on Microwave Theory and Techniques,* vol. 64, no. 7, pp. 2207–2225.

National Instruments. (December, 2016). National Instruments. [Online]. www.ni.com/sdr/usrp/.

NetWorld2020. (September, 2014). 5G: Challenges, Research Priorities, and Recommendations.

NITOS. (August, 2016). FED4FIRE [Online]. www.fed4fire.eu/nitos/.

Norhan Mohsen and Khaled S. Hassan. (2015). C-RAN simulator: A tool for evaluating 5G cloud-based networks system-level performance. In *2015 IEEE 11th International Conference on Wireless and Mobile Computing, Networking and Communications (WiMob),* Abu Dhabi, United Arab Emirates, pp. 302–309.

PhantomNet. (December, 2016). PhantomNet [Online]. http://phantomnet.org/.

Remcom. (2016). 5G and MIMO Using Wireless InSite. [Online]. www.remcom.com/5g-mimo.

Robert Ricci, Jonathon Duerig, Leig Stoller, Gary Wong, Srikanth Chikkulapelly, and Woojin Seok. (June 11–13, 2012). Designing a federated testbed as a distributed system. In *Testbeds and Research Infrastructure. Development of Networks and Communities: 8th International ICST Conference, TridentCom 2012,* Thessanoliki, Greece, pp. 321–337. Revised Selected Papers. Springer, Berlin Heidelberg.

Simone Redana, Cinzia Sartori, and Eiko Seidel. (July, 2015). 5G NORMA: A NOvel Radio Multiservice Adaptive Network Architecture for 5G networks. *In European Conference and Networks (EuCNC),* Paris, France.

Sourjya Dutta, Marco Mezzavilla, Menglei Zhang, Mustafa Riza Akdeniz, and Sundeep Rangan. (2015). 5G MmWave module for the ns-3 network simulator. *In Proceedings of the 18th ACM International Conference on Modeling, Analysis and Simulation of Wireless and Mobile Systems,* New York, pp. 283–289.

TWIST. (December, 2016). Cognitive Radio Experimentation World, Twist. [Online]. www.crew-project.eu/twist.

Virginia Tech. (December, 2016). LTE-Enhanced Cognitive Radio Testbed. [Online]. www.cornet.wireless.vt.edu/lte.html.

VSN. (December, 2016). Cognitive Radio Experimentation World, Vsn. [Online]. www.crew-project.eu/vsn.

WINS Lab. (December, 2016). WINS Lab Test-bed. [Online]. http://wins.ceng.metu.edu.tr.

Wireless Innovation. (December, 2016). What is Software Defined Radio. [Online]. www.wirelessinnovation.org/assets/documents/SoftwareDefinedRadio.pdf.

Wolfgang Braun and Michael Menth. (2014) Software-defined networking using OpenFlow: Protocols, applications and architectural design choices. *Future Internet,* vol. 6, pp. 302–336.

Yang Zhang, Ruoyu Jin, Xin Zhan, and Dacheng Yang. (October, 2011). Study on link-level simulation in multi-cell LTE downlink system. *In 4th IEEE International Conference on Broadband Network and Multimedia Technology (IC-BNMT),* Shenzhen, pp. 168–172.

5G Massive MIMO Testbed, 2016. (December, 2016). 5G Massive MIMO Testbed [Online]. www.ni.com/white-paper/52382/en/.

Part III

Efficient Solutions for Future Heterogenous Networks

11 A Fuzzy Logic–Based QoS Evaluation Method for Heterogeneous Networks

Farnaz Farid, Seyed Shahrestani, and Chun Ruan

CONTENTS

11.1 INTRODUCTION

Decision-making processes related to telecommunication networks are becoming extremely complex due to the increasing complexity of the environments, fast-growing technological evolution, rapid changes in service availability, market structure, and growing customer demands (Figueira et al., 2005). Involvement of multiple communication technologies in a single network is making the QoS evaluation of any network more complex than ever as each technology has its own characteristics and the applications running over them have their own QoS requirements. These multidimensional QoS issues in the new technological platforms lay down the ground for multi-criteria analysis methods. Multi-criteria decision-making (MCDM) or multi-attribute decision-making (MADM) algorithms have been widely used for QoS evaluation of heterogeneous networks (Alshamrani et al., 2011; Augestad and Lindsetmo, 2009; Kantubukta Vasu et al., 2012; Karimi and Fathy, 2010; Lusheng and Binet, 2009; Qingyang and Jamalipour, 2005; Sgora et al., 2010; Shu-ping et al., 2011; Wei and Weihua, 2012).

However, the QoS analyses applying these various methods show that there could be situations in which the QoS level of an application or access network overlaps two consecutive measurement metrics (Farid et al., 2013a). For example, if any application or access network has a QoS level value between 0.6 and 0.8, the application or the access network is regarded as having a medium level of QoS. If the application or radio access network (RAN) has a QoS level value between 0.8 and 1, then the QoS level is considered good. Now, there are cases in which an application or RAN could have a QoS level value of 0.79 or 0.8. In such overlapping cases, using the fixed and dynamic weight-based methods, it is hard to determine the QoS of that application or RAN precisely. For instance, it is hard to classify the QoS level of that application or RAN as "Good" or "Medium." Therefore, the evaluation process of any network can involve a certain degree of uncertainty, and the QoS level sometimes is not just a crisp value due to the multiple engaging factors. Fuzzy logic is suitable for dealing with these sorts of uncertain and imprecise parameters (Zadeh, 1988).

As fuzzy logic–based solutions use linguistic variables and inference rules, the solutions are closer to the way people think. As a result, these solutions can lead to more rational and dynamic outcomes than the conventional algorithms. Using fuzzy logic, the accumulated operator and user knowledge about the QoS evaluation can be easily accommodated. Another advantage of using fuzzy logic is its simplicity. QoS evaluation of heterogeneous networks itself is complex in nature. Therefore, the proposed solutions for this type of network should be simple enough and should avoid complex mathematical analytics. The fuzzy logic–based mathematical reasoning in this regard is very simple and easy to understand. This simplicity makes fuzzy logic a perfect candidate for providing simple QoS evaluation solutions (Alkhawlani and Ayesh, 2008). Therefore, this chapter outlines a fuzzy logic–based QoS evaluation method (FLQEM). A hierarchical fuzzy-logic system is used in designing the proposed method as there are several subsystems involved. The application QoS evaluation subsystems derive the application QoS metrics, the RAN QoS evaluation subsystems derive the RAN QoS metrics, and the network configuration subsystem derives the configuration QoS metric.

11.2 BACKGROUND

The communication technologies present in a heterogeneous network have different characteristics. These technologies have their offered bandwidths, coverage area, and operating frequencies. Their QoS characteristics, such as delay, throughput, and packet loss, as well as usage and implementation costs, also differ from each other. As a result, the adaptation of heterogeneous network–based architecture for the provision of different applications, especially multimedia applications, faces significant challenges. Among these challenges, QoS-related issues such as effective QoS evaluation, management, and monitoring still top the list (Carmen Lucas-Estañ and Gozalvez, 2013).

The methods for QoS evaluation of heterogeneous wireless networks have been extensively studied. The motivations for these studies can be categorized as access network selection (ANS), joint admission control (JAC), joint scheduling control (JSC), and vertical handover (VHO) in heterogeneous networks. These studies mainly focus on two aspects. First, the QoS evaluation of a single application or a single RAN in a heterogeneous environment must be able to handover to a better network. Second, the strategies to maintain the QoS of the current network while new call admission.

Most of these studies take into account the effects of a single access network on application performance rather than considering the impacts of the access networks on the other side of the connection. To measure the QoS in a heterogeneous environment, the conventional methods do not consider the performance of all the applications running on a network. For example, if there are voice and video conferencing applications running on a UMTS network, these methods do not include the performance analysis of both of these applications to quantify the network QoS. In addition, there is no unified metric to quantify the QoS of a network, which considers the performance of all the access networks present in it. For example, in a heterogeneous environment, there are three access networks, such as UMTS, WiMAX, and LTE. At present, no unified metric can represent the performance of this whole configuration using the QoS-related parameters of these access networks.

Many previous works in this context have also analyzed traffic characteristics to recommend QoS evaluation processes for heterogeneous networks. However, precise evaluation of the overall characteristics of such network traffic is a challenging task. The presence of different types of communication technologies and a varying number of users make that evaluation even more complex. The same user scenario can suggest varying performance results based on the measurement of packet loss and end-to-end delay. For example, with twenty voice clients and one streaming client, in some networks, the end-to-end delay of voice calls shows an acceptable value. However, in terms of packet loss, the voice calls do not achieve an acceptable performance value for the same number of users (Farid et al., 2013a). Therefore, in some cases, even though the end-to-end delay may be at an acceptable level for some applications, packet loss may simply be too high. In addition, the effects of different communication technologies on the performance of various applications, say, voice and video, must be accounted for in an efficient and methodical manner. In such situations, a unified QoS metric that would quantify the performance of the whole network configuration is more helpful than analyzing the resultant value of each QoS parameter separately.

To evaluate the QoS of any heterogeneous network, some studies have ranked each access network by combining different QoS metrics. The most common parameters, which are considered during this ranking process, are mainly service, network, and user related, for instance, received signal strength (RSS), type of service (e.g., conversational, streaming, interactive, background), minimum bandwidth, end-to-end delay, throughput, packet loss, bit error rate, cost, transmit power, traffic load, current battery status of the mobile terminal, and the user's preferences. To combine these parameters into a single value, at first, a weight is assigned to each of these parameters according to its relative importance.

The weights are both subjective and objective in nature. For example, the importance of network-related parameters such as RSS and bandwidth are objective in nature. Application-related parameters such as end-to-end delay, packet loss, and jitter seem objective in nature; some studies reveal that they could be subjective in nature too. For example, a study conducted in Tanzania (Sedoyeka et al., 2009) shows that to evaluate the quality of a network, the users give moderate importance to end-to-end delay over packet loss. Another research study is conducted by the European Telecommunication Standards Institute (ETSI) (ETSI, 2006), which reveals that the users give strong importance to end-to-end delay over packet loss. Therefore, the importance of application parameters can vary based on the context, for example, between home and an industrial environment or between developed and developing countries.

The primary issue with the conventional weight calculation methods is that the weights have been assigned to QoS-related parameters in each application according to separate objective functions, and the context-based importance of these parameters has been ignored. Additionally, the application significance is not considered in these methods. However, as with the QoS-related parameters, the importance of any application can also vary with the changing context. For instance, an application for the education service would have more significance than an application for the entertainment service. Without considering this information, the weights will not reflect the proper importance attached to the considered parameter, application, or network.

QoS analysis that considers different aspects of a heterogeneous network is necessary to evaluate and enhance the service-based. To address this necessity, several studies (Alshamrani et al., 2011; Karimi and Fathy, 2010; Shu-ping et al., 2011) have examined the service-centric performance of these networks. However, the establishment of an integrated QoS metric for this type of configuration as a whole is still challenging (Luo and Shyu, 2011). Most of the existing research discussed previously focuses on the partial QoS evaluation of a heterogeneous network by deriving the QoS level of a single access network and a single application present within a heterogeneous environment. Also, different studies have come up with different performance metrics for QoS evaluation of these networks. In this chapter, the focus is on combining the different performance metrics together to come up with a unified QoS metric to evaluate the QoS of heterogeneous network configuration as a whole. The proposed method in this chapter integrates the application QoS level and the network QoS level in heterogeneous wireless networks under various traffic and mobility scenarios.

The VHO process usually has three phases: system discovery phase, decision phase, and execution phase (McNair and Fang, 2004; Wen-Tsuen, Jen-Chu, and Hsieh-Kuan et al., 2004). In Kantubukta Vasu et al. (2012), a QoS-aware fuzzy rule–based algorithm is proposed for vertical handoff. The algorithm takes a multi criteria–based decision considering various parameters of different traffic classes, such as end-to-end delay, jitter, and packet loss. The weights for each of these parameters are considered based on their relative importance in relation to that specific traffic class. In Sgora et al. (2010), a network selection method based on analytical hierarchy process (AHP) and fuzzy Total Order Preference by Similarity to the Ideal Solution (TOPSIS) is proposed. The fuzzy TOPSIS method is used to rank each access network according to application performance. Table 11.1 shows the details of the results from this method.

In Qingyang and Jamalipour (2005), the AHP and Grey relational analysis (GRA) are combined to calculate the final network QoS. The AHP breaks down the network selection criteria into several subcriteria and assigns a weight to each criterion and subcriterion. Then GRA is applied to rank the candidate access networks. Then the MADM algorithm is applied again to rank the candidate networks.

In Bari and Leung (2007a), the authors have applied elimination and choice translating reality (ELECTRE) for network selection. In Wei and Qing-An (2008), a cost function–based access network selection model is proposed which considers the available bandwidth and received signal strength for network selection. An application-based vertical handoff algorithm is proposed

TABLE 11.1

Network Ranking Using Fuzzy TOPSIS Method

Applications	Technology	Requirements
Web browsing	UMTS	0.62
	WiMAX	0.67
	WLAN	0.65
VOIP	UMTS	0.6
	WiMAX	0.61
	WLAN	0.615
Streaming media	UMTS	0.58
	WiMAX	0.66
	WLAN	0.62

in Wen-Tsuen and Yen-Yuan (2005). A QoS level for each candidate network is quantified by the application performance. Then this QoS value is compared with the current QoS value for that specific application before a mobile terminal initiates any handoff. In Alkhawlani and Ayesh (2008), an access network selection scheme is proposed based on fuzzy logic, genetic algorithms (GAs), and the MCDM algorithm. The scheme has used GA to assign suitable values for weights of the considered criteria. Three parallel fuzzy logic subsystems are used to score each of the available RAN. The scores and the weights are then passed on to the MCDM system to take the final decision for VHO. For weight assignment, the available literature for QoS evaluation in network selection has mostly used the AHP method, which is primarily developed by Saaty (Bari and Leung, 2007b; Kantubukta Vasu et al., 2012; Qingyang and Jamalipour, 2005; Sgora et al., 2010; Stevens-Navarro et al., 2008; Wenhui, 2004). Some studies have also assigned fixed weights to these parameters based on their importance to service performance (Wen-Tsuen and Yen-Yuan, 2005). Both the AHP and fixed-weight methods are unable to handle the subjective and ambiguous factors related to weight determination, such as context-based significance. To deal with this situation, in Charilas et al. (2009), the authors have used the fuzzy AHP (FAHP) method to assign these weights and the ELECTRE method to select the most suitable network from a range of available candidate networks.

11.3 APPLICATIONS OF FUZZY LOGIC

Fuzzy logic is viewed as a multi-valued logic that deals with the approximate mode of reasoning rather than the precise mode (Zadeh, 1988). It has been described as computing with words (Zadeh, 1996). Primarily it was introduced to model the uncertainty of natural language and since then has been widely used as a means to support intelligent systems (Zadeh, 1965). Although the history of fuzzy logic goes back to 1965, the usage of fuzzy logic in computer networking started to become popular during the 1990s (Yen, 1999). It has been extensively used in QoS evaluation of networks and web-based services (Diao et al., 2002; Munir et al., 2007; Vuong Xuan and Tsuji, 2008).

11.3.1 FUZZY LOGIC PRINCIPLES

A brief overview of the basic principles of fuzzy logic is outlined in this section. Using fuzzy logic, it is possible to map the imprecise values or the word expressions to a crisp number. It also provides the options to use natural language for rule classification. As a result, complex algorithms of a mathematical model could be explained easily and be understandable by nonexperts (Zadeh, 1973).

11.3.2 FUZZY VERSUS NON-FUZZY SET THEORY

The traditional sets are defined by crisp boundaries. Fuzzy sets, on the other hand, are defined by ambiguous boundaries (Ross, 2010). For example, in today's world, to classify whether a certain product is made in Australia or imported would be challenging if the classical set is used, because these days it is hard to classify a locally made product as some parts of that product may be imported from outside. Therefore, the fuzzy theory can be used in such situations as they do not have clear boundaries. The age of people is also an example of fuzzy sets. Using the crisp set theory, a person 50 or older is regarded as old. So into what group would a person who is 49 years old fall? On his or her 50th birthday, he or she will suddenly become old. However, such a sudden change is against common sense.

Another example could be given in relation to the application QoS evaluation. For example, 400 msec is regarded as the upper boundary of end-to-end delay for a quality voice call. Now, if any voice call experiences a 395-msec delay, how is the quality going to be evaluated in this case? Would this call be evaluated as poor or acceptable? Hence, in many situations, when a person analyses a statement, there is always natural fuzzification. Therefore, a membership curve can describe to what degree an element does or does not belong to a specific set.

As stated earlier, a fuzzy set has uncertain boundaries. A set is represented using a membership function defined on a space called universe of discourse. The fuzzy sets apply the idea of degree of membership. Therefore, the membership function maps the elements of the universe of discourse onto numerical values between 0 and 1. A membership function value of 0 implies that the corresponding element is definitely not an element of the fuzzy set, while a value of 1 means that the element fully belongs to the set. A value close to 0 means it has lesser belonging to the set, and a value close to 1 means it has greater belonging to the set.

The advantage of fuzzy set theory over crisp set theory is that a fuzzy set can include elements with only a partial degree of membership (a membership degree value between 0 and 1). In the example of the end-to-end delay stated previously, it was hard to classify the delay of 395 msec. By applying fuzzy theory, it would be easier to classify these types of values as fuzzy sets do not have sharply fixed boundaries. Therefore, a delay of 395 msec can be an acceptable delay with a degree of 0.3. As it has a lower degree of membership, the evaluated QoS level of the voice call will show the impact.

11.3.3 FUZZY MEMBERSHIP FUNCTIONS

A membership function is the graphical representation of a fuzzy set. A membership function for a fuzzy set X on the universe of discourse U is defined as $\mu_X : U \rightarrow [0,1]$, where each element of U is mapped to a value between 0 and 1. This value, called the membership value or the degree of membership, quantifies the grade of membership of the element in U to the fuzzy set X. The membership function in the example of a locally made product is the list of all the products. The term *locally made* would correspond to a graphical curve that would define the degree to which any product would be considered locally made.

There are different types of membership functions (Jin and Bose, 2002). These functions are illustrated as follows:

- ***Linear functions:*** These are the simplest types of membership functions. There are mainly two types of linear membership functions. They are defined as follows:
  - **Triangular function**: A triangular membership function is defined by three points, a lower limit, a, an upper limit, b, and a value, m, where $a < m < b$. It can be defined as

$$f(m,a,b,c) = \max\left\{\min\left(\frac{m-a}{b-a}, \frac{c-m}{c-b}\right), 0\right\} \tag{11.1}$$

  - **Trapezoidal function**: A trapezoidal function has a flat top. Four points are used to define this function; they are a lower limit, a, an upper limit, b, a lower support limit, b, and an upper support limit, c.
    A trapezoid function for a fuzzy set μ is defined by

$$f(m,a,b,c,d) = \max\left\{\min\left(\frac{m-a}{b-a}, 1, \frac{d-m}{d-c}\right), 0\right\} \tag{11.2}$$

    There are two special cases for a trapezoidal function, which are called the R function and L function. In the case of R functions, $a = b = -\infty$, and for L functions, this is $c = d = +\infty$.
- ***Gaussian functions:*** A Gaussian function is defined by a central value m and a standard deviation $\sigma > 0$. A Gaussian function is expressed as

$$f(m,\sigma,c) = e^{\frac{-(m-c)^2}{2\sigma^2}} \tag{11.3}$$

where the parameter c is the distance from the origin, and σ is the standard deviation or indicates the width of the curve. The Gaussian membership functions have the features of being smooth and nonzero at all points.

- **Bell-shaped functions:** The bell-shaped functions have a symmetrical shape. They can be expressed as

$$f(m,a,b,c) = -\frac{1}{1 + \left| \frac{m-c}{a} \right|^{2b}} \qquad (11.4)$$

where the b parameter is usually positive, the parameter c indicates the center of the curve, and a represents the width of the curve. Bell-shaped functions also have the feature of being smooth and nonzero at all points as with Gaussian functions.

- **Sigmoidal function:** A sigmoidal function is usually open to the right or left depending on the sign of the steepness of the function. It is expressed as

$$f(m,a,c) = -\frac{1}{1 + e^{-a(m-c)}} \qquad (11.5)$$

where c locates the distance from the origin, and a determines the steepness of the function. If a is positive, the membership function is open to the right, and if it is negative, it is open to the left. This type of construction gives them the advantage to represent "very large" or "very negative."

- **Polynomial-based functions:** The polynomial functions can be categorized as polynomial-Z (*zmf*), polynomial-S (*smf*), and polynomial-PI (*pimf*). Polynomial-Z and polynomial-S are always asymmetric.

$$y = zmf(m,[a,b]) \qquad (11.6)$$

where the parameters a and b locate the extremes of the sloped portion of the curve.

$$y = smf(m,[a,b]) \qquad (11.7)$$

This spline-based curve is a mapping on the vector x and is named because of its "S" shape. The parameters a and b locate the extremes of the sloped portion of the curve.

$$y = pimf(m,[a,b,c,d]) \qquad (11.8)$$

11.3.4 LINGUISTIC VARIABLES

If a variable takes a word in natural languages as its values instead of numerical values, it is called a linguistic variable (Zimmermann, 1996). Linguistic variables are used to define the uncertainly. For example, the speed of a car can be defined as "fast," "very fast," "slow," etc. The variables are defined through a quintuple (X,L,U,G) (Herrera and Herrera-Viedma, 2000). X is the name for the variable; L is the set of labels for X; U is the universe of discourse, which contains all possible values obtained by X; G is the rule defined in the form of a grammar; and M is the membership function for each label L. There are two ways to choose an appropriate linguistic term set and its semantic (Herrera and Herrera-Viedma, 2000).

It can be defined by using a context-free grammar. The semantic of the linguistic terms is illustrated using fuzzy numbers, which are represented by membership functions. For example, a context-free grammar G is a 4-tuple structure, (V_N, V_T, I, P), where V_N is the non-terminal, V_T is the terminal, I is the starting rule, and P is the production rule. Now if G has the primary terms (high, medium, low), hedges (not, much, very, . . .), relations (higher, lower, . . .), and connectives (and, but, or), producing I as any term, the linguistic term set H = (high, very high, low, very low, . . .) is generated by using P.

The second approach, on the other hand, defines the linguistic term set using an ordered structure, and the semantic of this linguistic term set is derived from the ordered structure, which may be either symmetrically distributed on an interval of [0,1] or not.

11.3.5 Fuzzy Rules

A fuzzy rule takes the form of IF-THEN conditional statements (Wang, 1997). It is expressed as

IF <fuzzy proposition> Then <fuzzy proposition>

Fuzzy propositions can be two types: atomic fuzzy propositions and compound fuzzy propositions. An atomic fuzzy proposition is a single fuzzy statement. For example,

Delay Is Low

where *delay* is the linguistic variable and *low* is the linguistic value of delay. A compound fuzzy preposition, on the other hand, consists of more than one atomic fuzzy proposition and uses the connectives "and," "or," and "not" representing fuzzy intersection, fuzzy union, and fuzzy complement, respectively. For example,

IF Delay Is Low and Jitter Is Low but Packet Loss Is Not Low

There are many ways to interpret the fuzzy rules. They are the Dienes–Rescher implication, the Lukasiewicz implication, the Zadeh implication, the Gödel implication, and the Mamdani implication. In this chapter, the Mamdani implication method is used for its simplicity and efficiency. In the Mamdani implication, the fuzzy if-then rules are interpreted as a fuzzy relation R_{MM} in $U \times V$ with the membership function:

$$\mu_{R_{MM}}(x, y) = \min\left[\mu_{FP_1}(x), \mu_{FP_2}(y)\right] \tag{11.9}$$

11.3.6 Fuzzy Inference System

A fuzzy inference system (FIS) maps the fuzzy input parameters to the output parameters using fuzzy theory. This mapping provides the ability to decision making or pattern detection (Jang et al., 1997). The strength of FISs depends on their twofold identities. First, they have the ability to handle linguistic concepts. Second, their universal approximators are able to perform non-linear mappings between inputs and outputs (Guillaume, 2001). There are mainly two types of fuzzy inference systems: the Mamdani fuzzy inference method and the Sugeno inference method. The Mamdani FIS is used due to its simplicity, widespread acceptance, and the vibrancy of output membership functions as compared to Sugeno systems. A Mamdani inference system has the following steps:

- **Fuzzification of the input variables:** The first step is to take the crisp inputs and determine the degree to which these inputs belong to each of the appropriate fuzzy sets. This step involves simple calculation by applying graphical analysis to a crisp input value (X axis) versus its membership value (Y axis). In other words, fuzzification is the process of taking actual real-word data and converting them into fuzzy input. For example, an end-to-end delay input value of 400 msec may be converted to a fuzzy input value of high. This process produces multiple fuzzy inputs for every real-word input based on the number of membership functions defined for the system. The goal of the fuzzification step is to produce inputs that can be processed by the rule evaluation step.

- **Rule evaluation:** The second step is to take the fuzzified inputs and apply them to the antecedents of the fuzzy rules. If a given fuzzy rule has multiple antecedents, AND or OR are used to obtain a single number that represents the result of the antecedent evaluation. This number is then applied to the consequent membership function. For example, if delay is low, jitter is low, and packet loss is low, then the QoS is good. The result of the rule evaluation step is a set of rule strengths. The rule strength is determined from the numeric values of its input labels.
- **Aggregation of the rule outputs:** Aggregation is the process of unification of the outputs of all rules. In this step, the membership functions of all rule consequents previously scaled are combined into a single fuzzy set. In other words, the input of the aggregation process is the list of scaled consequent membership functions, and the output is one fuzzy set for each output variable.
- **Defuzzification:** Fuzziness helps to evaluate the rules, but the final output of a fuzzy system has to be a crisp number. The input for the defuzzification process is the aggregate output fuzzy set. The output is a single number. There are several defuzzification methods, but probably the most popular one is the centroid technique. This technique works by finding the point at which a vertical line would slice the aggregate set into two equal masses.

11.4 E-MODEL

The E-model is a well-known transmission-planning tool that provides a forecast of the expected voice quality of the considered connection (ITU, 2008). It is applicable for quality assessment of the voice quality of both wireline and wireless scenarios, based on circuit-switched and packet-switched technology. The quality that is predicted represents the perception of a typical telephone user for a complete end-to-end (i.e., mouth-to-ear) telephone connection under different conversational conditions. To evaluate the user satisfaction and communication quality, the model uses several parameters; these are basic signal-to-noise ratio, delay impairment factor, equipment impairment factor, and an advantage factor. The equipment impairment includes the influence of packet loss. The advantage factor represents an advantage of access, which certain systems may provide in comparison to conventional systems. The primary output of the E-model is a scalar quality rating value called the transmission rating factor, R. R can be transformed into other quality measures, such as mean opinion score (MOS), percentage good or better (GoB), and percentage poor or worse (PoW).

The E-model is established on an impairment factor–based mathematical algorithm. This algorithm is responsible for the transformation of the individual transmission parameters into different individual impairment factors. These impairment factors are assumed to be additive on a psychological scale (ITU, 2008). The algorithm also considers the combined effects of the simultaneous occurrence of those impairment factors at the connection level and the masking effects. E-model predictions may be inaccurate when the combined effect is not considered proper. The relation between the different impairment factors and R is given by the following equation:

$$R = R_0 - I_s - I_d - I_e, \mathit{eff} + A \qquad (11.10)$$

where:

The term R_0 expresses the basic signal-to-noise ratio (received speech level relative to the circuit and acoustic noise).

The term I_s represents all impairments that occur more or less simultaneously with the voice signal, such as too loud speech level (non-optimum OLR), non-optimum sidetone (STMR), quantization noise (qdu), etc.

The term I_d sums all impairments that take place due to delay and echo effects.

The term I_e, eff is an effective equipment impairment factor, which represents impairments caused by low bit-rate codecs. It also includes impairment due to packet losses of random distribution. The values of I_e are transformed into I_e, eff in the case of random packet loss, using the E-model algorithm. The values of I_e are taken from (I).

TABLE 11.2
E-Model Details

R	MOS	User Satisfaction
90 < R < 100	4.34 < MOS < 4.5	Very satisfied
80 < R < 90	4.03 < MOS < 4.34	Satisfied
70 < R < 80	3.60 < MOS < 4.03	Some users dissatisfied
60 < R < 70	3.10 < MOS < 3.60	Many users dissatisfied
50 < R < 60	2.58 < MOS < 3.10	Nearly all users dissatisfied

The term A is an advantage factor, which allows for an advantage of access for certain systems relative to conventional systems, trading voice quality for convenience. While all other impairment factors are subtracted from the basic signal-to-noise ratio, R_0, A is added to compensate for other impairments to a certain amount. It takes into account the fact that the user will tolerate some decrease in transmission quality in exchange for the advantage of access. Examples of such advantages are cordless and mobile systems and connections into hard-to-reach regions via multi-satellite hops.

A value between 90 and 100 for the R factor leads to higher user satisfaction. Table 11.2 shows the interpretations of the R factor for different values, the interpretations of MOS, and the satisfaction measurement based on MOS.

11.5   THE PROPOSED EVALUATION METHOD

In the previous section, the concepts related to fuzzy logic are discussed in detail. It is viewed as a multi-valued logic, which deals with the approximate mode of reasoning rather than the precise mode. It can be described as computing with words as well. It maps the imprecise values or the expression of words to a crisp number and provides the options to use natural language for rule classification. As a result, a sophisticated algorithm of a mathematical model could be easily explained and understandable by non-experts. A fuzzy inference system or fuzzy rules–based system has five main functional blocks; these are a rule base, which contains the if-then rules, the input membership functions, a fuzzification interface to transform the crisps input parameters to match linguistic variables, and a defuzzification interface, which converts the fuzzy inputs into a crisp number. The process is as follows:

- The first step involves the formulation of the problem and definition of the input and the output variables. The universe of discourse and the linguistic terms for each variable are also defined, for example, a fuzzy set A in the universe of discourse U.
- When designing the structure of the system, for instance, when there are multiple subsystems present, the input and the output variables should be defined clearly. The input variable of one subsystem could be the output variable of another subsystem.
- Define fuzzy membership functions for each variable. For example, a linguistic variable x in the universe of discourse U is defined by $T(x) = \left\{T_x^1, T_x^2, ..., T_x^l\right\}$ and $\mu(x) = \left\{\mu_x^1, \mu_x^2, ..., \mu_x^l\right\}$, where $T(x)$ is a term set of x.
- Define the rules for the system. If there are multiple subsystems, then rules are set for each of the subsystem blocks.
- Perform defuzzification to derive a crisp value from the fuzzy values.

A hierarchical fuzzy logic system is used as there are several subsystems involved in this method. The application QoS evaluation subsystems derive the application QoS metrics, the RAN QoS evaluation subsystems derive the RAN QoS metrics, and the network configuration subsystem derives the configuration QoS metric. Figure 11.1 shows this model. The first level has the application

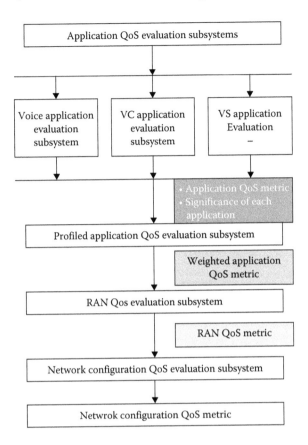

FIGURE 11.1 The proposed model.

QoS evaluation subsystem, and under this, there are three modules. They are the voice, VC, and VS application-based modules. The input parameters for the first level of subsystems are delay, jitter, and packet loss. The second tier profiles the applications depending on their significance and the evaluated QoS outcome. The third level has several RAN QoS evaluation subsystems. The input parameters for these subsystems are the output parameters from the previous subsystems. The fourth level is the subsystem for network configuration QoS evaluation. The input for this subsystem is the output from the previous subsystem, which is the RAN QoS metric. The output of the network configuration subsystem is the network configuration QoS metric. A MATLAB® tool is used to design this conceptual model.

11.6 FUZZY MODELLING OF APPLICATION QoS EVALUATION SUBSYSTEM

There are three application-based modules in the first level of the fuzzy system. These are the voice, VC, and VS modules. The specifications of these modules are described as follows.

11.6.1 VARIABLES

The following tasks are related to the variables:

- Define the input and the output variables.
- Define the universe of discourse for each variable.
- Choose the number of linguistic terms that states each variable.

Most of the fuzzy logic–based systems use three, five, or seven terms for each linguistic variable. Fewer than three terms are rarely used as most concepts in human language consider at least two extremes and the middle ground. On the other hand, one rarely uses more than seven terms because humans interpret technical figures using their short-term memory. In general, human short-term memory can compute only up to seven symbols at a time (Von Altrock, 1995). Therefore, the odd cardinality of the linguistic set is used. According to the discussions in Farnaz Farid and Ruan (2012), the key parameters to evaluating the QoS of any network-based, real-time application are delay, jitter, and packet loss. Therefore, these QoS-related parameters are used as input variables of the application QoS evaluation subsystems. The justifications for choosing these variables are as follows:

- Delay: Delay is one of the important parameters for measuring the performance of real-time applications such as voice and video conferencing. It impacts the timeliness of the delivery of information and maintains the QoS.
- Jitter: The value of jitter is one of the best indicators for observing service performance, especially real-time or delay-sensitive applications. Jitter can be used to model the packet loss of a network.
- Packet loss: Packet loss is another important parameter for evaluation of network QoS. Packet-based metric is quite promising for evaluation of quality. It also contributes to the calculation of peak signal-to-noise ratio (PSNR).

The variables that are used in the application-based subsystems are as follows:

$QP_{k,A}$: Input QoS-related parameters in an application, A, where k is the index for the QoS-related parameters, and $k = \{1, 2,..., p\}$
$QoSAM_A$: The QoS metric for any application, A, where A = Voice, VC, etc.
U: Universe of discourse for the QoS-related input variables
V^A: Universe of discourse for the application QoS metric

In Farnaz Farid and Ruan (2016), the dynamic weight-based QoS evaluation method (DWQEM) uses an acceptable range for QoS-related parameters based on different contexts. In the proposed fuzzy logic–based approach, these ranges are used as the universe of discourse for each of the QoS-related input parameters. Tables 11.3 through 11.5 show the universe of discourse for the QoS-related parameters of each application. The universe of discourse for each input variable is expressed as

$$U = \left[U^{\min}, U^{\max} \right] \tag{11.11}$$

The universe of discourse for the output variable application QoS metric $QoSAM_A$ is set as [0, 1]. It is expressed as

$$V^A = \left[V^{A,\min}, V^{A,\max} \right] \tag{11.12}$$

TABLE 11.3

Universe of Discourse for VC Application

Metrics	Universe of Discourse
End-to-end delay (sec)	0–400
Jitter (sec)	0–75
Packet loss (%)	0–5

TABLE 11.4

Universe of Discourse for Voice Application

Metrics	Universe of Discourse
End-to-end delay (sec)	0–400
Jitter (sec)	0–75
Packet loss (%)	0–12

TABLE 11.5

Universe of Discourse for VS Application

Metrics	Universe of Discourse
End-to-end delay (sec)	0–10
Packet loss (%)	0–5

11.6.2 Membership Functions

The ranges of the universe of discourse, which are defined in the previous section, are used to determine the fuzzy membership functions for the application QoS evaluation subsystems. Each input variable has three different membership functions: Low (L), Medium (M), and High (H). The membership functions are designed based on the empirical and simulation analysis results illustrated in Farnaz Farid and Ruan (2012). The design follows the conditions that each membership function overlaps only with the closest neighbouring membership function, and for any possible input data, its membership values in all relevant fuzzy sets should sum to 1 or close to 1 (Yen and Langari, 1999). Each membership function is expressed through a set of linguistic terms. The input variables associated with the membership functions are also termed as the linguistic variable.

Recall from Equation 11.1 that if X is a collection of objects denoted graphically by x, then a fuzzy set A in X is a set of ordered pairs.

$$A = \left\{ (x, \mu_A(x) \mid x \in X) \right\} \tag{11.13}$$

Using this concept, the fuzzy set of the QoS-related parameters in the universe of discourse U is defined as

$$Q = \left\{ QP_{k,A}, \mu_Q(QP_{k,A}) \mid QP_{k,A} \in U \right\} \tag{11.14}$$

The term set for the QoS-related parameter is defined as

$$T(QP_{k,A}) = \left\{ T^1_{QP_{k,A}}, T^2_{QP_{k,A}}, T^3_{QP_{k,A}} \right\} \tag{11.15}$$

where $T^1_{QP_{k,A}} = Low$, $T^2_{QP_{k,A}} = Medium$, and $T^3_{QP_{k,A}} = High$.

For example, the fuzzy set for end-to-end delay in a voice application is expressed as

$$Q = \left\{ QP_{d,V}, \mu_Q(QP_{d,V}) \mid QP_{d,V} \in U \right\}$$

where d denotes end-to-end delay, and V denotes voice application.

The term set for the end-to-end delay for voice application is expressed as

$$T(QP_{d,v}) = \left\{ Low, Medium, High \right\}$$

The fuzzy set for the application QoS metric $QoSAM_A$ in the universe of discourse V^A is defined as

$$A = \left\{ QoSAM_A, \mu_A(QoSAM_A) \mid QoSAM_A \in V^A \right\} \quad (11.16)$$

The term set for the application QoS metric $QoSAM_A$ is defined as

$$T(QoSAM_A) = \left\{ T^1_{QoSAM_A}, T^2_{QoSAM_A}, T^3_{QoSAM_A} \right\} \quad (11.17)$$

where $T^1_{QoSAM_A} = Poor$, $T^2_{QoSAM_A} = Average$, and $T^3_{QoSAM_A} = Good$.

For example, the term set for the voice application QoS metric is as follows:

$$T(QoSAM_V) = \left\{ Poor, Average, Good \right\}$$

Table 11.6 shows the ranges of these membership functions.

The justifications for choosing the ranges for the membership functions of the QoS-related parameters are described as follows:

- Voice application (V): For voice applications, three input variables are considered; these are delay, jitter, and packet loss. Table 11.7 shows the ranges of the membership functions for these variables. For each of these variables, three membership functions are defined. The justifications for choosing these ranges are as follows:
  - End-to-end delay (E2ED): For this metric, the membership functions, which are designed for the rural area, have the ranges L: 0 to 320, M: 300 to 400, and H: 380 to 600. The low for a rural area is chosen based on the simulation results in Farnaz Farid and Ruan (2012). The simulation results show that with an average delay of around 300 msec, most of the calls achieve an expected QoS level. The recommendations from the ITU and 3GPP (3GPP, 2002; ETSI, 2006) also confirm the same ranges. The recommendations state that an end-to-end delay of around 400 msec is considered acceptable, including the constraints in the current technology. The medium and high fuzzy sets are derived based on Yen and Langari (1999) using an overlap of 20.

TABLE 11.6

Membership Functions for the Application QoS Metric

Membership Function	Range
Poor	0.0–0.6
Average	0.6–0.8
Good	0.8–1.0

TABLE 11.7

Membership Functions for Voice Application

Metrics		Membership Functions		
		Low	Medium	High
End-to-end delay (msec)	Rural	0–320	300–400	380–600
	Urban	0–150	130–300	280–600
Jitter (msec)	Rural	0–50	45–75	70–100
	Urban	0–30	25–50	45–75
Packet loss (%)	Rural	0–5	4–9	8–20
	Urban	0–3	2–5	4–20

The membership functions of the urban area are designed based on the recommendations from industrial environments such as Cisco. Therefore, the low range is set to 0 to 150 msec. Cisco recommends that for enterprise networks, delay should be no more than 150 msec to ensure quality voice calls. For high-quality speech, this should be no more than 100 msec.

- **Jitter (J):** The values of the membership functions of jitter are also defined using the same method. In the rural area, L: 0 to 50, M: 45 to 75, and H: 70 to 100. For instance, Cisco requires the jitter for an audio call to be less than 30 msec (Szigeti, 2005), and the requirement for jitter stated in ITU-T REC. Y.541 is less than 50 msec (Stankiewicz et al., 2011). The limit for low is defined according to these recommendations, the simulation results, and other relevant studies stated in (Farnaz Farid and Ruan, 2012). The medium and high fuzzy sets are derived using an overlap of 5. In the urban area, low is set to 0 to 30 msec. This is based on the recommendation from Cisco that jitter should be less than 30 msec for quality voice calls.
- **Packet loss (PL):** The fuzzy sets of packet loss of the rural area are also defined according to the simulation results studied in (Farnaz Farid and Ruan, 2012), the recommendations of ITU-T, Cisco, 3GPP, and a study that has gathered data from the African continent (3GPP, 2002; ETSI, 2006; Zennaro et al., 2006; Szigeti, 2005). For rural areas, L: 0 to 5, M: 4 to 9, and H: 8 to 20, and for the urban area, this is L: 0 to 3, M: 2 to 5, and H: 4 to 20.
- **Video conferencing (VC):** For VC, three input variables are considered, which are delay, jitter, and packet loss. For each of these variables, three membership functions are defined as well. The ranges are outlined in Table 11.8. The justifications for choosing these ranges are as follows:
  - End-to-end delay (E2ED): The ranges of membership functions for VC application in a rural area network are defined based on the simulation results in Farnaz Farid and Ruan (2012) and the recommendations from the ITU and 3GPP (3GPP, 2002; ETSI, 2006). The ranges of the urban area are defined mainly based on the recommendations for enterprise networks, such as Cisco recommendations (Szigeti, 2005).
  - Jitter (J): The ranges of jitter for VC also use the same recommendations as voice applications. The ranges of membership functions for the rural area are determined according to the experimental simulation results in Farnaz Farid and Ruan (2012), and recommendations from the ITU and 3GPP (3GPP, 2002; ETSI, 2006). The ranges of the urban area are defined based on the recommendations from Cisco (Szigeti, 2005).
  - Packet loss (PL): The fuzzy sets for packet loss are defined for the rural area based on the experimental results in Farnaz Farid and Ruan (2012) and, the recommendations from the ITU, Cisco, 3GPP, and PingER research, which has gathered data from the African continent (3GPP, 2002; ETSI, 2006; Zennaro et al., 2006; Szigeti, 2005). For

---

TABLE 11.8
Membership Functions for VC

Metrics		Low	Medium	High
		Low	**Medium**	**High**
End-to-end delay (msec)	Rural	0–200	180–400	380–600
	Urban	0–100	50–200	180–600
Jitter (msec)	Rural	0–50	45–75	70–100
	Urban	0–30	25–50	45–75
Packet loss (%)	Rural	0–2.5	1.5–5	4–12
	Urban	0–2	1–3	2–12

rural areas, L: 0 to 2.5, M: 1 to 5, and H: 4 to 12, and for the urban area, this is L: 0 to 2, M: 1 to 2.5, and H: 2 to 12. The experimental results from PingER show that for a VC session, packet loss of less than 1% is regarded as the most satisfactory, and 1% to 2.5% of loss is regarded as acceptable. However, if the loss is greater than 2.5% and the sessions experience a loss of 2.5% to 5%, then the VC quality is poor. If this loss is between 5% and 12%, then the quality is very poor, and a loss of greater than 12% is unacceptable. The results also indicate that if the loss is between 4% and 6%, then it becomes impossible for non-native speakers to communicate.

- **Video streaming (VS):** Delay and packet loss are considered the input variables for the VS application subsystem. Jitter is not seen as an essential parameter for VS application performance evaluation. For each of these two variables, three membership functions are defined. The justifications for the ranges of these membership functions are as follows:
  - End-to-end delay (E2ED): For the VS application, ITU and Cisco recommend different end-to-end delay. The experimental results in Farnaz Farid and Ruan (2012) also show the effect of different environments on packet loss of VS sessions. For example, when the VS server is in an urban outdoor environment and the client is in the rural outdoor environment, the VS client experiences a 314.5-msec delay and 5.72% packet loss. On the other hand, when the server is placed in a suburban environment, the same users experience 6.21% packet loss, and the delay almost remains the same. Based on these observations and the available recommendations, the ranges for the end-to-end delay are defined.
  - Packet loss (PL): The membership functions for packet loss in the context of rural and urban areas are also defined according to the same recommendations and simulation results in Farnaz Farid and Ruan (2012). Table 11.9 shows the values of these membership functions.

The proposed QoS evaluation method applies Gaussian membership functions for several reasons. Primarily, these membership functions are flexible in nature. They can be easily modified by simply adjusting the mean and the variance of the membership functions. Second, the boundless increase of the number of statistically independent samples takes the probability distribution of the sample mean to a Gaussian shape. This is stated in the central limit theorem. Since many real-life, random phenomena are a sum of many independent fluctuations, it can be expected that a Gaussian process will perform well. Third, the single sigmoid function does not represent a closed class interval. Finally, the triangular function is unable to ensure that all inputs are fuzzified in some classes (Hallani, 2007). Figures 11.2 through 11.4 show the Gaussian membership functions for voice application, which are designed based on the limits of rural areas. Figure 11.5 shows the membership function for the application QoS metric.

TABLE 11.9
Membership Functions for VS Application

Metrics		Membership Functions		
		Low	Medium	High
End-to-end delay (sec)	Rural	0–9	8–12	11–20
	Urban	0–5	4–8	7–10
Packet loss (%)	Rural	0–3	2.5–4	3.5–12
	Urban	0–3	1–5	4–10

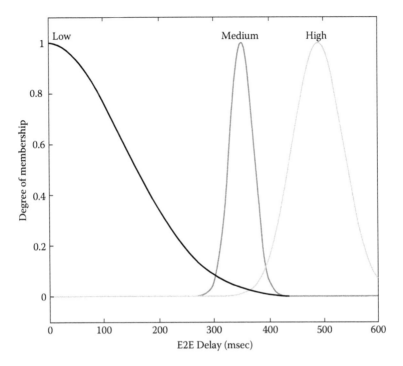

FIGURE 11.2   Membership function for E2ED (rural area).

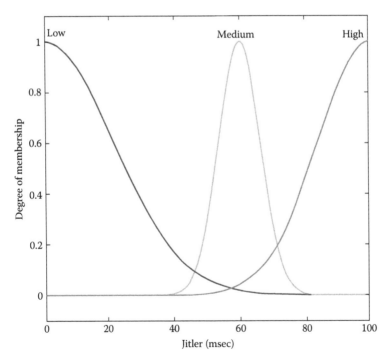

FIGURE 11.3   Membership function for jitter (rural area).

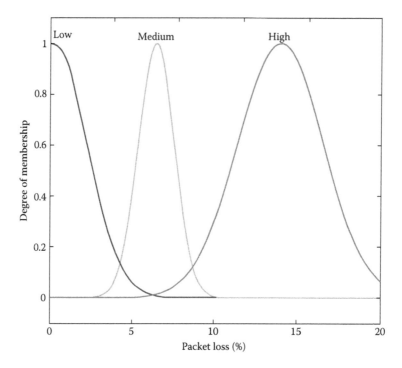

FIGURE 11.4   Membership function for packet loss (rural area).

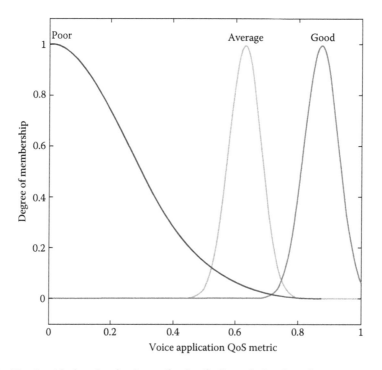

FIGURE 11.5   Membership function for the application QoS metric (rural area).

11.6.3 Rules

The definition of fuzzy rules is followed by the definition of the input and output membership functions of each subsystem. The fuzzy rules are in the form of if-then statements. These statements look at the system inputs and determine the desired output. If there are R rules, each with l premises in the system, the rth rule has the following form:

Ru_r: **If** $QP_{k,A}$ **is** Q_r^l **then** $QoSAM_A$ **is** A_r^l

The Mamdani fuzzy implication method is used in this model for its simplicity. Recall from Equation 11.13 that for two fuzzy sets, A and B, the Mamdani implication can be expressed as

$$\mu_{R_{MM}}(x,y) = \min\left[\mu_A(x),\mu_B(y)\right] \qquad (11.18)$$

or it can be expressed as

$$\Phi\left[\mu_A(x),\mu_B(y)\right] \equiv \mu_A(x)\cap\mu_B(y) \qquad (11.19)$$

The antecedent of the previously discussed general rule is interpreted as a fuzzy relationship, which is obtained from the intersection of the fuzzy sets Q_r^l and A_r^l.

$$FR_l = Q_r^l \cap A_r^l \qquad (11.20)$$

The membership function for this relationship is defined as

$$FR_l(QP_{k,A},QoSAM_A) = Q_r^l(QP_{k,A}) \wedge A_r^l(QoSAM_A) \qquad (11.21)$$

Using the Mamdani method, the generalized rule can be denoted as

$$\Phi\left[\mu_{Q_r^l}(QP_{k,A}),\mu_{A_r^l}(QoSAM_A)\right] \equiv \mu_{Q_r^l}(QP_{k,A})\cap\mu_{A_r^l}(QoSAM_A) \qquad (11.22)$$

Using the *max* operator, the aggregation is denoted as

$$Ag(z) = \bigcup_{q=1}^{R} (\mu_{Q_r^l}(QP_{k,A})\cap\mu_{A_r^l}(QoSAM_A)) \qquad (11.23)$$

The VC- and voice application–based subsystem has 3^3 rules depending on the number of input variables and membership functions. For the VS application–based subsystem, the number of rules is 3^2. The rules are designed according to the behavior analysis of the considered parameters. These studies are conducted using various experimental data available in the literature.

- **Voice application (V):** For voice applications, the rules are designed based on the performance analysis of voice applications in different conditions. The experimental results from the E-model and other studies are considered to define the rules. For example, the first rule is:

If Voice Delay Is Low, Voice Jitter Is Low, and Voice Packet Loss Is Low, Then Voice Application QoS Is Good.

No how it is defined that if all these three parameters are low, then the application QoS for voice is usually good? The behavioral analyses of delay, jitter, and packet loss in the voice application show that the voice application has a higher level of QoS if these parameters have lower values. The ranges regarded as low for these parameters have been discussed in Farnaz Farid and Ruan (2012).

These ranges are determined based on the various experimental results, which are rendered with numerical values.

The E-model has been discussed in the previous section. It is a well-known model established by ITU, which estimates the communication quality and the user satisfaction. To evaluate the user satisfaction and communication quality, the model uses several parameters; these are basic signal-to-noise ratio, delay impairment factor, equipment impairment factor, and an advantage factor. The equipment impairment includes the influence of packet loss. The advantage factor represents an advantage of access, which particular systems may provide, compared to conventional systems. The primary output of the E-model is a scalar quality rating value called the transmission rating factor, R. R can be transformed into other quality measures such as MOS, GoB, and PoW. The relationship between the R factor and the other parameters has been discussed in the previous section.

A value between 90 and 100 for the R factor leads to higher user satisfaction. Table 11.10 shows the interpretations of the R factor for different values, the interpretations of MOS, and the satisfaction measurement based on MOS. Some of the experimental results show that with a packet loss of 0% and a delay of 185.8 mesc, the R factor is 90. The E-model does not consider jitter as one of its parameters to evaluate the voice transmission quality. However, other experiments show that for larger jitter values, user satisfaction in terms of MOS decreases. Most of the test results in the literature outline that regardless of codec and other factors, low delay, jitter, and packet loss lead to a good QoS level.

The experimental results also show that if the packet loss is high (12%), even with a smaller delay (28.8 ms), the MOS value of the network decreases dramatically. According to this analysis, the second rule is:

If Voice Delay Is Low, Voice Jitter Is Low, and Voice Packet Loss Is High, Then Voice Application QoS Is Poor.

Table 11.11 outlines the conditions for the output QoS level of the applications running on any network. As shown in the table, a rule with a poor QoS level evaluates if any of the considered

TABLE 11.10

R Factor, Mean Opinion Score, and Satisfaction

R	MOS	User Satisfaction
$90 < R < 100$	$4.34 < MOS < 4.5$	Very satisfied
$80 < R < 90$	$4.03 < MOS < 4.34$	Satisfied
$70 < R < 80$	$3.60 < MOS < 4.03$	Some users dissatisfied
$60 < R < 70$	$3.10 < MOS < 3.60$	Many users dissatisfied
$50 < R < 60$	$2.58 < MOS < 3.10$	Nearly all users dissatisfied

TABLE 11.11

Conditions for Voice Application QoS Level

Fuzzy Consequent	Fuzzy Conditions
Poor	Any of the evaluated parameters has a high value.
Average	All the evaluated parameters have a medium value.
	At least two parameters have a medium value, and the other parameter has a value other than high.
Good	At least one has a medium value, and the other two have low values.
	All the evaluated parameters have a low value.

parameters within the fuzzy rule have a high value. The application that undergoes this performance level triggers prompt action from the service providers for root cause analysis of the poor performance. This application performance can affect the whole network's QoS level.

A rule with an average QoS level evaluates if at least two of the parameters within the fuzzy rule have a medium value, and the other parameter has a value other than high. The rule can also have an average value if all the parameters are evaluated as having medium values. The application that undergoes this performance level can have a long-term investigation from service providers for root cause analysis of the performance.

A rule with a good QoS level evaluates if all the parameters within the fuzzy rule have a lower value, or at least one parameter has a medium value, while other two have low value. The application that undergoes this performance level can be a model case for service operators to assess other networks. Figures 11.6 through 11.8 show the relationship of the voice application QoS metric to E2ED, jitter, and packet loss. It is apparent from the graphs that the voice application QoS metric decreases when the values of these parameters increase. In Figure 11.6, if the jitter value goes around 100 and the E2ED value is around 600, the voice application QoS metric is 0.

- **Video conferencing (VC):** The rules for VC applications are derived by analyzing the impact of various ranges of delay, jitter, and packet loss on their performance. Real-time applications are in general delay sensitive, but loss-tolerant and nonreal-time applications

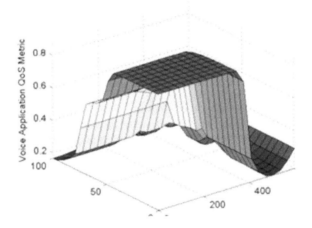

FIGURE 11.6   Variation of the voice application QoS metric in relation to jitter and E2ED.

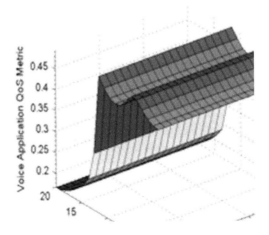

FIGURE 11.7   Variation of the voice application QoS metric in relation to E2ED and packet loss.

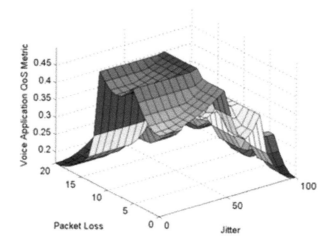

FIGURE 11.8   Variation of the voice application QoS metric in relation to packet loss and jitter.

TABLE 11.12

Conditions for VS Application QoS Level

Fuzzy Consequent	Fuzzy Conditions
Poor	Any of the evaluated parameters has a high value.
Average	All the evaluated parameters have a medium value.
Good	All the evaluated parameters have a low value.

are delay tolerant but sensitive to packet loss. In Farnaz Farid and Ruan (2012), some data have been presented to back up these theories. For instance, based on a PingER report, for a VC session, packet loss of less than 1% is regarded as satisfactory, 1% to 2.5% of loss is acceptable, 2.5% to 5% is poor, 5% to 12% is very poor, and loss greater than 12% is unacceptable. The observations indicate that when any VC session experiences 4% to 6% packet loss, it is hard for non-native speakers to communicate properly.

For conversational and videophone applications, ITU, Cisco, and 3GPP agree that it is preferred to have an end-to-end delay of no greater than 150 msec. However, according to ITU-T, this refers to a long-term achievable value. Given the current technology, an end-to-end delay of around 400 msec is considered acceptable by both ITU-T and 3GPP. The ranges also vary between developed and developing countries or an industry and a home environment. All these analyses have been presented in Farnaz Farid and Ruan (2012). The rules use the same pattern as the voice application that is outlined in Table 11.11. The first rule of the VC application subsystem is as follows:

If VC Delay Is Low, VC Jitter Is Low, and VC Packet Loss Is Low, Then VC Application QoS Is Good.

- **Video streaming (VS):** In the case of VS applications, two parameters are considered to form the rules; these are delay and packet loss. The reasons for considering only these two parameters have been discussed in Farnaz Farid and Ruan (2012). In this case, the rules take the pattern outlined in Table 11.12.

11.7   FUZZY MODELLING OF PROFILED APPLICATION QoS SUBSYSTEM

After analyzing the QoS level of the applications running on a network, the application significances are applied to assess the overall QoS of the application.

11.7.1 Variables

This subsystem takes two input parameters. One of them is the output from the previous application-based subsystems, which is the application QoS metric. The second input parameter is the significance of those applications depending on the particular context. Therefore, the input variables are as follows:

$QoSAM_A$: Application QoS metric. S_A: Application significance.
The output variable is: $QoSWAM_A$ (profiled application QoS metric)
The other notations that are used in this subsystem are as follows:
V^A: Universe of discourse for the input variable application QoS metric
V^S: Universe of discourse for the input variable application significance
WV^A: Universe of discourse for the profiled application QoS metric

The universe of discourse for the significance of the application, S_A, is set to [0, 1]. The fuzzy set for the significance of the application, ^{Ru}, in the universe of discourse V^S is defined as

$$SG = \left\{ S_A, \mu_{SG}(S_A) \mid S_A \in V^s \right\} \tag{11.24}$$

The term set for this variable is defined as

$$T(S_A) = \left\{ T_{S_A}^1, T_{S_A}^2, T_{S_A}^3 \right\} \tag{11.25}$$

where $T_{S_A}^1 = Low$, $T_{S_A}^2 = Medium$, and $T_{S_A}^3 = High$.
For example, the fuzzy set for the input variable voice application significance is expressed as

$$SG = \left\{ S_V, \mu_{SG}(S_V,) \mid S_V \in V^s \right\} \tag{11.26}$$

The universe of discourse for the profiled application QoS metric output variable WV^A is [0, 1]. It is expressed as

$$WV^A = \left\{ WV_1^A, WV_2^A, ..., WV_l^A \right\} \tag{11.27}$$

The fuzzy set for the profiled application QoS metric output variable $QoSWAM_A$ in the universe of discourse WV^A is defined as

$$WA = \left\{ QoSWAM_A, \mu_{WA}(QoSWAM_A) \mid QoSWAM_A \in WV^A \right\} \tag{11.28}$$

$$T(QoSWAM_A) = \left\{ T_{QoSWAM_A}^1, T_{QoSWAM_A}^2, T_{QoSWAM_A}^3 \right\} \tag{11.29}$$

where $T_{QoSWAM_A}^1 = Low$, $T_{QoSWAM_A}^2 = Medium$, and $T_{QoSWAM_A}^3 = High$.

11.7.2 Membership Functions

In this case, the membership functions also are formed as Gaussian functions. The reason for using Gaussian membership functions is illustrated in Subsection 11.3.3. The ranges of the input and output variables of the membership functions are a scale between zero and one. The details of the application QoS metric are outlined in Subsection 11.6.2. The input variable application significance takes a scale between zero and one. The output variable profiled application QoS metric also uses a scale between zero and one, with three membership functions of good, average, and poor. Table 11.13 shows the ranges for the application significance metric.

TABLE 11.13

Membership Functions for the Application Significance Parameter

Membership Function	Range
Low	0–0.6
Medium	0.6–0.75
High	0.75–1.0

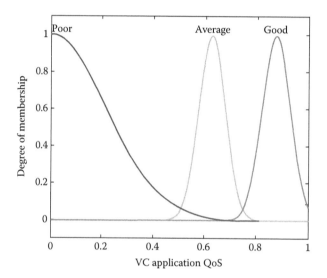

FIGURE 11.9 VC application QoS metric.

The membership function for the voice application QoS metric is presented in Figure 11.5. The membership functions for VC and VS application are described in the following figures. The membership functions for the other input variable application significance and the output variable profiled application QoS metric are also outlined in Figures 11.9 through 11.12.

11.7.3 RULES

Each of the profiled application QoS evaluation subsystems has 3^2 rules depending on the number of membership functions and input variables. The rules are designed based on the simulation analysis results in Farnaz Farid and Ruan (2015). In those studies, the impact of application significance on the overall network performance has been outlined. In one of the simulation analysis results, the network QoS demonstrates a good performance level with a voice application QoS level being average and a VS application QoS level being good when the voice and VS applications both have equal importance. When the significance of a voice application is changed to extremely important compared to a VS application, the network's QoS comes down to an average QoS level. Although the performance of the VS application was good, because of having a lower importance, it had a minimal effect on the network's QoS level. On the other hand, the voice application having greater importance affected the network QoS level

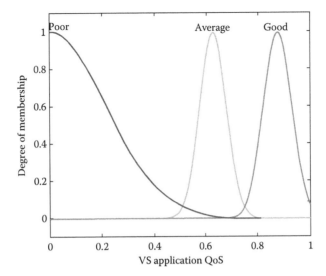

FIGURE 11.10   VS application QoS metric.

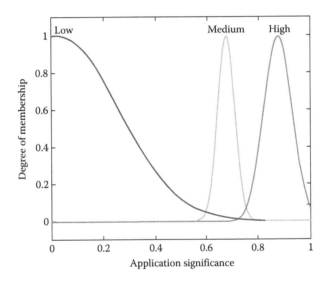

FIGURE 11.11   Application significance.

to a greater extent. Based on such analysis results, the rules are defined for the profiled application subsystem.

Table 11.14 illustrates the conditions for the output QoS level of the profiled application QoS evaluation subsystems. As shown in the table, a rule with a poor QoS level evaluates if any of the considered applications with a higher importance have a poor QoS level. The profiled application QoS adopts a value of "Good" if one of the high importance applications has good QoS. The medium or low importance application, in this case, gets a lower priority compared to the high importance application. Therefore, even if these applications have a medium QoS level, they affect the overall application QoS level to a limited extent. The average QoS level is also determined based on similar criteria. The example of one of the rules from the profiled application QoS evaluation subsystem is as follows:

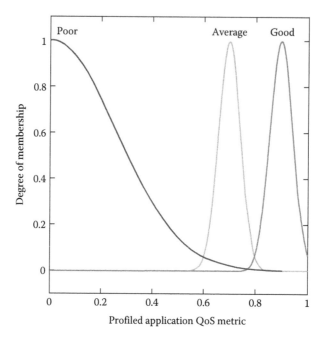

FIGURE 11.12   Profiled application QoS metric.

TABLE 11.14

Conditions for the Profiled Application QoS Level

Fuzzy Consequent	Fuzzy Conditions
Poor	The application with high/medium importance has a poor QoS level.
Average	The application with high/medium importance has an average QoS level.
	The application with low importance has a poor QoS level.
Good	The application with high/medium/low importance has a good QoS level.
	The application with low importance has an average QoS level.

If Voice Application Significance Is Low and the Voice Application QoS Is Poor, Then the Profiled Application QoS Is Average.

If there are R rules, each with l premises in the system, the rth rule has the following form:

$Ru_r :$ **If** S_A **is** SG_r^l **and** $QoSWAM_A$ **is** A_r^l **then** $QoSWAM_A$ **is** WA_r^l

The antecedent of this general rule can be interpreted as a fuzzy relationship, which is obtained through the intersection of the fuzzy sets SG_r^l, A_r^l, and WA_r^l.

$$FR_l = SG_r^l \bigcap A_r^l WA_r^l \tag{11.30}$$

The membership function for this relationship is defined as

$$FR_l(S_A, QoSAM_A, QoSWAM_A) = SG_r^l(S_A) \wedge A_r^l(QoSAM_A) \wedge WA_r^l(QoSWAM_A) \tag{11.31}$$

Using the *max* operator, the aggregation is denoted as

$$Ag(z) = \bigcup_{q=1}^{R} (\mu_{SG_r^l}(S_A) \cap \mu_{A_r^l}(QoSAM_A) \cap \mu_{WA_r^l}(QoSWAM_A)) \qquad (11.32)$$

11.8   FUZZY MODELLING OF A RADIO ACCESS NETWORK QoS SUBSYSTEM

In this section, the fuzzy modelling of a RAN QoS evaluation subsystem is outlined. One network can have multiple RANs. As a result, the QoS level of each RAN is assessed separately. The number of RAN subsystems is defined based on the number of RANs present in the network. The specifications of the RAN subsystem are described as follows.

11.8.1   VARIABLES

The input parameter for this subsystem is the output parameter from the previous subsystem, which is the profiled application QoS metric.

$QoSWAM_A^R$: Profiled application QoS metric
$QoSRM_R$: RAN QoS metric
W^R: Universe of discourse for the RAN QoS metric

The universe of discourse for the RAN QoS metric output variable W^R is [0, 1]. It is expressed as

$$W^R = \left\{ W_1^R, W_2^R, \ldots, W_l^R \right\} \qquad (11.33)$$

The fuzzy set for the RAN QoS metric output variable $QoSRM_R$ in the universe of discourse W^R is defined as

$$RQ = \left\{ QoSRM_R, \mu_{RQ}(QoSRM_R) \mid QoSRM_R \in W^R \right\} \qquad (11.34)$$

The term set for this $QoSRM_R$ is defined as

$$T(QoSRM_R) = \left\{ T_{QoSRM_R}^1, T_{QoSRM_R}^2, T_{QoSRM_R}^3 \right\} \qquad (11.35)$$

where $T_{QoSRM_R}^1 = Poor$, $T_{QoSRM_i}^2 = Average$, and $T_{QoSRM_R}^3 = Good$.

11.8.2   MEMBERSHIP FUNCTIONS

In this case, the membership functions are also defined as Gaussian functions. The reason for using Gaussian membership functions is illustrated in the previous sections. The ranges of the input and output variables of the membership functions use a scale between zero and one. The details for the profiled application QoS metric are outlined in previous sections. The output variable RAN QoS metric adapts a scale between zero and one, with three membership functions of good, average, and poor. Table 11.15 illustrates the ranges for each membership function in the RAN QoS evaluation subsystem. The output variable RAN QoS metric is presented in Figure 11.13.

TABLE 11.15

Membership Functions for the Network RAN QoS Metric

Membership Function	Range
Poor	0.0–0.50
Average	0.50–0.75
Good	0.75–1.0

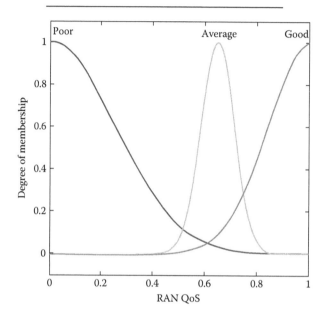

FIGURE 11.13 Membership functions for the RAN QoS metric.

11.8.3 RULES

The number of rules for each RAN QoS evaluation subsystem depends on the number of membership functions and input variables. For example, if there are three applications in an access network, the number of rules for that access network subsystem is 3^3. In Farnaz Farid and Ruan (2012), the simulation analysis results demonstrate that the applications with lower importance have a minimal effect on the network's QoS level. On the other hand, the applications that have greater importance affected the network's QoS level to a greater extent. As the application significance parameters are already embedded into the profiled QoS level, the rules defined in the RAN subsystems are straightforward.

Table 11.16 outlines the conditions for the output QoS level of RANs. As shown in the table, a rule with a poor QoS level evaluates if any of the profiled application QoS levels are poor. This rule is in synchronization with the rules of the profiled application QoS evaluation subsystem. According to the rule patterns of those subsystems, if any of the applications with high or medium importance has poor QoS, then the profiled application QoS for that application is evaluated as poor.

The RAN QoS level is set to good if any of the profiled applications are labeled as having good QoS. According to the rule in the previous subsystem, the profiled QoS of any application is labeled as good if any of the applications, regardless of importance, has a good QoS level or any application with low importance has an average QoS level. The low importance applications, in this case, get lower priority as they do not affect the overall network QoS level that much. An example of one of the rules from the RAN QoS evaluation subsystem is as follows:

TABLE 11.16
Conditions for the RAN QoS Level

Fuzzy Consequent	Fuzzy Conditions
Poor	The profiled QoS level of any application is poor.
Average	All, the maximum number, or an equal number of profiled application QoS are average.
	No profiled application QoS is poor.
Good	All or the maximum number of profiled application QoS levels are good.
	No profiled application QoS is poor.

If in Network R the Profiled Voice Application QoS Is Poor and If the Profiled VC Application QoS Is Good and the Profiled VC Application QoS Is Poor, Then the RAN QoS Metric Is Poor.

If there are R rules, each with l premises in the system, the rth rule has the following form:

Ru_q: **If** $QoSWAM_{A_i}$ **is** WA_r^l **then** $QoSRM_R$ **is** RQ_r^l

where i denotes the number of applications, and $i = \{1,2,\dots,m\}$. The antecedent of this general rule can be interpreted as a fuzzy relationship, which is obtained from the intersection of the fuzzy sets WA_r^l and RQ_r^l:

$$FR_l = WA_r^l \cap RQ_r^l \tag{11.36}$$

The membership function for this relationship is defined as

$$FR_l(QoSWAM_{A_i}, QoSRM_R) = WA_r^l(QoSWAM_{A_i}) \wedge RQ_r^l(QoSRM_R) \tag{11.37}$$

Using the *max* operator, the aggregation is denoted as

$$Ag(z) = \bigcup_{q=1}^{R} (\mu_{WA_r^l}(QoSWAM_{A_i}) \cap \mu_{RQ_r^l}(QoSRM_R)) \tag{11.38}$$

11.9   FUZZY MODELLING OF A NETWORK CONFIGURATION QoS SUBSYSTEM

This subsystem produces the final network configuration QoS metric by utilizing the RAN QoS metrics.

11.9.1   VARIABLES

The output variables from RAN subsystems are the input variable of this subsystem.

$QoSRM_R^N$: Input variable RAN QoS metric
$QoSCM_N$: Output variable network configuration QoS metric
X^N: Universe of discourse for network configuration QoS metric

The universe of discourse for the configuration QoS metric $QoSCM_N$ is [0, 1]. It is expressed as

$$X^N = \left\{ X_1^N, X_2^N, \dots, X_l^N \right\} \tag{11.39}$$

The fuzzy set for $QoSCM_N$ in the universe of discourse X^N is defined as

$$NC = \left\{ QoSCM_N, \mu_{NC}(QoSCM_N) \mid QoSCM_N \in X^N \right\} \tag{11.40}$$

The term set for this $QoSCM_N$ is defined as

$$T(QoSCM_N) = \left\{ T^1_{QoSCM_N}, T^2_{QoSCM_N}, T^3_{QoSCM_N} \right\} \tag{11.41}$$

where $T^1_{QoSCM_N} = Poor$, $T^2_{QoSCM_N} = Average$, and $T^3_{QoSCM_l} = Good$.

11.9.2 MEMBERSHIP FUNCTIONS

Figure 11.14 shows the membership function for the network configuration QoS metric. Gaussian membership functions are used for the same reasons as stated in Section 11.3.3. The output variable network configuration QoS metric takes a scale between 0 and 1, with three membership functions of good, average, and poor. Table 11.17 shows the ranges for the application significance metric. Figure 11.14 illustrates the Gaussian membership functions.

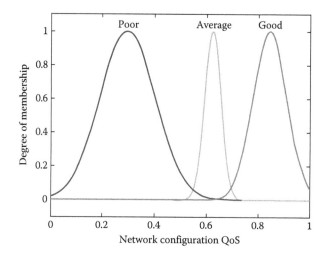

FIGURE 11.14   Membership function for the network configuration QoS metric.

TABLE 11.17
Membership Functions for the Network Configuration QoS Metric

Membership Function	Range
Poor	0–0.6
Average	0.6–0.75
Good	0.75–1.0

TABLE 11.18

Conditions for the Network Configuration QoS Level

Fuzzy Consequent	Fuzzy Conditions
Poor	At least one or all the RANs have a poor QoS level.
Average	All or the maximum number of RANs have an average QoS level. No RAN has a poor QoS level.
Good	All or the maximum number of RANs have a good QoS level. No RAN has a poor QoS level.

11.9.3 RULES

The number of access networks in a network determines the number of rules for this subsystem. If the network has two RANs, then the number of rules is 3^2. The rules are designed according to the simulation analysis results from Farid et al. (2013a, b) and Farnaz Farid and Ruan (2015) In those works, the network configuration QoS metric has been calculated by combining the performance metrics of different RANs in a network. The analyses from those calculations demonstrate that if all the RANs have average QoS, usually the overall network QoS is average. If one of the RANs has poor QoS, then the overall network QoS is usually poor. To achieve a good network QoS, all the RANs' QoS should be good. Based on such analysis results, the rules are defined for the RAN subsystem. Table 11.18 shows the rule patterns for this QoS evaluation subsystem.

If there are R rules, each with l premises in the system, the rth rule has the following form:

If $QoSRM_{R_j}^N$ is RQ_r^l then $QoSCM_N$ is NC_r^l

where j is the number of RANs present in the network, and $j = \{1,2,\dots, \text{n}\}$. The antecedent of this general rule can be interpreted as a fuzzy relationship, which is obtained by the intersection of the fuzzy sets RQ_r^l and NC_r^l.

$$FR_l = RQ_r^l \cap NC_r^l \tag{11.42}$$

The membership function for this relationship is defined as

$$FR_l(QoSRM_{R_j}^N, QoSCM_N) = RQ_r^l(QoSRM_{R_j}^N) \wedge NC_r^l(QoSCM_N) \tag{11.43}$$

If the subsystem has r number of rules, then using the *max* operator, the aggregation is denoted as

$$Ag(z) = \bigcup_{q=1}^{R} \left(\mu_{RQ_r^l}\left(QoSRM_{R_j}^N\right) \bigcap \mu_{NC_r^l}\left(QoSCM_N\right) \right) \tag{11.44}$$

11.10 DEFUZZIFICATION

The next step is the defuzzification process. Each rule produces a fuzzy output set. These sets are then mapped to a crisp value through the defuzzification method. There are many methods of defuzzification, such as the center-of-gravity (CoG), or centroid, method; the weighted-mean method; and the min-max method. The most commonly used strategy, the centroid method, is used for this chapter. The centroid calculation is denoted as

$$QoS_N^* = \frac{\int_{i=1}^{n} \mu_{NC_r^l}(QoSCM_N).QoSCM_N\, dQoSCM_N}{\int_{i=1}^{n} \mu_{NC_r^l}(QoSCM_N)\, dQoSCM_N} \qquad (11.45)$$

where n is the number of outputs.

11.11   CASE STUDY

In this section, a case study is outlined to evaluate the efficiency of the proposed QoS evaluation method. A service operator, Z, maintains a network, N, which is used mainly for voice applications. In the near future, this network will be heavily used for education services. As a result, video-conferencing and video-streaming applications will be two major applications in this network along with voice applications. The operator wants to investigate the performance of the network with the new configuration settings before deploying the final network. They also want to examine the required QoS for each application and each RAN in the network. The network has UMTS and LTE RANs along with WLANs. The proposed method discussed previously is used to carry out the analyses.

In the first phase, the current network for Z is simulated with the actual number of voice users. The network QoS is calculated for this stage. Then, in the second phase, the network with the upcoming changes is simulated with the newly added VC and VS users. The performance is analyzed for each application, access network, and the whole network configuration. The results of the first and the second phase are compared to provide recommendations for the upcoming upgrades. Table 11.19 shows the network settings in both phases. In the first phase, the whole network has a total of 46 users, and in the second phase, this number is increased to 113. The UMTS RAN has 52 users, and the LTE RAN has 51 users.

Table 11.20 shows the QoS analysis of the voice application with the current and the forthcoming changes in the network. The performance of the voice application on the UMTS network decreases

TABLE 11.19
Network Settings

Network	Number of Users in Voice Application	Number of Users in VC Application	Number of Users in VS Application
UMTS + WLAN (1st phase)	20	1	1
UMTS (2nd phase)	40	8	4
LTE (1st phase)	20	2	2
LTE (2nd phase)	40	6	5

TABLE 11.20
QoS Analysis of the Voice Application

RANs	Number of Users	Specifications			
		End-to-End Delay (msec)	Jitter (msec)	Packet Loss (%)	$QoSAM_A^R$
UMTS	20	240	45	6.25	0.836
	40	330	55	8.10	0.677
LTE	20	120	25	5.15	0.877
	40	200	45	6.23	0.837

TABLE 11.21

QoS Analysis of the VC Application

RANs	Number of Users	Specifications			
		End-to-End Delay (msec)	Jitter (msec)	Packet Loss (%)	$QoSAM_A^R$
UMTS	1	65	10	1.5	0.894
	8	320	50	4.1	0.516
LTE	2	30	12	1.12	0.898
	6	135	40	3.2	0.741

TABLE 11.22

QoS Analysis of the VS Application

RANs	Number of Users	Specifications		
		End-to-End Delay (msec)	Packet Loss (%)	$QoSAM_A^R$
UMTS	1	1	2	0.853
	4	4	3.8	0.559
LTE	2	0.30	0.60	0.898
	5	2	3	0.794

by 23.3% due to the increased numbers of voice, VC, and VS users. For the LTE network, this performance degradation is only 3.6%. Table 11.21 depicts the QoS analysis for a VC application. With the new settings on the UMTS network, the performance of the VC application degrades by around 20%. For the LTE network, this number is 12%. Table 11.22 illustrates the performance analysis results for the VS application. With three more users, the performance of the VS application on the UMTS network decreases by 29.1%. For the LTE network, the performance deterioration is only 5.6%.

After analyzing the data for the application level, the performance of each RAN is evaluated. For the first phase, the importance for the VC application is set to medium, the importance for the voice application is set to high, and the importance for the VS application is set to low. For the second phase, the significances for the VC, voice, and VS applications are set to high, medium, and high respectively. The analysis shows that for the newly added applications and users, the performance of the overall network drops by 54%. This is mainly due to the performance drop of the UMTS network. The analysis is outlined in Figure 11.15.

Using such analysis, it is easier for operator Z to figure out which part of the network can face potential performance issues when the new configurations are deployed. After detecting the performance issue in the UMTS network, operator Z makes some changes to its planned configurations. As a part of these changes, some users are moved from the UMTS to the LTE network. Among them, there are six VC users and two VS users. The performance is analyzed again after these changes. Table 11.23 shows the performance evaluation with the new user arrangements.

Figure 11.16 shows the comparison with the old upgrade configurations. In the UMTS network, the voice performance improves by 9%. For VC and VS applications, the improvements are 34.9% and 28.5%, respectively. In the case of LTE networks, although the performance decreases for all the applications, it is still within the acceptable level. Figure 11.17 outlines the comparisons of the overall performance of the network. With the newly planned settings, the performance of the whole network is improved by 50.9%, with a 54.7% improvement in the UMTS network and a 7.7% improvement in the LTE network. Although the performance of the LTE network decreases, it is still within the acceptable QoS level.

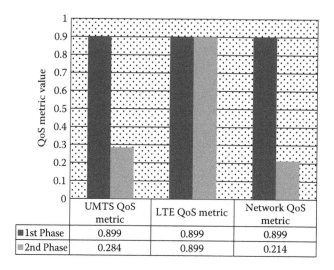

	UMTS QoS metric	LTE QoS metric	Network QoS metric
■ 1st Phase	0.899	0.899	0.899
▦ 2nd Phase	0.284	0.899	0.214

FIGURE 11.15   Performance evaluation of the network configuration.

TABLE 11.23

Performance Evaluation with the Newly Planned Settings

		Specifications			
RAN	Applications	End-to-End Delay (msec)	Jitter (msec)	Packet Loss (%)	$QoSAM_A^R$
UMTS	Voice	300	48	7.25	0.74
	VC	100	15	2.8	0.854
	VS	1.8	N/A	2	0.853
LTE	Voice	240	50	7.50	0.711
	VC	145	42	3.8	0.601
	VS	2.5	N/A	3	0.794

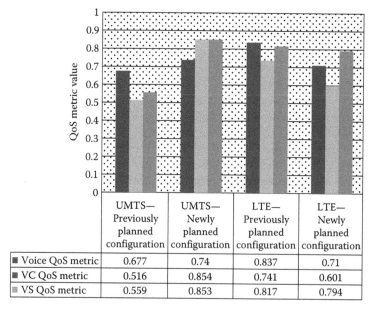

	UMTS—Previously planned configuration	UMTS—Newly planned configuration	LTE—Previously planned configuration	LTE—Newly planned configuration
■ Voice QoS metric	0.677	0.74	0.837	0.71
■ VC QoS metric	0.516	0.854	0.741	0.601
▦ VS QoS metric	0.559	0.853	0.817	0.794

FIGURE 11.16   Performance comparison.

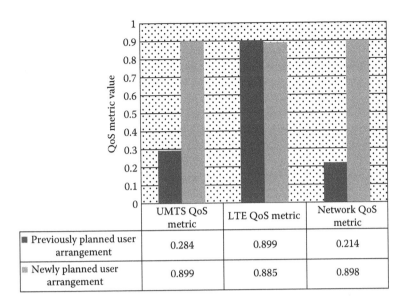

	UMTS QoS metric	LTE QoS metric	Network QoS metric
■ Previously planned user arrangement	0.284	0.899	0.214
▨ Newly planned user arrangement	0.899	0.885	0.898

FIGURE 11.17   Performance comparison of different settings.

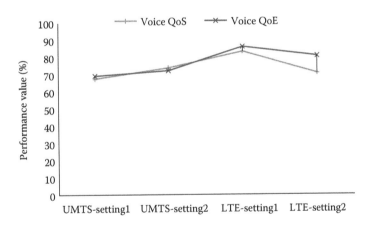

FIGURE 11.18   Comparisons of QoS and QoE values.

The analysis of this case study demonstrates that when using the proposed method, it is possible to evaluate the performance of heterogeneous networks efficiently. If the operators plan to upgrade any network configurations, the potential issues with the new settings in the post-deployment stage can be figured out beforehand. It is easier to detect the specific settings that could affect the whole network performance. Based on such analysis, the configurations can be adjusted for better performance.

Using this evaluation method, the QoE level of a network can be analyzed. The QoS metrics for the voice application of this case study are used to interpret the QoE, and the results are compared to the E-model of ITU. For effective comparison, the voice QoS metric values are converted to percentiles to match the QoE scale of E-model. Figure 11.18 outlines the comparisons. The QoS metric values are annotated as QoE; Setting-1 refers to the previously planned configurations, and Setting-2 refers to the newly planned configurations. The measured values from the E-model do not show much difference from the voice QoS metric values. Therefore, it is possible to measure the QoE of the voice application using the QoS metrics derived using this method.

11.12 COMPARISON WITH OTHER METHODS

In this section, the comparison results of the FLQEM with the fixed and DWQEMs are presented. Table 11.22 shows the comparison results with the dynamic weight-based method. Most of the time, the networks that are evaluated using both of these methods have demonstrated the same ranking. However, it differs in the case when the voice calls experience a delay of 224 msec, jitter of 65 msec, and packet loss of 3.28%. The dynamic weight-based method shows the QoS level as "Poor" in this situation. However, using the fuzzy logic approach, this is measured as "Average" because the fuzzy logic approach can handle an overlapping situation efficiently compared to the dynamic weight-based method. With a delay of 224 msec, jitter of 65 msec, and packet loss of 9.24%, the voice QoS takes a value of 0.6. Using the dynamic weight-based method, it is hard to define the QoS level of this value as the "Average" and the "Poor" QoS values overlap. The fuzzy logic method uses rules; therefore, firing the proper rule, it calculates the QoS level as "Average."

Table 11.24 shows the mapping of the QoE scale of the E-model to the proposed QoS metric scale. It should be noted that this is only for voice applications as the E-model can evaluate only the voice transmission quality. It uses a five-term scale to analyze the QoE of any voice transmission quality. The proposed fuzzy logic approach uses a three-term scale to measure the QoS of voice applications. This three-scale measurement is mapped to the five-scale QoE measurement scale of the E-model to evaluate the QoS level in terms of QoE. Figure 11.19 shows the comparisons of QoS

TABLE 11.24
QoS Metrics for Mixed Traffic

Number of Active Calls	E2ED (msec)	J (msec)	PL (%)	DWQEM Application QoS Metric	DWQEM QoS Level	FLQEM Application QoS Metric	FLQEM QoS Level
8	216	40	2.92	0.82	Good	0.895	Good
10	221	58	2.95	0.62	Average	0.712	Average
12	224	65	3.28	0.54	Poor	0.67	Average
14	228	68	4.41	0.49	Poor	0.588	Poor
16	230	70	5.34	0.45	Poor	0.563	Poor
18	233	99	6.46	0.38	Poor	0.281	Poor

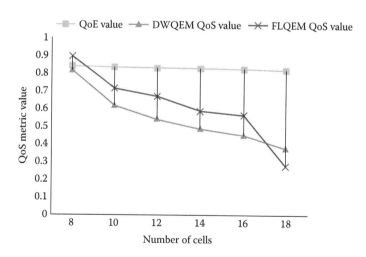

FIGURE 11.19 Comparisons of DWQEM and FLQEM metric values with QoE values.

TABLE 11.25
Mapping of the QoE Scale of E-Model

R Factor Scale	QoS Metric Scale	QoS Level	QoE Level
90 < R < 100	0.8–1	Good	Very satisfied
80 < R < 90			Satisfied
70 < R < 80	0.6–0.8	Average	Some users dissatisfied
60 < R < 70			Many users dissatisfied
50 < R < 60	0–0.6	Poor	Nearly all users dissatisfied

values of voice applications on some networks using the dynamic weight-based and fuzzy logic–based methods. It also shows their closeness to the measured QoE value using the E-model. The fuzzy logic approach shows a closer measurement result to the measured QoE value of the E-model than the DWQEM method (Table 11.25).

11.13 CONCLUSION

In this chapter, a novel QoS evaluation method based on fuzzy logic concepts has been proposed and evaluated. The fundamental idea behind this method is the consideration of uncertain situations in performance analysis of heterogeneous networks. There could be situations in which the QoS level of an application or access network overlaps two consecutive measurement metrics. In those cases, it is hard to assess the QoS level accurately. Through the simulation analysis, it has been demonstrated that without considering the overlapping values, the network QoS could be evaluated with a higher or lower QoS level than the expected level. The proposed method can effectively handle these situations. The method uses a set of rules for assessing the QoS of heterogeneous networks. Hence, the results can be categorized in a more efficient manner. The case study of a network-planning stage shows that using the fuzzy logic–based method, the effects of any new upgrade on application and network performance can be examined in a very efficient manner. For instance, the analysis results of this case study have outlined that the performance of voice applications on the UMTS network has decreased by 23.3% and 3.6% on the LTE network due to the newly added users. After finding out such performance issues, the operator has updated the previously planned configuration. The comparison results of the newly planned settings with the old ones demonstrate that, on the UMTS network, the voice performance improves by 9%. For VC and VS applications, the improvements are 34.9% and 28.5%, respectively. In the case of LTE networks, although the performance decreases for all the applications, it is still within the acceptable level. The comparisons of the overall performance of the network show that with the changed settings, the performance of the whole network is improved by 50.9%, with a 54.7% improvement in the UMTS network and a 7.7% improvement in the LTE network. After receiving such numerical analysis results, it is very easy for the service operator to go with the planned network upgrades.

The QoE level of a network has been also analysed using this method. The values of the QoS-related parameters have been utilized to measure the voice QoS and QoE metrics. The voice QoS metric has been measured using the fuzzy logic–based method, and the QoE metric has been measured using the E-model. The voice QoS metric values have been converted to percentiles to match with the QoE scale of E-model. The measured values from the E-model do not show much difference when compared to the voice QoS metric values. Therefore, it is possible to measure the QoE of the voice application using the QoS metrics that are derived using this method. This method is also compared with the dynamic weight-based evaluation. For instance, with a delay of 224 msec, jitter of 65 msec, and packet loss of 9.24%, the voice QoS on the UMTS network takes a value of 0.6. Using the dynamic weight-based method, it is hard to determine the QoS level of this value as the

"Average" and the "Poor" QoS values overlap. The fuzzy logic method uses rules; therefore, firing the proper rule, it calculates the QoS level as "Average." Therefore, the fuzzy logic–based method is more efficient in the cases when the QoS level overlaps.

REFERENCES

Alkhawlani, M., and Ayesh, A. (2008). Access network selection based on fuzzy logic and genetic algorithms. *Advances in Artificial Intelligence, 8*(1), 1–12. doi: 10.1155/2008/793058

Alshamrani, A., Xuemin, S., and Liang-Liang, X. (2011). QoS provisioning for heterogeneous services in cooperative cognitive radio networks. *IEEE Journal on Selected Areas in Communications, 29*(4), 819–830. doi: 10.1109/JSAC.2011.110413

Augestad, K., and Lindsetmo, R. (2009). Overcoming distance: Video-conferencing as a clinical and educational tool among surgeons. *World Journal of Surgery, 33*(7), 1356–1365. doi: 10.1007/s00268-009-0036-0

Bari, F., and Leung, V. (11–15 March 2007a). Application of ELECTRE to network selection in a hetereo-geneous wireless network environment. Paper presented at the *IEEE Wireless Communications and Networking Conference, WCNC.*

Bari, F., and Leung, V. C. M. (2007b). Automated network selection in a heterogeneous wireless network environment. *IEEE Network, 21*(1), 34–40. doi: 10.1109/mnet.2007.314536

Carmen Lucas-Estañ, M., and Gozalvez, J. (2013). On the real-time hardware implementation feasibility of joint radio resource management policies for heterogeneous wireless networks. *IEEE Transactions on Mobile Computing, 12*(2), 193–205. doi: 10.1109/TMC.2011.256

Charilas, D., Markaki, O., Psarras, J., and Constantinou, P. (2009). Application of fuzzy AHP and ELECTRE to network selection. In F. Granelli, C. Skianis, P. Chatzimisios, Y. Xiao, and S. Redana (Eds.), *Mobile Lightweight Wireless Systems* (Vol. *13*, pp. 63–73). Berlin, Heidelberg: Springer.

Diao, Y., Hellerstein, J., and Parekh, S. (2002). Optimizing quality of service using fuzzy control. In M. Feridun, P. Kropf, and G. Babin (Eds.), *Management Technologies for E-Commerce and E-Business Applications* (Vol. *2506*, pp. 42–53). Springer Berlin Heidelberg.

ETSI. (2006). Review of available material on QoS requirements of multimedia services. *Telecommunications and Internet Converged Services and Protocols for Advanced Networking (TISPAN)* (Vol. 1). ETSI.

Farid, F. (2015). Methodical evaluation of quality of service for heterogeneous networks.

Farid, F., Shahrestani, S., and Ruan, C. (2013b). A user-centric network architecture for sustainable rural areas. In V. Guyot (Ed.), *Advanced Infocomm Technology* (Vol. *7593*, pp. 138–150). Springer Berlin Heidelberg.

Farid, F., Shahrestani, S., and Ruan, C. (21–24 October 2013a). QoS analysis and evaluations: Improving cellular-based distance education. *Paper presented at the IEEE 38th Conference on Local Computer Networks Workshops (LCN Workshops).*

Farnaz Farid, S. S., and Ruan, C. (2012). Quality of service requirements in wireless and cellular networks: Application-based analysis. Paper presented at the *International Business Information Management Association Conference*, Barcelona, Spain.

Farnaz Farid, S. S., and Ruan, C. (2015). Application-based QoS evaluation of heterogeneous networks. Paper presented at the *Seventh International Conference on Wireless & Mobile Networks*, Sydney, Australia.

Figueira, J., Greco, S., and Ehrgott, M. (2005). *Multiple Criteria Decision Analysis: State of the Art Surveys.* Springer.3GPP R 26.937 version 1.2.0 RTP usage model (2002).

Guillaume, S. (2001). Designing fuzzy inference systems from data: An interpretability-oriented review. *IEEE Transactions on Fuzzy Systems, 9*(3), 426–443. doi: 10.1109/91.928739

Hallani, H. (2007). *Utilising Behaviour History and Fuzzy Trust Levels to Enhance Security in Ad-Hoc Networks.* University of Western Sydney.

Herrera, F., and Herrera-Viedma, E. (2000). Linguistic decision analysis: Steps for solving decision prob-lems under linguistic information. *Fuzzy Sets and Systems, 115*(1), 67–82. doi: http://dx.doi.org/10.1016/S0165-0114(99)00024-X

ITU. (30 May 2008). E-model Tutorial. Accessed on 16 February 2015. Retrieved from www.itu.int/ITU-T/studygroups/com12/emodelv1/tut.htm.

Jang, J. S. R., Sun, C. T., and Mizutani, E. (1997). *Neuro-Fuzzy and Soft Computing: A Computational Approach to Learning and Machine Intelligence.* Prentice Hall.

Jin, Z., and Bose, B. K. (5–8 November 2002). Evaluation of membership functions for fuzzy logic controlled induction motor drive. Paper presented at the *IECON 02 IEEE 28th Annual Conference of the Industrial Electronics Society.*

Kantubukta Vasu, S. M., Mahapatra, S., and Kumar, C. S. (2012). QoS-aware fuzzy rule-based vertical handoff decision algorithm incorporating a new evaluation model for wireless heterogeneous networks. *EURASIP Journal on Wireless Communications and Networking, 2012, 322.* doi: 10.1186/1687-1499-2012-322

Karimi, O. B., and Fathy, M. (2010). Adaptive end-to-end QoS for multimedia over heterogeneous wireless networks. *Computers & Electrical Engineering, 36*(1), 45–55. doi: http://dx.doi.org/10.1016/j.compeleceng.2009.04.006

Luo, H., and Shyu, M.-L. (2011). Quality of service provision in mobile multimedia—A survey. *Human-Centric Computing and Information Sciences, 1*(1), 1–15. doi: 10.1186/2192-1962-1-5

Lusheng, W., and Binet, D. (17–20 May 2009). MADM-based network selection in heterogeneous wireless networks: A simulation study. Paper presented at the *1st International Conference on Wireless Communication, Vehicular Technology, Information Theory and Aerospace & Electronic Systems Technology (Wireless VITAE).*

McNair, J., and Fang, Z. (2004). Vertical handoffs in fourth-generation multinetwork environments. *IEEE Wireless Communications, 11*(3), 8–15. doi: 10.1109/mwc.2004.1308935

Munir, S. A., Yu Wen, B., Ren, B., and Ma, J. (11–15 March 2007). Fuzzy logic based congestion estimation for QoS in wireless sensor network. Paper presented at the *IEEE Wireless Communications and Networking Conference, WCNC.*

Qingyang, S., and Jamalipour, A. (2005). Network selection in an integrated wireless LAN and UMTS environment using mathematical modeling and computing techniques. *IEEE Wireless Communications, 12*(3), 42–48. doi: 10.1109/mwc.2005.1452853

Ross, T. J. (2010). *Classical Sets and Fuzzy Sets: Fuzzy Logic with Engineering Applications* (pp. 25–47). Wiley.

Sedoyeka, E., Hunaiti, Z., and Tairo, D. (2009). Analysis of QoS requirements in developing countries. *International Journal of Computing and ICT Research, 3*(1), 18–31.

Sgora, A., Chatzimisios, P., and Vergados, D. (2010). Access network selection in a heterogeneous environment using the AHP and fuzzy TOPSIS methods. In P. Chatzimisios, C. Verikoukis, I. Santamaría, M. Laddomada, and O. Hoffmann (Eds.), *Mobile Lightweight Wireless Systems* (Vol. 45, pp. 88–98). Springer Berlin Heidelberg.

Shu-ping, Y., Talwar, S., Geng, W., Himayat, N., and Johnsson, K. (2011). Capacity and coverage enhancement in heterogeneous networks. *IEEE Wireless Communications, 18*(3), 32–38. doi: 10.1109/MWC.2011.5876498

Stankiewicz, R., Cholda, P., and Jajszczyk, A. (2011). QoX: What is it really? *IEEE Communications Magazine, 49*(4), 148–158. doi: 10.1109/mcom.2011.5741159

Stevens-Navarro, E., Yuxia, L., and Wong, V. W. S. (2008). An MDP-based vertical handoff decision algorithm for heterogeneous wireless networks. *IEEE Transactions on Vehicular Technology, 57*(2), 1243–1254. doi: 10.1109/tvt.2007.907072

Szigeti, T. H. C., Hattingh, C., Barton, R., and Briley, K. Jr. (2005). *End-to-End QoS Network Design: Quality of Service for Rich-Media & Cloud Network.* Indianapolis, IN: Cisco Press, pp. 26, 734, Hattingh, Christina (Translator).

Von Altrock, C. (1995). *Fuzzy Logic and NeuroFuzzy Applications Explained.* Prentice Hall PTR.

Vuong Xuan, T., and Tsuji, H. (20–22 October 2008). QoS based ranking for web services: Fuzzy approaches. Paper presented at the *4th International Conference on Next Generation Web Services Practices, NWESP '08.*

Wang, L.-X. (1997). *A Course in Fuzzy Systems and Control.* Prentice-Hall.

Wei, S., and Qing-An Z. (2008). Cost-function-based network selection strategy in integrated wireless and mobile networks. *IEEE Transactions on Vehicular Technology, 57*(6), 3778–3788. doi: 10.1109/tvt.2008.917257

Wei, S., and Weihua, Z. (2012). Performance analysis of probabilistic multipath transmission of video streaming traffic over multi-radio wireless devices. *IEEE Transactions on Wireless Communications, 11*(4), 1554–1564. doi: 10.1109/twc.2012.021512.111397

Wenhui, Z. (21–25 March 2004). Handover decision using fuzzy MADM in heterogeneous networks. Paper presented at the *IEEE Wireless Communications and Networking Conference, WCNC.*

Wen-Tsuen, C., and Yen-Yuan, S. (13–17 March 2005). Active application oriented vertical handoff in next-generation wireless networks. Paper presented at the *IEEE Wireless Communications and Networking Conference.*

Wen-Tsuen, C., Jen-Chu, L., and Hsieh-Kuan, H. (7–9 July 2004). An adaptive scheme for vertical handoff in wireless overlay networks. Paper presented at the *Tenth International Conference on Parallel and Distributed Systems, ICPADS Proceedings.*

Yen, J. (1999). Fuzzy logic-a modern perspective. *IEEE Transactions on Knowledge and Data Engineering*, *11*(1), 153–165.

Yen, J., and Langari, R. (1999). *Fuzzy Logic: Intelligence, Control, and Information*. Prentice Hall.

Zadeh, L. A. (1965). Fuzzy sets. *Information and Control, 8*(3), 338–353. doi: http://dx.doi.org/10.1016/S0019-9958(65)90241-X

Zadeh, L. A. (1973). Outline of a new approach to the analysis of complex systems and decision processes. *IEEE Transactions on Systems, Man and Cybernetics, 3*(1), 28–44. doi: 10.1109/TSMC.1973.5408575

Zadeh, L. A. (1988). Fuzzy logic. *Computer, 21*(4), 83–93. doi: 10.1109/2.53

Zadeh, L. A. (1996). Fuzzy logic computing with words. *IEEE Transactions on Fuzzy Systems, 4*(2), 103–111. doi: 10.1109/91.493904

Zennaro, M., Canessa, E., Sreenivasan, K. R., Rehmatullah, A. A., and Cottrell, R. L. (2006). Scientific measure of Africa's connectivity. *The Massachusetts Institute of Technology Information Technologies and International Development, 3*(1), 55–64.

Zimmermann, H. J. (1996). *Fuzzy Set Theory–and Its Applications* (3rd ed.). Boston: Kluwer Academic Publishers.

12 Network Virtualization for Next-Generation Computing and Communication Infrastructures

Scalable Mapping Algorithm and Self-Healing Solution

Qiang Yang

CONTENTS

12.1   INTRODUCTION

In recent years, network virtualization [1] has become an increasingly important research area in the computer science community. A virtual network (VN) consists of a set of virtual nodes and virtual links. The former are hosted on multiple substrate nodes, while the latter are the end-to-end physical paths connecting virtual nodes across the substrate network. The operational goal of VNs is to provide a collection of transparent packet delivery systems with diverse protocols and packet formats over the same substrate network for the delivery of a range of services to network users.

As one of the key tasks for setting up a VN, VN mapping is the process of selecting the appropriate physical nodes and link resources subject to a set of predefined constraints (e.g., topology constraints, bandwidth) while meeting certain service requirements (e.g., quality of service, security). While a variety of state-of-the-art algorithms [2–7] have been proposed and attempted to address this issue from different facets, the challenge still remains in the context of large-scale communication networks. These existing online mapping algorithms were typically designed in a centralized fashion using a central server, which needs to maintain up-to-date topology and operational information of the entire substrate network (e.g., bandwidth and network configuration). When the network scale becomes large, the performance of these algorithms will be significantly undermined due to poor scalability. The maintenance of the up-to-date information of the underlying large-scale substrate network can be difficult, even impractical. The available mapping algorithmic solutions were mainly assessed in the context of small-scale networks (e.g., a network with 40–100 nodes) [5–7], and their performance in terms of computational efficiency will become unacceptable along with growth of the network scale. In addition, the realistic large-scale substrate network is composed of a set of heterogeneous administrated domains belonging to multiple infrastructure providers (InPs), which would not like to share their secret topology and operational details. As a result, these existing solutions, which assume that they have complete information of the entire substrate network (e.g., network bandwidth and topology), are not applicable for solving the problem of VN mapping across multiple domains in this case. This chapter proposes a novel hierarchical algorithm in conjunction with a substrate network decomposition approach to cope with the complexity of network virtualization in large-scale networks as well as to address the problem of VN mapping across multiple domains.

VN mapping across multiple domains is illustrated in Figure 12.1. The substrate network is partitioned into three domains, with VN configuration brokers (VNCBs) located in individual domains which are responsible for managing physical resources within the domains (e.g., link bandwidth, node CPU capability). The substrate network also maintains a central server which carries out the task of VN creation through cooperation with the aforementioned VNCBs. By taking advantage of the hierarchical network strategy, the suggested VN mapping algorithm can be broken down into two parts: a local VN mapping algorithm (LVNMA) operating within each VNCB and a global VN mapping algorithm (GVNMA) functioning in the central server. The LVNMA will conduct intradomain path computation according to the request sent by the GVNMA and send back the computation result, while the GVNMA will carry out the VN mapping based on a mechanism motivated by the path-vector protocol.

Compared with the existing VN mapping algorithms, the contribution made in this chapter can be summarized as follows: (1) the approach simplifies network management, as the central server no longer needs to maintain the overall substrate network's information, and individual VNCBs need only to be aware of the network information of their own scope, which can greatly improve the scalability of the algorithm as well as keep the secrecy of substrate network information; and (2) the cooperation (between the GVNMA and a set of LVNMAs) and the local computation (in LVNMA) can significantly decrease the computational complexity for VN mapping, as well as support the parallel processing of multiple received VN mapping requests in different domains. Our experimental

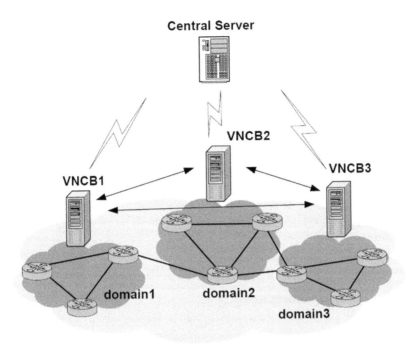

FIGURE 12.1   Hierarchical VN management environment.

evaluations show that the proposed hierarchical algorithm can be about 5 to 20 times faster than the traditional centralized approach with acceptable communication overhead between GVNCA and LVNCA while performing almost as well as the centralized solutions.

The remainder of this chapter is organized as follows. Section 12.2 overviews the most relevant work in the literature. Following that, Section 12.3 presents the network model and formulates the VN mapping problem. Section 12.4 proposes a network decomposition approach that is suitable for the hierarchical mapping algorithm, which is described in detail in Section 12.5. Section 12.6 presents the assessment of the proposed approach with a set of numerical simulation results. Finally, some concluding remarks are given in Section 12.7.

In parallel-network virtualization (e.g., [8–10]), supporting a set of embedded VNs to coexist in a shared infrastructure has been considered a promising approach for accommodating and promoting novel heterogeneous services and mechanisms toward the next-generation network architectures. Current large-scale networks (e.g., Internet) are facing "ossification" and many technical challenges which need to be addressed by more advanced networking architectures. Multiple VNs can share an underlying physical substrate to deliver a variety of tailored services by adopting heterogeneous network technologies in an isolated manner. In addition, the provision of standardized interfaces among VNs could enable seamless and transparent communications across multiple domains. From the network management perspective, network virtualization can greatly improve the global utilization efficiency of network resources and hence the capacity of network service provision.

However, it should be noted that VN invulnerability, which is one of the key aspects of network virtualization, has been long ignored, and little research effort has been made yet. The self-healing capability of VNs involves several aspects, which can be summarized in four parts: (1) proactively detecting the potential faults of physical components in the substrate to prevent service interruption due to reorganization of VNs (i.e., remapping of links and nodes); (2) keeping the load balance of the physical substrate to avoid or alleviate a network resource bottleneck; (3) restoring the service delivery in the failed VNs as quickly as possible upon physical network failures; and (4) protecting the services delivered by the VNs against a range of network attacks (e.g., denial of service [DoS]

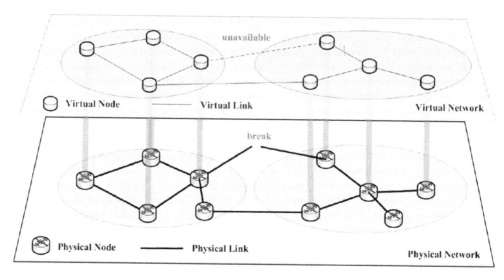

FIGURE 12.2   An example of VN mapping.

and network worm propagation). Figure 12.2 illustrates that a failed physical link can result in
the unavailability of the virtual link, and as a consequence, the overall VN can no longer deliver
expected services. In general, following to the physical substrate faults, the affected VNs need to
be remapped to establish a new set of VNs through network mapping algorithms, which may often
impose prohibitive complexity in network management, particularly in large-scale networks.

   In this chapter, a novel approach, MFP-VNMH, is exploited and presented by adopting the multi-
commodity flow problem (MFP) [11] based strategy to enable the self-healing capability in network
virtualization. More specifically, this approach aims to achieve fast service restoration upon VN
failures by improving the service restoration success ratio and minimizing the restoration time
while maintaining the load balance across the overall physical substrate.

12.2   LITERATURE REVIEW

12.2.1   Recent Studies on Network Virtualization Mapping

Lots of research effort has been made to address the VN mapping problem in recent years (e.g.,
[3–7,12]). A heuristic VN mapping algorithm is presented in [3] with the aim of minimizing mapping cost, which is in general determined by the required physical resources. [4] and [5] formulated
VN mapping as a multi-commodity flow problem. In [4], the authors used the maximum-flow model
to pre-allocate resources on each edge node and allocate these resources to VNs according to specific requirements. To improve physical resource utilization efficiency, [5] proposed an approach of
using multi-path mapping to map a virtual link to multiple physical links and resolved it as a multi-
commodity flow problem. The VN mapping problem is modeled in [6] with an integer programming
method and addressed by a fixed VN mapping algorithm (D-ViNE) as well as a random VN mapping algorithm. In [7], applying the isomorphism algorithm to VN mapping is suggested, aiming
to increase the VN request acceptance ratio. The authors in [8] proposed a distributed algorithm
that solves the mapping problem without any centralized controller but at the price of additional
communication overhead for disseminating network information to all network nodes via broadcasting. In the case of VN mapping across multiple domains, a policy-based, inter-domain VN
embedding (PolyViNE) framework is presented in [13] to heuristically address the inter-domain
VN mapping problem without optimizing the mapping performance and computational cost. A

hierarchical management framework for VN mapping across multiple domains is discussed in [14], where management elements are responsible for resource allocation within the scope of their own domains. However, no algorithmic solution or results are presented in this work.

In summary, the available algorithms with centralized operation are able to achieve acceptable VN mapping results, but their performance can degrade severely due to the scalability problem when applied to large-scale networks. While the parallelism of the distributed solution can significantly improve computational efficiency and enhance the robustness in coping with single-point failure, the communication overhead induced by cooperation among network nodes can be unacceptable with a massive number of network nodes. The proposed hierarchical solution aims to find the best trade-off between these two alternatives and to be scalable enough to be deployed in networks with large scale.

12.2.2 RECENT STUDIES ON THE VIRTUAL NETWORK HEALING MECHANISM

It is well known that many VN mapping problems have NP-complete (NPC) complexity. To address such computational intractability, a collection of heuristics- or linear programming–based proposals has been reported (e.g., [2,4,5,7,12–17]). In recent years, the fault-tolerant technology has been extensively studied (e.g., [18–21]). In [18], the authors considered two fault-management techniques in an IP-over-WDM network where network nodes employ optical cross-connects and IP routers to provide protection in the WDM layer and restoration in the IP layer. However, such an approach cannot be directly applied to the VN environment. An automatic pilot VN architecture, which is able to monitor the VN failures, together with the concept of self-healing for failed VNs is presented in [19]. However, the specific approach and algorithmic solution to heal the failed VNs is not given. In [20], the shared backup network is designed to protect the embedded VN. Each VN will receive some pre-allocated backup resources to avoid VN failure when the physical substrate suffers faults. Such a shared backup network is an offline approach to VN protection but with restricted flexibility and low efficiency of network resource utilization. In [21], the authors proposed a fault diagnosis mechanism in the network virtualization environment, which attempts to restore the VNs through fast fault detection and restoration of the physical network. Such an approach can be time consuming and suffer from an undeterminably long period of time. In fact, VN service restoration upon physical network failures is expected to be processed as soon as possible, as well as minimize the interruption of other VNs and guarantee the physical network's operational efficiency and flexibility.

In summary, current VN mapping solutions, either for random or specific topologies, focus merely on addressing the VN mapping problem and resource allocation without considering the fast service restoration upon network failures. Current studies on fault-tolerant networks focus either on fast detection and repairing of the physical substrate faults to protect the embedded VNs or on adopting the VN mapping algorithm (e.g., the overall VN remapping [2], remapping all virtual nodes and links of the failed VN, see Figure 12.3b) and the end-to-end link remapping [4] (i.e., remapping the failed virtual links upon physical-link failure, see Figure 12.3c). In fact, such VN restoration is not an easy task, calling for an efficient and flexible approach, which is scalable enough to be adopted in a large-scale VN environment. The key contribution made in this chapter can be summarized as follows: (1) a multi-commodity-based approach, MFP-VNMH, is proposed to restore the failed VNs through remapping of the virtual links and evaluating its performance against the existing overall VN remapping and end-to-end link remapping approaches; (2) through remapping the affected virtual links to available physical resources rather than the overall VN or end-to-end path due to the physical network failures, the scale of the VN remapping is minimized and hence promotes the efficiency of VN restoration and minimizes the interruption to other operational VNs.

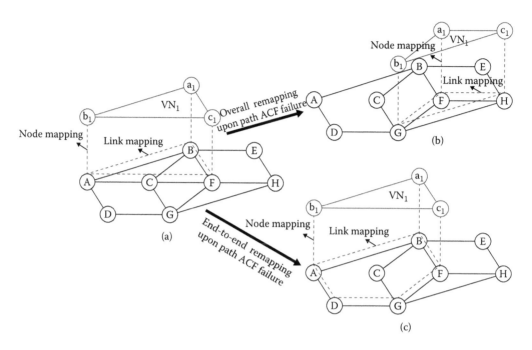

FIGURE 12.3 Overall VN remapping and end-to-end link remapping.

12.3 SCALABLE MAPPING ALGORITHM FOR LARGE-SCALE NETWORK VIRTUALIZATION

12.3.1 VN Mapping Problem Formulation

12.3.1.1 Network Model and Problem Description

The substrate network can be represented by a weighted undirected graph error, where V and E are the set of substrate nodes and the set of physical links, respectively. The network nodes are classified into two categories: core nodes and access nodes, where they are capable of packet forwarding, and the latter can act as sources or sinks. A VN request is denoted by $\{G^v(V^v,E^v),D\}$, where $G^v(V^v,E^v)$, V^v, E^v, and $D = \{d_{e^v} \mid e^v \in E^v\}$ are the VN topology, the set of virtual nodes, the set of virtual links, and the set of bandwidth requirements, d_{e^v}, of an arbitrary virtual link, e^v.

VN mapping consists of virtual node mapping and virtual link mapping. The former is to map a virtual node, $s^v \in V^v$, to a substrate node, $s \in V$, and the latter is to map a virtual link, $e^v = (s^v, t^v)$, to a substrate path from s to t. In reality, virtual nodes which act as end-users' access nodes are often predefined and mapped to access substrate nodes for many applications. Therefore, this chapter restricts its view to the task of virtual link mapping. In this case, the VN request can be expressed as a set of virtual links, denoted by $R = \{(s_1,t_1,d_1),...,(s_l,t_l,d_l)\}$, where (s,t,d) is a virtual link from s to t with the bandwidth requirement of d, and l is the number of virtual links.

12.3.1.2 Linear Program Formulation for VN Mapping

Assuming that the substrate network supports flexible splitting of virtual links over multiple substrate paths, as a result, the VN mapping problem can be formulated as a multi-commodity flow problem. Each virtual link, k, can be considered a commodity with source and destination nodes $1 \le i \le k$. Then the VN mapping problem becomes finding the paths for all commodities based on flow conservation subject to substrate link capacity constraints, which can be presented as follows:

VNM_LP (virtual network mapping linear program)

Inputs:

VN request: $R = \{(s_1, t_1, d_1), ..., (s_l, t_l, d_l)\}$

Substrate network: $G(V, E)$

The weight of the link: $\forall \overline{uv} \in E$, $w(u, v)$ *represents the cost of buying/renting a unit of link bandwidth*

The residual bandwidth of the link: $\forall \overline{uv} \in E$, $r(u, v)$ *denotes the residual bandwidth of link* $\overline{uv}$

Variables:

$f_{u,v}^i$ *is a flow variable denoting the total amount of flow from u to v on the link* $\overline{uv}$ *for the ith virtual link.*

Objective:

$$\text{Minimize} \sum_{\overline{uv} \in E} \sum_{i=1}^{l} f_{u,v}^i \cdot w(u, v) \tag{12.1}$$

Constraints:

$$\sum_{i=1}^{l} \left(f_{u,v}^i + f_{v,u}^i \right) \leq r(u, v), \ \forall u, v \in V \tag{12.2}$$

$$\sum_{w \in V} f_{u,w}^i - \sum_{w \in V} f_{w,u}^i = 0, \ \forall i, \forall u \in V \setminus \{s_i, t_i\} \tag{12.3}$$

$$\sum_{w \in V} f_{s_i,w}^i - \sum_{w \in V} f_{w,s_i}^i = d_i, \ \forall i \tag{12.4}$$

$$\sum_{w \in V} f_{t_i,w}^i - \sum_{w \in V} f_{w,t_i}^i = -d_i, \ \forall i \tag{12.5}$$

$$f_{u,v}^i \geq 0, \ \forall u, v \in V \tag{12.6}$$

The objective function of the **VNM_LP** (Equation 12.1) attempts to minimize the cost of mapping the VN request. The constraint set in Equation 12.2 is capacity constraints, summing up $f_{u,v}^i$, and $f_{v,u}^i$ ensures that the summation of flows on both directions of the undirected link $\overline{uv}$ remains within its available bandwidth. The constraint sets in Equations 12.3 through 12.5 can be combined to refer to the condition of flow conservation. The constraint set in Equation 12.6 denotes the real domain constraints on the variables.

12.3.2 Network Decomposition Approach

Before the proposed hierarchical mapping algorithm can be applied, the large-scale substrate network needs to be partitioned into a collection of subnetworks (i.e., domains). By properly applying

the historical statistical data obtained from previous VN mapping requests to the network partitioning, the efficiency of the hierarchical mapping algorithm executing in the partitioned domains can
potentially be improved. Since mapping a virtual link with its nodes in the same domain can be
simply carried out by using the local mapping algorithm proposed in this chapter (i.e., LVNMA), it
is suggested that the nodes be grouped into the same domain if a large number of virtual connections are between them to minimize the probability of VN mapping across multiple domains.

When decomposing the substrate network into a set of domains, the following statistics from
previous VN mapping requests are taken into account:

- $N_1(u_1, u_2)$: The total bandwidth of virtual links between nodes u_1 and u_2 within period T,
  including links that are operating and those that have been dismantled because of expiration of the VN.
- $N_2(u_1, u_2)$: The average bandwidth of operating virtual links between u_1 and u_2 within
  period T. $N_2(u_1, u_2) = \dfrac{1}{T} \displaystyle\int_0^T N_2(u_1, u_2, t)dt$, where $N_2(v_1, v_2, t)$ is the total bandwidth of operating virtual links between nodes u_1 and u_2 at the time of t.
- $\alpha(u_1, u_2)$: The amount of virtual links between nodes u_1 and u_2 within period T.

Although a number of network decomposition algorithms [22,23] are available aiming to decompose the network with a minimized number of links, this chapter takes a new look by taking the
historical VN mapping statistics into account during network decomposition as the statistics implies
the relation between access nodes in the substrate network. The rest of this section describes the
principle of the network decomposition algorithm that is suitable for the hierarchical mapping
algorithm.

- **Bandwidth property:** Nodes u_1 and u_2 with a large $N_1(u_1, u_2)$ are highly recommended
  to be partitioned into the same domain because a large $N_1(u_1, u_2)$ suggests plenty of communication between u_1 and u_2.
- **VN creation frequency property:** Nodes u_1 and u_2 with a large $\alpha(u_1, u_2)$ should be in the
  same domain. When u_1 and u_2 are grouped into the same domain, our algorithm for virtual link mapping needs only to carry out local mapping. Partitioning them into the same
  domain can improve the practical operational efficiency of our mapping algorithm.
- **Distance property:** Nodes u_1 and u_2 with a few hops should be grouped into the same
  domain.

The key idea behind the decomposition algorithm can be briefly described as follows: It first classifies the substrate network access nodes based on the previously stated properties, and the nodes
with the same class will be grouped into the same domain. Afterward, the substrate network core
nodes are distributed to different domains based on the maximum multi-commodity flow and the
node connectivity degree. It is assumed that $G(V, E)$ is decomposed into N domains, and V_a and
V_c denote the set of access nodes and core nodes in the substrate, respectively. L is the number of
access nodes. $B_{L \times L}^1$ and $B_{L \times L}^2$ are the $L \times L$ matrix. $B_{L \times L}^1(i, j) = \dfrac{N_1(u_i, u_j)}{\text{dist}(u_i, u_j)^2}$, where $\text{dist}(u_i, u_j)$ is the
hop between u_1 and u_2. $B_{L \times L}^2(i, j) = \alpha(u_i, u_j)$. $B_{L \times L}^1$ reflects the relation between nodes' similarity
weight about the bandwidth feature and the distance feature, and $B_{L \times L}^2$ reflects the relation between
nodes' similarity weight about the construction frequency feature. Both $B_{L \times L}^1$ and $B_{L \times L}^2$ are normalized. The decomposition algorithm is described in Algorithm 12.1.

To illuminate Algorithm 12.1, we take an example as shown in Figure 12.4. It is
assumed that nodes 1, 4, 7, 9, 14, and 16 are network access nodes (see Figure 12.4a), with
$N_2(1,4) = 5, N_2(7,9) = 5, N_2(14,16) = 3$, and all link bandwidths are set as 2 units. For example, the

ALGORITHM 12.1 NETWORK DECOMPOSITION ALGORITHM

Inputs: Substrate network $G(V,E)$, **Outputs:** N domains, $G(V_i,E_i)$, $1 \leq i \leq N$

1. Generate a similarity matrix $B_{L \times L} \leftarrow \alpha \cdot B_{L \times L}^1 + \beta \cdot B_{L \times L}^2$, and set $B_{L \times L}(i,i) \leftarrow 1, 1 \leq i \leq L$, where α and β are weight factors ($\alpha + \beta = 1$).
2. Based on similarity matrix $B_{L \times L}$ and the spectral clustering algorithm, V_a is partitioned into $\{V_1, V_2, \cdots, V_N\}$, and for $\forall i, j, i \neq j, V_i \cap V_j = \varnothing$.
3. For $\forall k \in \{1, \cdots, N\}$
4. Create a maximum multi-commodity flow model, which takes $\forall u_i, u_j \in V_k$ as the source and the sink of a commodity and takes $N_2(u_1, u_2)$ as its demand.
5. Obtain the paths, denoted by PATHs, in which all commodities can be simultaneously routed by using maximum multi-commodity flow algorithm. (PATHs is a set of nodes.)
6. $V_k \leftarrow V_k \cup$ PATHs.
7. For $\forall u_i \in V_c$
8. If u_i is included by several domains or is not included by any domains, it is assigned to a domain, V_j, with a maximum degree of node connectivity.

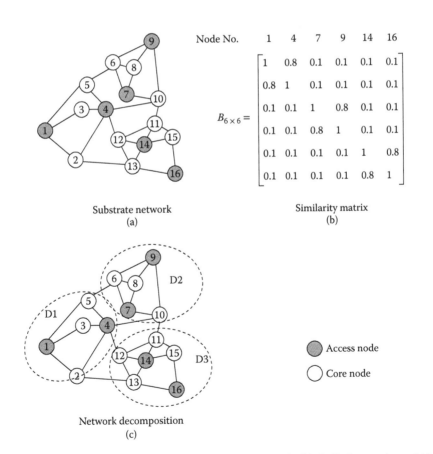

Node No.	1	4	7	9	14	16
1	1	0.8	0.1	0.1	0.1	0.1
	0.8	1	0.1	0.1	0.1	0.1
	0.1	0.1	1	0.8	0.1	0.1
$B_{6 \times 6} =$	0.1	0.1	0.8	1	0.1	0.1
	0.1	0.1	0.1	0.1	1	0.8
	0.1	0.1	0.1	0.1	0.8	1

Substrate network
(a)

Similarity matrix
(b)

Network decomposition
(c)

FIGURE 12.4 An example of network partition: (a) Substrate network, (b) similarity matrix, and (c) network decomposition.

substrate network is expected to be decomposed into 3 domains. In step 1, a similarity matrix $B_{6\times 6}$ is generated using $B^1_{6\times 6}$ and $B^2_{6\times 6}$, as shown in Figure 12.4b. Then, step 2 clusters the access nodes based on such a similarity matrix, and we obtain the clustering result as $V_1 = \{1,4\}, V_2 = \{7,9\}, V_3 = \{14,16\}$. In steps $3 \sim 6$, the core nodes are added into $V_i, 1 \leq i \leq 3$. Afterward, three domains are obtained: $V_1 = \{1,2,3,4,5\}, V_2 = \{6,7,8,9,10\}, V_3 = \{13,14,15,16\}$. Since some core nodes may be added into more than one set or never be added into any domain, these nodes are dealt with by steps 7 and 8. For example, nodes 12 and 11 can be added to D3 through steps 7 and 8 until all nodes are included in the domains, as shown in Figure 12.4c.

In this chapter, the proposed network decomposition algorithm will be carried out offline before the deployment of VNCBs and VN mapping, and hence the network decomposition can be implemented by using sophisticated algorithms, which will not introduce any additional complexity in VN mapping.

12.3.3 A Hierarchical VN Mapping Algorithm

Assuming that the substrate network $G(V,E)$ is partitioned into N domains, with the topology of individual domains denoted as $G_i(V_i,E_i)$, $1 \leq i \leq N$, where V_i denotes the set of nodes in domain i, V_i^{border} denotes the set of border nodes of domain i (the two nodes of a link connecting two different domains are called border nodes). E_i is the set of links within domain i, and the set of links connecting domain i and other domains is denoted by E_i^{border}. Each domain is considered a meta node, and the set of meta nodes is denoted as V_{meta}. Let $E_{i,j}$ be the set of border links that span from nodes in domain i to nodes in domain j. These links are collapsed into a single link (meta link). The set of meta links and the meta network are denoted as E_{meta} and $G_{\text{meta}}(V_{\text{meta}},E_{\text{meta}})$, respectively.

The proposed hierarchical VN mapping algorithm consists of a GVNMA and a LVNMA. The GVNMA can take use of information such as V_i^{border}, E_i^{border}, $r(u,v)$, and $w(u,v)$, $\forall \overline{uv} \in E_i^{\text{borde}}$, $1 \leq i \leq N$, and the LVNMA obtains relevant network domain information, including domain topology, $G_i(V_i,E_i)$, $r(u,v)$, and $w(u,v)$, $\forall \overline{uv} \in E_i \cup E_i^{\text{border}}$. Using their network information, the GVNMA and the LVNMA can obtain different views of the substrate network. The former (i.e., GVNMA) obtains two high-level views: meta network and path-vector network, where the meta network represents the connectivity among domains (Figure 12.5c), and the path vector is a logical connection of two nodes in the same domain of the substrate network (Figure 12.5d). The view of the latter (i.e., LVNMA) is a subgraph of the substrate network's topology, namely, its domain topology. For a VN request, the GVNMA uses **VNM_LP** to solve a VN mapping on a path-vector network. A challenging task for the GVNMA is to determine the bandwidth capacities and weights of logic connections in a path-vector network, as it has no knowledge of domain topology information. A solution to tackle this problem can be that the LVNMA allocates the bandwidth and calculates the weights for logical connections in individual domains. Since the solution of **VNM_LP** on a path-vector network is only the result of VN mapping onto a path-vector network, the result of VN mapping onto a substrate network can be derived by the LVNMA from the **VNM_LP** solution obtained by the GVNMA. This processing of the LVNMA is called path mapping, which implies that the LVNMA computes the paths on the substrate network for path vectors. The interaction between the GVNMA and the LVNMA is shown in Figure 12.5.

Given a substrate network (see Figure 12.5a), upon the receipt of a VN request at the GVNMA, the request will be fragmented into a set of intra-domain VN requests and an inter-domain VN request (see Figure 12.5b). The former refers to the request if the nodes on all virtual links are in the same domain; the latter refers to the request if the nodes of each virtual link belong to different domains. For example, Figure 12.5b illustrates a VN request, $\{(a1,a3),(a2,a3),(a3,a4),(a3,a5),(a4,a5)\}$, which is divided into an intra-domain request, $((a3,a4),(a3,a5),(a4,a5))$, and an inter-domain VN request, $(a1,a3),(a2,a3)$. For the intra-domain VN requests, the GVNMA sends them to the corresponding LVNMAs where the requests can be addressed by the VNM_LP solver. Therefore, we mainly look into the process of dealing with inter-domain VN requests. For the sake of clarity, three definitions are given as follows:

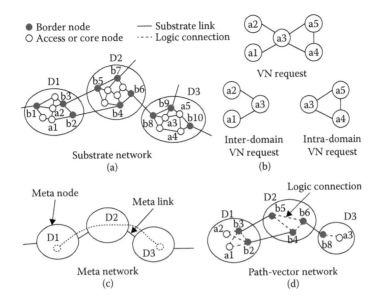

FIGURE 12.5 The concept of the hierarchical VN mapping algorithm: (a) Substrate network, (b) VN request, (c) meta network, and (d) path-vector network.

> **Definition 1, meta request:** *All virtual links are collapsed into a single connection if their source and destination nodes both are within the same domain. Such a single connection is defined as a meta request. For example, in Figure 12.3b, for virtual link $\overline{a1a3}$, $\overline{a2a3}$, a1, and a2 belong to D1 and a3 belongs to D3, so (D1,D3) is a meta request.*
> **Definition 2, available domain:** *A domain which is potentially involved in the VN mapping.*
> **Definition 3, available border node:** *The border node connecting two available domains. For example, in Figure 12.5d, if D1 and D2 are available domains, b2 and b3 are the available border nodes of D1, and b1 is not an available border node.*

For the inter-domain VN request, the computation at the GVNMA mainly includes the following steps:

- Find the k-shortest paths for every meta request in the meta network to determine available domains. In this case, the domains on the k-shortest paths are considered to be potentially used in VN mapping (i.e., available domains). The value of k needs to be properly selected to minimize the number of available domains while achieving desirable VN mapping.
- After determining the available domains, the GVNMA will generate a path-vector network, denoted by $G_{path}(V_{path}, E_{path})$, where V_{path} consists of the nodes of the inter-domain VN request and the available border nodes of all available domains, and E_{path} includes the edge of the mesh network constructed by available nodes and the edge connecting the nodes of the VN request and available nodes. For example, as shown in Figure 12.5d, $V_{path} = \{b2, b3, b4, b5, b6, b8, a1, a2, a3\}$ and $E_{path} = (b2, b3), (a1, b2), (a1, b3), (a2, b2), (a2, b3), (b4, b5), (b4, b6), (b5, b6), (a3, b8), (b3, b5), (b2, b4), (b6, b8)$.

- The GVNMA sends a message to the LVNMA of the corresponding domains to compute the weight and allocated bandwidth of the path vector. Upon receiving all feedback messages, the GVNMA solves VNM_LP based on the path-vector network and passes the solution to the LVNMA to compute path mapping in the domain.

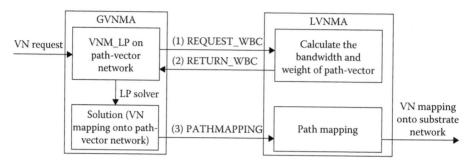

FIGURE 12.6 The interaction between the GVNMA and the LVNMA. (1) Weight and bandwidth computation request (*GVNMA → LVNMA*). (2) Computation return (*LVNMA → GVNMA*). (3) Path mapping (*GVNMA → LVNMA*).

The GVNMA takes an online VN request as inputs and maps it onto the substrate network through cooperation and coordination with the LVNMA. The messages between the GVNMA and the LVNMA are described as follows and as shown in Figure 12.6:

(1) For an intra-domain VN mapping request, one type of message is required:

INTRADOMAINVNM(i, R^i_{intra}): Denotes the message sent to the LVNMA of domain i by the GVNMA to map the VN onto the substrate network of domain i according to intra-domain VN request, R^i_{intra}.

(2) For an inter-domain VN mapping request, three types of messages are required:

REQUEST_WBC(i, E^i_{path}): Represents the message sent to the LVNMA of domain i by the GVNMA to get the weight and allocated bandwidth for each path vector, where E^i_{path} denotes a set of path vectors in domain i.

RETURN_WBC(i,W,B): A message sent from the LVNMA to the GVNMA. $W = \{w(u,v): \forall \overline{uv} \in E^i_{path}\}$ and $B = \{r(u,v): \forall \overline{uv} \in E^i_{path}\}$ are the weight and allocated bandwidth for each path vector, $\overline{uv} \in E^i_{path}$.

*PATHMAPPING(i, F^i_{path}): Denotes the message sent to the LVNMA of domain i by the GVNMA. This message passes the solution to the **VNM_LP** problem based on the path-vector network so that the LVNMA can carry out path mapping based on the solution. $F^i_{path} = \{f^j_{u,v} : \forall j, \forall \overline{uv} \in E^i_{path}\}$, where $f^j_{u,v}$ is a variable to the **VNM_LP** problem based on the path-vector network.*

The GVNMA is described as shown in Algorithm 12.2.

In Algorithm 12.2, steps 1 through 5 subdivide the VN request into a set of intra-domain VN requests, R^i_{intra}, and an inter-domain VN request, R^j_{inter}. For each intra-domain VN request, steps 6 through 8 send it to the corresponding LVNMA. For the inter-domain request, step 10 obtains the meta request, denoted by R_{meta}, and the sets of nodes of the VN request in each domain, denoted by $SD_Q, \forall Q \in V_{meta}$. Steps 11 and 12 use the k-shortest path algorithm to determine the set of available domains, D_a. Steps 13 through 16 generate the set of path vectors for each domain and send a message to the corresponding LVNMA to compute the weight and bandwidth of the path vector. Steps 18 and 19 solve the **VNM_LP** problem based on the path-vector network and send the solution to the LVNMA.

The LVNMA mainly executes two key tasks, namely the weight computation and the bandwidth allocation of the path vector and the path mapping in the domain.

ALGORITHM 12.2   GVNMA: GLOBAL VIRTUAL NETWORK MAPPING ALGORITHM

Inputs: A VN request $R = \{(s_k, t_k, d_k), 1 \le k \le l\}$

Require: Meta network $G_{\text{meta}}(V_{\text{meta}}, E_{\text{meta}})$, the access nodes of each domain, V_i^{border}, E_i^{border}, $r(u,v)$, and $w(u,v)$, $\forall \overline{uv} \in E_i^{\text{border}}, \forall i$

1. For $i = 1$ to l
   2. if s_i and t_i belong to the same domain j
      3. $R_{\text{intra}}^j = R_{\text{intra}}^j \cup \{s_i, t_i\}$
   4. else
      5. $R_{\text{inter}} = R_{\text{inter}} \cup \{s_i, t_i\}$
         6. For $\forall i \in \{1, \cdots, l\}$
7. if $R_{\text{intra}}^i \ne \varnothing$
   - send message INTRADOMAINVNM(i, R_{intra}^i)
     – if $R_{\text{inter}} = \varnothing$, stop.
8. $\forall (s,t,d) \in R_{\text{inter}}$, find S,T satisfying $s \in S, t \in T$, where $S,T \in V_{\text{meta}}$. Then $R_{\text{meta}} = R_{\text{meta}} \cup \{(S,T)\}, SD_S = SD_S \cup \{s\}, SD_T = SD_T \cup \{t\}$
9. $\forall (S,T) \in R_{\text{meta}}$, perform k-shortest path algorithm to find k path P_k from S to T in network $G_{\text{meta}}(V_{\text{meta}}, E_{\text{meta}})$, where P_k denotes the set of path.
10. $\forall P \in P_k$, $\forall Q \in P$, $D_a = D_a \cup \{Q\}$
11. $\forall Q \in D_a$, put available border nodes of Q into B_Q
12. $\forall S, T \in D_a$, if $\overline{ST} \in E_{\text{meta}}$, then put the border links between S and T into $E_{\overline{ST}}^{\text{border}}$
13. $\forall Q \in D_a$, Path Vector$_Q = \{\overline{uv} : \forall u, v \in B_Q, u \ne v\} \cup \{\overline{uv} : \forall u \in B_Q, v \in SD_Q\}$
14. $\forall Q \in D_a$, send message REQUEST_WBC(Q, Path Vector$_Q$) to domain Q. Wait for RETURN_WBC.
15. Generate path-vector network $G_{\text{path}}(V_{\text{path}}, E_{\text{path}})$:

$$V_{\text{path}} = \left(\bigcup_{\forall Q \in D_a} B_Q \right) \bigcup SD_Q$$

$$E_{\text{path}} = \left(\bigcup_{\forall Q \in D_a} \text{Path Vector}_Q \right) \bigcup \left(\bigcup_{\forall S, T \in D_a} E_{\overline{ST}}^{\text{border}} \right)$$

16. Perform **VNM_LP** solver

Input: $G_{\text{path}}(V_{\text{path}}, E_{\text{path}})$, $r(u,v), w(u,v), \forall \overline{uv} \in E_{\text{path}}$

$\forall Q \in D_a$, send message PATHMAPPING(Q, F_{path}^i)

Optimal bandwidth allocation and weight can hardly be obtained to ensure optimal VN mapping as LVNMAs only have the local network information (no global information). This chapter adopts the optimal max-min fair multi-commodity (OPT_MMF) [24] to fairly allocate bandwidth to path vectors through taking a path vector as a commodity with a demand of 1 unit. After obtaining the bandwidth allocation for each commodity (path vector), the average weight can be expressed as

$$\text{average weight } w\left(\overline{st}\right) = \frac{\sum_{\forall \overline{uv} \in E_i} w(u,v) f_{u,v}^{\overline{st}}}{\lambda_{\overline{st}}} \tag{12.7}$$

ALGORITHM 12.3   LVNMA: LOCAL VIRTUAL
NETWORK MAPPING ALGORITHM

Require:$G_j(V_j,E_j)$, $r(u,v)$, and $w(u,v)$, $\forall \overline{uv} \in E_j$

1. Block and wait for message from GVNMA
2. If message is INTRADOMAINVNM(j,R_{intra}^j)
   3. perform **VNM_LP** solver.
      Input:$G_j(V_j,E_j)$, $r(u,v)$ and $w(u,v)$, $\forall \overline{uv} \in E_j$
   4. go to step 1.
5. If message is REQUEST_WBC(j,E_{path}^j)
   6. perform OPT_MMF algorithm.
      Input: $G_j(V_j,E_j)$, E_{path}^j. Take $\overline{st} \in E_{\text{path}}^j$ as a commodity $(s,t,1),s$, and t is the source and the sink of a commodity with the demand of 1 unit.
      Return: $\lambda_{\overline{st}}$, $f_{u,v}^{\overline{st}}$, $\forall \overline{st} \in E_{\text{path}}^j$, $\forall \overline{uv} \in E_j$
   7. $\forall \overline{st} \in E_{\text{path}}^j$, compute the weight according to Equation 12.7.
   8. $W = \{w(\overline{st}): \forall \overline{st} \in E_{\text{path}}^j\}, B = \{\lambda_{\overline{st}}: \forall \overline{st} \in E_{\text{path}}^j\}$, send RETURN_WBC($j,W,B$) to GVNMA.
   9. go to step 1.
10. If message is PATHMAPPING(j, F_{path}^j)
    11. perform PM_LP solver.
    Input: $G_j(V_j,E_j)$, F_{path}^j, $r(u,v)$, $w(u,v)$
12. Go to step 1.

where $\overline{st}$ denotes a path vector in domain i, and $\overline{uv}$ is a substrate link of domain i. $f_{u,v}^{\overline{st}}$ is the total amount of flow on the substrate link $\overline{uv}$ for the path vector $\overline{st}$, and $\lambda_{\overline{st}}$ is the bandwidth allocation for $\overline{st}$.

Another task is to compute path mapping in the domain. Once the GVNMA computes a feasible solution to **VNM_LP** on the path-vector network, the remaining problem is to derive a feasible solution to **VNM_LP** on the substrate network from the feasible solution on the path-vector network. A simple approach to solve this problem is, for a solution $f_{u,v}^i$ to a virtual link i on the path-vector $\overline{uv}$, if $f_{u,v}^i > 0$, we associate with node u a supply $f_{u,v}^i$ and associate with node v a demand $f_{u,v}^i$ and then use the minimum cost flow on the substrate network to obtain the path mapping onto the substrate network. But this approach cannot obtain the optimal path mapping in the domain. We formulate the optimal path mapping in the domain j as PM_LP, presented as follows:

PM_LP (path mapping linear program)
Inputs:
Domain topology: $G(V_j,E_j)$ GVNMA's solution about domain j: $\forall i, \forall \overline{uv} \in E_j, f_{u,v}^i$
The weight of the link: $\forall \overline{uv} \in E_j$
The residual bandwidth of the link: $\forall \overline{uv} \in E_j$
Variables:
$x_{u,v}^i$ is a flow variable denoting the total amount of flow from u to v on the link $\overline{uv}$ for the ith virtual link.
Objective:

$$\text{Minimize} \sum_{\overline{uv} \in E_j} \sum_{i=1}^{l} x_{u,v}^i \cdot w(u,v) \tag{12.8}$$

Constraints:

$$\sum_{i=1}^{l}\left(x_{u,v}^{i}+x_{v,u}^{i}\right)\le r(u,v),\ \forall u,v\in V_{j} \tag{12.9}$$

$$\sum_{w\in V_j}x_{u,w}^{i}-\sum_{w\in V_j}x_{w,u}^{i}=0,\ \forall i,\forall u\in V_{j}\setminus V_{\text{path}}^{i,j} \tag{12.10}$$

$$\sum_{w\in V_j}x_{s,w}^{i}-\sum_{w\in V_j}x_{w,s}^{i}=\sum_{v\in V_{\text{path}}^{i,j}}f_{s,v}^{i}-\sum_{v\in V_{\text{path}}^{i,j}}f_{v,s}^{i},\ \forall i,\ \forall s\in V_{\text{path}}^{i,j} \tag{12.11}$$

$$x_{u,v}^{i}\ge 0,\ \forall u,v\in V_{j} \tag{12.12}$$

The objective function of the **PM_LP**, Equation 12.8, is the same as Equation 12.1, which attempts to minimize the path-mapping cost in a domain, where l denotes the number of virtual links. The constraint set in Equation 12.9 is a capacity constraint. The constraint sets in Equations 12.10 and 12.11 are combined to refer to the condition of flow conservation, where $V_{\text{path}}^{i,j}$ denotes the set of available border nodes and the nodes of virtual link i in domain j. The constraint set in Equation 12.12 denotes the real domain constraints.

The LVNMA is described in Algorithm 12.3, which performs according to the received messages from the GVNMA.

12.3.4 PERFORMANCE EVALUATION

This section first describes the assessment environment, followed by the presentation of a set of key numerical results. The objective of the simulation experiments falls into two categories: first, we evaluate the performance of the proposed hierarchical VN mapping algorithm in terms of runtime, cost, and communication overhead (in terms of the number of exchanged messages) for a range of substrate networks and VN mapping requests using the centralized algorithm based on VNM_LP (e.g., [5]) as a comparison benchmark, which gives the optimal VN mapping solution; second, we assess the performance of the proposed algorithm in the scenarios with VN dynamics (i.e., VN creation and termination) in comparison with the PolyViNE algorithm [13].

The substrate network topologies in the simulation experiments are randomly generated using the BRITE [25] tool, and the substrate networks are partitioned into 20 domains for all experiments. The open-source linear programming library GLPK [26] is used to implement the NM_LP algorithm. All the simulations are carried out 20 times, and the statically averaged values are presented as the results. The substrate link bandwidth is a real number which is uniformly distributed between 100 and 120 units. In all experiments, we randomly generated an inter-domain VN request whose bandwidth requirements of virtual links are uniformly distributed between 1 and 5 units, and the source and destination nodes are randomly selected from substrate network nodes. It is also assumed that the sum of the communication time of the message exchange between the GVNMA and the LVNMA is set as 2 seconds.

Here we present a set of key simulation results obtained from the numerical experiments. Figure 12.7 gives the runtime performance results of the hierarchical VN mapping algorithm in substrate networks with different network sizes against the centralized (optimal) approach. In this experiment, we randomly generated 7 networks with the number of nodes increased from 600 to 1200, and an average degree of inter-domain connectivity of 4. The simulated VNs to be mapped to the substrate network all have 50 virtual links. It is clearly demonstrated that the runtime of the centralized approach increased significantly along with the growth of network size, whereas the runtime of the proposed hierarchical solution almost remains unchanged. The difference between

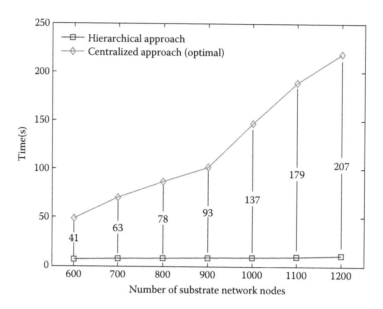

FIGURE 12.7   Runtime performance for different-sized substrate networks.

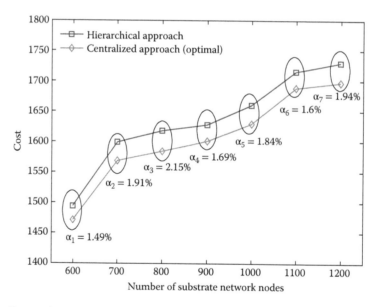

FIGURE 12.8   Cost performance for different-sized substrate networks.

hierarchical and centralized in terms of runtime is presented in Figure 12.7 (e.g., 137 s for the sub-strate network with 1000 nodes). This is mainly due to two reasons: (1) while the size of the overall network increases from 600 to 1200 nodes, the size of individual domains increases very gradually (20 domains in total); and (2) parallel computation for VN mapping can be carried out in individual domains. Therefore, the hierarchical solution exhibits superiority in terms of runtime performance over centralized solutions.

Figure 12.8 illustrates the VN mapping performance in terms of mapping cost for both the hier-archical and the centralized algorithm against the number of substrate network nodes (from 600 to 1200). Due to the nature of the centralized algorithm, it gives the optimal VN mapping in terms of

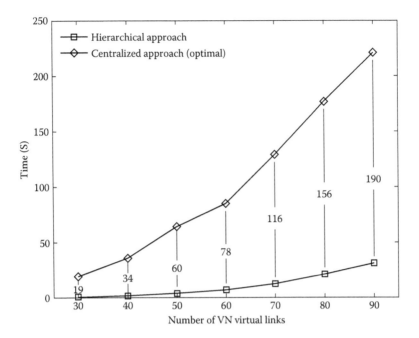

FIGURE 12.9 Runtime performance for different-sized substrate networks.

cost (minimal cost). The gap between the hierarchical and optimal solutions, $\alpha_1 - \alpha_7$, is presented in Figure 12.8, which shows that for all simulated substrate networks, the maximum gap is less than 3%. Here the gap $= \dfrac{|C_1 - C_2|}{C_2}$, where C_1 and C_2 denote the mapping cost of the hierarchical and optimal approaches, respectively.

Figure 12.9 shows the runtime performance results of the hierarchical approach for VNs with different size against the centralized (optimal) approach. In this experiment, the substrate network has 800 nodes, and the average degree of inter-domain connectivity is 4. We randomly generate the VN request, with the number of virtual links increased from 30 to 90. It is illustrated that the runtimes of both the hierarchical and the centralized approach grow with increasing number of VN virtual links; however, the runtime of the centralized approach increased significantly compared to the hierarchical solution. The difference between hierarchical and centralized in terms of runtime is also presented in Figure 12.9 (e.g., 190 seconds for the VN with 90 virtual links). This is mainly due to the fact that the scale of VNM_LP increases by the size of the substrate network when the number of virtual links increases by one. The size of individual domains is far less than the substrate network, so the increase in the runtime of the hierarchical approach is less than the centralized approach.

Figure 12.10 shows the VN mapping performance in terms of mapping cost for both the hierarchical and the centralized algorithm with different numbers of VN virtual links (from 30 to 90). The gap between the hierarchical and the optimal solution, $\beta_1 - \beta_7$, is presented in Figure 12.10. It illustrates that for all simulated substrate networks, the maximum gap is also less than 3%.

The communication overhead performance (the number of exchanged messages) for the hierarchical solution is given in Figure 12.11, with the 95% confidence interval. In this experiment, we generate a random substrate topology with 800 nodes and an average degree of inter-domain connectivity of 4. The VN requests all have 50 virtual links. The upper bound of the number of exchanged messages is 60. The upper bound is determined by $N_{message} = 3 \cdot N_{domain}$, where $N_{message}$ denotes the number of exchanged messages, and N_{domain} denotes the number of domains involved in

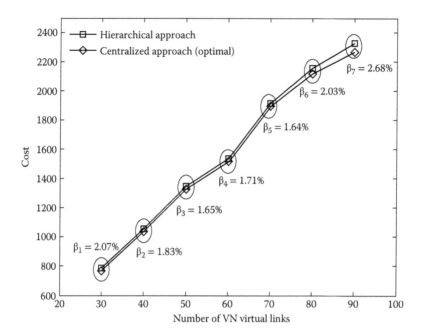

FIGURE 12.10 Cost performance for different-sized substrate networks.

the VN mapping. As there are three messages that need to be exchanged between the GVNMA and the LVNMA to complete an inter-domain VN mapping and the simulated substrate networks are partitioned into 20 domains. When the number of domains is 20, the overhead reaches the maximum value. But if the VN requests involve only part of the domains, the communication overhead can decrease, as shown in Figure 12.11 (e.g., about 31 messages for the VN request involved in 4 domains). This is because we compute the k-shortest path to determine the domains involved when

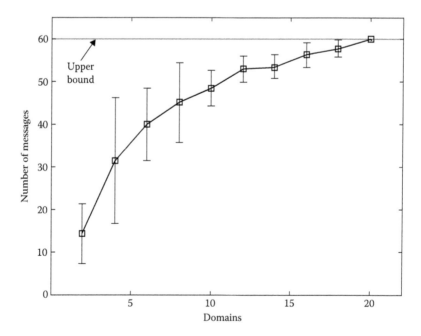

FIGURE 12.11 Communication overhead of the hierarchical VN mapping algorithm.

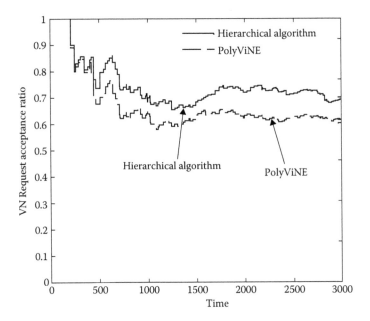

FIGURE 12.12   VN request acceptance ratio over time.

mapping the VN, rather than use all domains in the VN mapping. In this case, the average number of exchanged messages is 46.

In the following experiment, we study the dynamic performance of the proposed algorithm in terms of the VN request acceptance ratio in comparison with the average acceptance ratio, A, defined as $A = \dfrac{N_a}{N_r}$, where N_a denotes the number of accepted VN requests, and N_r denotes the total number of VN requests. All simulations are carried out for 3000 time units to obtain steady-state performance measurements, and the performance evaluation metrics are described as follows. The VN request is randomly generated, and its size is uniformly distributed from 20 to 30 in terms of virtual links. The average bandwidth requirements of the virtual links are $E[BW = 25]$. It is assumed that the arrival pattern of the VN requests follows a Poisson distribution with the mean inter-arrival time of 15 time units, and the VN lifetimes are exponentially distributed with the mean of 80 time units.

Figure 12.12 depicts the algorithm performance in terms of the average acceptance ratio. The results demonstrate that our algorithm outperforms and leads to a higher acceptance ratio as well as more revenue than the PolyViNE algorithm. Since the PolyViNE algorithm maps each virtual link onto a single path in the substrate network, it becomes difficult to find a single path which could satisfy the bandwidth capacity of the corresponding virtual links with a larger bandwidth requirement in the VN requests. While our algorithm allows path splitting, the substrate network resources can be efficiently utilized. As a result, our algorithm could accept more VN requests and hence generates more revenue than PolyViNE in the long run.

12.4   MULTI-COMMODITY–BASED SELF-HEALING APPROACH FOR VIRTUAL NETWORKS

12.4.1   Network Definition

This section firstly gives the network definition before presents the proposed multi commodity based self-healing solution.

12.4.1.1 Physical Network

The physical network is modeled as an undirected graph, denoted as $G^P = (N^P, L^P, C_N^P, C_L^P)$, where N^P and L^P are the set of physical nodes and physical links, respectively. We consider the capacity, C_N^P, as the key node attribute for an arbitrary node, $n \in N^P$, and the bandwidth, C_L^P, as the key link attribute for an arbitrary link, $l \in L^P$.

12.4.1.2 Virtual Network

The VN is also modeled as an undirected graph, denoted as $G^V = (N^V, L^V, C_N^V, C_L^V)$, where N^V and L^V are the set of virtual nodes and virtual links, respectively. We consider the required capacity, C_N^V, as the key node attribute for an arbitrary virtual node, $n \in N^V$, and the required bandwidth, C_L^V, as the key link attribute for an arbitrary virtual link, $l \in L^V$.

12.4.1.3 Virtual Network Mapping

The mapping problem upon a VN request, $G^V = (N^V, L^V, C_N^V, C_L^V)$, onto a substrate network, $G^P = (N^P, L^P, C_N^P, C_L^P)$, can be formulated as follows: find a subgraph, $G^{P'} = (N^{P'}, L^{P'})$, of G^P such that all virtual nodes are mapped onto different physical nodes, and virtual links are mapped onto loop-free paths concatenated by substrate links, denoted as

$$M : M(G^V) = (N^{P'}, P^{P'}) \tag{12.13}$$

where $N^{P'}$ and $P^{P'}$ denote the subset of substrate nodes and loop-free paths in G^P, respectively. M can be resolved into node and link mapping as follows:

$$M_N : M_N(N^V) = N^{P'} \tag{12.14}$$

$$M_L : M_L(L^V) = P^{P'} \tag{12.15}$$

Here, M is considered a valid mapping only if $M : M(G^V)$ meets the constraints given in Equations 12.16 and 12.17.

$$C_N^V \leq C_N^P, \forall n \in N^V, M_N(n) \in N^{P'} \tag{12.16}$$

$$C_L^V \leq C_L^P, \forall l \in L^V, M_L(l) \in P^{P'} \tag{12.17}$$

12.4.2 THE MFP-VNMH OVERVIEW

This section presents and discusses the MFP-VNMH design considerations and procedure, together with the description of control protocols, which are used to synchronize the VN restoration process between the infrastructure provider (InP) and the service provider (SP).

12.4.2.1 The MFP-VNMH Mechanism

MFP-VNMH is designed as a self-healing VN restoration approach with high efficiency and an acceptable success rate. The operational procedure of MFP-VNMH can be briefly summarized as follows. Following the unavailability of VNs due to substrate network failures, the end users detect that their accessed VN applications become unavailable, they inform their affiliated SPs to restore the service provision, and the SPs forward the requests to the InPs by sending a set of messages. After the messages are received, the InPs identify the VN failures and restore the failed VN and hence the service, as soon as possible, through the use of a suggested VN restoration approach.

12.4.2.2 The MFP-VNMH Control Protocols

Apart from a standard procedure for SPs and InPs to synchronize the whole VN restoration process, which originates from the SP side and terminates at the InP side, a set of control protocols is also needed to synchronize the sessions and exchange messages during the VN restoration process, as illustrated in Figure 12.13. More specifically, the control protocols are designed based on five messages, RESUME, SUCCESS, FAILURE, WAIT, and ACK, and two key functions, restore() and check(), which are carried out with different functionalities at certain stages and described as follows:

- **RESUME(fail_id, M):** This message is sent from the SP to the InP to inform the InP to restore the failed VN, M, with an exclusive identification, denoted as fail_id, which initiates the VN restoration process.
- **SUCCESS(fail_id, M):** Upon the restoration process's completion, the InP feedbacks to the SP with the updated VN mapping, M.
- **FAILURE(fail_id , error):** If the InP is not able to restore the failed VN mapping, M, due to certain reasons (e.g., resource constraints or topological limitations), this message will be sent to the SP informing the reasons denoted as error (e.g., no appropriate physical link for remapping is found).
- **WAIT(fail_id, M):** After the FAILURE message is acknowledged by the SP, this message is sent to the InP informing it that the SP is waiting for the failed VN mapping, M, restoration, and M will be added into a priority queue. The InP will attempt to resume the uncompleted VN mapping process based on the request waiting in the queue.
- **ACK(fail_id):** When the SP receives the SUCCESS messages from the InP, it will reply this message to terminate the overall VN restoration process.

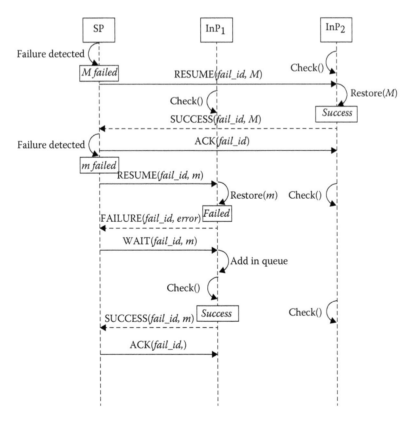

FIGURE 12.13 The operation of control protocols.

- **restore()**: The InP applies this function to restore the failed VNs due to different physical component failures (i.e., physical link failure and physical path failure), which is discussed in detail in Sections 12.4.3.5 and 12.4.3.6.
- **check()**: The unrestored VN mapping requests are added into a priority queue (introduced in Section 12.4.3.2), and the InP uses this function to process these VN requests in the queue periodically based on their priority.

As shown in Figure 12.13, the *SP* provides the services to its end users based on InP_1 and InP_2, respectively (e.g., the VNs are mapped onto InP_1 and InP_2). The control protocols enable the *SP* to communicate with both InPs. When the *SP* receives a request from its end user with the message indicating that the established VN mapping, M, becomes unavailable, it sends a RESUME message to InP_2 with the failed VN mapping, M, and the unique fail_id. After this message is received, InP_2 restores M successfully and sends the SUCCESS message to the *SP*. The *SP* replies back the ACK message to finish the overall restoration process. Similarly, the *SP* receives a request from its end user with the message that the VN mapping, m, is no longer unavailable due to failures or dynamics that occurred in the physical substrate, and it informs the InP_1 to update this VN mapping to restore the VN services. However, InP_1 may not be able to update the VN mapping, m, immediately due to certain reasons. As a result, it returns a FAILURE message to the *SP*. The *SP* replies with a WAIT message to notify InP_1 to attempt to restore m later. Meanwhile, InP_1 adds this mapping update request, m, into its priority queue. The InP could process the requests in its own priority queue based on the assigned priorities, depending on the InP's resource availability and the nature of the services (e.g., real-time services, critical data transmission, or best-effort services) by its own check() function, which executes periodically with a predefined cycle. The VN mapping, m, is updated after the second execution of InP_1's check() function, and the SUCCESS message is sent to the *SP* informing it that the VN mapping, m, is successfully restored.

12.4.3 MORE DETAILS

This section introduces the model of the InP control manager and presents two key components in detail: priority queue and physical substrate reliability estimation. The former is used to process the unsuccessfully restored VN mapping requests following the restore() function, and the latter is used to estimate the reliability of the physical node and link. Based on these two mechanisms, the proposed VN restoration approach, MFP-VNMH, is discussed.

12.4.3.1 InP Control Manager

The InP control manager is one of the key components for managing the InP and enhancing its self-healing capability, which communicates with the SPs through a set of control protocols, administrates the priority queue, which deals with the unprocessed VN mapping restoration requests,

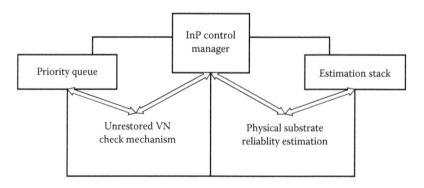

FIGURE 12.14 Infrastructure provider control manager.

and updates the current status (failures and other unexpected dynamics) of the physical substrate's components (i.e., routers and communication channels). The model of the InP control manager consisting of two key components is illustrated in Figure 12.14.

12.4.3.2 Priority Queue and Check Function

This work adopts the management strategy of "restorable first" to manage the priority queue for maintaining the list of uncompleted VN mapping update requests. During each InP checking cycle, it scans the VN requests in sequence. If a VN mapping request is identified as restorable, then the request is processed immediately. Otherwise, the request is moved to the queue's rear. This mechanism could enable fast identification of the VN mapping update request, which can be implemented with the available network resources, and hence the overall VN restoration efficiency is significantly promoted.

This restorable first mechanism is illustrated through an example in Figure 12.15. Suppose that the InP attempts to restore two requests in the queue during an InP checking cycle, T, and there are four requests (from 1st to 4th) in the queue initially, denoted by m_1, m_2, m_3, and m_4. In the first checking cycle, m_1 is popped from the queue as it can be restored successfully, and m_2 is pushed to the rear of the queue as it cannot be restored with the current resource availability. During the second cycle, m_3 and m_4 are both removed from the queue consequently upon their successful restoration, and a new VN mapping update request, m_5, arrives, and hence it is added to the queue's rear.

12.4.3.3 Physical Substrate Reliability Estimation

To enhance the robustness of the restored VNs, not only the failed physical components including the physical nodes and links should be recorded, but their reliability needs to be estimated. A stack is adopted to assess their reliability based on the criteria of "least recently failed," motivated by the locality principle. This stack maintains and records all the physical components which have ever failed by imposing the most venerable component at the top of the stack. Once a failed physical component is detected in the InP, its identification (ID) will be pushed into the top of the stack.

As shown in Figure 12.16, the stack is initialized with three physical components, which have failed, denoted as physical node N_1, physical link L_1, and physical link L_2, with the position to the top of the stack set as $N_1 > L_1 > L_2$, which means the latest failed physical component is L_2, the next one is L_1, and the last one is N_1. N_1 meets another failure at the time slot t_1, and as a result, it is moved to the top of the stack. At the time slot t_2, the physical node N_2 is moved to the top since it is the latest failed physical node. The situation at the time slot t_3 is similar to the one at t_1, and L_2 occupies the top position. The reliability of the physical node, n, or link, l, is defined as follows:

$$r_n = \frac{\text{pos}[n \mid l]}{|V| + |E| + 1} \tag{12.18}$$

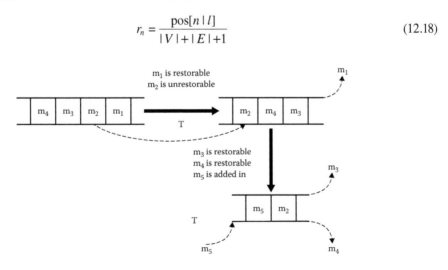

FIGURE 12.15 The priority queue for unprocessed VN restoration request.

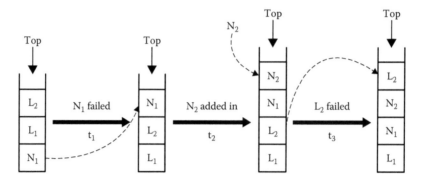

FIGURE 12.16 The stack structure for reliability estimation.

where $|V|$ and $|E|$ denote the number of physical nodes and links in the physical substrate, and $pos[n \mid l]$ represents the stack index of the physical node, n, or link, l. Taking an example shown in Figure 12.16, the stack index of L_2 is 1, denoted as $pos[L_2] = 1$. If the node, n, or link, l, has never appeared in the stack, it implies that the n or l has never failed and the reliability of n or l is fixed to $r_{n|l} = 1$.

12.4.3.4 The Restore Function
This section discusses the function for restoring the failed VN mapping in detail. In general, a VN mapping failure can be due to the physical substrate's failure, which leads to current VN mapping unavailable (i.e., virtual node or virtual link mapping). The key issue addressed in this chapter is to update the failed VN mapping and hence restore the VN services as fast as possible and avoid the overall VN remapping, which can often be time consuming and even intractable. For the sake of simplicity, the physical nodes in the physical network shown in Figures 12.17 and 12.18 are assumed with sufficient CPU capacities to support their delivered virtual nodes, while the physical link bandwidths in the physical network are the key resource constraints needed to be met. Meanwhile, the reliability of both physical nodes and links are not considered.

12.4.3.5 Physical Link Failure
Some VN mappings will be unavailable inevitably when the underlying physical links become available. A straightforward solution under such circumstances is to find all the failed VN mapping caused by this failed physical link and remap all these failed VN requests, including all the affiliated virtual nodes and virtual links regardless of the still available and already completed mapping. Therefore, the failed VN mapping can be updated. It should be noted that, under certain circumstances, the majority of virtual node mapping and virtual link mapping in a failed VN mapping is still functional, and hence it is wise to maintain them. Very often, the virtual link is mapped onto the physical path, which is concatenated by a set of physical links. This implies that a virtual link can be partially mapped onto a physical link. The physical link failure is likely to be a disconnected segment of a physical path which is supporting some virtual links. In fact, a large portion of virtual links' mapping information in a failed VN can be reused, which provides the possibility of improving the efficiency of the VN mapping update and service restoration.

Suppose that there is a failed physical link, $L^P = \overline{AB}$, supporting k virtual links, denoted as $L_i^V (1 \leq i \leq k)$, which are involved in r VN mappings, denoted as $M_i (1 \leq i \leq r \leq k)$. The physical link $\overline{AB}$ will lead to k virtual links' mapping being unavailable and r VN mapping failures when it fails, implying that k network flows, each sized f_i^V, are reduced between physical nodes A and B if all virtual links' capacities, C_i^V, are considered network flows (i.e., $f_i^V = C_i^V$, $1 \leq i \leq k$). The restoration of r failed VN mappings is equivalent to solving a MFP between nodes A and B with the flow size, F, subject to

$$
\left\{
\begin{array}{rcl}
F & = & \displaystyle\sum_{i=1}^{k} f_i^V \\
f_1^V & = & C_1^V, f_2^V = C_2^V, \dots, f_k^V = C_k^V
\end{array}
\right.
\tag{12.19}
$$

and adding the physical paths that the k flows traversed into their corresponding failed VNs, respectively.

Thus, the VN mapping restoration problem is effectively transformed into the MFP, which can be addressed by various state-of-art algorithmic solutions. These MFP solutions are able to promote the load balance and success rate of the VN mapping update as a failed virtual link can be remapped onto multiple available physical links [5]. The restoration of k failed VNs needs only to solve the MFP once and at the price of incorporating schemes to deal with the out-of-order packets due to service interruption, where existing solutions are available (e.g., [27–29]). The flow value, φ, for a commodity flow between two physical nodes, A and B, is defined as follows:

$$
\varphi = \prod_{i=1}^{m} r_{n_i} \cdot \prod_{j=1}^{n} r_{l_j}
\tag{12.20}
$$

where n_i and l_j represent the physical node and the physical link traversed by the commodity flow between A and B, and r_{n_i} and r_{l_j} represent their reliability. φ is used as a selection criteria if multiple

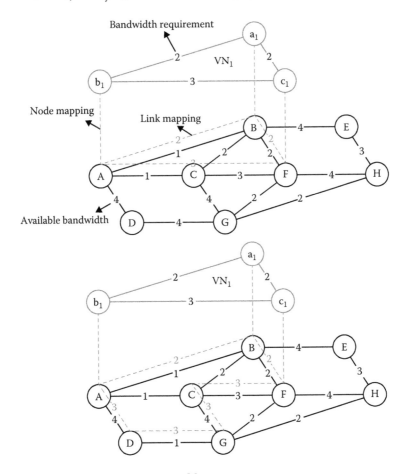

FIGURE 12.17   The restoration of VN mapping M_1.

candidate parts are generated by the MFP approach, and the solution with the highest φ is selected. It is noticed that any physical node or link carrying the flow between A and B can lead to the failure of flow delivery, which means the fewer physical nodes or links a flow has traversed, the better reliability of the flow delivery.

The pseudocode of Algorithm 12.4 is given as follows:

As shown in Figure 12.17, the physical link $\overline{AC}$ fails, resulting in the unavailability of virtual link mapping $M_1(\overline{b_1c_1}) = \overline{ACF}$, which is part of the VN mapping, M_1. We restore the flow between the physical nodes A and C with the capacity $C^V_{\overline{b_1c_1}}$ by the use of the MFP solution and simply find a physical path, $\overline{ADGC}$, to support, which fundamentally equals remapping the virtual link $\overline{b_1c_1}$ onto the physical path $\overline{ADGCF}$ (i.e., $M_1(\overline{b_1c_1}) = \overline{ADGCF}$). The updated virtual link mapping is added into the VN mapping, M_1, eventually to finish the whole restoration process.

12.4.3.6 Physical Path Failure

Under certain conditions, the failure occurs on the end-to-end physical path (concatenated by a collection of physical links) during network operations. In this work, physical path failure is defined as the links in the end-to-end physical path simultaneously failing. Such a condition can happen when physical network reinforcement or network maintenance is carried out. Restoring the physical path is similar to that under physical link failure but needs to restore the individually affected virtual links to the available physical links.

Suppose that a failed physical path exists, denoted as $P^P = \overline{A_1A_2...A_n}$, which supports k virtual links, denoted as $L^V_i (1 \le i \le k)$. These k virtual links are included in r VN mappings, noted by $M_i (1 \le i \le r \le k)$. If the physical path meets some faults, it will lead to k virtual links' mapping being unavailable and r VN mapping failures. This circumstance is more complex than a single physical link failure. We assume that these k virtual links $L^V_i (1 \le i \le k)$ are mapped onto m segments of the failed physical path $P^P = \overline{A_1A_2...A_n}$, respectively, noted by $\overline{A_{s_i} A_{s_{i+1}}...A_{t_i}} (1 \le i \le m \le k, 1 \le s_i \le t_i \le n)$, which means that each segment, $\overline{A_{s_i} A_{s_{i+1}}...A_{t_i}}$, of the physical path, P^P, supports a part of these k virtual links, which can consist of a set denoted as $S_i (1 \le i \le m)$. According to the experience of restoring the failure caused by the physical link, the failure of the physical path, P^P, indicates that each sub-physical path, $\overline{A_{s_i} A_{s_{i+1}}...A_{t_i}}$, cannot deliver the network flow traversing it with F_i size. The update of r failed VN mappings is equivalent to restoring the total i flows with each sized F_i on each segment of P^P, based on the MPF solution, which is subject to Equation 12.21, and adding the physical paths that each flow traversed into its corresponding failed VN, respectively.

$$
\begin{cases}
F_i & = & \displaystyle\sum_{L \in S_i} f^V_L \\[2ex]
f^V_L & = & C^V_L, L \in S_i \\[2ex]
\displaystyle\bigcup_{i=1}^{m} \overline{A_{s_i} A_{s_{i+1}}...A_{t_i}} & = & \overline{A_1A_2...A_n} \\[2ex]
\displaystyle\bigcup_{i=1}^{m} S_i & = & \{L^V_1, L^V_2, ..., L^V_k\} \\[2ex]
S_i \bigcap S_j & = & \phi, 1 \le i,j \le m, i \ne j
\end{cases}
\tag{12.21}
$$

$$1 \le m \le k, 1 \le s_i \le t_i \le n$$

Figure 12.18 illustrates an example in which the virtual link $\overline{b_1c_1}$ is mapped onto the physical path $\overline{ACF}$, and the virtual link $\overline{b_2c_2}$ is partially mapped onto the physical link $\overline{CF}$. Once the path

**ALGORITHM 12.4   VN MAPPING RESTORATION
ALGORITHM UNDER PHYSICAL LINK FAILURE (RLF)**

1. Find all k virtual links $L_i^V (1 \le i \le k)$ mapped or partially mapped onto the failed physical link $L^P = \overline{AB}$.
2. Solve the MFP problem with k flows, $f_i^V (1 \le i \le k)$, between the source node, A, and the destination node, B, with the constraints in Equation 12.18, and select the solution with the highest flow value φ.
3. Join the physical paths that each commodity flow traversed to its corresponding failed VN mapping.

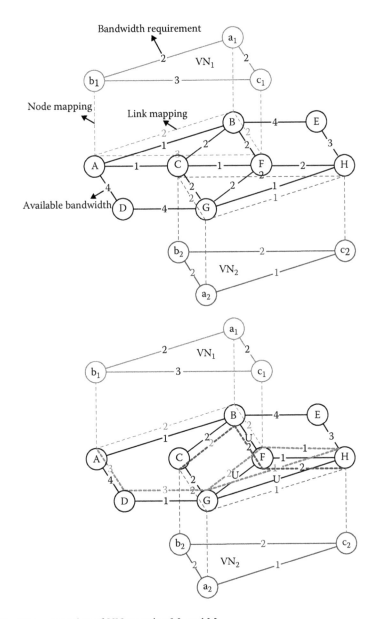

FIGURE 12.18   The restoration of VN mapping M_1 and M_2.

ALGORITHM 12.5   VN MAPPING RESTORATION
ALGORITHM UNDER PHYSICAL PATH FAILURE (RPF)

1. Find all the k virtual links $L_i^V (1 \le i \le k)$ mapped or partially mapped onto the failed physical path $P^P = \overline{A_1 A_2 ... A_n}$.
2. **For** i from 1 to m **do**
3. execute the RLF function with $| S_i |$ flows with each flow size $f_L^V = C_L^V (L \in S_i)$ between the source node A_{s_i} and the destination node A_{t_i} with the constraints in Equation 12.21.
4. **End for.**

$\overline{ACF}$ is unavailable, we can restore the flow F_1 between nodes A and F, and the flow F_2 between nodes C and F to restore the failed VN mappings M_1 and M_2, where $F_1 = C_{\overline{b_1 c_1}}^V$ and $F_2 = C_{\overline{b_2 c_2}}^V$ according to Equation 12.21. The InP updates the former virtual link mapping $M_1(\overline{b_1 c_1}) = \overline{ACF}$ to $M_1(\overline{b_1 c_1}) = \{\overline{ACGF}, \overline{ACGHF}\}$ according to the MPF solution. Similarly, the virtual link $\overline{b_2 c_2}$ mapped onto the physical path $\overline{CF}$ is remapped onto $\overline{CBF}$. As a result, the virtual link mapping $M_2(\overline{b_2 c_2}) = \overline{CFH}$ is updated to $M_2(\overline{b_2 c_2}) = \overline{CBFH}$.

The restoration algorithm for the physical path failure (RPF) is given in Algorithm 12.5.

12.4.4   PERFORMANCE ASSESSMENT AND NUMERICAL RESULTS

12.4.4.1   Experiment Setting and Performance Metrics

In this section, we assess the performance of the MFP-VNMH algorithm through extensive simulation experiments with a particular focus on three performance metrics: (1) the VN mapping restoration success ratio; (2) efficiency; and (3) load balance throughout the VN restoration process. The MFP-VNMH algorithm is evaluated using the overall VN remapping approach and the end-to-end link remapping approach as the comparison benchmarks. The environment is given in Table 12.1.

Two different physical network topologies are used in the experiment: a random network topology (see Figure 12.19) and a realistic university network topology (see Figure 12.20). The former is generated by the GT-ITM tool [30], which is organized by 300 physical nodes with an average node degree of 6, resulting in 891 physical links, while the latter consists of 18 routers and 28 departments (each department has 10–30 connected end users), which is used to validate the simulation results obtained from the random network topology. The physical node CPU capacity and physical link bandwidth on both topologies are generated following a uniform distribution in the range [4,26]. In respect to the failure pattern in the physical substrate, it is assumed that the arrival of failures follows the Poisson distribution with the mean inter-arrival time of 10 time units, and the lifetime of failures follows the exponential distribution with the mean time of 100 time units. All simulations are carried out for 2000 time units to obtain steady-state performance measurements.

In the valuated scenarios, we assume that there are 50 VNs in operation sharing these two physical substrates. The CPU and bandwidth requirements of virtual nodes and links are uniformly distributed in the range [3,5]. In the experimental process, we access the MFP-VNMH strategy for coping with these VNs against the physical failures. The performance of MFP-VNMH is assessed with three key metrics, described as follows:

- **Success ratio:** Measured by proportion of successfully restored VNs
- **Efficiency:** Measured by the mean time cost for restoring a failed VN
- **Load balance:** Measured by the standard deviation between each physical link load and average physical link load

TABLE 12.1

Simulation Environment Setup

CPU	Intel Core2 i7 920 (four cores and 2.67 GHZ)
MEMORY	6G (three 2G DDR3 1333 HZ memories)
OS	Linux Ubuntu (version 11.04)
COMPILER	GCC (version 4.5.0)

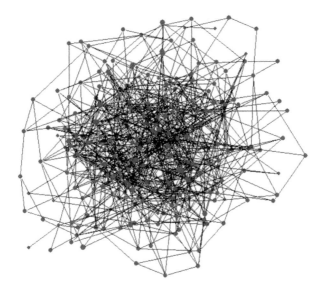

FIGURE 12.19   The random network topology.

The success ratio, *SR*, for VN restoration is defined as

$$SR = \frac{S_v}{F_v} \tag{12.22}$$

where F_v denotes the number of failed VNs caused by physical faults, and S_v represents the number of VNs which have been successfully restored. Also, we assess the efficiency for MFP-VNMH, overall VN remapping, and end-to-end link remapping by calculating the mean time consumption spent in restoring a single VN. The load balance degree of the physical network denoted as Ω merely considers the physical link load balance, which is defined as

$$\Omega = \sqrt{\frac{\sum_{i=1}^{l} (D_{L_i} - \overline{D_L})^2}{l}} \tag{12.23}$$

where D_{L_i} is the remaining bandwidth of physical link i, and $\overline{D_L}$ is the average residual bandwidth of the total l physical links. According to Equation 12.23, it can be seen that with a smaller D_{L_i}, the virtual links are allocated to the physical links more averagely which implies that the physical network is more balanced.

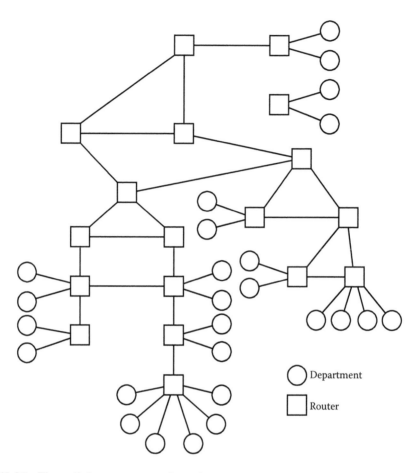

FIGURE 12.20 The realistic campus network topology.

12.4.4.2   Numerical Results

Figure 12.21 presents the results of the VN restoration performance in terms of the success ratio upon the physical link failures. The results show that the restoration success ratio of MPF-VNMH outperforms in comparison with the two existing solutions: the overall VN remapping and the end-to-end link remapping. The MPF-VNMH strategy is able to keep the success ratio of the restoration at more than 70%, while the other two have relatively lower restoration success ratios. This is due to the fact that the MPF-VNMH takes the multi-commodity strategy to remap a failed virtual link onto multiple available physical links, which makes full use of the residual bandwidth in the physical network and hence improves the remapping acceptance ratio. A similar result to the restoration success ratio is obtained from the realistic university network topology, which validates the effectiveness of the proposed algorithm.

The efficiency of the MPF-VNMH, overall VN remapping, and end-to-end link remapping approaches are measured in terms of average time cost for a single VN restoration, as shown in Figure 12.22. It can be seen that the efficiency of MPF-VNMH is almost the same as that of end-to-end link remapping as the latter approach is based on the Dijkstra algorithm, which has the same time complexity as MPF-VNMH. Also, the efficiency of MFP-VNMH is nearly 1.5 times more than that of overall VN remapping, which is mainly due to the fact that when MPF-VNMH is executed once, it enables multiple failed virtual links based on the failed physical link to be restored to an available physical link, while the overall VN remapping approach merely restores one failed VN.

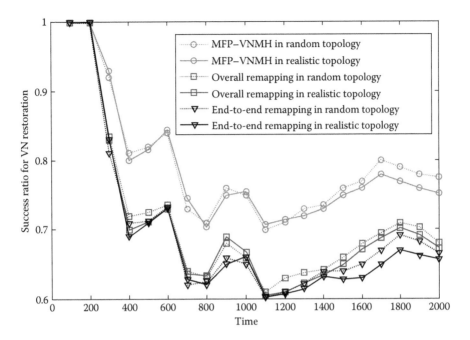

FIGURE 12.21 The VN restoration success ratio.

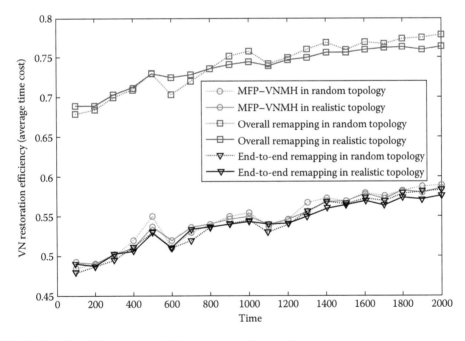

FIGURE 12.22 The VN restoration efficiency (average time cost).

Again, the result of the efficiency obtained from random network topology is similar to that from a realistic university network.

The result of the overall physical network load balance (i.e., Ω) after the VN's restoration by MPF-VNMH, the overall VN remapping, and the end-to-end link remapping are compared and shown in Figure 12.23. The results show a significant difference among MPF-VNMH, overall VN

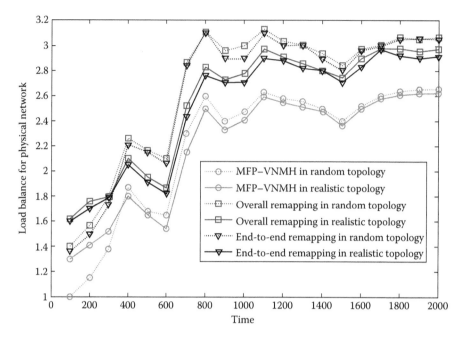

FIGURE 12.23   The load balance of the physical network.

remapping, and end-to-end link remapping. MPF-VNMH achieves better performance in maintaining the load balance of the physical network, which is due to the fact that MPF-VNMH can remap a failed virtual link to multiple physical links or paths to distribute the network demands across the overall network. However, the other two approaches are only able to carry out the one-to-one mapping (i.e., a virtual link can only be mapped onto a physical link). Thus, the proposed MFP-VNMH can utilize the physical network's resources in a more efficient manner.

12.5   CONCLUSION

In this chapter, we presented a novel hierarchical VN-mapping algorithm based on a substrate network decomposition approach. The hierarchical algorithm executes through a GVNMA and a LVNMA operated in the network's central server and within individual subnetworks, respectively, and the GVNMA cooperates with the LVNMA to finish VN mapping tasks. We implemented a hierarchical algorithm and evaluated the performance in comparison with the algorithm presented in [5] and [13] through simulation experiments, and the key findings can be summarized as follows:

- The hierarchical approach has significantly improved the VN mapping performance in terms of runtime; it is about 5 to 20 times faster than the conventional centralized approach.
- The hierarchical approach is close to optimized performance in terms of mapping cost, within 3% for all simulated networks.
- The communication overhead is mainly determined by the number of domains in the substrate network. If the VN requests involve partial domains, the communication overhead can decrease by about 20%. The hierarchical algorithm increases about 10% in terms of the VN request acceptance ratio compared to the PolyViNE algorithm in the dynamic network scenarios.

In addition, this chapter presented a novel VN restoration algorithmic solution, MFP-VNMH, which could enhance VN mapping and service restoration subject to link failures in the physical substrate while avoiding the remapping of the overall VNs. This is achieved by using the sessions between the SP and the InP supervised by a set of control protocols. The suggested method addresses the VN mapping based on the MFP with the primary aim of improving the VN restoration success ratio, maintaining the load balance of the whole physical network, and minimizing the time cost. The performance of MFP-VNMH is assessed through extensive numerical experiments for a range of network scenarios in comparison with the overall VN remapping and end-to-end link remapping approaches. The result is encouraging and demonstrates its effectiveness and efficiency for VN mapping update and service restoration upon physical substrate failures.

ACKNOWLEDGMENT

This work is supported by the National Grand Fundamental Research 973 Programme of China (No. 2012CB315903) and the National Natural Science Foundation of China (No. 51107113). The author also would like to thank Dr. Min Zhang and Dr. Yuting Miao for their contributions to conducting the simulation experiments and invaluable input to this work.

REFERENCES

1. Eason G., Noble B., Sneddon I. N. Network virtualization: A view from the bottom. In *Proceedings of the 1st ACM Workshop on Virtualized Infrastructure Systems and Architectures*, 2009, pp. 73–80, Barcelona, Spain.
2. Zhu Y., Ammar M. Algorithms for assigning substrate network resources to virtual network components. In *Proceedings of IEEE INFOCOM*, 2006, pp. 1–12, Barcelona, Spain.
3. Zhou L., Sen A. Topology design of service overlay network with a generalized cost model. In *Proceedings of IEEE GLOBECOM*, 2007, pp. 75–80, Washington, DC.
4. Szeto W., Iraqi Y., Boutaba R. A multi-commodity flow based approach to virtual network resource allocation. In *Proceedings of IEEE Global Telecommunications Conference (GLOBECOM '03)*, 2003, pp. 3004–3008, San Francisco, CA.
5. Yu M., Yi Y. et al. Rethinking virtual network embedding: Substrate support for path splitting and migration. *ACM SIGCOMM Computer Communication Review*, 2008, 38(2): 17–29.
6. Mosharaf N. M., Chowdhury K. et al. Virtual network embedding with coordinated node and link mapping. *In Proceedings of IEEE INFOCOM*, 2009, pp. 783–791, Rio de Janeiro, Brazil.
7. Lischka J., Karl H. A virtual network mapping algorithm based on subgraph isomorphism detection. In *Proceedings of ACM SIGCOMM*, 2009, pp. 81–88, Barcelona, Spain.
8. Anderson T., Peterson L, Shenker S., Turner J. Overcoming the internet impasse through virtualization. *IEEE Computer Magazine*, 2005, 38(4): 34–41.
9. GENI (Global Environment for Network Innovations). [online]. Available at: www.geni.net/ (accessed on 28 June 2017).
10. Bavier A., Feamster N., Huang M., Peterson L., Rexford J. In VINI Veritas: Realistic and controlled network experimentation. In *Proceedings of ACM SIGCOMM*, 2006, Pisa, Italy.
11. Ahuja R. K., Magnanti T. L., Orilin J. B. *Network Flows: Theory, Algorithms, and Applications*. Upper Saddle River, NJ: Prentice Hall, Inc., 1993.
12. Houidi I., Louati W. et al. A distributed virtual network mapping algorithm. In *Proceedings of IEEE ICC*, 2008, pp. 5634–5640, Beijing, China.
13. Chowdhury M., Samuel F., Boutaba R. PolyViNE: Policy-based virtual network embedding across multiple domains. In *Proceedings of the Second ACM SIGCOMM Workshop on Virtualized Infrastructure Systems and Architectures*, New Delhi, India 2010, pp. 49–56.
14. Lv B., Wang Z., Huang T., Chen J., Liu Y. A hierarchical management architecture for virtual network mapping. In *Proceedings of Internet Technology and Applications*, Wuhan, China 2010, pp. 1-4.
15. Fan J., Ammar M. Dynamic topology configuration in service overlay networks—A study of reconfiguration policies. In *Proceedings of IEEE INFOCOMM 06*, 2006, pp. 1–12, Barcelona, Spain.
16. Chowdhury N. M. M. K., Rahman M. R., Boutaba R. Virtual network embedding with coordinated node and link mapping. In *Proceedings of IEEE INFOCOMM 09*, 2009, pp. 783–791, Rio de Janeiro, Brazil.

17. Ilhem F., Nadjib A. S. et al. VNE-AC: Virtual network embedding algorithm based on ant colony meta-heuristic. In *Proceedings of IEEE ICC*, Kyoto, Japan, 2011.

18. Sahasrabuddhe L., Ramamurthy S., Mukherjee B. Fault management in IP-over-WDM networks: WDM protection versus IP restoration. *IEEE Journal on Selected Areas in Communications*, 2002, 20(1) : 21–33.

19. Fajjari I., Aiari M., Braham O. Towards an automatic piloting virtual network architecture. In *Proceedings of IFIP International Conference on New Technologies, Mobility and Security (NTMS)*, 2011, Paris, France.

20. Guo T., Wang N., Moessner K. Shared backup network provision for virtual network embedding. In *Proceedings of IEEE ICC*, 2011, Kyoto, Japan.

21. Yalian P., Xuesong Q., Shunli Z. Fault diagnosis in network virtualization environment. In *Proceedings of 18th Conference on Telecommunications (ICT)*, 2011, Ayia Napa, Cyprus.

22. Khandekar R., Rao S., Vazirani U. Graph partitioning using single commodity flows. In *Proceedings of the Thirty-Eighth Annual ACM Symposium on Theory of Computing*, 2006, pp. 385–390, Seattle.

23. Lang K., Mahoney M. W., Orecchia L. Empirical evaluation of graph partitioning using spectral embeddings and flow. In *Proceedings of the 8th International Symposium on Experimental Algorithms*, 2009, pp. 197–208, Dortmund, Germany.

24. Allalouf M., Shavitt Y. Centralized and distributed algorithms for routing and weighted max-min fair bandwidth allocation. *IEEE/ACM Transactions on Networking*, 2008, 16(5): 1015–1024.

25. Medina A., Lakhina A., Matta I., Byers J. BRITE: An approach to universal topology generation. *MASCOTS 2001, Proceedings Ninth International Symposium on Modeling, Analysis and Simulation of Computer and Telecommunication Systems*, Cincinnati, OH, 2001, pp. 346–353. [Online]. Available at: www.cs.bu.edu/brite/.

26. GNU Linear Programming Kit. [Online]. Available at: www.gnu.org/software/glpk/ (accessed on 28 June 2017).

27. Avramopoulos I., Syrivelis D., Rexford J., Lalis S. Secure availability monitoring using stealth probes. Princeton University, Technical Report TR-769-06, 2006.

28. Feldmann A., Greenberg A., Lund C., Reingold N., Rexford J. Netscope: Traffic engineering for IP networks. *IEEE Network Magazine*, 2000, 14(2): 11–19.

29. Augustin B., Cuvellier X. et al. Avoiding traceroute anomalies with Paris traceroute. In *Proceedings of Internet Measurement Conference*. New York: ACM Press, 2006, pp. 153–158.

30. Zegura E. W., Calvert K. L., Bhattacharjee S. How to model an internetwork. *Proceeding of INFOCOM '96*, pp. 594–602, San Francisco, California, 2016.

13 Maximizing the Lifetime of Wireless Sensor Networks by Optimal Network Design

Keqin Li

CONTENTS

13.1 INTRODUCTION

The Internet of Things (IoT) has been defined in Recommendation ITU-T Y.2060 (06/2012) as a global infrastructure for the information society, enabling advanced services by interconnecting physical and virtual things based on existing and evolving interoperable information and communication technologies [1]. The IoT is the network of physical objects (e.g., goods, products, vehicles, buildings) embedded with electronics, sensors, software, and network connectivity, which enable objects to collect and process data. The IoT allows objects to be sensed and controlled remotely through existing network infrastructure, creating opportunities for tight integration of the physical world into computer and communication systems. Each thing is uniquely identifiable through its embedded devices and is able to interoperate within the existing Internet infrastructure. It is estimated that the IoT will consist of 50 billion objects by 2020 [2] and contribute 19 trillion USD in the global economy [3].

One of the major enabling technologies for the IoT is low-energy wireless sensor networks [4–6]. When IoT is augmented with sensors and actuators, the technology becomes cyber-physical systems, such as smart grids, smart homes, smart cities, and intelligent transportation systems. A wireless sensor network (WSN) consists of spatially distributed autonomous sensors which are able to monitor physical and environmental conditions and to cooperatively transmit their sensed data through the network to a base station. Originally motivated by military applications such as battle-field surveillance, WSNs are now deployed and used widely in various applications, such as environmental and earth monitoring (air and water quality and pollution monitoring, forest fire, landslide, and natural disaster detection and prevention); industrial monitoring (machine health monitoring, data logging, and industrial sense and control applications); agriculture (accurate agriculture, irrigation management, greenhouses); passive localization and tracking; smart home monitoring; and IoT [7,8].

Due to very limited power supply and severe energy constraint in sensors [9], the lifetime of a WSN has gained substantial research attention in recent years [10]. Energy consumption in WSNs contains two components, namely, the energy required for data sensing and the energy used for data transmission. Research in lifetime maximization of WSNs has been focused on the first component only [11–14], the second component only [15–20], and both components [21–23]. We believe that the lifetime maximization problem of WSNs should be studied by taking both components of energy consumption into consideration [24].

Several methods have been proposed to increase the lifetime of a WSN, including redundant sensors [25], nonuniform sensor distributions [26], and aggregation and forwarding nodes for data transmission [27,28]. All these methods are based on the observation that sensors at different locations consume their battery power at different speeds. In particular, sensors close to a base station consume energy much faster than sensors far away from the base station [29,30]. Therefore, the most effective way to maximize the lifetime of a WSN is to allocate initial energy to sensors such that they exhaust their energy at the same time [21,31–33].

It has been found that the lifetime of a WSN and an optimal initial energy allocation are determined by a network design. Network lifetime maximization is a two-stage process, namely, optimal network design and optimal energy allocation. In reality, a WSN design includes the locations, sensing ranges, communication ranges, and data generation rates of all sensors, energy consumption for both data sensing and data transmission, as well as a routing algorithm for data transmission to a base station (i.e., a sink). All these factors have an impact on sensor and network lifetime as well as optimal energy allocation. For a given network design, an optimal initial energy allocation which maximizes the network lifetime can be determined [31]. Hence, the lifetime of a WSN is essentially determined by a network design.

It has been known that the lifetime of a WSN can be maximized by an optimal network design. In [34], the network lifetime obtained by optimal energy allocation is represented as a function of the number m of annuli, and it is shown that m has a significant impact on network lifetime. It is proved that for annuli of identical widths, if the energy consumed by data transmission is proportional to $d^\alpha + c$, where d is the distance of data transmission, and α and c are some constants, then for a circular area of interest with radius R, the optimal number of annuli that maximizes the network lifetime is

$$m = R\left(\frac{\alpha - 1}{c}\right)^{1/\alpha} \tag{13.1}$$

for an arbitrary sensor density function.

The investigation in [34] assumes that all annuli have identical widths based on the observation that the energy consumption of a data transmission is minimized when all hops have the same distance [32,35]. However, whether identical annulus widths give the maximum network lifetime is worth further investigation. In this chapter, we show that it is indeed the case that an optimal network design has different widths of annuli. The main contribution of the present chapter is to develop an algorithm to find an optimal network design which maximizes the lifetime of a WSN obtained by optimal initial energy allocation for an arbitrary sensor distribution. We show that the optimal WSN design problem can be formulated as a nonlinear system of equations. Our results reveal the fact that an optimal network design has different widths of annuli. In particular, in an optimal network design, an annulus closer to a sink has larger width. Compared with a network design with identical annulus widths, a network design with variable annulus widths can lead to a noticeable increment of the network lifetime.

The chapter is organized as follows. In Section 13.2, we present preliminary information, including the network design model, the sensor distribution model, and the energy consumption model used in our study. In Section 13.3, we give an example to motivate our investigation. In Section 13.4, we develop analytical forms of network lifetime and optimal energy allocation. In Section 13.5, we give an algorithm to find an optimal network design which maximizes the network lifetime obtained by optimal energy allocation for a uniform sensor distribution, where the optimal network design problem is formulated as a nonlinear system of equations. In Section 13.6, we give an algorithm to find an optimal network design for a nonuniform sensor distribution. In Section 13.7, we demonstrate numerical examples. In Section 13.8, we mention some future work. We conclude the chapter in Section 13.9.

13.2 PRELIMINARIES

In this section, we provide preliminary information, including the network design model, the sensor distribution model, and the energy consumption model. We also give a motivational example which inspires our research.

13.2.1 THE NETWORK DESIGN MODEL

Let us consider a circular area of interest A which has radius R meters (see Figure 13.1). Assume that A is divided into m annuli (also called coronae) $A_1, A_2, ..., A_m$ by m circles with radii $r_1, r_2, ..., r_m$ centered at a sink, where $0 < r_1 < r_2 < \cdots < r_m = R$ [32]. For convenience, we assume that there is A_0 with width $r_0 = 0$ which contains a sink. All sensors report sensory data to the sink. Annulus A_j has width $r_j - r_{j-1}$, where $1 \le j \le m$. For a fixed R, the number m of annuli as well as the sequence of

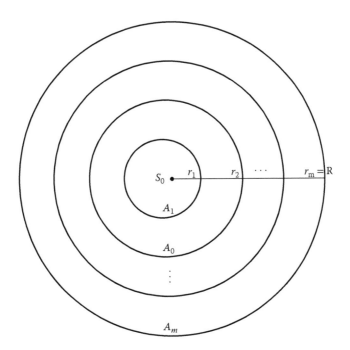

FIGURE 13.1   A circular area of radius R with m annuli.

values $(r_1, r_2, ..., r_{m-1})$ is called a *network design* or a *network configuration*, which has significant impact on energy consumption and network lifetime.

Assume that there are N sensors, $s_1, s_2, ..., s_N$, randomly distributed in A. We use s_0 to represent a sink. All sensors in A_j are designed in such a way that they have the same transmission range, $r_j - r_{j-1}$. All sensors also have a certain sensing range. It is assumed that N is sufficiently large such that a WSN is connected. Furthermore, it is assumed that the sensing range is sufficiently large such that A is well covered. Let N_j be the number of sensors in A_j.

13.2.2   THE SENSOR DISTRIBUTION MODEL

Let $f(r)$ be any *sensor density function* (or *sensor distribution function*) in a circular area of interest, A, with radius R, where $0 \le r \le R$. In other words, the number of sensors in a small area, z, with distance r to the sink is $f(r)z$. The function $f(r)$ should satisfy

$$\int_0^R 2\pi r f(r) dr = N \tag{13.2}$$

The number of sensors in A_j is

$$N_j = \int_{r_{j-1}}^{r_j} 2\pi r f(r) dr \tag{13.3}$$

for all $1 \le j \le m$. For instance, for a uniform distribution, we have

$$f(r) = \frac{N}{\pi R^2} \tag{13.4}$$

$$\int_0^R 2\pi r f(r) dr = \frac{N}{R^2} \int_0^R 2r dr = \left(\frac{N}{R^2}\right) r^2 \Big|_0^R = N \tag{13.5}$$

and

$$N_j = \frac{N}{R^2} \int_{r_{j-1}}^{r_j} 2r dr = \left(\frac{N}{R^2}\right) r^2 \Big|_{r_{j-1}}^{r_j} = \left(\frac{r_j^2 - r_{j-1}^2}{R^2}\right) N \tag{13.6}$$

for all $1 \le j \le m$.

As an example of nonuniform sensor distribution, let us consider

$$f(r) = \left(\frac{N}{\pi \ln(1 + 1/u)}\right)\left(\frac{1}{r^2 + uR^2}\right) \tag{13.7}$$

where $u > 0$. It is easy to see that

$$\int_0^R 2\pi r f(r)\,dr = \int_0^K 2\pi r \left(\frac{N}{\pi \ln\,(1+1/u)}\right)\left(\frac{1}{r^2+uR^2}\right)dr \tag{13.8}$$

$$= \frac{N}{\ln\,(1+1/u)} \int_0^R \left(\frac{1}{r^2+uR^2}\right) d\left(r^2+uR^2\right)$$

$$= \frac{N}{\ln\,(1+1/u)} \int_{uR^2}^{(u+1)R^2} \frac{dx}{x} \quad \left(\text{letting } x=r^2+uR^2\right)$$

$$= \frac{N}{\ln\,(1+1/u)} \ln x \,\vert_{uR^2}^{(u+1)R^2}$$

$$= \frac{N}{\ln\,(1+1/u)}\Big(\ln\big((u+1)R^2 - \ln\,(uR^2)\big)\Big)$$

$$= N$$

Notice that the ratio of the largest density (when $r=0$) to the smallest density (when $r=R$) is $(1+1/u)$. Thus, the parameter u indicates uniformity of sensor distribution. For a small u, sensors are more densely distributed in the area closer to the sink. As $u \to 0$, the sensor density near the sink can be arbitrarily large. One the other hand, as u increases, sensors are more evenly distributed in A. For a very large u, we have $\ln\,(1+1/u) \approx 1/u$, and

$$f(r) \approx \left(\frac{N}{\pi(1/u)}\right)\left(\frac{1}{r^2+uR^2}\right) = \frac{N}{\pi\left(r^2/u+R^2\right)} \approx \frac{N}{\pi R^2} \tag{13.9}$$

That is, as $u \to \infty$, $f(r)$ approaches a uniform distribution.

The previous $f(r)$ gives rise to

$$N_j = \frac{N}{\ln\,(1+1/u)} \ln\left(\frac{r_j^2+uR^2}{r_{j-1}^2+uR^2}\right) \tag{13.10}$$

for all $1 \le j \le m$. For a very large u, we have

$$\ln\left(\frac{r_j^2+uR^2}{r_{j-1}^2+uR^2}\right) = \ln\left(1+\frac{r_j^2-r_{j-1}^2}{r_{j-1}^2+uR^2}\right) \approx \frac{r_j^2-r_{j-1}^2}{r_{j-1}^2+uR^2} \tag{13.11}$$

and

$$N_j \approx \frac{N}{(1/u)} \cdot \frac{r_j^2-r_{j-1}^2}{r_{j-1}^2+uR^2} \approx \left(\frac{r_j^2-r_{j-1}^2}{R^2}\right) N \tag{13.12}$$

which is the N_j for a uniform distribution.

13.2.3   THE ENERGY CONSUMPTION MODEL

The amount of energy consumed by a sensor to sense and receive data in one unit of time is p mJ/second.

The amount of energy needed to transmit one bit over distance d meters is $\left(a_1 d^\alpha + a_2\right)$ pJ, where a_1 is the energy required to run a transmitter amplifier, a_2 is the energy used to activate a transmitter circuitry, and $2 \leq \alpha \leq 6$ is a constant [36]. The previous expression has a significant implication in minimizing the energy cost of data transmission in WSNs.

Assume that each datum has size b bytes $= 8b$ bits. Then, the amount of energy needed to transmit one datum over distance d meters is

$$q = 8b\left(a_1 d^\alpha + a_2\right) \mathrm{pJ} = 8a_1 b\left(d^\alpha + \frac{a_2}{a_1}\right)\mathrm{pJ} = \frac{8a_1 b}{10^6}\left(d^\alpha + c\right)\mathrm{mJ} = a\left(d^\alpha + c\right)\mathrm{mJ} \quad (13.13)$$

where $a = a_1 b / 125{,}000$ mJ/m$^\alpha$ and $c = a_2 / a_1$ m$^\alpha$. (Notation: pJ = pico Joule, nJ = nano Joule, mJ = micro Joule, m$^\alpha$ = meter raised to the power α.) For instance, when $b = 25$ bytes, $a_1 = 10$ pJ/bit/m$^\alpha$, $a_2 = 50$ nJ / bit $= 50{,}000$ pJ / bit, we have $a = 0.002$ mJ/m$^\alpha$, $c = 5000$ m$^\alpha$, and $q = 0.002 d^\alpha + 10$ mJ. Based on the previous discussion, we know that the amount of energy consumed by a sensor in A_j to transmit a datum is

$$q_j = a\left(\left(r_j - r_{j-1}\right)^\alpha + c\right)\mathrm{mJ} \quad (13.14)$$

for all $1 \leq j \leq m$.

13.3 A MOTIVATIONAL EXAMPLE

Before we proceed, let us consider the following motivational example. Consider $(k+1)$ sensors $s_0, s_1, s_2, \ldots, s_k$ along a line, where the distance between s_{j-1} and s_j is d_j, for all $1 \leq j \leq k$, and $d = d_1 + d_2 + \cdots + d_k$ (see Figure 13.2). Each sensor s_j needs to send a bit to s_0 along the path $\left(s_j, s_{j-1}, \ldots, s_0\right)$ with j hops, for all $1 \leq j \leq k$. Then, for fixed d, the energy consumed by the above k data transmissions is a function of $d_1, d_2, \ldots, d_{k-1}$ (since $d_k = d - d_1 - d_2 - \cdots - d_{k-1}$):

$$E\left(d_1, d_2, \ldots, d_{k-1}\right) = \sum_{j=1}^{k}(k-j+1)\left(a_1 d_j^\alpha + a_2\right) = a_1 \sum_{j=1}^{k}(k-j+1)d_j^\alpha + a_2 \cdot \frac{k(k+1)}{2} \quad (13.15)$$

It is clear that to minimize $E\left(d_1, d_2, \ldots, d_{k-1}\right)$, we need only to minimize

$$f\left(d_1, d_2, \ldots, d_{k-1}\right) = \sum_{j=1}^{k}(k-j+1)d_j^\alpha = \sum_{j=1}^{k-1}(k-j+1)d_j^\alpha + \left(d - \sum_{j=1}^{k-1}d_j\right)^\alpha \quad (13.16)$$

Consider the case when $\alpha = 2$. Since the terms in $f\left(d_1, d_2, \ldots, d_{k-1}\right)$ that include d_j are

$$(k-j+2)d_j^2 + 2d_j \sum_{j' \neq j} d_{j'} - 2dd_j \quad (13.17)$$

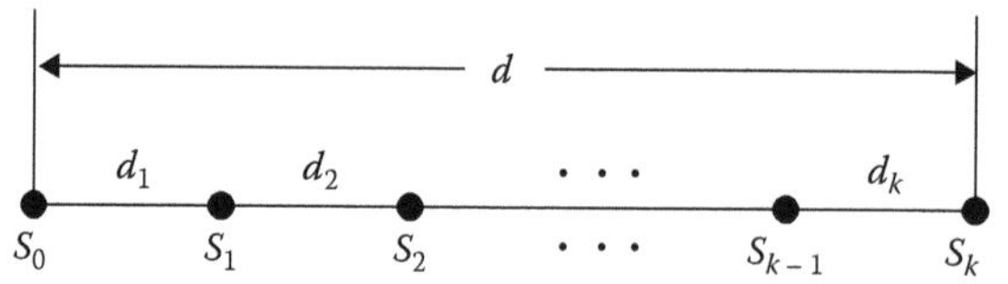

FIGURE 13.2 A data transmission path.

we have

$$\frac{\partial f(d_1, d_2, ..., d_{k-1})}{\partial d_j} = 2(k-j+2)d_j + 2\sum_{j' \neq j} d_{j'} - 2d \tag{13.18}$$

for all $1 \leq j \leq k-1$. Hence, we get a linear system of equations:

$$(k-j+2)d_j + \sum_{j' \neq j} d_{j'} = d \tag{13.19}$$

for all $1 \leq j \leq k-1$. It is straightforward to verify that when $k=2$, we have $d_1 = \left(\frac{1}{3}\right)d$ and $d_2 = \left(\frac{2}{3}\right)d$; when $k=3$, we have $d_1 = \left(\frac{2}{11}\right)d$, $d_2 = \left(\frac{3}{11}\right)d$, and $d_3 = \left(\frac{6}{11}\right)d$; and when $k=4$, we have $d_1 = \left(\frac{3}{25}\right)d$, $d_2 = \left(\frac{4}{25}\right)d$, $d_3 = \left(\frac{6}{25}\right)d$, and $d_4 = \left(\frac{12}{25}\right)d$. We observe that the d_j's are different, a quite different conclusion from the fact that for a single data transmission, energy consumption is minimized when all hops have the same distance [34]. Such phenomenon inspires the optimal network configuration problem solved in this chapter.

13.4   NETWORK LIFETIME AND OPTIMAL ENERGY ALLOCATION

In this section, we develop analytical forms of network lifetime and optimal energy allocation, which are necessary to formulate our optimization problems (i.e., network lifetime maximization and optimal network design).

When a datum is transmitted from a sensor, s_j, in A_j to the sink, s_0, the datum is sent along a path $(s_j, s_{j-1}, s_{j-2}, ..., s_1, s_0)$ from s_j to s_0, where $s_i \in A_i$ for all $j \geq i \geq 0$. Assume that each sensor senses and transmits μ data to a sink per second. This implies that sensors in A_j contribute $N_j\mu$ data transmissions per second to all A_i's, where $1 \leq i \leq j$. It is also assumed that all sensors in A_j are treated equally such that they all perform the same amount of data transmission. Since there are $(N_j + N_{j+1} + \cdots + N_m)\mu$ data transmissions to be performed by N_j sensors in A_j per second, a sensor in A_j performs β_j data transmissions in one unit of time, where

$$\beta_j = \frac{1}{N_j}(N_j + N_{j+1} + \cdots + N_m)\mu = \left(\frac{r_m^2 - r_{j-1}^2}{r_j^2 - r_{j-1}^2}\right)\mu \tag{13.20}$$

for all $1 \leq j \leq m$.

A sensor in A_j is equipped with E_j of initial energy. Let E denote the total energy budget, that is,

$$E = \sum_{j=1}^{m} N_j E_j \tag{13.21}$$

Once a sensor is deployed, E_j is not renewable or replenishable. The network lifetime is determined by the *initial energy allocation* $(E_1, E_2, ..., E_m)$, which is determined by a network design, that is, $(r_1, r_2, ..., r_{m-1})$.

The energy consumed by a sensor in A_j in one unit of time is $p + \beta_j q_j$, which implies that the lifetime of a sensor in A_j is

$$L_j = \frac{E_j}{p + \beta_j q_j} \tag{13.22}$$

A reasonable definition of *network lifetime* L is $L = \min(L_1, L_2, ..., L_m)$, since when all sensors in A_j run out of battery power, a WSN becomes disconnected and inoperational. It is clear that L is maximized if and only if $L_1 = L_2 = \cdots = L_m$, that is, all sensors die at the same time; otherwise, we can initially move energy from sensors which work longer to sensors which die sooner so that the network lifetime is increased.

To have an identical lifetime L for all the sensors, that is, $E_j / (p + \beta_j q_j) = L$, we need $E_j = L(p + \beta_j q_j)$, for all $1 \le j \le m$. Since $N_1 E_1 + N_2 E_2 + \cdots + N_m E_m = E$, that is,

$$\sum_{j=1}^{m} N_j L(p + \beta_j q_j) = E \tag{13.23}$$

we get

$$L = \frac{E}{\sum_{j=1}^{m} N_j (p + \beta_j q_j)} \tag{13.24}$$

and

$$E_j = \frac{(p + \beta_j q_j) E}{\sum_{j=1}^{m} N_j (p + \beta_j q_j)} \tag{13.25}$$

for all $1 \le j \le m$. An initial energy allocation $(E_1, E_2, ..., E_m)$ that results in $L_1 = L_2 = \cdots = L_m = L$ is called an *optimal energy allocation*.

We notice that

$$N_j + N_{j+1} + \cdots + N_m = \int_{r_{j-1}}^{r_m} 2\pi r f(r) dr \tag{13.26}$$

Hence, the network lifetime is

$$
\begin{aligned}
L &= \frac{E}{Np + \sum_{j=1}^{m} (N_j + N_{j+1} + \cdots + N_m) \mu q_j} \\
&= \frac{E}{Np + \mu a \sum_{j=1}^{m} \left(\int_{r_{j-1}}^{r_m} 2\pi r f(r) dr \right) \left((r_j - r_{j-1})^\alpha + c \right)}
\end{aligned}
\tag{13.27}
$$

Since

$$p + \beta_j q_j = p + \frac{1}{N_j}\left(N_j + N_{j+1} + \cdots + N_m\right)\mu a\left((r_j - r_{j-1})^\alpha + c\right) \tag{13.28}$$

$$= p + \frac{\int_{r_{j-1}}^{r_m} 2\pi r f(r)dr}{\int_{r_{j-1}}^{r_j} 2\pi r f(r)dr} \cdot \mu a\left((r_j - r_{j-1})^\alpha + c\right)$$

the optimal energy allocation is

$$E_j = E \cdot \frac{p + \left(\int_{r_{j-1}}^{r_m} 2\pi r f(r)dr\right)\left(\int_{r_{j-1}}^{r_j} 2\pi r f(r)dr\right)^{-1}\mu a\left((r_j - r_{j-1})^\alpha + c\right)}{Np + \mu a \sum_{j=1}^{m}\left(\int_{r_{j-1}}^{r_m} 2\pi r f(r)dr\right)\left((r_j - r_{j-1})^\alpha + c\right)} \tag{13.29}$$

for all $1 \le j \le m$.

13.5 OPTIMAL NETWORK DESIGN FOR UNIFORM DISTRIBUTIONS

In this section, for a uniform sensor distribution, we define our network lifetime maximization problem as a multivariable minimization problem. Then, we give an algorithm to find an optimal network design which maximizes the network lifetime obtained by optimal energy allocation.

13.5.1 THE OPTIMIZATION PROBLEM

Notice that for a uniform distribution of sensors, we have

$$N_j = \left(\frac{\pi r_j^2 - \pi r_{j-1}^2}{\pi R^2}\right)N = \left(\frac{r_j^2 - r_{j-1}^2}{R^2}\right)N \tag{13.30}$$

and $N = N_1 + N_2 + \cdots + N_m$. Then, we have

$$\sum_{j=1}^{m} N_j\left(p + \beta_j q_j\right) = \sum_{j=1}^{m} N_j p + \sum_{j=1}^{m} N_j \beta_j q_j$$

$$= Np + \sum_{j=1}^{m}\left(N_j + N_{j+1} + \cdots + N_m\right)\mu q_j$$

$$= Np + \sum_{j=1}^{m}\left(\frac{r_m^2 - r_{j-1}^2}{R^2}\right)N\mu a\left((r_j - r_{j-1})^\alpha + c\right) \quad C \tag{13.31}$$

$$= Np + \frac{N\mu a}{R^2}\sum_{j=1}^{m}(r_m^2 - r_{j-1}^2)\left((r_j - r_{j-1})^\alpha + c\right)$$

$$= N\left(p + \frac{\mu a}{R^2}\sum_{j=1}^{m}(r_m^2 - r_{j-1}^2)\left((r_j - r_{j-1})^\alpha + c\right)\right)$$

Hence, the network lifetime is

$$L = \frac{E}{N\left(p + \dfrac{\mu a}{R^2} \displaystyle\sum_{j=1}^{m} (r_m^2 - r_{j-1}^2)\big((r_j - r_{j-1})^\alpha + c\big) \right)} \tag{13.32}$$

Since

$$
\begin{aligned}
p + \beta_j q_j &= p + \frac{1}{N_j}\big(N_j + N_{j+1} + \cdots + N_m\big)\mu a\big((r_j - r_{j-1})^\alpha + c\big) \\
&= p + \left(\frac{r_m^2 - r_{j-1}^2}{r_j^2 - r_{j-1}^2}\right)\mu a\big((r_j - r_{j-1})^\alpha + c\big)
\end{aligned}
\tag{13.33}
$$

we obtain the optimal energy allocation

$$E_j = \frac{E}{N} \cdot \frac{p + \left(\dfrac{r_m^2 - r_{j-1}^2}{r_j^2 - r_{j-1}^2}\right)\mu a\big((r_j - r_{j-1})^\alpha + c\big)}{p + \dfrac{\mu a}{R^2}\displaystyle\sum_{j=1}^{m}(r_m^2 - r_{j-1}^2)\big((r_j - r_{j-1})^\alpha + c\big)} \tag{13.34}$$

for all $1 \le j \le m$.

It is clear that the network lifetime L is a function of $r_1, r_2, \ldots, r_{m-1}$. To maximize the network life-time, we need to minimize the following function:

$$F(r_1, r_2, \ldots, r_{m-1}) = \sum_{j=1}^{m}(r_m^2 - r_{j-1}^2)\big((r_j - r_{j-1})^\alpha + c\big) \tag{13.35}$$

where $r_0 = 0$ and $r_m = R$. Notice that $F(r_1, r_2, \ldots, r_{m-1})/R^2$ gives the average number of m^α taken by a single data transmission. It is clear that $F(r_1, r_2, \ldots, r_{m-1})$ is a quantity determined by a net-work design $(r_1, r_2, \ldots, r_{m-1})$ and $F(r_1, r_2, \ldots, r_{m-1})$ determines the energy expenditure of data trans-mission.

13.5.2 Optimal Network Design

Now, we develop an algorithm to solve our optimization problem. Our main idea is to formulate the optimal network design problem as a nonlinear system of equations. (For technically less-experienced readers, this section can be skipped.)

To minimize the function

$$F(r_1, r_2, \ldots, r_{m-1}) = \sum_{j=1}^{m}(R^2 - r_{j-1}^2)\big((r_j - r_{j-1})^\alpha + c\big) \tag{13.36}$$

we should have

$$\frac{\partial F}{\partial r_j} = f_j(r_1, r_2, \ldots, r_{m-1}) = 0 \tag{13.37}$$

for all $1 \le j \le m-1$. Since the terms in $F(r_1, r_2, ..., r_{m-1})$ that include r_j are

$$(R^2 - r_{j-1}^2)((r_j - r_{j-1})^\alpha + c) + (R^2 - r_j^2)((r_{j+1} - r_j)^\alpha + c) \tag{13.38}$$

we get

$$f_j(r_1, r_2, ..., r_{m-1}) = \alpha(R^2 - r_{j-1}^2)(r_j - r_{j-1})^{\alpha-1} - 2r_j((r_{j+1} - r_j)^\alpha + c) - \alpha(R^2 - r_j^2)(r_{j+1} - r_j)^{\alpha-1} \tag{13.39}$$

Therefore, we have a nonlinear system of equations, that is,

$$\begin{aligned} f_1(r_1, r_2, ..., r_{m-1}) &= 0 \\ f_2(r_1, r_2, ..., r_{m-1}) &= 0, \\ &\vdots \\ f_{m-1}(r_1, r_2, ..., r_{m-1}) &= 0 \end{aligned} \tag{13.40}$$

By using vector notation to represent the variables $r_1, r_2, ..., r_{m-1}$, we write

$$\mathbf{r} = (r_1, r_2, ..., r_{m-1}) \tag{13.41}$$

and $f_j(r_1, r_2, ..., r_{m-1}) = f_j(\mathbf{r})$, where $f_j : \mathbb{R}^{m-1} \to \mathbb{R}$ maps $(m-1)$-dimensional space $\mathbb{R}^{m-1}$ into the real line $\mathbb{R}$. By defining a function $\mathbf{F} : \mathbb{R}^{m-1} \to \mathbb{R}^{m-1}$ which maps $\mathbb{R}^{m-1}$ into $\mathbb{R}^{m-1}$

$$\mathbf{F}(r_1, r_2, ..., r_{m-1}) = (f_1(r_1, r_2, ..., r_{m-1}), f_2(r_1, r_2, ..., r_{m-1}), ..., f_{m-1}(r_1, r_2, ..., r_{m-1})) \tag{13.42}$$

namely,

$$\mathbf{F}(\mathbf{r}) = (f_1(\mathbf{r}), f_2(\mathbf{r}), ..., f_{m-1}(\mathbf{r})) \tag{13.43}$$

then our nonlinear system of equations is

$$\mathbf{F}(\mathbf{r}) = 0 \tag{13.44}$$

where $\mathbf{0} = (0, 0, ..., 0)$.

The previous nonlinear system of equations can be solved by using Newton's method. To this end, we need the Jacobian matrix $J(\mathbf{r})$ defined as

$$J(\mathbf{r}) = \begin{bmatrix} \dfrac{\partial f_1(\mathbf{r})}{\partial r_1} & \dfrac{\partial f_1(\mathbf{r})}{\partial r_2} & \cdots & \dfrac{\partial f_1(\mathbf{r})}{\partial r_{m-1}} \\[2ex] \dfrac{\partial f_2(\mathbf{r})}{\partial r_1} & \dfrac{\partial f_2(\mathbf{r})}{\partial r_2} & \cdots & \dfrac{\partial f_2(\mathbf{r})}{\partial r_{m-1}} \\[2ex] \vdots & \vdots & \ddots & \vdots \\[2ex] \dfrac{\partial f_{m-1}(\mathbf{r})}{\partial r_1} & \dfrac{\partial f_{m-1}(\mathbf{r})}{\partial r_2} & \cdots & \dfrac{\partial f_{m-1}(\mathbf{r})}{\partial r_{m-1}} \end{bmatrix} \tag{13.45}$$

where

$$\frac{\partial f_j(\mathbf{r})}{\partial r_{j-1}} = -2\alpha r_{j-1}(r_j - r_{j-1})^{\alpha-1} - \alpha(\alpha-1)(R^2 - r_{j-1}^2)(r_j - r_{j-1})^{\alpha-2} \tag{13.46}$$

ALGORITHM 13.1   AN ALGORITHM FOR FINDING
AN OPTIMAL NETWORK DESIGN

Algorithm: Optimal Network Design

Input: Parameters α, c, R, m.

Output: An optimal network design $\mathbf{r} = (r_1, r_2, ..., r_{m-1})$ which satisfies $\mathbf{F}(\mathbf{r}) = \mathbf{0}$.

$\mathbf{r} \leftarrow (R/m, 2R/m, 3R/m, ..., (m-1)R/m)$;	(1)
Repeat	(2)
Calculate $\mathbf{F}(\mathbf{r}) = (f_1(\mathbf{r}), f_2(\mathbf{r}), ..., f_{m-1}(\mathbf{r}))$;	(3)
Calculate $J(\mathbf{r})$, where $J(\mathbf{r})_{j,k} = \partial f_j(\mathbf{r})/\partial r_k$ for $1 \le j, k \le m-1$;	(4)
Solve the linear system of equations $J(\mathbf{r})\mathbf{x} = -\mathbf{F}(\mathbf{r})$;	(5)
$\mathbf{r} \leftarrow \mathbf{r} + \mathbf{x}$;	(6)
until $\|\mathbf{x}\| \le \epsilon$.	(7)

for all $2 \le j \le m-1$, and

$$\frac{\partial f_j(\mathbf{r})}{\partial r_j} = \alpha(\alpha - 1)\left(R^2 - r_{j-1}^2\right)(r_j - r_{j-1})^{\alpha-2} - 2\left((r_{j+1} - r_j)^\alpha + c\right)$$

$$+ 4\alpha r_j(r_{j+1} - r_j)^{\alpha-1} + \alpha(\alpha - 1)\left(R^2 - r_j^2\right)(r_{j+1} - r_j)^{\alpha-2} \tag{13.47}$$

for all $1 \le j \le m-1$, and

$$\frac{\partial f_j(\mathbf{r})}{\partial r_{j+1}} = -2\alpha r_j(r_{j+1} - r_j)^{\alpha-1} - \alpha(\alpha - 1)\left(R^2 - r_j^2\right)(r_{j+1} - r_j)^{\alpha-2} \tag{13.48}$$

for all $1 \le j \le m-2$, and

$$\frac{\partial f_j(\mathbf{r})}{\partial r_k} = 0 \tag{13.49}$$

for all $1 \le j \le m-1$ and $k \ne j-1, j, j+1$.

Our algorithm for finding an optimal network design $\mathbf{r} = (r_1, r_2, ..., r_{m-1})$ which satisfies the nonlinear system of equations $\mathbf{F}(\mathbf{r}) = \mathbf{0}$ is given in Algorithm 13.1. This is essentially Newton's standard iterative method ([37], p. 451). Our initial approximation of $\mathbf{r}$ is $r_j = jR/m$ for all $1 \le j \le m-1$ [line (1)]. The value of $\mathbf{r}$ is then repeated modified as $\mathbf{r} + \mathbf{x}$ [line (6)], where $\mathbf{x}$ is the solution to the linear system of equations $J(\mathbf{r})\mathbf{x} = -\mathbf{F}(\mathbf{r})$ [line (5)]. Such modification is repeated until $\|\mathbf{x}\| \le \epsilon$ [line (7)], where

$$\|\mathbf{x}\| = \sqrt{x_1^2 + x_2^2 + \cdots + x_{m-1}^2} \tag{13.50}$$

and ϵ is a sufficiently small constant, say, 10^{-10}. The linear system of equations in line (5) can be solved by using the classic Gaussian elimination with backward substitution algorithm ([37], pp. 268–269).

13.6 OPTIMAL NETWORK DESIGN FOR NONUNIFORM DISTRIBUTIONS

In this section, for an arbitrary nonuniform sensor distribution, we define our network lifetime maximization problem as a multivariable minimization problem. We also give an algorithm to find an optimal network design which maximizes the network lifetime obtained by optimal energy allocation by extending our method in Section 13.5.

13.6.1 THE OPTIMIZATION PROBLEM

We follow the method in Section 13.5.1.

For the nonuniform sensor distribution in Section 13.2.2, since

$$N_j + N_{j+1} + \cdots + N_m = \frac{N}{\ln(1+1/u)} \ln \left(\prod_{i=j}^{m} \frac{r_i^2 + uR^2}{r_{i-1}^2 + uR^2} \right)$$

$$= \frac{N}{\ln(1+1/u)} \ln \left(\frac{r_m^2 + uR^2}{r_{j-1}^2 + uR^2} \right)$$

$$(13.51)$$

we have

$$L = \frac{E}{N \left(p + \dfrac{\mu a}{\ln(1+1/u)} \displaystyle\sum_{j=1}^{m} \ln \left(\dfrac{r_m^2 + uR^2}{r_{j-1}^2 + uR^2} \right) \left((r_j - r_{j-1})^\alpha + c \right) \right)} \qquad (13.52)$$

Because

$$p + \beta_j q_j = p + \frac{1}{N_j} \left(N_j + N_{j+1} + \cdots + N_m \right) \mu a \left((r_j - r_{j-1})^\alpha + c \right)$$

$$= p + \frac{\ln \left(\dfrac{r_m^2 + uR^2}{r_{j-1}^2 + uR^2} \right)}{\ln \left(\dfrac{r_j^2 + uR^2}{r_{j-1}^2 + uR^2} \right)} \cdot \mu a \left((r_j - r_{j-1})^\alpha + c \right) \qquad (13.53)$$

we obtain

$$E_j = \frac{E}{N} \cdot \frac{p + \ln \left(\dfrac{r_m^2 + uR^2}{r_{j-1}^2 + uR^2} \right) \left(\ln \left(\dfrac{r_j^2 + uR^2}{r_{j-1}^2 + uR^2} \right) \right)^{-1} \mu a \left((r_j - r_{j-1})^\alpha + c \right)}{p + \dfrac{\mu a}{\ln(1+1/u)} \displaystyle\sum_{j=1}^{m} \ln \left(\dfrac{r_m^2 + uR^2}{r_{j-1}^2 + uR^2} \right) \left((r_j - r_{j-1})^\alpha + c \right)} \qquad (13.54)$$

for all $1 \le j \le m$.

To maximize the network lifetime, we need to minimize the following function:

$$F(r_1, r_2, \ldots, r_{m-1}) = \sum_{j=1}^{m} \ln \left(\frac{r_m^2 + uR^2}{r_{j-1}^2 + uR^2} \right) \left((r_j - r_{j-1})^\alpha + c \right) \qquad (13.55)$$

where $r_0 = 0$ and $r_m = R$.

13.6.2 OPTIMAL NETWORK DESIGN

Now, we follow the method in Section 13.5.2. (For technically less-experienced readers, this section can be skipped.)

To minimize the previous function, we should have

$$\frac{\partial F}{\partial r_j} = f_j(r_1, r_2, ..., r_{m-1}) = 0 \tag{13.56}$$

for all $1 \le j \le m-1$. Since the terms in $F(r_1, r_2, ..., r_{m-1})$ that include r_j are

$$\ln\left(\frac{r_m^2 + uR^2}{r_{j-1}^2 + uR^2}\right)\left((r_j - r_{j-1})^\alpha + c\right) + \ln\left(\frac{r_m^2 + uR^2}{r_j^2 + uR^2}\right)\left((r_{j+1} - r_j)^\alpha + c\right) \tag{13.57}$$

we get

$$f_j(r_1, r_2, ..., r_{m-1}) = \ln\left(\frac{r_m^2 + uR^2}{r_{j-1}^2 + uR^2}\right)\alpha(r_j - r_{j-1})^{\alpha-1}$$
$$-\left(\frac{2r_j}{r_j^2 + uR^2}\right)\left((r_{j+1} - r_j)^\alpha + c\right) - \ln\left(\frac{r_m^2 + uR^2}{r_j^2 + uR^2}\right)\alpha(r_{j+1} - r_j)^{\alpha-1} \tag{13.58}$$

Therefore, we have a nonlinear system of equations, that is,

$$\begin{aligned}
f_1(r_1, r_2, ..., r_{m-1}) &= 0 \\
f_2(r_1, r_2, ..., r_{m-1}) &= 0, \\
&\vdots \\
f_{m-1}(r_1, r_2, ..., r_{m-1}) &= 0
\end{aligned} \tag{13.59}$$

Again, by using vector notation to represent the variables $r_1, r_2, ..., r_{m-1}$, we write

$$\mathbf{r} = (r_1, r_2, ..., r_{m-1}) \tag{13.60}$$

and $f_j(r_1, r_2, ..., r_{m-1}) = f_j(\mathbf{r})$. By defining a function $\mathbf{F} : \mathbb{R}^{m-1} \to \mathbb{R}^{m-1}$

$$\mathbf{F}(r_1, r_2, ..., r_{m-1}) = \left(f_1(r_1, r_2, ..., r_{m-1}), f_2(r_1, r_2, ..., r_{m-1}), ..., f_{m-1}(r_1, r_2, ..., r_{m-1})\right) \tag{13.61}$$

namely,

$$\mathbf{F}(\mathbf{r}) = \left(f_1(\mathbf{r}), f_2(\mathbf{r}), ..., f_{m-1}(\mathbf{r})\right) \tag{13.62}$$

then our nonlinear system of equations is

$$\mathbf{F}(\mathbf{r}) = \mathbf{0} \tag{13.63}$$

By using the same algorithm in Section 13.5.2, the previous nonlinear system of equations can be solved by using Newton's method, where the Jacobian matrix $J(\mathbf{r})$ is defined as

$$J(\mathbf{r}) = \left[\frac{\partial f_j(\mathbf{r})}{\partial r_k} \right]_{1 \leq j,k \leq m-1} \tag{13.64}$$

with

$$\frac{\partial f_j(\mathbf{r})}{\partial r_{j-1}} = -\left(\frac{2r_{j-1}}{r_{j-1}^2 + uR^2} \right) \alpha(r_j - r_{j-1})^{\alpha-1} - \ln\left(\frac{r_m^2 + uR^2}{r_{j-1}^2 + uR^2} \right) \alpha(\alpha-1)(r_j - r_{j-1})^{\alpha-2} \tag{13.65}$$

for all $2 \leq j \leq m-1$, and

$$\frac{\partial f_j(\mathbf{r})}{\partial r_j} = \ln\left(\frac{r_m^2 + uR^2}{r_{j-1}^2 + uR^2} \right) \alpha(\alpha-1)(r_j - r_{j-1})^{\alpha-2}$$

$$- 2\left(\frac{uR^2 - r_j^2}{(r_m^2 + uR^2)^2} \right) \left((r_{j+1} - r_j)^{\alpha} + c \right) + \left(\frac{4r_j}{r_j^2 + uR^2} \right) \alpha(r_{j+1} - r_j)^{\alpha-1} \tag{13.66}$$

$$+ \ln\left(\frac{r_m^2 + uR^2}{r_j^2 + uR^2} \right) \alpha(\alpha-1)(r_{j+1} - r_j)^{\alpha-2}$$

for all $1 \leq j \leq m-1$, and

$$\frac{\partial f_j(\mathbf{r})}{\partial r_{j+1}} = -\left(\frac{2r_j}{r_j^2 + uR^2} \right) \alpha(r_{j+1} - r_j)^{\alpha-1} - \ln\left(\frac{r_m^2 + uR^2}{r_j^2 + uR^2} \right) \alpha(\alpha-1)(r_{j+1} - r_j)^{\alpha-2} \tag{13.67}$$

for all $1 \leq j \leq m-2$, and

$$\frac{\partial f_j(\mathbf{r})}{\partial r_k} = 0 \tag{13.68}$$

for all $1 \leq j \leq m-1$ and $k \neq j-1, j, j+1$.

13.7 NUMERICAL EXAMPLES

In this section, we demonstrate numerical examples for both uniform and nonuniform sensor distributions.

13.7.1 UNIFORM DISTRIBUTIONS

To show a numerical example of optimal network design for a uniform distribution of sensors, we set $c = 5000$ and $R = 200$. We notice that Newton's algorithm can only find $\mathbf{r} = (r_1, r_2, ..., r_{m-1})$ for m not exceeding a certain limit, m^*. The value of m^* is 3 for $\alpha = 2$, 15 for $\alpha = 3$, 32 for $\alpha = 4$, 48 for $\alpha = 5$, and 63 for $\alpha = 6$. In fact, m^* is the optimal number of annuli when all the annuli have the same width [34].

In Table 13.1, we show the optimal network design $(r_1, r_2, ..., r_m)$ when $\alpha = 3$ and $m = 15$. We also compare r_j with $j(R/m)$, that is, the value of r_j when all the annuli have the same width. It is easily

TABLE 13.1

Optimal Network Design

j	r_j	$j(R/m)$
0	0.00	0.00
1	13.74	13.33
2	27.44	26.67
3	41.12	40.00
4	54.76	53.33
5	68.38	66.67
6	81.96	80.00
7	95.51	93.33
8	109.01	106.67
9	122.47	120.00
10	135.86	133.33
11	149.18	146.67
12	162.39	160.00
13	175.41	173.33
14	188.12	186.67
15	200.00	200.00

Note: $\alpha = 3$, $m = 15$, uniform distribution.

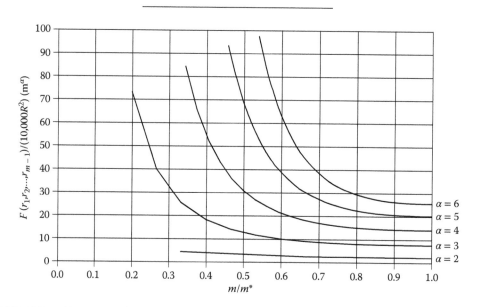

FIGURE 13.3 $F(r_1, r_2, ..., r_{m-1})$ versus number of annuli m (uniform distribution).

observed that $F(r_1, r_2, ..., r_{m-1})$ is minimized when annuli have different widths. Furthermore, from A_1 to A_m, annuli have decreasing widths, that is, $r_1 - r_0 > r_2 - r_1 > \cdots > r_m - r_{m-1}$.

In Figure 13.3, we display the value of $F(r_1, r_2, ..., r_{m-1})$ (actually,

$$F(r_1, r_2, ..., r_{m-1}) / (10{,}000R^2)$$ (13.69)

is shown) for $\alpha = 2, 3, 4, 5, 6$, where $1 \leq m \leq m^*$. We observe that for all $\alpha > 2$, as m increases, $F(r_1, r_2, ..., r_{m-1})$ decreases rapidly. $F(r_1, r_2, ..., r_{m-1})$ reaches its minimum value at m^*. Thus, the energy used for data transmission is very sensitive to the choice of m.

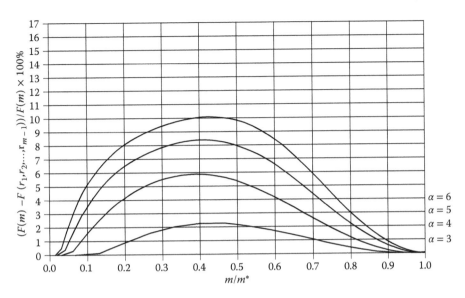

FIGURE 13.4   Relative reduction of $F(r_1, r_2, ..., r_{m-1})$ versus number of annuli m (uniform distribution).

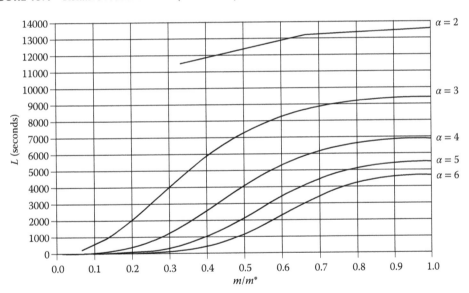

FIGURE 13.5   Network lifetime L versus number of annuli m (uniform distribution).

Let $F(m)$ denote $F(r_1, r_2, ..., r_{m-1})$ when all the annuli have the same width, that is, $r_j = jr$ with $r = R/m$, for all $1 \le j \le m$. In Figure 13.4, we show the relative reduction of $F(r_1, r_2, ..., r_{m-1})$ when compared with $F(m)$, that is,

$$\left(\frac{F(m) - F(r_1, r_2, ..., r_{m-1})}{F(m)} \right) \times 100\% \tag{13.70}$$

For $\alpha = 2$, the relative reduction is small (less than 0.7%) and not shown in the figure. For $\alpha > 2$, the relative reduction is noticeable, except for m close to 1 and m^*.

In Figure 13.5, we demonstrate network lifetime L as a function of m, where $1 \le m \le m^*$, and show the effect of α on L. We set $p = 6$, $a = 0.002$, $c = 5000$, $\mu = 0.03$, $R = 200$, $E/N = 100$ J, and $\alpha = 2, 3, 4, 5, 6$. We observe that for all α, as m increases, L increases rapidly. L reaches its maximum value at m^*. Thus, the network lifetime is very sensitive to the choice of m. Furthermore, the value of α also has a noticeable impact on the network lifetime.

Let L^* denote network lifetime when all the annuli have the same width. In Figure 13.6, we show the relative increment of L when compared with L^*, that is,

$$\left(\frac{L - L^*}{L^*}\right) \times 100\% \tag{13.71}$$

For $\alpha = 2$, the relative increment is small (less than 0.13%) and not shown in the figure. For $\alpha > 2$, the relative increment is noticeable, except for m close to 1 and m^*.

In Figure 13.7, we show the normalized optimal energy allocation $E_j / (E / N) / E_1$, where $1 \le j \le m$. We set $p = 6$, $a = 0.002$, $c = 5000$, $\mu = 0.03$, $R = 200$, and $\alpha = 2$ with $m = 3$, $\alpha = 3$ with $m = 15$, $\alpha = 4$ with $m = 32$, $\alpha = 5$ with $m = 48$, and $\alpha = 6$ with $m = 63$. Each m is the optimal choice

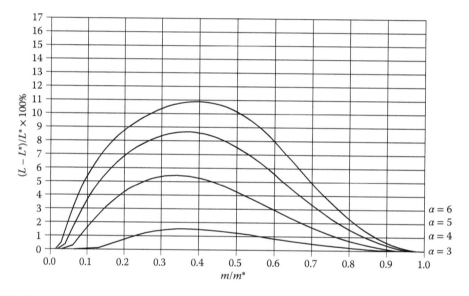

FIGURE 13.6 Relative increment of L versus number of annuli m (uniform distribution).

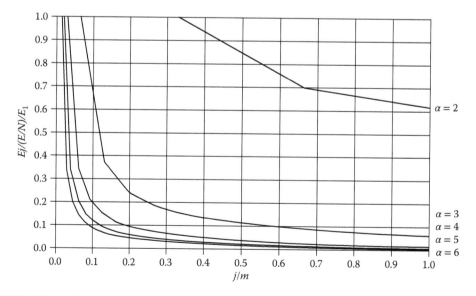

FIGURE 13.7 Optimal energy allocation E_j versus j (uniform distribution).

for the corresponding α. It is observed that an optimal energy allocation is not balanced. In particular, we have $E_1 > E_2 > \cdots > E_m$. Sensors closer to a sink receive significantly more energy than sensors far away from the sink. Such an imbalance increases as α increases.

13.7.2   NONUNIFORM DISTRIBUTIONS

To show a numerical example of optimal network design for a nonuniform distribution of sensors, we consider the following nonuniform sensor distribution function:

$$f(r) = \left(\frac{N}{\pi \ln\,(1+1/u)} \right)\left(\frac{1}{r^2 + uR^2} \right) \tag{13.72}$$

In Figure 13.8, we display the above $f(r)$, where $0 \le r \le R$, assuming that $N = 10,000$, $R = 200$, and $u = 0.125, 0.250, 0.500, 1.000, 2.000, 4.000, 8.000$. It can be seen that as u increases, $f(r)$ approaches the uniform distribution $f(r) = N/(\pi R^2) = 0.0795774$.

In the following, we continue to use the previous nonuniform sensor density function, where $u = 0.5$, that is, the ratio of the largest density to the smallest density is 3. Again, we set $c = 5000$ and $R = 200$. We notice that the value of m^*, that is, the optimal number of annuli when all the annuli have the same width, is identical to that of the uniform distribution [34].

In Table 13.2, we show the optimal network design $(r_1, r_2, ..., r_m)$ for a nonuniform distribution of sensors when $\alpha = 3$ and $m = 15$. We also compare r_j with $j(R/m)$, that is, the value of r_j when all the annuli have the same width. It is observed that $F(r_1, r_2, ..., r_{m-1})$ is minimized when the annuli have different widths. Furthermore, from A_1 to A_m, annuli have decreasing widths, that is, $r_1 - r_0 > r_2 - r_1 > \cdots > r_m - r_{m-1}$.

In Figure 13.9, we display the value of $F(r_1, r_2, ..., r_{m-1})$ (actually,

$$F(r_1, r_2, ..., r_{m-1})/(10,000\ln\,(1+1/u)) \tag{13.73}$$

is shown for comparison with Figure 13.3) for $\alpha = 2, 3, 4, 5, 6$, where $1 \le m \le m^*$. As expected, the behavior of $F(r_1, r_2, ..., r_{m-1})$ is similar to and less than that of the uniform distribution in Figure 13.3.

In Figure 13.10, we show the relative reduction of $F(r_1, r_2, ..., r_{m-1})$ when compared with $F(m)$, that is, $F(r_1, r_2, ..., r_{m-1})$ when all the annuli have the same width. For $\alpha = 2$, the relative reduction is small (less than 0.6%) and not shown in the figure. As expected, the behavior of the relative

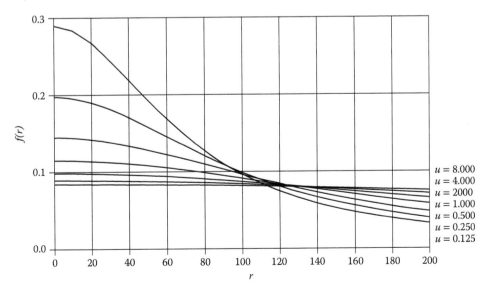

FIGURE 13.8   Nonuniform sensor distribution functions.

TABLE 13.2
Optimal Network Design

j	r_j	$j(R/m)$
0	0.00	0.00
1	13.72	13.33
2	27.39	26.67
3	41.00	40.00
4	54.57	53.33
5	68.09	66.67
6	81.58	80.00
7	95.02	93.33
8	108.43	106.67
9	121.79	120.00
10	135.11	133.33
11	148.38	146.67
12	161.58	160.00
13	174.66	173.33
14	187.56	186.67
15	200.00	200.00

Note: $\alpha = 3$, $m = 15$, nonuniform distribution.

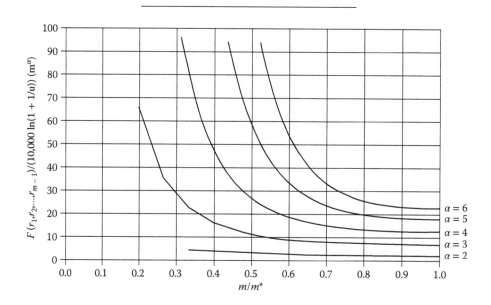

FIGURE 13.9 $F(r_1, r_2, \ldots, r_{m-1})$ versus number of annuli m (nonuniform distribution).

reduction is similar to that of a uniform distribution in Figure 13.4. Furthermore, the relative reduction is greater than that of the uniform distribution in Figure 13.4, which is less obvious.

In Figure 13.11, we demonstrate network lifetime L as a function of m, where $1 \le m \le m^*$, and show the effect of α on L. We set $p = 6$, $a = 0.002$, $c = 5000$, $\mu = 0.03$, $R = 200$, $E/N = 100$ J, and $\alpha = 2,3,4,5,6$. As expected, the behavior of L is similar to and greater than that of the uniform distribution in Figure 13.5.

In Figure 13.12, we show the relative increment of L when compared with L^*, that is, the network lifetime when all the annuli have the same width. For $\alpha = 2$, the relative increment is small (less than 0.1%) and not shown in the figure. As expected, the behavior of the relative increment is similar to that of the uniform distribution in Figure 13.6. Furthermore, the relative increment is greater than

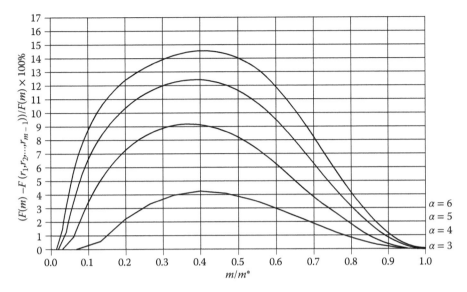

FIGURE 13.10   Relative reduction of $F(r_1, r_2, ..., r_{m-1})$ versus number of annuli m (nonuniform distribution).

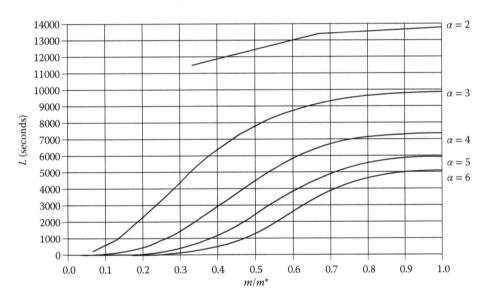

FIGURE 13.11   Network lifetime L versus number of annuli m (nonuniform distribution).

that of the uniform distribution in Figure 13.6, which is less obvious but a direct consequence of Figure 13.10.

In Figure 13.13, we show the normalized optimal energy allocation, that is, $E_j / (E/N) / E_1$, where $1 \leq j \leq m$. We set $p = 6$, $a = 0.002$, $c = 5000$, $\mu = 0.03$, $R = 200$, and $\alpha = 2$ and $m = 3$, $\alpha = 3$ and $m = 15$, $\alpha = 4$ and $m = 32$, $\alpha = 5$ and $m = 48$, and $\alpha = 6$ and $m = 63$. Each m is the optimal choice for the corresponding α. As expected, the optimal energy allocation is similar to but more balanced than that in Figure 13.7.

Finally, in Figure 13.14, we demonstrate network lifetime L as a function of u, where $0 < u \leq 5$, and show the impact of u on L. We set $p = 6$, $a = 0.002$, $c = 5000$, $\mu = 0.03$, $R = 200$, $E/N = 100$, and $\alpha = 2$ with $m = 3$, $\alpha = 3$ with $m = 15$, $\alpha = 4$ with $m = 32$, $\alpha = 5$ with $m = 48$, and $\alpha = 6$ with $m = 63$. We observe that when $u < 1$, the network lifetime can be increased noticeably by using a nonuniform sensor distribution. As u increases, the network lifetime approaches that of a uniform distribution.

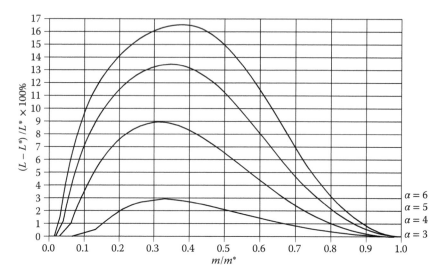

FIGURE 13.12   Relative increment of L versus number of annuli m (nonuniform distribution).

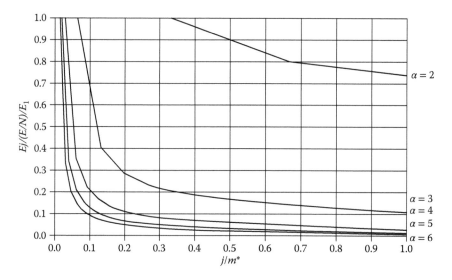

FIGURE 13.13   Optimal energy allocation E_j versus j (nonuniform distribution).

13.8   FUTURE WORK

It has been observed that an optimal energy allocation that results in the maximum network lifetime is not balanced. Sensors close to a sink bear much heavier data transmission loads and consume much more energy than sensors far away from the sink. If such uneven energy depletion and allocation cause problems in real implementation and applications of WSNs, additional effort should be made to produce balanced energy allocation. Fortunately, just as with network lifetime, an optimal energy allocation is determined by a network design. This means that a WSN can be designed in such a way that an optimal energy allocation that yields the maximum network lifetime also satisfies certain balance constraints. For instance, the ratio of the maximum energy allocated to a sensor to the minimum energy allocated to a sensor does not exceed a certain limit. Future research efforts can be directed toward solving the problem of optimal network design with energy balance constraints, namely, to design a WSN such that an optimal energy allocation satisfies a given balance constraint.

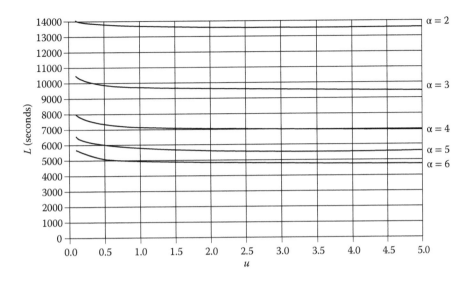

FIGURE 13.14   Network lifetime L versus u (nonuniform distribution, varying α).

13.9   CONCLUSIONS

We have developed an algorithm to find an optimal network design which maximizes the lifetime of a wireless sensor network obtained by optimal initial energy allocation for an arbitrary sensor distribution. Our strategy is to solve a nonlinear system of equations. We have shown that an optimal network design has different widths of annuli. Furthermore, for the same number of annuli, an optimal network design with variable annulus widths can yield a noticeably longer network lifetime than a network design with identical annulus widths.

REFERENCES

1. Available at: www.itu.int/en/ITU-T/gsi/iot/Pages/default.aspx. (Accessed on 16 September 2016).
2. D. Evans. *The Internet of Things: How the Next Evolution of the Internet Is Changing Everything.* White Paper, Cisco, 2011.
3. Available at: www.bloomberg.com/news/articles/2014-01-08/cisco-ceo-pegs-internet-of-things-as-19-trillion-market (Accessed on 16 September 2016).
4. C. Alcaraz, P. Najera, J. Lopez, and R. Roman. Wireless sensor networks and the Internet of things: Do we need a complete integration? *1st International Workshop on the Security of the Internet of Things (SecIoT10),* 2010.
5. D. Christin, A. Reinhardt, P. S. Mogre, and R. Steinmetz. Wireless sensor networks and the Internet of things: Selected challenges. Available at: www.ti5.tu-harburg.de/events/fgsn09/proceedings/fgsn_031.pdf (Accessed on 16 September 2016).
6. L. Mainetti, L. Patrono, and A. Vilei. Evolution of wireless sensor networks towards the Internet of things: A survey. *19th International Conference on Software, Telecommunications and Computer Networks (SoftCOM),* pp. 1–6, 2011.
7. M. Srivastava, D. Culler, and D. Estrin. Special section on sensor network applications. *IEEE Computer,* vol. 37, no. 8, 41–78, 2004.
8. D. E. Culler and W. Hong, eds. Special section on wireless sensor networks. *Communications of the ACM,* vol. 47, no. 6, 2004.
9. G. Anastasi, M. Conti, M. Di Francesco, and A. Passarella. Energy conservation in wireless sensor networks: A survey. *Ad Hoc Networks,* vol. 7, no. 3, 537–568, 2009.
10. Y. Gu, Y. Ji, and B. Zhao. Maximize lifetime of heterogeneous wireless sensor networks with joint coverage and connectivity requirement. *Proceedings of International Conference on Scalable Computing and Communications* and *Eighth International Conference on Embedded Computing,* pp. 226–231, 2009.
11. M. Cardei and D.-Z. Du. Improving wireless sensor network lifetime through power aware organization. *ACM Wireless Networks,* vol. 11, no. 3, 333–340, 2005.

12. M. Cardei, J. Wu, M. Lu, and M. O. Pervaiz. Maximum network lifetime in wireless sensor networks with adjustable sensing ranges. *Proceedings of IEEE International Conference on Wireless and Mobile Computing, Networking and Communications*, 2005.

13. A. Dhawan, C. T. Vu, A. Zelikovsky, Y. Li, and S. K. Prasad. Maximum lifetime of sensor networks with adjustable sensing range. *Proceedings of the 7th ACIS International Conference on Software Engineering, Artificial Intelligence, Networking, and Parallel/Distributed Computing*, pp. 285–289, 2006.

14. S. Slijepcevic and M. Potkonjak. Power efficient organization of wireless sensor networks. *Proceedings of IEEE International Conference on Communications*, vol. 2, 472–476, 2001.

15. J.-H. Chang and L. Tassiulas. Maximum lifetime routing in wireless sensor networks. *IEEE/ACM Transactions on Networking*, vol. 12, no. 4, 609–619, 2004.

16. A. Giridhar and P. R. Kumar. Maximizing the functional lifetime of sensor networks. *Proceedings of the 4th International Symposium on Information Processing in Sensor Networks*, pp. 5–12, 2005.

17. R. Kacimi, R. Dhaou, and A.-L. Beylot. Load balancing techniques for lifetime maximizing in wireless sensor networks. *Ad Hoc Networks*, vol. 11, no. 8, 2172–2186, 2013.

18. K. Kalpakis, K. Dasgupta, and P. Namjoshi. Efficient algorithms for maximum lifetime data gathering and aggregation in wireless sensor networks. *Computer Networks*, vol. 42, no. 6, 697–716, 2003.

19. J. Kim, X. Lin, N. B. Shroff, and P. Sinha. Minimizing delay and maximizing lifetime for wireless sensor networks with anycast. *IEEE/ACM Transactions on Networking*, vol. 18, no. 2, 515–528, 2010.

20. Z. Nutov and M. Segal. Improved approximation algorithms for maximum lifetime problems in wireless networks. *Fifth International Workshop on Algorithmic Aspects of Wireless Sensor Networks* (*Lecture Notes in Computer Science*), vol. 5804, pp. 41–51, 2009.

21. A. Alfieri, A. Bianco, P. Brandimarte, and C. F. Chiasserini. Maximizing system lifetime in wireless sensor networks. *European Journal of Operational Research*, vol. 181, 390–402, 2007.

22. M. Bhardwaj and A. P. Chandrakasan. Bounding the lifetime of sensor networks via optimal role assignments. *Proceedings of the 21st Annual Joint Conference of the IEEE Computer and Communications Societies (INFOCOM)*, vol. 3, pp. 1587–1596, 2002.

23. Q. Zhao and M. Gurusamy. Maximizing network lifetime for connected target coverage in wireless sensor networks. *Proceedings of IEEE International Conference on Wireless and Mobile Computing, Networking and Communications*, pp. 94–101, 2006.

24. Y. Chen and Q. Zhao. On the lifetime of wireless sensor networks. *IEEE Communications Letters*, vol. 9, no. 11, 976–978, 2005.

25. F. Ye, G. Zhong, J. Cheng, S. Lu, and L. Zhang. PEAS: A robust energy conserving protocol for long-lived sensor networks. *Proceedings of the 23rd International Conference on Distributed Computing Systems*, 2003.

26. D. Wang, B. Xie, and D. P. Agrawal. Coverage and lifetime optimization of wireless sensor networks with Gaussian distribution. *IEEE Transactions on Mobile Computing*, vol. 7, no. 12, 1444–1458, 2008.

27. Y. T. Hou, Y. Shi, H. D. Sherali, and S. F. Midkiff. Prolonging sensor network lifetime with energy provisioning and relay node placement. *Proceedings of the 2nd IEEE Communications Society Conference on Sensor and Ad Hoc Communications and Networks*, pp. 295–304, 2005.

28. Y. Yang, M. I. Fonoage, and M. Cardei. Improving network lifetime with mobile wireless sensor networks. *Computer Communications*, vol. 33, no. 4, 409–419, 2010.

29. Z. Hu and B. Li. On the fundamental capacity and lifetime limits of energy-constrained wireless sensor networks. *Proceedings of the 10th IEEE Real-Time and Embedded Technology and Applications Symposium*, pp. 2–9, 2004.

30. J. Lian, K. Naik, and G. B. Agnew. Data capacity improvement of wireless sensor networks using non-uniform sensor distribution. *International Journal of Distributed Sensor Networks*, vol. 2, no. 2, 121–145, 2006.

31. K. Li and J. Li. Optimal energy allocation in heterogeneous wireless sensor networks for lifetime maximization. *Journal of Parallel and Distributed Computing*, vol. 72, no. 7, 902–916, 2012.

32. S. Olariu and I. Stojmenović. Design guidelines for maximizing lifetime and avoiding energy holes in sensor networks with uniform distribution and uniform reporting. *Proceedings of the 25th IEEE International Conference on Computer Communications (INFOCOM)*, 2006.

33. V. Rai and R. N. Mahapatra. Lifetime modeling of a sensor network. *Proceedings of the Conference on Design, Automation and Test in Europe*, vol. 1, pp. 202–203, 2005.

34. K. Li. Optimal number of annuli for maximizing the lifetime of sensor networks. *Journal of Parallel and Distributed Computing*, vol. 74, no. 1, 1719–1729, 2014.

35. M. Bhardwaj, T. Garnett, and A. P. Chandrakasan. Upper bounds on the lifetime of sensor networks. *Proceedings of IEEE International Conference on Communications*, vol. 3, pp. 785–790, 2001.

36. W. B. Heinzelman, A. P. Chandrakasan, and H. Balakrishnan. An application-specific protocol architecture for wireless microsensor networks. *IEEE Transactions on Wireless Communications*, vol. 1, no. 4, 660–670, 2002.

37. R. L. Burden, J. D. Faires, and A. C. Reynolds. *Numerical Analysis*, 2nd edition. Boston: Prindle, Weber & Schmidt, 1981.

14 Bandwidth Allocation Scheme with QoS Provisioning for Heterogeneous Optical and Wireless Networks

Siti H. Mohammad, Nadiatulhuda Zulkifli,
Sevia Mahdaliza Idrus, and Arnidza Ramli

CONTENTS

14.1 INTRODUCTION

Bandwidth scarcity due to the increasing number Internet subscribers and the emergence of future services, such as Internet of Things (IoT)–based applications, are among the key factors in the growth of broadband access networks. This development is particularly intended to construct a network with wider coverage, higher capacity, advanced reliability, and flexibility. In addition, network service providers are aiming for much more cost-effective networks in terms of construction and maintenance. Thus, it is crucial for researchers in the field to develop a network which can benefit both the subscribers and the service providers.

Through the years, broadband access networks have been dominated by optical and wireless access networks. Not long after the introduction of optical fiber communications in the core long-haul network, it was expanded close to the users through fiber-to-the-home (FTTH) technology, which is considered a superior alternative for broadband access networks given its large bandwidth capacity, sufficiently small size, lightness, and invulnerability to electromagnetic interference. FTTH in its early stages has been fabricated in light of the point-to-point Ethernet convention.

The development of point-to-point Ethernet optical networks is not economic that outweighs it benefits. As the outcome, FTTH based on passive optical networks (PONs) has been developed. Among PONs' multiple access techniques, time-division multiplexing (TDM) PON is the current deployment which allows *n* users to share in time the bandwidth offered by a single wavelength (Hernandez et al., 2012; Alvizu et al., 2012). Three standardized versions of TDM PON are Ethernet PON (E-PON), broadband PON (B-PON), and gigabit PON (G-PON), which are mainly distinguished by their available data rates (Shaddad et al., 2014).

Currently, the integration of optical and wireless technologies to create a single access network has been one of the topics among researchers because of advantages in the mobility and flexibility of the network compared to the fixed line. Wireless access networks are ubiquitous, with major advantages in mobility and flexible penetration to end users (Kazovsky et al., 2012; Mohammad et al., 2011). The technology is broadly recognized by its ability to remove the restrictions of time and place over high data transfer capacity and demonstrated dependability of optical systems. There are multiple wireless access network technologies which have been standardized, and the technologies are vastly improving year to year. Among the main wireless access technologies are wireless fidelity (Wi-Fi), Worldwide Interoperability of Microwave Access (WiMAX), Long-Term Evolution (LTE), and wireless 5G.

In this chapter, PON technologies are discussed in Section 14.2. The future of PON is also included in this section. Section 14.3 describes wireless access networks technologies and the future of the network, including 5G and beyond. The related work of heterogenous broadband access networks is discussed in Section 14.4. Bandwidth allocation schemes in heterogeneous PON-wireless networks are discussed in Section 14.5. Section 14.6 concludes this chapter.

14.2 PASSIVE OPTICAL ACCESS NETWORKS

The accelerating demands for high-speed data access have been the main factor in developing and upgrading the broadband access network. For decades, PON has been among the solutions for the deployment of efficient broadband access networks. It is reflected by the standards introduced by the Full-Service Access Network (FSAN) group of ITU-T, which covers PON solutions operating at gigabit rates, especially G-PONs (Leligou et al., 2006). As the network uses only passive elements in the distribution network, PON had its name founded on that function (Lee et al., 2006). In this way, the operating and maintenance costs of the access network can be lessened with the nonexisting active element.

Three basic arrangements of PON are tree, bus, and ring. The preferred arrangement is the tree topology, in which PON can expand its coverage, utilizing less network splitters so that optical power loss can be minimized (Gutierrez et al., 2005). The simple architecture of PON has the optical line terminal (OLT) as the main element at the local exchange, while optical network units (ONUs) are deployed on the subscriber's side as the network interface (Nowak and Murphy, 2005). Optical fiber and an optical distribution node (ODD) connect those elements, as shown in Figure 14.1.

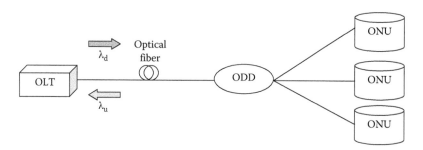

FIGURE 14.1   Basic architecture of PON.

Several multiple-access techniques have been used to share a single PON architecture; among those techniques are TDM, wavelength-division multiplexing (WDM), subcarrier multiple access (SCMA), code-division multiple access (CDMA), and orthogonal frequency-division multiplexing (OFDM). Among these techniques, TDM-PON, WDM-PON, and OFDM-PON are quite popular and mostly used. The comparisons of TDM-PON, WDM-PON, and OFDM-PON multiple-access techniques are summarized in Table 14.1 (Banerjee et al., 2005; Davey et al., 2006; Choi et al., 2007; Grobe and Elbers, 2008; Akanbi et al., 2006; Khairi et al., 2009; and Shaddad et al., 2014). In the following section, the widely used and matured TDM-PON–based technologies are discussed further.

14.2.1 TDM-PON–BASED TECHNOLOGIES

TDM-PON displays simple and cost-effective PON (Choi et al., 2007). The architecture of TDM-PON is shown in Figure 14.2. In the downstream operation, TDM-PON is a point-to-multipoint network in which data is broadcast to all ONUs, while in the upstream transmission, each ONU will have to send the data in the TDM manner. To control the upstream bandwidth transmission between ONUs, a bandwidth allocation algorithm is used in the OLT so that the collision of the data between ONUs can be avoided. Three standardized versions of TDM-PON are B-PON, E-PON, and G-PON. The most used PON variations found today are E-PON and G-PON.

TABLE 14.1
TDM-PON versus WDM-PON versus OFDM-PON

Features	TDM-PON	WDM-PON	OFDM-PON
Signal Transmission	Broadcast and select in devoted time slots	Dedicated wavelength to each user	Broadcast and select (frequency)
	Point to multipoint	Point to point	Point to multipoint
Data Rate	2.5 Gb/s	10 Gb/s per wavelength	100 Gb/s
Splitting Loss	Insertion loss of passive splitter/coupler	Insertion loss of arrayed waveguide grating (AWG)	Insertion loss of passive splitter/coupler
	10 log (N) dB	3–5 dB	10 log (N) dB
Security in Physical Layer	Not guaranteed	Guaranteed	Not guaranteed
Remarks	Low data rate at large number of users, but simple and cost-effective	High data rate and high scalability, but high cost	High data rate and low cost
Implementation	B-PON, E-PON, and G-PON	Coarse WDM-PON (CWDM-PON) and dense WDM-PON (DWDM-PON)	Bandwidth scalable OFDM-PON (BSOFDM-PON)

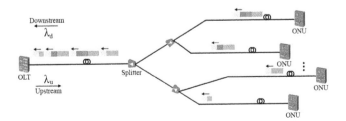

FIGURE 14.2 TDM operation concept in the upstream transmission of PON.

TABLE 14.2

E-PON versus G-PON according to IEEE 802.3ah and ITU-T G.984 Standards

Features	E-PON	G-PON
Standard Body	IEEE 802.3ah	ITU-T G.984
Maximum Speed	1 Gb/s	2.5 Gb/s
Framing	Ethernet	G-PON encapsulation method (GEM)/asynchronous transfer mode (ATM)
Frame Size	64–1518 bytes	5 bytes of GEM header ≤ 1518 bytes of frame fragment
Splitting Ratio	1:16	1:64
Control Unit	Logical-link identifier (LLID)	Transmission container (T-CONT)

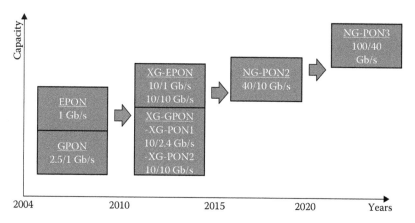

FIGURE 14.3 PON generations. (From H. S. Abbas and M. A. Gregory. [2016]. *Journal of Network and Computer Applications*, 67, pp. 53–74.)

The E-PON and G-PON standards have similar general rules as far as their network systems and applications, yet their operations are distinctive because of the executions of the physical and data-link layers (Olmos et al., 2011). E-PON is approved by the Institute of Electrical and Electronics Engineers (IEEE), standardized as IEEE 802.3, while G-PON is defined by the FSAN group that was certified as ITU-T G.984. E-PON is widely employed in Asia, while G-PON is deployed in North America, Europe, the Middle East, and Australasia (Skubic et al., 2009). The general comparison of E-PON and G-PON is shown in Table 14.2. High-speed TDM-PONs standards, 10G-EPON and XG-PON, were ratified by the IEEE and ITU-T in 2009 and 2010, respectively (IEEE std 802.3av, 2009; ITU-T Recommendation G.987 series, 2010).

14.2.2 FUTURE OF PON

Abbas and Gregory in 2016 stated that the progression of PON technology is characterized into three generations: the first generation (implemented PON), next-generation stage 1 (NG-PON1), and next-generation stage 2 (NG-PON2). The evolution of the PON architectures and their corresponding capacities are shown in Figure 14.3.

According to ITU-T, the next-generation technologies should be of lower cost, easy to maintain, and adapted to local environments (ITU-T, 2016). The inauguration of broadband technologies, community antennas, optical fiber, satellite, and fixed and mobile wireless has realized traditional and

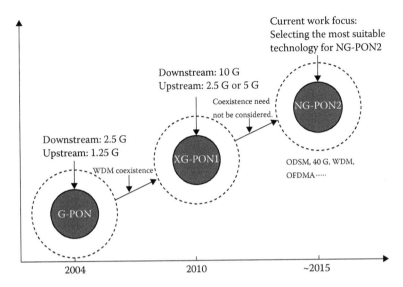

FIGURE 14.4 FSAN roadmap of G-PON. (From A. Srivastava. (2013). Next generation PON evolution. Proceedings In Broadband Access Communication Technologies VII. Vol. 8645, pp 1-15. San Francisco, California, USA: SPIE.)

new methods of telecom transportations all over the world (ITU-T, 2016). As shown in Figure 14.4, FSAN has outlined a roadmap for G-PON technology where current work focuses on selecting the most suitable technology for the NG-PON standard (Srivastava, 2013). The newest member of the ITU-T family of PON standards in force is 10 GB–capable symmetric PON (XGS-PON), which allows bit rates reaching 10 Gb/s (ITU-T G.9807.1). The related work in progress is leading to upgrade the bit rates up to 40 Gb/s for upstream and downstream transmission.

Dissimilar PONs can be hybrid and joint to build high-potential heterogeneous PONs. For example, WDM and TDM can be combined in a heterogeneous network, such as WDM/TDM-PON or TWDM-PON. The hybrid network was proposed to optimize network performance and bandwidth utilization. It benefits from high-speed TDM-PONs and the huge amount of wavelengths of WDM-PONs. The FSAN community chose TWDM-PON as a key resolution to NG-PON2 in its April 2012 meeting (Luo et al., 2013). The aggregate PON rate can be increased by assembling XG-PONs through several pairs of wavelengths. Another recent effort on hybrid PONs involved long-reach optical access (LROA) networks, which mostly integrate TDM-PON with DWDM-PON (Mohamed and Ab-Rahman, 2015).

As the telecommunication technologies are rapidly progressing, ubiquitous broadband access machineries have become accessible, particularly wireless, as it offers performance that is comparable to, or expands upon, wired access networks (ITU-T, 2016). There is an additional functional prerequisite to combining wired and wireless networks to employ a heterogeneous broadband access network. Network service providers also demand dynamic resource sharing and the ability to provide backhaul support to next-generation wireless technologies, such as 3G and LTE (Chen et al., 2013).

14.3 WIRELESS ACCESS NETWORKS

Wireless access networks are ubiquitous, with major advantages in mobility and flexible penetration to end users (Kazovsky et al., 2012; Mohammad et al., 2011). There are multiple wireless access network technologies which have been standardized, and the technologies are vastly improving year to year. Among the main wireless access technologies are Wi-Fi, WiMAX, LTE, and wireless 5G.

14.3.1 WIRELESS ACCESS NETWORK TECHNOLOGIES

Wi-Fi is based on IEEE 802.11 standards and classified into several standards, including IEEE 802.11a, IEEE 802.11b, IEEE 802.11g, and IEEE 802.11n. It is famous because of its low cost, high-speed data rate from 54 to 600 Mb/s according to classified standards, and easy deployment. Two construction approaches of Wi-Fi are infrastructure mode and ad-hoc mode (Sakarindr and Ansari, 2007; Akyildiz et al., 2005). In the infrastructure mode, an AP is required to control and run the network, while the ad-hoc mode permits self-management for all subscribers and does not assign any user as an administrator. Due to the high and wide dispersion of Wi-Fi, it is very useful in cities where a large number of users have personal electronic devices.

WiMAX is built on IEEE 802.16 standards (Shaw et al., 2007). It is intended to address the requirements for wireless metropolitan area networks (WMANs). A WiMAX base station can support a total data rate reaching 75 Mb/s within a 5-km range. Two configuration modes of IEEE 802.16 networks are P2MP and mesh mode. An adaptive modulation with varying levels of the forward error correction is used in the network to boost transmission rate and error performance (Mohammad et al., 2012). There are also different modulation systems using binary phase-shift keying (BPSK), QPSK, 16-, 64-, and 256-QAM, which are defined by the WiMAX OFDM standard. There are two standards that have been developed for WiMAX: 802.16a (fixed applications) and 802.16e (mobile applications) at speeds up to 120 km/h (Shaddad et al., 2014). There is also an NG standard, 802.16m, which occupies both fixed and mobile applications (Yu, 2011).

LTE is a mobile telecommunication technology standardized by 3GPP with striving needs for data rates, capacity, and latencies. There is also an innovative form of LTE, called LTE-Advanced, correspondingly a 4G recognized mobile technology, founded on 3GPP UMTS Release 10 in 2011 (ITU, 2012). LTE is best for bursty data traffic with good VoIP support. It uses OFDMA for downlink transmission and single-carrier FDMA (SC-FDMA) for uplink transmission. SC-FDMA diminishes peak-to-average power ratio (PAPR) by 3 to 5 dB, which improves the coverage or throughputs of cell-edge users. Legacy LTE and LTE-Advanced use licensed IMT-2000 bands at bands such as 700, 900, 1800, 2100, and 2600 MHz, which are usually accessible at favored low-frequency bands, giving it more coverage gain than WiMAX (ITU, 2012). LTE can hold flexibility reaching 350 km/h, similar to that of WiMAX 2.0. LTE is undergoing preparation on its Release 12 and Release 13 to improve coverage, deliver much higher stable data rates, and encounter the anticipated high-traffic demand (Aldmour, 2013).

The term *5th-generation mobile technology (5G)* is not yet legitimately used for any certain specification or in any certified document made public by telecommunication companies or standardization bodies. It is expected to be standardized and deployed in the next 5 to 10 years (Thompson et al., 2014). Li et al. in 2014 said that a 5G network should achieve a 1000-fold capacity increase compared to the current generation of wireless network deployments. The main driver for achieving within the estimated 90- to 160-times range of capacity growth is expected to come from the network side (Li et al., 2014), with key enabling techniques to include heterogeneous networks, cell densification, device-to-device (D2D) communication, and multi-radio access technology (RAT) operation. 5G Nanocore was introduced by Patil et al. in 2012. It is a convergence of nanotechnology, cloud computing, and all IP platforms. Beam-division multiple access (BDMA) is proposed to be used as a radio interface for 5G networks (KAIST IITC, 2008). BDMA splits an antenna beam based on the locations of the mobile stations to permit the mobile stations to provide multiple accesses, thus significantly growing the capacity of the system.

14.3.2 FUTURE OF WIRELESS ACCESS TECHNOLOGY

5G is realized as one of the qualifiers of what is denoted as "networked society" (i.e., a future with user- and machine-centric communications where access to information and sharing of data is ubiquitous) (Fiorani et al., 2015). To further increase the data rates, the 3GPP Release 12 in June 2014 introduced new technologies such as network function virtualization (NFV), software-defined networks (SDNs), and heterogeneous networks (HetNets) (Mitchell, 2014). NFV enables the separation of hardware from software or "function," whereas SDN is an extension of NFV wherein software can perform dynamic reconfiguration of network topology to adjust to load and demand. On the other hand, small-cell deployments are a key feature of the HetNets approach as they allow flexibility as to where they are positioned. Although such technologies can potentially address the greedy 5G capacity requirements and reduced RAN-related capital and operational expenditures (CapEX and OpEX), a new challenge has nonetheless emerged: the 5G backhaul (Jaber et al., 2016).

The backhaul is the network that connects the base stations to the core network. This network could be of copper, fiber, or microwave technology. In fact, the use of satellite technology as a backhaul is expected in 6G in which it enables the global coverage area. This will increase performance, efficiency, and reliability and allow limitless expansion because airwaves cannot be overcrowded (Karki and Garia, 2016). On the other hand, the next-generation wireless network, which is 7G, will be the most advanced compared to its predecessors. 7G systems can be supported by global navigational satellite systems using techniques such as OFDM and FEC for the speed of the communication process (Karki and Garia, 2016).

Traditionally, the copper cable and microwave link are often used as wireless backhaul. However, the cost of deployment for both technologies scales linearly with the bandwidth. Although advanced copper-based technologies (e.g., G.fast) are able to carry signals with rates of a few Gb/s, they can cover only short distances (Fiorani et al., 2015). Thus, this technology is suitable to be used for the network with low-bit options or short-distance applications. Since 3G applications, the deployment of cost-effective solutions has been discussed for backhaul networks to accommodate the higher demand in data rates as well as to prepare for LTE networks. Due to this, optical fiber is favored as the backhaul solution because it offers cost reduction, huge capacity, and high reliability in which most of the existing research have focused in GPON network.

Meanwhile, the wireless backhaul network is an attractive option for locations in which it is difficult to deploy fiber infrastructure. Due to the ability to achieve high transmission capacity over short and medium distances (i.e., up to few km), the advanced wireless technologies (i.e., millimeter wave [mmwave] and free-space optics [FSO]) are deployed in LTE backhauling. In 5G, the deployment of small cells is a viable way to provide high capacity (Chen et al., 2016; Larrabeiti et al., 2014). More details on backhaul technologies for wireless networks can be found in Jaber et al. (2016). However, this approach imposes a financial challenge because of its high cost of deployment and the maintenance of a large number of active devices connected at gigabit speeds scattered over metropolitan areas (Larrabeiti et al., 2014). Again, the optical network receives considerable attention to serve as a backhaul of 5G wireless networks, and several works have been reported for this approach. For example, Chen et al. (2016) have leveraged the TWDM-PON, which offers a cost-minimized design of a mobile backhaul network for 5G systems. Meanwhile, Draxler and Karl (2015) have adopted the WDM-PON in the 5G mobile backhaul system due to its flexibility and capacity.

14.4 HETEROGENEOUS BROADBAND ACCESS NETWORKS

Over the years, as the number of people who rely on the Internet increases daily, the research on heterogeneous networks also increases. From previous research (Kuznetsov et al., 2000; Martinez et al., 2001; Luo et al., 2006, 2007; Shaw et al., 2007; Yeh and Chow, 2011; Fadlullah et al., 2013; Stöhr et al., 2014; De Andradea et al., 2014), the heterogeneous terminology is defined by two meanings:

1. Heterogeneous, in a single network, is used to describe the hybrid/integration/converged technologies, such as multiple traffic formats, in that single-network system.
2. Heterogeneous, in a two-channel network, is used to define the integration/hybrid/converged architectures of the system.

In this work, the term *heterogeneous* is used as the latter, which involves integration between optical networks and wireless networks. A heterogeneous optical and wireless access network is an ideal convergence of an optical backhaul and a wireless front end for a proficient access network. There are certain advantages and features obtainable by the heterogeneous network. It is cost effective, flexible, and robust. Furthermore, the converged network is able to construct "self-healing" if there is any failure in the system or the infrastructure. Significantly, governments are motivated to provide incentives to the operators and contractors to implement heterogeneous network infrastructure in expectation to manipulate the reliability and high capacity of the network.

14.4.1 HETEROGENEOUS OPTICAL AND WIRELESS BROADBAND ACCESS NETWORKS

Sarkar et al. in 2007 projected a multi-domain hybrid network, namely the wireless optical broadband access network (WOBAN) (Sarkar et al., 2007). The architecture of WOBAN is shown in Figure 14.5. The network comprises a wireless network at the front end, and an optical network supports its backhaul. Multiple ONUs are fed by multiple OLTs, which are positioned in a telecom central office (CO). WOBAN is associated by fiber links from the ONU to the CO. The ONUs are wirelessly linked to the subscribers in single-hop or multi-hop fashion. The tree topology is used in the optical part, while a mesh is envisaged in its wireless front end, formed by the wireless gateways and routers. In this multi-domain construction, several wireless routers can be connected with a single gateway, in which the gateways are the key aggregation points. As numerous gateways can connect to a single ONU, the ONUs are the higher aggregation points. Thus, OLT is the highest aggregation point for the network before the traditional metro/core aggregation befalls for the rest of the network. More details on the WOBAN technology are found in Sarkar et al. (2008), Chowdury et al. (2009), and Mukherjee (2008).

Another WOBAN-like architecture was proposed by Shaw et al. in 2007, namely GROW-Net. As shown in Figure 14.6, the backhaul of the architecture is built by a reconfigurable optical network, while its front end consists of a wireless mesh network (WMN). The existence of a WMN in the system enables the network to support the user everywhere in a cost-effective manner. Several wireless

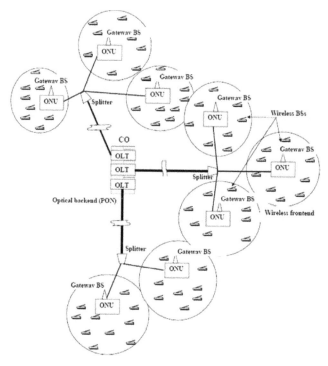

FIGURE 14.5 WOBAN architecture. (From S. Sarkar et al. [2007]. *Journal of Lightwave Technology*, 25, 3329–3340.)

mesh routers are installed to exchange traffic via gateways, which are linked to the web through the PON. Because of its restricted flexibility, lower power requirements, and upgraded computational abilities, the WMN of GROW-Net is infrastructure based. Traffic management in the mesh network is exploited to enhance dependability and load balancing in the WMN of GROW-Net. The authors likewise anticipated that deep review on the operability of wireless innovations is required to enhance the capacity, reliability, upkeep, and portability of WMN in GROW-Net 9 (Shaw et al., 2007).

In 2010, Qiao et al. presented optical wireless integration (OWI) technology, which is an expected phase from fixed-mobile convergence (FMC). In OWI architecture, the wireless front end consists of WiMAX technology to deliver the last-mile broadband, while PON is deployed as the optical backhaul. The capacity of the wireless front end in OWI architecture is enhanced by merging the point-to-multipoint WiMAX with multiple-input, multiple-output (MIMO) technology (see Figure 14.7). Several spatially disseminated component arrays are utilized to exploit the space range of

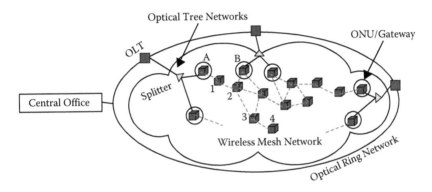

FIGURE 14.6 GROW-Net architecture. (From W.-T. Shaw et al. [2007]. *Journal of Lightwave Technology*, 25, pp. 3443–3451.)

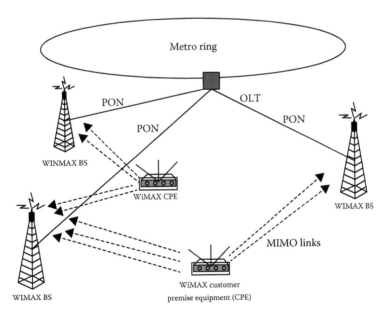

FIGURE 14.7 OWI network architecture. BS, base station. (From C. Qiao et al. [2010]. Integrated optical and wireless access/metro networks. *Proceedings of Conference on Optical Fiber Communications (OFC)/ Collocated National Fiber Optic Engineers Conference*, pp. 1–3.)

wireless channels in MIMO. Furthermore, it exploits multipath spread with no additional spectrum while its connections are used to synchronize antenna resource assignment among WiMAX base stations. If there should arise an occurrence of base station failures, a centralized antenna assignment procedure at the OLT will exploit MIMO diversity to accomplish improved throughput and adaptability (Qiao et al., 2010).

14.4.2 QUALITY OF SERVICES IN HETEROGENEOUS OPTICAL AND WIRELESS BROADBAND ACCESS NETWORKS

Obele and Kang presented a converged GEPON and WiMAX architecture in 2008 (see Figure 14.8). It was the main endeavor at unequivocally accounting for both wired and wireless users in designs for integrating WiMAX and GEPON. In the projected architecture, a single piece of customer premise equipment (CPE) replaces the GEPON ONU and WiMAX BS, termed ONU-BS. As a member of the IEEE family of standards (Kramer and Mukherjee, 2002), GEPON ought to comply with the QoS components characterized in the IEEE 802.1D standard. Seven traffic classes distinguished by the standard are network control, voice, video, controlled load, excellent effort, best effort, and background (Obele and Kang, 2008). The QoS provisioning for GEPON is coarse grained and assigned per user/terminal. On the other hand, WiMAX QoS provisioning is fine grained, connection oriented, and assigned per service flow per subscriber/terminal. According to WiMAX standard IEEE 802.16e, five traffic classes are differentiated: unsolicited grant service (UGS), real-time polling service (rtPS), extended real-time polling service (ertPS), non-real-time polling service (nrtPS), and best effort (BE) (Obele and Kang, 2008). As well as a different number of QoS classes, GEPON and WiMAX also differ in terms of frame arrangements and addressing structures. Additionally, the author proposed the aggregation in the number of service

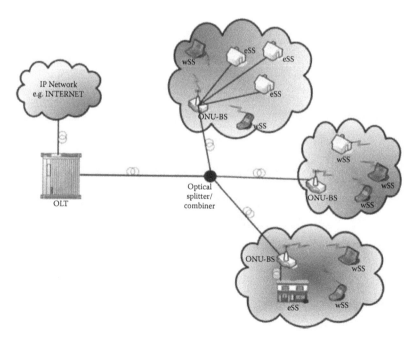

FIGURE 14.8 GEPON-WiMAX architecture. (From B. O. Obele and M. Kang, Fixed mobile convergence: A self-aware QoS architecture for converging WiMAX and GEPON access networks. *The Second International Conference on Next Generation Mobile Application, Services and Technologies [NGMAST]*, pp. 411–418, 2008.)

differentiation queues in the routine GEPON ONU from seven to four to match the WiMAX QoS queues (Obele and Kang, 2008).

Yang et al. (2009) classified three priorities of EPON services, which are BE, assured forwarding (AF), and expedited forwarding (EF), while WiMAX service classes follow IEEE 802.16 standards. The five WiMAX QoS classes are accumulated only into three types: UGS, rtPS (includes ertPS), and BE (includes nrtPS). As the UGS class in WiMAX has similar characteristics to the EF class of E-PON, the UGS service is mapped to the EF queue. The same goes to the rtPS class of WiMAX, which is mapped to E-PON AF, and WiMAX BE is mapped to E-PON BE. The 1-Gb/s bandwidth in both downstream and upstream supported by E-PON is shared by a group of distant ONUs. For example, for a group of 16 ONUs, each ONU holds about 1/16 Gb/s = 65 Mb/s bandwidth on average, matching the total capacity presented by an 802.16 BS, which is around 70 Mb/s over a 20-MHz channel (Yang et al., 2009).

An integrated OFDMA-PON and LTE network was proposed by Lim et al. in 2013. Due to OFDMA-PON's adaptability in establishing virtual transmission pipes of variable bandwidths at the PHY layer, it is a sufficient choice for backhauling high-capacity wireless networks such as LTE (Lim et al., 2013). Each E-UTRAN enhanced NodeB (eNB) is attended by a devoted OFDMA-PON ONU, called an ONU/eNB pair. The QoS model for the evolved packet system (EPS) for LTE standardized by 3GPP is based on the logical concept of an EPS bearer and categorized by detailed QoS parameters (e.g., capacity, delay, and packet loss) (Bhat et al., 2012). An individual QoS class identifier (QCI) is assigned to two types of bearers, defined as guaranteed bit rate (GBR) (QCIs1-4) and non-GBR (QCIs5-9). Following a DiffServ viewpoint, OFDMA-PON combined several flows into virtual queues (e.g., XG-PON T-CONTs). A mapping controller element (MCE) is assigned to conduct the mapping of these queues. The MCE exists either between the eNB and the ONU (in uplink) or between the EPC and the OLT (in downlink) (see Figure 14.9).

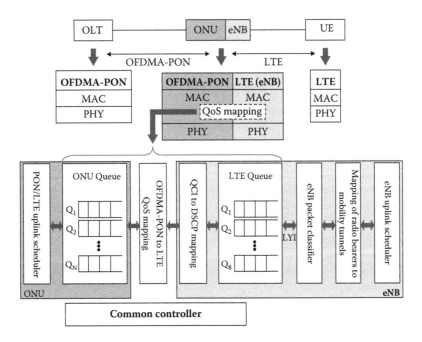

FIGURE 14.9 QoS mapping process conducted by MCE. (From W. Lim et al., Modeling of LTE back-hauling through OFDMA-PONs. In *17th International Conference on Optical Network Design and Modeling [ONDM]*, pp. 240–245, IEEE, 2013.)

14.5   QoS-AWARE BANDWIDTH ALLOCATION SCHEME FOR HETEROGENEOUS OPTICAL AND WIRELESS NETWORKS

A system must have an appropriate incorporation of a DBA algorithm, an intra ONU-BS scheduling algorithm, and a QoS mapping machinery to achieve optimal QoS performance (Ranaweera et al., 2011). In traditional PON structures, DBA is described as the method of delivering numerical multiplexing among ONUs (McGarry et al., 2008). In the case of heterogeneous optical and wireless access networks, the offered statistical multiplexing is also extended to the wireless part of the network (i.e., the subscriber station [SS]). Thus, the main goal of DBA is to share network resources through statistical multiplexing between the subscribers on the same network using either optical or wireless crossing points (Raslan et al., 2015; Sarigiannidis et al., 2015).

The bandwidth sharing algorithms in heterogeneous optical and wireless networks could be classified as centralized, decentralized, or hierarchical based on where in the network the bandwidth allocation decisions are performed. In the centralized bandwidth allocation algorithms, the OLT takes full responsibility of performing bandwidth allocation decisions. In contrast, the bandwidth allocation decisions are applied at ONUs in the decentralized bandwidth allocation algorithms. These techniques require supporting peer-to-peer communication between ONUs so that the ONU bandwidth requests can be received by the neighbouring ONUs. Last, in the hierarchical bandwidth allocation algorithms, the OLT allocates the bandwidth to the ONUs and then the ONUs reallocate the granted bandwidth for its interior queue of different traffic types. There is another classification of the bandwidth allocation algorithms that is either prediction based or nonprediction based on the traffic prediction (Shyu et al., 2012; Ramantas et al., 2014; Mirahmadi and Shami, 2012).

Several bandwidth allocation schemes have been proposed in the research community aiming to smoothly converge both optical and wireless networks. Ou et al. (2008) focused on video-on-demand (VOD) services where just enough bandwidth is provided to VOD clients to upsurge global bandwidth consumption. Figure 14.10 shows the procedure of dynamic bandwidth control for the proposed integrated G-PON and WiMAX networks by (Ou et al., 2008). The method is appropriate for the video clips fixed at CBR. Otherwise, for the VBR videos, steps 3 through 5 in Figure 14.10 will be summoned occasionally to encounter the video profile. Bandwidth utilization is affected by the frequency of the bandwidth-resizing mechanism. Frequent resizing attains higher bandwidth consumption. Nevertheless, frequent resizing could impact framework signalling overhead.

In 2010, Ou et al. also presented slotted DBA (S-DBA) for converged PON (EPON or G-PON) and WiMAX networks, as shown in Figure 14.11. The term *slot* is used as a volume of allocatable bandwidth. In a heterogeneous network, the allocatable bandwidth is separated into a quantity of slots. A server station assigns bandwidth to its user stations in the element of slots. The proposed S-DBA is intended to operate with both OLT and ONU-BS. The users' stations receive their bandwidth from this server station (OLT or ONU-BS) in a preservice-type manner. Ou et al. highlighted that the drive of S-DBA is not to substitute a specific bandwidth provision approach, but it was planned to cooperate with the present DBA approaches to lessen signaling overhead in converged PON and WiMAX networks. The S-DBA scheme is demonstrated to work competently as far as signalling overhead reduction, blocking probability, and uplink channel utilization (Ou et al., 2010).

In 2013, Moradpoor et al. proposed an inter-channel and intra-channel dynamic wavelength-/bandwidth-allocation algorithm (IIDWBA) for converged hybrid WDM/TDM EPON with wireless technologies for next-generation broadband access networks. The evaluation of the IIDWBA algorithm in OPNET Modeler shows improved performance as far as average queuing delays, utilizations, and throughput when it is related with the situation in which it has not been deployed. It is verified over simulation experimentations that the projected algorithm is proficient at accumulating the excess bandwidth from the lightly loaded channels and disseminating it across the severely loaded user stations, which may be dispersed over different channels (Moradpoor et al., 2013).

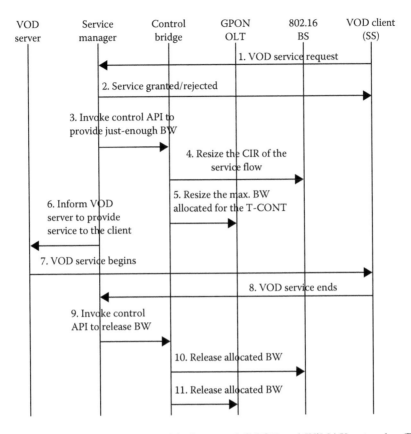

FIGURE 14.10 Dynamic bandwidth control in integrated G-PON and WiMAX networks. (From S. Ou et al., A control bridge to automate the convergence of passive optical networks and IEEE 802.16 [WiMAX] wireless networks. *5th International Conference on Broadband Communications, Networks and Systems [BROADNETS]*, pp. 514–521, 2008.)

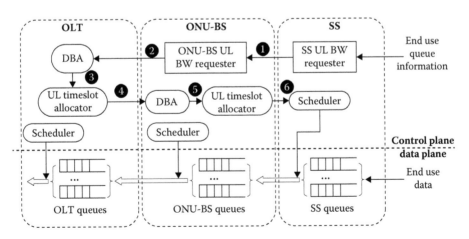

FIGURE 14.11 Dynamic bandwidth allocation in converged PON and WiMAX networks. (From S. Ou et al., *IEEE Systems Journal*, 4, 467–476, 2010.)

In 2014, Radzi et al. proposed two DBA algorithms, named status reporting excessive bandwidth (SREB) and non-status reporting excessive bandwidth (NSREB), for a combined G-PON and Wi-Fi network. In a SREB algorithm, the requested bandwidth is transmitted first to the ONU by wireless routers and then the ONU will collect all the bandwidth from all the wireless routers associated with it. Meanwhile, in NSREB, due to its non-status reporting behavior, the ONU does not request the bandwidth from the OLT. Thus, the OLT has to make an approximation for the requested bandwidth according to the former bandwidth demanded by an individual ONU. The OLT separates the ONUs into underloaded and overloaded before granting them the guaranteed bandwidth. Then, each wireless router receives the allocated bandwidth from the ONU using the carrier-sense multiple access with collision avoidance (CSMA/CA) algorithm. The SREB algorithm is proven to be better than NSREB as far as bandwidth consumption because it contributes bandwidth per request. On the contrary, NSREB outperforms SREB in terms of delay as it eliminates the waiting time for requested bandwidth from the ONUs to allocate the bandwidth (Radzi et al., 2014).

In 2015, Hwang et al. introduced a hierarchical DBA with delay-constraint admission control for G-PON-long-term evolution (G-PON-LTE) converged networks to assure the QoS requirements (Hwang et al., 2015). The first is called limited dynamic bandwidth assignment (LDBA) with a prediction scheme implemented at the OLT to enhance the overall system bandwidth consumption. Another is bandwidth reassignment at the ONU-eNB to reallocate bandwidth assigned from the OLT for respective a user by a service-level agreement (SLA) to accomplish the unified heterogeneous system. The anticipated DBA system can enhance the throughput approximately 3% to 10% and upsurge about 60 to 210 requests for subscribers while the predefined QoS delay is encountered.

14.6 CONCLUSION

This chapter discussed the heterogeneous broadband access network and its potential technological candidates for future applications. Optical and wireless networks are generally described in terms of developed network technologies. The future of PON and wireless access networks is also briefly deliberated. Related works on heterogeneous PON and wireless networks are explained, compelling more on QoS in the heterogeneous network. QoS-aware bandwidth-allocation schemes for heterogeneous PON and wireless networks are also included. In conclusion, DBA is a very significant feature in heterogeneous optical and wireless networks as it is needed to transmit the heterogeneous traffic efficiently in converged networks. DBA plays an important role in ensuring QoS parameters, especially delay, throughput, packet loss, and bandwidth utilization.

REFERENCES

H. S. Abbas and M. A. Gregory. (2016). The next generation of passive optical networks: A review. *Journal of Network and Computer Applications*, Vol. 67, pp. 53–74.

O. Akanbi, J. Yu, and G. K. Chang. (2006). A new scheme for bidirectional WDM-PON using upstream and downstream channels generated by optical carrier suppression and separation technique. *IEEE Photonics Technology Letters*, Vol. 18, no. 2, pp. 340–342.

F. Akyildiz, X. Wang, and W. Wang. (March 2005). Wireless mesh networks: A survey. *Computer Networks*, Vol. 47, no. 4, pp. 445–487.

I. Aldmour. (November 2013). LTE and WiMAX: Comparison and future perspective. *Communications and Network*, Vol. 5, no. 4, pp. 360–368.

R. Alvizu, A. Alfredo, H. Maybemar, H. Monica, and T. M. Idelfonso. (2012). Hybrid WDM-XDM PON architectures for future proof access networks. *International Journal on Advances in Systems and Measurements*, Vol. 5, no. 3 and no. 4, pp. 139–153.

A. Banerjee, Y. Park, F. Clarke, H. Song, S. Yang, G. Kramer, K. Kim, and B. Mukherjee. (2005). Wavelength-division-multiplexed passive optical network (WDM-PON) technologies for broadband access: a review [Invited]. *Journal of Optical Networking*, Vol. 4, no. 11, pp. 737–758.

P. Bhat, S. Nagata, L. Campoy, I. Berberana, T. Derham, G. Liu, X. Shen, P. Zong, and J. Yang. (February 2012). LTE-advanced: An operator perspective. *IEEE Communication Magazines*, Vol. 50, no. 2, pp. 104–114.

H. Chen, Y. Li, S. K. Bose, S. Member, W. Shao, and L. Xiang. (2016). Cost-minimized design for TWDM-PON-based 5G mobile backhaul networks. *Journal of Optical Communications and Networking*, Vol. 8, no. 11, pp. B1–B11.

X. Chen, Z. Zhang, and X. Hu. (2013). The evolution trends of PON and key techniques for NG-PON. *Proceedings in 9th International Conference on Information, Communications and Signal Processing (ICICS) 2013*, Tainan, Taiwan, pp. 1–6. IEEE.

K.-M. Choi, S.-M. Lee, M.-H. Kim, and C.-H. Lee. (May 2007). An efficient evolution method from TDM-PON to next generation PON. *IEEE Photonics Technology Letters*, Vol. 19, no. 9, pp. 647–649.

P. Chowdury, B. Mukherjee, S. Sarkar, G. Kramer, and S. Dixit. (2009). Hybrid wireless-optical broadband access network (WOBAN): Prototype development and research challenges. *IEEE Network*, Vol. 23, no. 3, pp. 41–48.

R. P. Davey, P. Healey, I. Hope, P. Watkinson, D. B. Payne, O. Marmur, J. Ruhmann, and Y. Zuiderveld. (2006). DWDM reach extension of a GPON to 135 km. *Journal of Lightwave Technology*, Vol. 24, no. 1, pp. 29–31.

M. De Andradea, M. Maier, M. P. McGarry, and M. Reisslein. (2014). Passive optical network (PON) supported networking. *Optical Switching and Networking, SI: PON Supported Networking*, Vol. 14, Part 1, pp. 1–10.

M. Draxler and H. Karl. (2015). Dynamic Backhaul Network Configuration in SDN-Based Cloud RANs, *arXiv Prepr.* arXiv1503.03309, 2015.

Z. M. Fadlullah, H. Nishiyama, N. Kato, H. Ujikawa, K. Suzuki, and N. Yoshimoto. (2013). Smart FiWi networks: Challenges and solutions for QoS and green communications. *IEEE Intelligent Systems*, Vol. 28, no. 2, pp. 86–91.

M. Fiorani, B. Skubic, J. Mårtensson, L. Valcarenghi, P. Castoldi, L. Wosinska, and P. Monti. (2015). On the design of 5G transport networks. *Photonic Network Communications*, Vol. 30, no. 3, pp. 403–415.

K. Grobe and J. P. Elbers. (2008). PON in adolescence: from TDMA to WDM-PON. *IEEE Communications Magazine*, Vol. 46, no. 1, pp. 26–34.

D. Gutierrez, K. S. Kim, S. Rotolo, F.-T. An, and L. G. Kazovsky. (2005). FTTH standards, deployments and research issues. *Proceedings of 8th Joint Conference on Information Sciences*, JCIS, Salt Lake City, UT, USA, pp. 1358–1361.

M. Hernandez, A. Alfredo, A. Rodolfo, and H. Monica. (2012). A review of XDMA-WDM-PON for next generation optical access networks. *Proceedings in Global Information Infrastructure and Networking Symposium (GIIS)*, IEEE.Choroni, Venezuela, pp. 1–6.

M. Kuznetsov, N. M. Froberg, S. R. Henion, H. G. Rao, J. Korn, K. A. Rauschenbach, A. H. Modiano, and V. W. S. Chan. (2000). A next-generation optical regional access network. IEEE Communications Magazine, Vol. 38, Issue:

I. S. Hwang, B. J. Hwang, A. F. Pakpahan, and A. Nikoukar. (2015). Adaptive delay constraint admission control with hierarchical DBA over GPON-LTE converged network. In *Proceedings of the International Multi Conference of Engineers and Computer Scientists*, Vol. 2. Hong Kong, China, IAENG.

IEEE Std 802.3av. (2009). Physical layer specifications and management parameters for 10 Gb/s passive optical networks. (*Amendment to IEEE Std 802.3-2008*). Approved 11 September 2009, IEEE-SA Standards Board.

ITU-T. (2016). Next Generation Networks. www.itu.int/en/ITU-D/Technology/Pages/NextGenerationNetworks.aspx. Retrieved on November 8, 2016.

ITU Press Release. (2012). IMT-Advanced Standards Announced for Next-Generation Mobile Technology. www.itu.int/net/pressoffice/press_releases/2012/02.aspx. (Accessed on 6th August, 2017).

ITU-T Recommendation G.987 series. (2010). 10-gigabit-capable passive optical network (XG-PON) systems: Definitions, abbreviations and acronyms. *Digital sections and digital line system: Optical line systems for local and access networks*.

M. Jaber, M. A. Imran, R. Tafazolli, and A. Tukmanov. (2016). 5G backhaul challenges and emerging research directions: A survey. *IEEE Access*, Vol. 4, pp. 1743–1766.

R. S. Karki and V. B. Garia. (2016). Next generations of mobile networks. *Proceedings of International Conference on Advances in Information Technology and Management (ICAIM)*, Cochin, India: McGraw Hill Education, pp. 13–17.

L. Kazovsky, S. Wong, T. Ayhan, K. Albeyoglu, M. Ribeiro, and A. Shastri. (May 2012). Hybrid optical–wireless access networks. *Proceedings of the IEEE*, Vol. 100, no. 5, pp. 1197–1225.

KAIST IITC. (2008). The Korean IT R&D program of MKE/IITA: 2008-F-004-01. 5G mobile communication systems based on beam-division multiple access and relays with group cooperation. KAIST IITC.

G. Kramer and B. Mukhejee. (August/September 2002). Supporting differentiated classes of service in Ethernet passive optical networks. *Journal of Optical Networking*, Vol. 1, no. 8 and no. 9. pp. 280–298

K. Khairi, Z. A. Manaf, D. Adriyanto, M. S. Salleh, Z. Hamzah, and R. Mohamad. (2009). CWDM PON system: next generation PON for access network. Proceedings of 2009 *IEEE 9th Malaysia International Conference on Communications (MICC)*, Kuala Lumpur, Malaysia: IEEE, pp. 765–768.

M. Kuznetsov, N. M. Froberg, S. R. Henion, H. G. Rao, J. Korn, K. A. Rauschenbach, A. H. Modiano, and V. W. S. Chan. (2000). A next-generation optical regional access network. *IEEE Communications Magazine*, Vol. 38, no. 1, pp. 66–72.

D. Larrabeiti, M. Umar, R. Sanchez, and J. A. Hernandez. (2014). Heuristics for PON-based 5G backhaul design. *Proceedings of International Conference on Transparent Optical Networks*, Graz, Austria, IEEE, pp. 5–6.

C.-H. Lee, W. V. Sorin, and B. Y. Kim. (December 2006). Fiber to the home using a PON infrastructure. *Journal of Lightwave Technology*, Vol. 24, no. 12, pp. 4568–4583.

H. C. Leligou, Ch. Linardakis, K. Kanonakis, J. D. Angelopoulos, and Th. Orphanoudakis. (2006). Efficient medium arbitration of FSAN-compliant GPONs. *International Journal of Communication Systems*, Vol. 19, pp. 603–617. doi:10.1002/dac.761.

Q. Li, H. Niu, A. Papathanassiou, and G. Wu. (March 2014). 5G network capacity: Key elements and technologies. *IEEE Vehicular Technology Magazine*, Vol. 9, no. 1, pp. 71–78.

W. Lim, P. Kourtessis, K. Kanonakis, M. Milosavljevic, I. Tomkos, and J. M. Senior. (2013). Modeling of LTE back-hauling through OFDMA-PONs. In 17th *International Conference on Optical Network Design and Modeling (ONDM)*, pp. 240–245. IEEE. Brest, France.

Y. Luo, S. Yin, T. Wang, Y. Suemura, S. Nakamura, N. Ansari, and M. Cvijetic. (2007). QoS-aware scheduling over hybrid optical wireless networks. *Proceedings of Conference on Optical Fiber Communication and the National Fiber Optic Engineers Conference (OFC/NFOEC)*, IEEE, Anaheim, CA, USA. pp. 1–7.

Y. Luo, T. Wang, S. Weinstein, M. Cvijetic, and S. Nakamura. (2006). Integrating optical and wireless services in the access network. *Proceedings of Optical Fiber Communication Conference, 2006 and the 2006 National Fiber Optic Engineers Conference (OFC)*. OSA, Anaheim, CA, USA.

Y. Luo, X. Zhou, F. Effenberger, X. Yan, G. Peng, Y. Qian, and Y. Ma. (February 2013). Time- and wavelength-division multiplexed passive optical network (TWDM-PON) for next-generation PON stage 2 (NG-PON2). *Journal of Lightwave Technology*, Vol. 31, no. 4, pp. 587–593.

A. Martinez, V. Polo, and J. Marti. (2001). Simultaneous baseband and RF optical modulation scheme for feeding wireless and wireline heterogeneous access networks. *IEEE Transactions on Microwave Theory and Techniques,* Vol. 49, no. 10, pp. 2018–2024.

M. P. McGarry, M. Reisslein, and M. Maier. (2008). Ethernet passive optical network architectures and dynamic bandwidth allocation algorithms. *IEEE Communications Surveys & Tutorials*, Vol. 10, no. 3, pp. 46–60.

A. B. Mohammad, R. Q. Shaddad, and S. A. Al-Gailani. (2012). Enabling optical and wireless broadband access technologies. *In Proceedings of 2012 IEEE 3rd International Conference on Photonics*. IEEE, Penang, Malaysia.

M. Mohamed and M. S. B. Ab-Rahman. (2015). Options and challenges in next-generation optical access networks (NG-OANs). *Optik-International Journal for Light and Electron Optics*, Vol. 126, no. 1, pp. 131–138.

S. H. Mohammad, N. Zulkifli, and S. M. Idrus. (October 2011). A review on the network architectures and quality of service algorithms for integrated optical and wireless broadband access networks. *Proceedings of IEEE 2nd International Conference on Photonics (ICP)*, IEEE, Kota Kinabalu, Malaysia, pp. 1–5.

M. Mirahmadi and A. Shami. (2012). Traffic-prediction-assisted dynamic bandwidth assignment for hybrid optical wireless networks. *Computer Networks*, Vol. 56, no. 1, pp. 244–259.

J. E. Mitchell. (2014). Integrated wireless backhaul over optical access networks. *Journal of Lightwave Technology*, Vol. 32, no. 20, pp. 3373–3382.

N. Moradpoor, G. Parr, S. McClean, and B. Scotney. (August 2013). An inter-channel and intra-channel dynamic wavelength/bandwidth allocation algorithm for integrated hybrid PON with wireless technologies for next generation broadband access networks. *Proceedings of AFIN 2013: The Fifth International Conference on Advances in Future Internet*, Barcelona, Spain, IARIA, pp. 9–14.

B. Mukherjee. (2008). Hybrid wireless-optical broadband access networks. *Proceedings of 21st Annual Meeting of the IEEE Lasers and Electro-Optics Society (LEOS)*, Acapulco, Mexico, IEEE, pp. 794–795.

D. Nowak and J. Murphy. (July 2005). FTTH: Overview of the existing technology. *Proceedings of SPIE Opto-Ireland 2005: Optoelectronics, Photonic Devices, and Optical Networks*, Dublin.SPIE

B. O. Obele and M. Kang. (2008). Fixed mobile convergence: A self-aware QoS architecture for converging WiMAX and GEPON access networks. *The Second International Conference on Next Generation Mobile Application, Services and Technologies (NGMAST)*, pp. 411–418.Cardiff, UK. IEEE, pp. 411–418.

J. J. V. Olmos, J. Sugawa, H. Ikeda, and K. Sakamoto. (2011). GPON and 10G-EPON coexisting systems and filtering issues at the OLT. *Proceedings of the 16th Opto-Electronics and Communications Conference (OECC)*, pp. 828–829. Kaohsiung, Taiwan. IEEE, pp. 828–829.

S. Ou, K. Yang, and H. H. Chen. (2010). Integrated dynamic bandwidth allocation in converged passive optical networks and IEEE 802.16 networks. *IEEE Systems Journal*, Vol. 4, no. 4, pp. 467–476.

S. Ou, K. Yang, M. P. Farrera, C. Okonkwo, and K. M. Guild. (2008). A control bridge to automate the convergence of passive optical networks and IEEE 802.16 (WiMAX) wireless networks. *5th International Conference on Broadband Communications, Networks and Systems (BROADNETS)*, London, UK: IEEE, pp. 514–521.

S. Patil, V. Patil, and P. Bhat. (January 2012). A review on 5G technology. *International Journal of Engineering and Innovative Technology (IJEIT)*, Vol. 1, no. 1, pp. 26–30.

C. Qao, J. Wang, and T. Wang. (2010). Integrated optical and wireless access/metro networks. *Proceedings of Conference on Optical Fiber Communications (OFC)/Collocated National Fiber Optic Engineers Conference*, S an Diego, CA, USA: IEEE, pp. 1–3.

N. A. M. Radzi, N. M. Din, N. I. M. Rawi, F. Abdullah, A. Ismail, and M. H. Al-Mansoori. (2014). A new dynamic bandwidth allocation algorithm for fiber wireless network. *Proceedings of 2014 IEEE 2nd International Symposium on Telecommunication Technologies (ISTT)*, pp. 301–304. IEEE. Langkawi, Malaysia.

K. Ramantas, K. Vlachos, A. N. Bikos, G. Ellinas, and A. Hadjiantonis. (2014). New unified PON-RAN access architecture for 4G LTE networks. *Journal of Optical Communications and Networking*, Vol. 6, no. 10, pp. 890–900.

C. Ranaweera, E. Wong, C. Lim, and A. Nirmalathas. (2011). Quality of service assurance in EPON-WiMAX converged network. *International Topical Meeting on Microwave Photonics 2011 & Asia-Pacific, MWP/APMP Microwave Photonics Conference*, Singapore: IEEE, pp. 369–372.

E. I. Raslan, H. S. Hamza, and R. A. El-Khoribi. (2016). Dynamic bandwidth allocation in fiber-wireless (FiWi) networks. *World Academy of Science, Engineering and Technology, International Journal of Computer, Electrical, Automation, Control and Information Engineering*, Vol. 9, no. 12, pp. 2374–2378.

P. Sakarindr and N. Ansari. (October 2007). Security services in group communications over wireless infrastructure, mobile ad hoc, and wireless sensor networks. *IEEE Wireless Communications*, Vol. 14, no. 5, pp. 8–20.

G. Sarigiannidis, M. Iloridou, P. Nicopolitidis, G. Papadimitriou, F.-N. Pavlidou, P. G. Sarigiannidis, M. D. Louta, and V. Vitsas. (2015). Architectures and bandwidth allocation schemes for hybrid wireless-optical networks. *IEEE Communications Surveys & Tutorials*, Vol. 17, no. 1, pp. 427–468.

S. Sarkar, H.-H. Yen, S. Dixit, and B. Mukherjee. (2008). Hybrid wireless-optical broadband access network (WOBAN): Network planning and setup. *IEEE Journal on Selected Areas in Communications*, Vol. 26, no. 6, pp. 12–21.

S. Sarkar, S. Dixit, and B. Mukherjee. (2007). Hybrid wireless-optical broadband access network (WOBAN): A review of relevant challenges. *Journal of Lightwave Technology*, Vol. 25, no. 11, pp. 3329–3340.

R. Q. Shaddad, A. B. Mohammad, S. A. Al-Gailani, A. M. Al-hetar, and M. A. Elmagzoub. (2014). A survey on access technologies for broadband optical and wireless networks. *Journal of Network and Computer Applications*, Vol. 41, pp. 459–472, ELSEVIER.

W.-T. Shaw, S.-W. Wong, N. Cheng, K. Balasubramanian, X. Zhu, M. Maier, and L. G. Kazovsky. (November 2007). Hybrid architecture and integrated routing in a scalable optical–wireless access network. *Journal of Lightwave Technology*, Vol. 25, no. 11, pp. 3443–3451.

Z. D. Shyu, J. Y. Lee, and I. S. Hwang. (2012). A novel hybrid dynamic bandwidth allocation and scheduling scheme for the integrated EPON and WiMAX architecture. *Proceedings of 2012 Fourth International Conference on Ubiquitous and Future Networks (ICUFN)*, IEEE. Phuket, Thailand: IEEE, pp. 330–335.

B. Skubic, J. Chen, J. Ahmed, L. Wosinska, and B. Mukherjee. (2009). A comparison of dynamic bandwidth allocation for EPON, GPON, and next-generation TDM PON. *IEEE Communications Magazine*, Vol. 47, no. 3, pp. S40–S48.

A. Srivastava. (2013). Next generation PON evolution. *Proceedings In Broadband Access Communication Technologies VII*. Vol. 8645, pp 1–15. San Francisco, California, USA: SPIE.

A. Stöhr, J. E. Mitchell, and Y. Leiba. (2014). Transparent wireless access to optical WDM networks using a novel coherent radio-over-fiber (CRoF) approach. *Proceedings of 16th International Conference on Transparent Optical Networks (ICTON)*, pp. 1–2. Graz, Austria. IEEE.

J. Thompson, X. Ge, H.-C. Wu, R. Irmer, H. Jiang, G. Fettweis, and S. Alamouti. (May 2014). 5G wireless communication systems: Prospects and challenges part 2 [Guest Editorial]. *IEEE Communications Magazine*, Vol. 52, no. 5, pp. 24–26.

C. H. Yeh and C. W. Chow. (2011). Heterogeneous radio-over-fiber passive access network architecture to mitigate Rayleigh backscattering interferometric beat noise. *Optics Express*, Vol. 19, no. 7, pp. 5735–5740.

S. Yu. (2011). IEEE Approves IEEE802:16m™: Advanced Mobile Broadband Wireless Standard. http://standards.ieee.org/news/2011/80216m.html. (Retrieved on 6th August, 2017).

K. Yang, S. Ou, K. Guild, and H.-H. Chen. (2009). Convergence of ethernet PON and IEEE 802.16 broadband access network. *IEEE Journal on Selected Areas in Communications*, Vol. 27, no. 2, pp. 101–116.

15 Energy Conservation Techniques for Passive Optical Networks

Rizwan Aslam Butt, Sevia Mahdaliza Idrus,
and Nadiatulhuda Zulkifli

CONTENTS

15.1 INTRODUCTION

The evolution of the Internet has drastically increased the demand for Information Communication Technology (ICT) facilities, which has led to the massive expansion of telecommunication and ICT infrastructure throughout the world. A recent ITU report has mentioned that from the year 2010 to 2015, the percentage of individuals using the Internet have risen by 46% with a 10.8% increase in fixed broadband and 47% increase in mobile services subscriptions [1]. However, this massive expansion of telecommunication networks in response to the continuous traffic growth has increased the electric power generation and power consumption of this sector. This is worrying because the increasing power consumption leads to increased maintenance costs for the network operators. Moreover, most of the electric generation to date is through thermal power plants that require burning of some kind of fuel such as oil or natural gas [2]. This burning of fuel in the emission of greenhouse gases (GHGs) into the atmosphere, which deteriorates the environment.

The consumption of power in the ICT sector was around 20 GW from the year 2010 to 2015, which was equal to approximately 1% of the total global electricity supply of 2 TW in the year 2010 [3]. It has been estimated that the ICT sector consumes 10% of the global power [4]. Other studies have also shown this figure to be up to 37% [5,6]. Resultantly, the carbon footprint of this sector matches that of the airline industry or 25% of the world-wide CO_2 emissions caused by cars [7]. It has been forecasted that by the year 2020, this power consumption will reach to 40 GW and by the year 2025, it is expected to reach to 0.13 TW or approximately 7% of the global electricity supply [3]. These reports indicate that this trend of power consumption is likely to increase further in the future. Specifically, the power consumption of the telecommunication sector is expected to increase further in the coming years [8]. This trend will lead to more electric power generation and thus will further aggravate the emission of GHGs in future [9].

In the ICT sector, the most critical area consuming approximately 70% of the total ICT power consumption is the access network [5]. The main reason for this was the huge network of copper cables causing significant power losses. In recent years, most of the service providers in the world have replaced copper cables with optical fiber–based access networks. These broadband networks are deployed in either fiber to the area (FTTA) or fiber to the building (FTTB) topology. In both topologies, synchronous digital hierarchy/synchronous optical network (SDH/SONET) is used to transport broadband and voice services between optical line terminal (OLT) and optical network units (ONUs). However, this architecture also requires electric power at the termination point in the ONU to convert optical signals to electrical and vice versa. Therefore, this architecture is termed an active optical network (AON). Nevertheless, in an AON from the ONU onwards copper cables are used to transport voice and broadband services to users within a range of 10–500 meters. Therefore, the electrical power consumption of an AON is also high and the maintenance costs of sites containing an ONU is still a headache. However, it is relatively energy efficient compared to older copper-based access networks. Passive optical networks (PONs) present a complete optical fiber–based access network solution that is passive in nature, except for the OLT at the central office (CO) and ONU at the user end, as shown in Figure 15.1. The only drawback of these networks is the broadcast nature of the downstream (DS) link, which results in the ONU being continuously ON and processing incoming frames even if they are not related to it. Therefore, it consumes the least power compared to both copper-based access networks and AONs [10]. However, there is still room for further improvement to make PON a really green communication network. To date the most deployed PON networks are EPON and GPON, which are IEEE and ITU standards, respectively. 10-EPON and XGPON have also been standardized

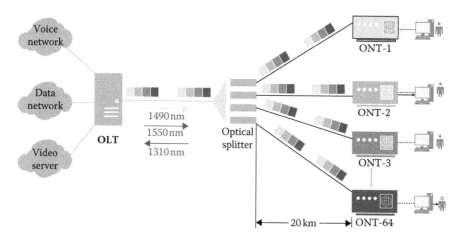

FIGURE 15.1   Showing a typical GPON network with 64 ONUs.

and are expected to be deployed soon. Both the physical layer and MAC layer for 40 Gbps NGPON2 (TWDM PON) have also been standardized [11]. This chapter is focused on energy conservation techniques for ITU PONs such as GPON, XGPON, and TWDM PON.

A PON typically comprises of a central unit at the CO termed the OLT and is linked to 32–128 ONUs through a single optical fiber branched passively through a power splitter. Both upstream (US) and DS communication use different wavelengths of 1310 nm and 1490 nm, respectively. The OLT is connected to the voice, video, and data servers through the backbone network. The DS traffic from the OLT to ONU in a TDMA PON is broadcast and select whereas US communication is TDMA. Due to this broadcast nature, an ONU is required to keep its receiver continuously ON and process the incoming frames. Up to 97% of the frames are processed and discarded as being unrelated, resulting in energy wastage. Due to this reason, 60% of PON power is consumed in the ONUs [12] whereas the OLT consumes only 7% of the total power consumption [13].

In this chapter, we will review the existing techniques to conserve energy of PONs with a focus on ITU PONs (GPON and XGPON). We will discuss all the relevant techniques, describe their merits and demerits, and finally compare all on the basis of energy saving, communication delays, complexity, and cost. We will particularly review ONU sleep mode techniques in detail as being the most economical, simple, and convenient technique to reduce the power consumption of a PON network. Finally, the critical factors impacting sleep mode performance and their optimum values are discussed. In Section 15.2, we broadly discuss the energy conservation techniques for PON. In Section 15.3, we talk about the energy conservation techniques for a PON OLT. Section 15.4 will describe all the energy saving techniques for a PON ONU. In Section 15.5, we present a comparative analysis of all ONU-based techniques and finally, in Section 15.6, we discuss the critical factors that impact the cyclic sleep and watchful sleep mode performance.

15.2 ENERGY CONSERVATION TECHNIQUES FOR PON

Many techniques have been proposed to conserve electrical energy in TDM-based PONs. When talking about TDM systems, the first alternative that comes to mind is a wavelength division multiplexing (WDM)-based PON system [14,15] that uses separate wavelengths for DS and US transmissions. This decoupling of US and DS links obviously eliminates extra processing of unrelated frames at the ONUs and thus, when ONUs are idle, they can be easily switched to low power mode to conserve energy. However, extra cost for ODN and thermoelectric cooling (TEC) required to stabilize the wavelengths diminishes such energy savings. Similarly, Orthogonal Frequency division Multiplexing (OFDMA)-based PON [16] appears to be another very attractive option to replace TDM PONs as it uses multiple orthogonal sub-carriers on a single fiber to provide higher bandwidth and separate communication links between ONUs and the OLT. The problem with this technology is its complexity and additional power requirements for advanced digital processing and DSP operations such as FFT and IFFT [17]. Similarly, Code Division Multiple Access (CDMA)-based PON has also been explored by the research community [18]. However, it has to major drawbacks. First, it suffers from the problem of inter-channel interference and second, it requires the components to be able to handle signal rate much higher than the line rate, which leads to very costly transceivers. Therefore, TDMA PONs are the most economical and commercially deployed optical fiber–based broadband networks to date. Power reduction techniques for a TDMA PON can be classified as OLT based or ONU based. Most of the research focuses on ONU power reduction as it contributes more to overall network power consumption. There is also little room to save energy consumption at the OLT as it is mostly busy in managing all the ONUs from a single PON port. In the following sections, we will discuss each of these issues separately.

15.2.1 OLT Power Saving Techniques

As the OLT is the main administrator that controls and manages communication among all the connected ONUs in the PON network, energy conservation at the OLT requires some special arrangements so that network communication is not disturbed. To the best of our knowledge, three techniques have been proposed in the literature to make an OLT energy efficient. The first technique proposed in reference [19] requires an additional OLT to work in a master-slave configuration using a switch box comprising of optical switches. This optical box monitors the traffic load and selects the OLT with lesser traffic as the master OLT and lets the other OLT sleep to save energy. In this scheme, when the traffic load of OLT-1 is below 10% and the OLT-2 is less than 40%, then OLT-2 can take over the load of OLT-1 and let it go to deep sleep to save energy. OLT-2 continues to serve until it has a traffic load of more than 90% in which case it brings OLT-1 to normal mode to take its load back. This technique requires an additional OLT, firmware changes in the OLT MAC, and an additional hardware setup to enable the master-slave configuration.

The second technique proposed in reference [20] is based on the idea of OLT PON line card sharing by multiple PON networks using an optical switch. Instead of powering of all the OLT line cards serving a group ONUs, only the minimum "m" line cards required to serve the "n" active ONUs should be turned on. However, this technique requires a centralized MAC layer controlling all OLT line cards, which could be computationally expensive. A similar idea using a wavelength router at the OLT has been described in references [21,22]. Only the minimum "m" ports are turned on and ONUs are assigned a wavelength corresponding to that port. Other ports are kept in sleep mode to conserve energy. This technique is suitable for TWDM PON and cannot be used in TDM PONs with a single non-tunable transceiver at the ONUs.

The third technique proposed in reference [23] presents the idea of an "elastic OLT" to save optical power by optimizing the transmission distance and split ratio according to the service requirements. The basic idea is to fully utilize the underutilized power budget that would be otherwise wasted at the receiving ONU. This technique can be useful at the time of a PON network deployment and cannot be applied to an already deployed functional network.

15.2.2 ONU-BASED ENERGY CONSERVATION TECHNIQUES

In a PON network as shown in Figure 15.2, from a single PON port "**r**" ONUs are served, where "**r**" is the split ratio. Typical values of "**r**" for ITU PONs are 32/64/128. Therefore, an energy conservation technique applied to an ONU saving "**m**" watts will actually result in an overall power saving of "**r x m**" watts. To understand the power consumption behavior of an ONU, an ONU power model was developed from the studies in [24–26] as shown in Figure 15.3. An ONU comprises of many optical and electrical function blocks with different power consumption. Based on this model, a GPON ONU consumes 10.6 Watts whereas an XGPON ONU consumes 14.5 Watts. The ONU power consumption can be categorized into four major sections as follows:

1. User section (P_{USER})
2. DC–DC power converter (P_{DC})
3. TDMA section (P_{TDMA})
4. Back end digital circuit (P_{BEDC})

The P_{USER} represents the power consumption of the user modules and their respective interfaces. Normally, this section comprises the ethernet and subscriber line interface card (SLIC) modules. The SLIC module provides the basic voice services such as ringing tones and on-hook and off-hook detections. The DC–DC power consumption, represented by P_{DC}, is consumed by the electronic circuit required to convert and regulate the input DC power to the desired voltage levels. The P_{TDMA} represents the power consumption of all the TDMA sections, which includes optical transceivers, local amplifiers (LAs), transimpedance amplifiers (TIAs), clock and data recovery (CDR) circuits, and serializer/de-serializer (SerDesr) blocks. Finally, the P_{BEDC} represents the MAC layer processing of the incoming frames and buffering in memory banks.

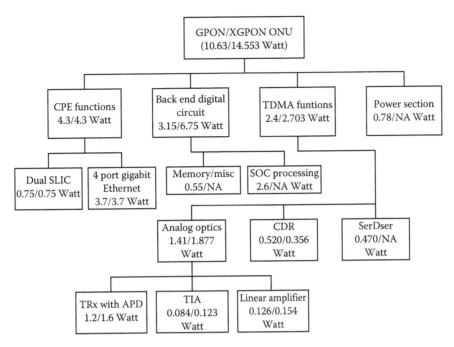

FIGURE 15.2 Average power consumption of different sections of a typical GPON/XGPON ONU.

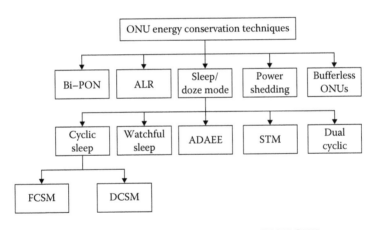

FIGURE 15.3 Power consumption of different sections of a typical GPON ONU.

The P_{DC} remains almost constant when ONU is powered on and can only be fully saved if ONU is turned off. Therefore, all the ONU-based energy conservation techniques try to reduce either or all of P_{USER}, P_{TDMA}, and P_{BEDC}. The study in reference [27] proposes to eliminate buffering at the ONUs to reduce P_{BEDC}. However, this leads to an increase of around 800 microseconds in communication delays and requires buffering at the user terminals/end nodes. The dynamic power component of the P_{TDMA}, P_{USER}, and P_{BEDC} depends on the line rate of the incoming signal. Therefore, the power consumption of these sections can be reduced by operating at lower line rates as in ALR and Bi-PON techniques [28]. The most commonly used technique is the ONU sleep mode. In this technique, an ONU completely turns off the user interfaces and partially or completely turns off the optical transceivers to save P_{TDMA} and P_{USER} when there is no traffic. This technique is termed ONU Doze/Sleep mode. Figure 4.2 lists all the ONU-based energy conservation techniques for an ITU PON. In the following sections, each technique is described in detail.

15.3   ONU SLEEP MODE TECHNIQUES

Sleep mode is the most commonly used energy conservation technique in all wired and wireless networks. In PON, up until now, it is the only technique that has so far been adopted by the PON standardizing bodies [29,30] or studied widely by the research communities. The basic idea is to switch off some components such as the optical transmitter, photo detector (PD), limiting amplifier (LA), CDR, DEMUX, and CPE functions when there is no available uplink or downlink traffic to/from that ONU to save P_{USER} and P_{TDMA} completely and P_{BEDC} partially as ONU turns off optical transceiver and UNI interfaces but the MAC layer processor is working. The idea of sleep mode for a PON was first proposed in reference [31] for a GPON network. In this work, the sleep mode mechanism is negotiated between ONUs and OLT, based on traffic queue length with the help of PLOAM messages. In this work, only two ONU power states—active and sleep—were considered. A similar proposal for EPON has also been presented in reference [32]. Many variations of ONU sleep modes have been proposed in the literature but most of these techniques have only been tested in an IEEE PON-based network environment. Only a few studies have studied sleep mode for ITU PONs. In this section, we will discuss all the sleep mode techniques that have been proposed for ITU PONs or those which are compatible with ITU PONs. In general, when using sleep mode, there has to be a trade-off between the network latency and the power consumption of the network [33].

15.3.1   ONU ARCHITECTURES

Before discussing sleep modes, first we discuss the ONU architecture as it plays very important role for an efficient sleep mode. A normal ONU receiver uses an avalanche photo diode (APD/PIN) to receive an optical signal that after passing thorough TIA and LA amplifier stages goes to the CDR unit. In this architecture, continuous mode CDR (CM-CDR) is used to recover 8 KHz clock from the received serialized GPON/XGPON frames and a SerDesr to output the frame bits in nibbles or bytes to the MAC layer processing module. This architecture consumes 3.85 Watts when active and only 750 mWatts when in sleep mode, but it requires 2–5 ms for clock recovery ($T_{Recovery}$) and achieving synchronization (T_{SYNC}) with the OLT [34]. The study in reference [34] proposes two improved ONU architectures to support sleep mode better. Option-1 uses a burst mode CDR (BM-CDR) unit and a local oscillator to keep track of the OLT clock when asleep. It keeps BM-CDR on during sleep and consumes 1.08 Watts in sleep and requires only 125 us for T_{SYNC} and 1–10 ns for $T_{Recovery}$. To avoid the cost of BM-CDR, option-2 suggests the use of CM-CDR and turns off the SerDesr module and instead uses a sleep control block to utilize the recovered clock to time the sleep mode with a counter module. This option consumes 1.28 Watts with the same recovery times. Therefore, option-2 is the most suitable and efficient ONU architecture to support sleep modes with very fast recovery times, high energy savings, and lower hardware cost.

At the transmitter side, ONUs normally use a DFB laser-based optical transmitter. In reference [35] it has been demonstrated that using VCSEL instead of a DFB laser at the ONU can reduce transmitter power from 0.7 Watts to 0.134 Watts. A saving of 12.8% in active mode and 14% in doze mode has been demonstrated in reference [36] using VCSEL instead of a DFB laser transmitter. A detailed discussion on ONU hardware design and its limitations is available in reference [3].

15.3.2   ADAPTIVE DELAY AWARE ENERGY EFFICIENT SLEEP (ADAEE)

A novel sleep mode mechanism is presented in reference [23]. The core of this mechanism is a sleep interval measurement and control (SIMC) unit at OLT. The SIMC unit computes the minimum (T_{min}) and maximum (T_{max}) sleep periods for an ONU based on average traffic arrival rate and target delays provided by the service provider. The sleep periods are calculated by the OLT and exponentially increased after each completed sleep cycle starting from T_{min}. A customized sleep process with asleep (AS), transmission mode (TM), active mode (AM), and doze mode (DM) is presented. In TM mode, the ONU can only transmit and in DM mode it can only listen. In AS mode, a mode controller logic block inside the ONU decides to put the ONU in light sleep mode (LSM) or deep sleep mode (DSM). It selects a LSM if the sleep interval (T_j) is less than a threshold ($T_{\text{Threshold}}$) value computed as in Equation 15.1; otherwise it selects DSM. During LSM, the option-2 ONU architecture described in Section 5.1 is assumed whereas in DSM, ONU is completely turned off. However, this technique does not consider the US traffic at an ONU and assumes that OLT is allocating the desired upstream bandwidths perfectly. The sleep start and end decision are made only at the OLT, which can result in increase of US communication delays and frame loss due to the bursty nature of traffic.

$$T_{\text{Threshold}} = T_j \quad \text{if } \frac{E_{T_j}^{\text{LSM}}}{E_{T_j}^{\text{DSM}}} = 1 \tag{15.1}$$

where $E_{T_j}^{\text{LSM}}$ and $E_{T_j}^{\text{DSM}}$ are the LSM and DSM mode energies during sleep interval T_j.

This technique does not consider the dynamic bandwidth assignment (DBA) mechanism for US transmissions and also does not follow any standard PON MAC layer protocol. Therefore, the results of this work need further validation in a more practical network environment.

15.3.3 Cyclic Sleep/Doze Mode

Cyclic sleep/doze modes are standardized energy conservation techniques for ITU PONs as defined in references [37,38]. If only the transmitter is turned off this half sleep state is termed as dozing. Four kind of states and power levels are associated with an ONU observing cyclic sleep termed active held (AH), active free (AF), sleep aware (SA), and asleep (AS) with corresponding OLT states; awake forced (AFD), awake free (AFR), low power sleep (LPS), and alerted sleep (ALS). If P_A, P_H, P_W, and P_S represent the ONU power level in AH, AF, SA and AS states, then $P_A = P_H > P_W > P_S$ as shown in Figure 15.4. ONU is fully active in AH and AF states but is allowed to exercise sleep mode only in the AF state. The SA state is like an observation state in which the ONU prepares to switch to/recover from the AS state. The choice of dwell time in the SA state (T_{SA}) and AS state (T_S) and DBA service interval (SI) time has a profound effect on the performance of the cyclic sleep operation, discussed in Section 15.12.

Figure 15.5 shows the OLT and ONU state machines for cyclic sleep. Various timers and events used to control these transitions are defined as follows;

- **Local Wake Up Indication (LWI):** This is a wake-up event to bring an ONU back to active state. The OLT calls this event on the arrival of DS frames for an ONU whereas an ONU calls this event on US traffic.
- **Local Sleep/Doze Indication (LSI/LDI):** This is an event that indicates permission for an ONU to avail sleep/doze mode.

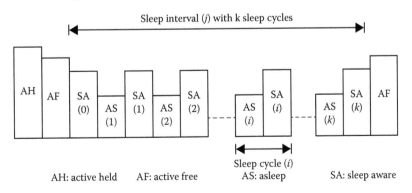

FIGURE 15.4 Cyclic sleep state transitions.

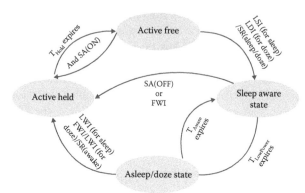

FIGURE 15.5 ONU state transition diagram for cyclic sleep/doze mode operation.

- **SA_OFF/ON:** This is a sleep termination/resumption call from the OLT for an ONU. The ONU maintains **SA (ON/OFF)** variables accordingly.
- T_{LowPower}, T_{SA} and T_{Hold}: ONU timers for monitoring sleep/doze, SA, and AF states.
- T_{Alerted} / T_{Eri} **Timers:** The time OLT waits before setting FWI = 1 for an ONU to wake up after sending an awake message/expected time for an ONU to send US frame before declaring a handshake violation.
- **/SR (Sleep/Awake):** Sleep start and sleep end messages exchanged between the OLT and ONUs to coordinate while exercising cyclic sleep mode.

Assuming that the DS and US traffic for an ONU follows the Poisson arrivals, the average power consumption and delay during a sleep interval can be computed using Equations 15.2 and 15.3 as worked out in [39]:

$$P_{\text{cyclic}} = \sum_{n=0}^{\infty} (P_{SA}.T_{SA} + n(P_{SA}.T_{SA} + P_S.T_S)) \bullet P_{FA}(X = n) / (T_{SA} + n(T_{Aware} + T_S)) \tag{15.2}$$

$$D_{\text{cyclic}} = \sum_{n=0}^{\infty} P_{FA}(X = n). \, T_S / 2 \tag{15.3}$$

where $P_{FA}(X = n)$ is the probability of frame arrival during the nth sleep cycle. If total observation time is T after LSI = 1, then during this time, the ONU will be in one of the AF, SA or AS states. Therefore, we can write $T = T_{AF} + T_{SA} + T_S$. Based on this we can say that in Equation 15.2, P_{cyclic} will be lower if $T_S > T_{SA} > T_{AF}$ and thus energy savings will be higher. However, the impact of awake time T_{Awake} from AS to SA has been disregarded in this equation. The work in reference [40] shows that as the ratio of T_{Awake} to T_S increases from 0 to 0.6, P_{cyclic} increases by 60%. The work in reference [41] also shows similar results for long reach PONs. This study also shows that increasing the reach also impacts the energy efficiency if the delay constraint is smaller like 5 ms, but if the delay constraint is larger like 100 ms then the energy efficiency is not affected.

15.3.4 Sleep Transmit Mode (STM)

During the cyclic sleep, the ONU has to buffer the frames arriving during the sleep period, which requires additional storage and increases communication delay. To avoid this, during the cyclic sleep process, an ONU is allowed to quit the AS state immediately if there is an US or DS frame arrival. AN LWI event at the ONU signals for immediate sleep cycle termination. This obviously

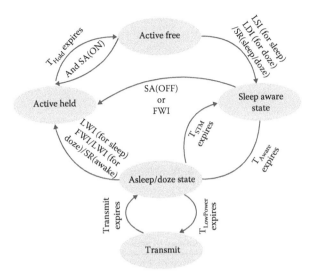

FIGURE 15.6   ONU state transition diagram for sleep transmit mode operation.

reduces energy savings due to shorter sleep cycles. Therefore, the authors of reference [42] propose a modified cyclic sleep, termed sleep transmit mode (STM); the process adds an extra transmit state during the sleep cycle to check US traffic arrivals during the sleep cycle and thus prolong AS state time. The addition of this extra transmit state for US transmission leads to less frequent interruption of sleep cycles due to US traffic arrivals thus enabling longer sleep intervals. An increased saving of 11% and 29% in STM compared to cyclic sleep were demonstrated with this approach with T_{SA} times of 2 ms and 0.1 ms. However, a T_{SA} time of 0.1 ms is not practically possible as it is even less than the duration of a single GPON/XGPON frame duration of 125 us, discussed further in Section 15.9. Figure 15.6 shows the state transition process for STM. It introduces two new timers, $T_{Transmit}$ and T_{STM}, to accommodate the additional transmit state.

15.3.5   DUAL CYCLIC SLEEP (DCS)

Another approach to decouple US and DS traffic handling to minimize sleep cycle interruptions has been presented in reference [43]. In this approach, an optical receiver is turned on every T_S period like in cyclic sleep. However, the optical transmitter is turned on only after "**m**" cycles resulting in more power savings compared to cyclic sleep and doze modes. However, the US delay slightly increases due to lesser ON frequency of the optical transmitter. This technique has been practically tested with an XGPON-based ETRI testbed and thus it is fully compatible with ITU PONs. However, this technique has only been tested at a traffic load of 1 Mbps with a frame size of 64 bytes and traffic arrivals were assumed to follow the Poisson distribution. Therefore, with a bursty traffic arrival and longer packet sizes of 1500 bytes, US delays might be even higher and result in upstream packet loss.

15.3.6   WATCHFUL SLEEP/UNIFIED SLEEP MODE

Cyclic sleep definitely offers more energy savings compared to doze mode but at a cost of higher US and DS delays; US are especially more affected. However, a careful integration of both techniques can result in high energy savings with moderate delays. If the traffic intensity is low, cyclic sleep can be used, otherwise doze mode can be selected. A unifying approach [44] demonstrated 67% energy savings. In this study, the switching criteria was based on traffic queue level (Q) compared to high (Q_{High}) and low (Q_{Low}) levels. LSI is set if the current queue size (Q) is less than Q_{Low} and LDI is set

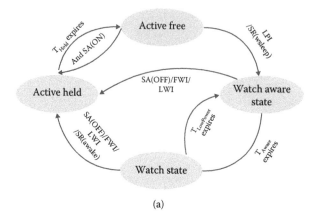

(a)

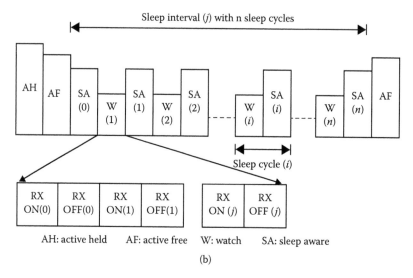

AH: active held   AF: active free   W: watch   SA: sleep aware

(b)

FIGURE 15.7 Watchful sleep mode: (a) state transition diagram; (b) power levels.

if $Q_{High} < Q > Q_{Low}$. But this study only assumes a DS traffic queue and does not discuss the switching policy if a US traffic queue is also considered. A better integration method has been described in reference [31] in which an ONU switches to doze mode after the sleep cycle instead of going to AF state. The ONU quits sleep mode if LWI is raised on the OLT asserts the FWI. Recently, ITU has also adopted this idea and included this mechanism as "watchful sleep" mode in the GPON and XGPON standards. Figure 15.7a and b shows the state transitions and power levels of watchful sleep mode states at the ONU.

During the Watch (W) state, the optical receiver is periodically turned ON/OFF. From the work in [37], the average power consumption ($P_{watchful}$) in a sleep interval can be given as

$$P_{watchful} = \sum_{n=0}^{\infty}(P_{SA}.T_{SA} + n(P_{SA}.T_{SA} + P_W.T_W)) . P_{FA}(X=n) / (T_{SA} + n(T_{SA} + T_W)) \quad (15.4)$$

and the average state transition delay in watchful sleep ($D_{watchful}$) can be computed as in Equation 15.5,

$$D_{watchful} = \sum_{n=0}^{\infty} P_{FA}(X=n) . T_{OFF}/2 \quad (15.5)$$

where P_w and T_w are the average power consumption and the total ONU time spent during the W state. Further, in terms of ON/OFF periods we can express P_w and T_w as in Equations 15.6 and 15.7:

$$P_{wt} = \left(P_{ON} + P_{OFF}\right) / 2 \qquad (15.6)$$

$$T_{wt} = \sum_{i=0}^{j}\left(P_{ON}\left(i\right) + P_{OFF}\left(i\right)\right) \qquad (15.7)$$

It has been shown in reference [38] that watchful sleep mode is more energy efficient than the cyclic and doze modes with slightly higher values of DS delay. The biggest advantage of watchful sleep is that its process can be converted to doze and cyclic sleep modes by defining $T_{watch} = T_{listen}$ and $T_{listen} = 0$ and by only configuring $T_{watch} = T_{sleep}$, respectively.

15.3.7   POWER SHEDDING

In this mode, an ONU turns off all the non-essential functions and user network interfaces (UNIs) to conserve power but keeps the optical link operational and fully synchronized. Only the basic voice service is provided during this mode and broadband services are not available to the user as shown in Figure 15.8. This mode is typically used during AC power failures to prolong the life of backup batteries [13,45]. During normal operation, this technique can be used to turn off non-essential function and user or network interfaces [26] as follows:

1. ONU monitors the status of all the UNIs based on reliable indicators such as physical link activity (e.g. loss of signal, loss of carrier) independent of the service layer.
2. UNI related core functions are switched off in case the UNI is not in use.
3. GEM/XGEM ports (excluding OMCI port) related to non-functional UNIs can be disconnected, stopping the flow into the ONU core functions and reducing traffic processing cycles. The mapping information of the disconnected ports to respective UNIs is preserved and can be activated again on UNI request.

The work in reference [45] shows that a combination of power shedding and sleep mode for a GPON ONU can result in 80% energy savings at lower traffic intensity. However, when using power

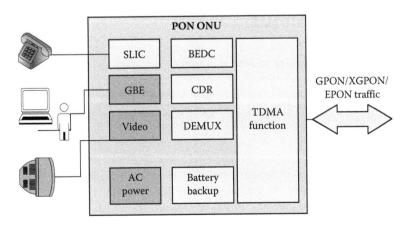

FIGURE 15.8   Power consumption of different sections of a typical GPON ONU.

shedding, all the traffic frames arriving at the ONU UNI are lost. Therefore, it is more suitable during long periods of ONU inactivity or when there is an electric power failure.

15.4 BI-PON

To save processing power due to unrelated frame processing in DS broadcast and reduce higher dynamic power consumption of electronic components at the ONU in TDM PONs, a new idea of bit interleaved frame structure was proposed in reference [21] termed bit interleaved PON (Bi-PON). The new frame structure has been tested with an FPGA-based testbed at 1 Gbps line rate showing 30% processing power savings versus the XGPON frame structure. The basic idea is the early decimation of bits from the DS frame received at line rate just after the CDR section, so that further frame processing functions such as de-scrambling, de-framing, and error decoding are performed at user rates instead of line rate. This leads to reduced power consumption in accordance to Equation 15.8 due to lower activity (A) and frequency (F). Furthermore, it has been pointed out in Bi-PON studies [46] that up to 97–99% of frames received by an ONU are unrelated leading to continuous wasting of MAC layer processing power in the ONUs. Therefore, significant processing energy is wasted in processing and finally discarding these unrelated frames.

$$P_{\text{Dynamic}} = \text{ACV}^2\text{F} \qquad (15.8)$$

To achieve this objective, the Bi-PON frame structure organizes all the ONU bits in a bit interleaved pattern in three sections: Sync, Header, and Payload. In the Sync and Header sections, the lane assignment is static and the offset of each lane from the first bit is equal to the ONU-ID. The lane assignment in the Payload section is dynamic and the lane offset and decimation rate is provided in the header section. Each ONU can tune to its own bits and avoid reading unrelated bits thus saving processing power and reducing unrelated processing activity. The detailed design and working of the Bi-PON decimator has been described in reference [47].

A detailed description of the Bi-PON scheme and its frame structure is explained in reference [46] with an experimental demonstration using FPGA and ASIC showing up to 35% processing energy savings. The negative side of Bi-PON is its disruptive approach, as it requires a decimator instead of DEMUX. The decimator function requires strict synchronization to only tune to related bits. Moreover, Bi-PON can only work for a fixed number of ONUs. In the current reported study, a fixed numbers of 256 ONUs was assumed. All the reported Bi-PON studies are on an FPGA-based laboratory testbed comprised of only a single OLT and ONU. Therefore, there is still a need for testing this concept in a network environment with multiple ONUs.

15.5 ADAPTIVE LINK RATE (ALR)

Similar to Bi-PON, the ALR technique tries to reduce dynamic power component of P_{TDMA} and P_{BEDC} by reducing the Frequency (F) factor in Equation 15.1. It achieves energy savings of the network by reducing the line rate when traffic is lower than a pre-decided threshold. At a lower frequency, the dynamic power of electronic components is lower because of reduced capacitive effects. The power consumption of optical functional blocks also reduces due to lower dispersion at lower line rates. Figure 15.9 shows the concept of ALR in a PON network. The ONU has both 1G and 10G receivers whereas the OLT has 1 Gbps and 10 Gbps transmitters. The idea is to have a dual rate adaptive communication between the OLT and ONU. The ALR function at the MAC layer in the OLT decides to choose between 1 Gbps and 10 Gbps line rates and informs the ONU via a downstream message. In case of EPON, this message will be through a GATE message frame whereas in case of ITU PONs this will be though a PLOAM message. The ALR function at the ONU at its MAC layer selects between the 1 Gbps receiver or 10 Gbps receiver based on OLT instructions.

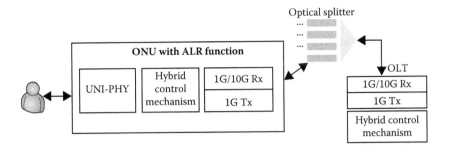

FIGURE 15.9 A PON network supporting the ALR function [50].

The switching from one line rate to another is a three-step process. First, the OLT ALR function needs to decide when to switch. This switching is performed on a pre-defined bandwidth threshold B_{Th}. The second step is ONU synchronization with OLT, requiring a synchronization time T_{SYNC}. The third step is the physical link rate switching at the hardware level. All these three steps require a transition time T_T. Selection of T_T also requires use of a suitable guard band, keeping in view the laser turn ON/OFF times. A large value of T_T may lead to buffer overflows and subsequent packet loss and increase of mean delays.

The ALR technique was first presented for an ethernet-based network in reference [48]. This study demonstrates an energy savings of up to 80%. Since a 10 G-EPON and the 1 G-EPON system can easily coexist by having separate wavelengths for both line rates as 1577 nm and 1490 nm, Kubo et al. [49,50] have successfully demonstrated an ALR technique using 10 G-EPON and 1 G-EPON transceivers in a single OLT in addition to sleep mode. It has been shown in reference [51] that the ALR technique is more effective than sleep mode when the power consumption in the 1 Gbps state is less than twice that of the sleep states of 10 Gbps ONUs.

However, up to now, no demonstration has been made for ITU PONs and all the studies have only chosen IEEE PONs (EPON and 10 G-EPON) in their analysis probably due to the incompatibility of higher and lower rate ITU PONs i.e., GPON and XGPON. GPON can operate at 1.2 Gbps as well as 2.5 Gbps downstream rates cannot be integrated with XGPON that operates at 10 Gbps downstream rate due to quite different MAC layer structure. For example, a GPON frame can carry only a single PLOAM message in a frame whereas XGPON can carry multiple PLOAM messages. Moreover, GPON can operate at variable data rates whereas XGPON downstream is fixed to 10 Gbps. Therefore, for XGPON additional changes at the MAC layer will be required to support slower data rate operation as the standard only specifies a fixed 10 Gbps downstream line rate [52]. To apply an ALR technique to an ITU PON XGPON, the MAC layer of the XGPON needs modifications to support lower line rates.

15.6 MOST SUITABLE ENERGY CONSERVATION TECHNIQUE FOR ITU PONS

Based on the comparative studies especially in references [28,53], a comparison of all the above discussed techniques has been made in Table 15.1 on the basis of energy savings, US/DS delays, new investment requirements, and computational complexity. It is clear that cyclic sleep and watchful sleep modes seem to be the best choices based on our comparison criteria. Although watchful sleep mode has been shown to be superior in energy conservation with slightly more US delays in references [37] and [38], in all of these analysis fixed bandwidth allocation has been assumed. Therefore, further research is needed to explore the merits and demerits of both these approaches and to determine the most suitable energy conservation technique for ITU PONs.

TABLE 15.1

Table Showing a Comparative Analysis of Energy Conservation Techniques for ITU PONs

Method	Energy Savings	US Delay	DS Delay	Requires New Investment	Changes Required	Complexity Level
ALR	High	Slight increase	Slight increase	Yes	Dual-rate transceivers	High
ALR with Sleep	Higher	High	High	Yes	Dual-rate transceivers	High
Bi-PON	High	Remains same	Remains same	Yes	New MAC layer	High
Bi-PON with Cyclic Sleep	Highest	High	High	Yes	New MAC layer	Highest
Doze Mode	Low	High	Low	No	No	Low
Cyclic Sleep	High	High	High	No	No	Low
Watchful Sleep	Higher	Higher	High	No	No	High

15.7 FACTORS IMPACTING CYCLIC/WATCHFUL SLEEP MODE PERFORMANCE

The performance of cyclic sleep depends upon four parameters of the DBA algorithm, T_S, *LWI*, *FWI* and service interval (*SI*), which are discussed separately in the following subsections.

15.7.1 Sleep Time (T_S)/Watch Time (T_W)

There are two approaches in selecting the sleep periods. The first approach is fixed cycle sleep mode (FCSM) in which sleep cycle length is fixed in each cycle unless interrupted by LWI or FWI events. Most of the sleep studies use FCSM. In this approach, a longer T_S provides more energy savings but may result in more buffering of incoming traffic packets and may result in frame loss and increase of communication delays. Sleep time of 20–50 ms has been shown to be a suitable choice in according to reference [54] whereas in reference [39] 50 ms has been shown to be the most optimistic value to maximize ONU energy savings and minimize delays. Dynamic cyclic sleep mode (DCSM) increases T_S according to some functions as in reference [29] or based on traffic types as in reference [55]. A dynamic sleep period calculation method based on the OLT queue length using a feedback controller has been described in reference [56]. However, all DCSM approaches increase computation load of the OLT and require reconfiguration of all sleep mode timers at the OLT and ONU every time, before starting a sleep cycle. This also increases messaging overhead between the OLT and ONUs. The study in [57] has shown FCSM to be more suitable than DCSM and showing up to 50% energy savings with a traffic load up to 31 Mbps. Therefore, FCSM is better than DCSM technique and can provide significant energy savings while maintaining QOS provided LWI indicators and T_S are set carefully.

15.7.2 Local Wake Up/Forced Wake Up Indication (LWI/FWI)

An LWI event at an OLT or ONU acts as a red signal of a traffic light to stop sleep mode. The OLT informs the ONU about its LWI event by first sending an awake message. However, if the ONU does not acknowledge within a period of $T_{alerted}$, it asserts FWI = 1 in every DS frame in the BWmap field of the GPON/XGPON frame. Typically in most of the studies, LWI is asserted by the OLT or ONU if there is even a single packet like in reference [24]. This surely reduces energy savings due to frequent interruption of sleep mode. Another approach in reference [38] proposes to hold LWI at ONU for a fixed time period of 40 ms and LWI at the OLT for a fixed time period of 50 ms after a traffic frame arrival during the sleep cycle with $T_S = 10ms$. This approach enhances energy savings but leads to higher queuing delays at higher traffic load.

15.7.3   Sleep Aware Time (T_{SA}) and Service Interval (SI)

In the cyclic sleep, the ONU temporarily comes out of the sleep mode for a time T_{SA} for three purposes:

1. To check if LWI at the ONU/OLT is set or FWI =1. If any one is high, then it quits sleep mode and switches to AF state.
2. To receive bandwidth allocation for PLOAM messages and send pending PLOAM messages to the OLT at the granted US time slots.
3. To receive US bandwidth allocation for sending its queue report for type 1 (T1) to type 4 (T4) traffic classes.
4. To receive US bandwidth allocation for US transmissions for all T1 to T4 traffic classes.

Therefore, T_{SA} should be long enough to let an ONU finish all the above tasks. SI plays a key role here for the selection of T_{SA} as in the worst case an ONU might have to wait for a time period equal to SI to receive US bandwidth allocation for sending PLOAM messages and queue reports for T2, T3, and T4 traffic classes to OLT. Therefore, T_{SA} > SI is a necessary condition to keep US communications intact. Sleep mode studies mostly choose values between 1 and 5 ms. A very short T_{SA} of 0.25 ms has been proposed in reference [58] but the analysis does not consider DBA algorithm. Typical values of SI used in DBA algorithms for ITU PONs are 5 or 10 [59,60]. Therefore, T_{SA} should be at least 1.3 ms and thus a choice of 2 ms is a suitable choice.

15.8   RECOMMENDATIONS

For the existing PON network, the sleep mode technique is the most suitable technique if its control parameters are carefully configured. Most of the present research on sleep modes does not consider all the traffic classes and the DBA algorithm. An ONU awakened from the AS state requires bandwidth for sending pending messages US to the OLT and then needs bandwidth allocation for all of the traffic classes to send buffer occupancy reports. Present research work assumes it to be perfectly working during sleep mode analysis. However, this coordination between cyclic sleep and the DBA algorithm needs further study. Although cyclic sleep has been widely studied, there is still a need to develop a cyclic sleep framework to guide on the selection/configuration of optimum values of T_S, LWI, and T_{SA} and all ONU and OLT sleep mode timers. The Bi-PON technique appears to be another attractive energy conservation technique for future PONs. It requires further validation of results in a network environment and at higher line rates. A combination of the sleep mode technique with Bi-PON could be the most energy-efficient solution in future PONs.

15.9   CONCLUSION

This chapter reviewed the energy conservation techniques compatible with ITU PONs. Most of the PON network energy is wasted at the ONU and comparatively little at the OLT. Energy conservation at the OLT requires special arrangements as all ONUs are connected to it. Therefore, it is more convenient and beneficial to reduce the energy consumption of a PON ONU. The energy consumption of an ONU is due to four main power components: P_{DC}, P_{USER}, P_{TDMA}, and P_{BEDC}. P_{DC} cannot be saved until ONU is completely powered off. However, P_{USER}, P_{TDMA}, and P_{BEDC} can be partially or fully saved. The Bi-PON study proposes a disruptive approach that changes the existing byte-oriented frame structure of ITU PONs to a bit interleaved pattern so that every ONU only tunes to its own bits. This results in reduction of 97–99% unrelated frame processing by the ONUs and thus reduces P_{BEDC}. Moreover, this technique also reduces the dynamic power component of P_{TDMA} by early decimation of data and OAM bits from the incoming frame at the line rate. The ALR technique tries to reduce the overall dynamic power consumption of all ONU modules by switching between a 1 Gbps and 10 Gbps line rates. Both OLT and ONU have a ALR control function in the MAC layer that controls the switching between higher and lower line rates based on traffic threshold B_{Th}. However, this technique requires additional optical transmitters at the OLT and receivers at the ONU. Therefore, it is costly and not easy to adapt especially in ITU PONs. ONU sleep mode is the most widely used technique because it is easier to adapt as it only requires changes in MAC layer. Doze mode is a half sleep in which only the optical transmitter of ONU is turned off and the optical receiver remains on. Many sleep mode techniques have been proposed in literature but cyclic sleep has been adapted by the PON standardizing bodies. Generally, sleep mode causes substantial increase in US and DS delays. US communication is more affected by sleep mode as the US bandwidth is shared between ONUs using a DBA algorithm. Doze mode is therefore more useful but it reduces energy savings.A combination of sleep and doze mode is more useful in balancing the energy savings and US delays. Several unifying techniques have been proposed in literature and now finally ITU has standardized this technique as watchful sleep mode. In one research study, it has been shown that watchful sleep provides higher energy savings compared to cyclic sleep but another study shows slightly lesser savings compared to cyclic sleep. The increase in communication delays with sleep mode can be controlled by optimizing the three main sleep control parameters: T_S, LWI, and SI.

REFERENCES

1. ICT Data and Statistics Division, ITU-T Facts and Figures, The world in 2015, Geneva (2015). http://www.iamsterdam.com/en/media-centre/facts-and-figures.
2. Enerdata, Global energy trends, Grenoble (2016). http://www.enerdata.net/enerdatauk/press-and-publication/publications/peak-energy-demand-co2-emissions-2016-world-energy.php.
3. R.S. Tucker, Green optical communications—Part II : Energy limitations in networks, *IEEE J. Sel. Top. Quantum Electron.* 17 (2011) 261–274.
4. R.M. Morais, C. Pavan, A.N. Pinto, Estimating the energy consumption in survivable optical transport networks, in: *EUROCON - International Conference on Computer as a Tool* , Lisbon, Portugal (2011): pp. 1–4. doi:10.1109/EUROCON.2011.5929224.
5. J. Malmodin, Å. Moberg, D. Lundén, G. Finnveden, N. Lövehagen, Greenhouse gas emissions and operational electricity use in the ICT and entertainment & media sectors, *J. Ind. Ecol.* 14 (2010) 770–790. doi:10.1111/j.1530-9290.2010.00278.x.
6. K. Wang, A. Gavler, M. Du, M. Kihl, Power consumption analysis of energy-aware FTTH networks, in: *10th International Conference on Digital Telecommunications*, IARIA, Barcelona, Spain (2015): pp. 1–6.
7. M.P. Mills, The cloud begins with coal: Big data, big networks, big infrastructure, and big power (2013). http://www.tech-pundit.com/wp-content/uploads/2013/07/Cloud_Begins_With_Coal.pdf?c761ac.
8. C. Lange, D. Kosiankowski, C. Gerlach, F. Westphal, A. Gladisch, Energy consumption of telecommunication Networks, in: *European Conference on Optical Communication*, Vienna, Austria (2009): pp. 2–3.
9. U.S Energy Information Administration, International energy outlook (2016). http://www.eia.gov/forecasts/ieo/world.cfm.
10. R.S. Tucker, J. Baliga, R. Ayre, K. Hinton, W.V. Sorin, Energy consumption in IP networks, *J. Light. Technol.* 27 (2009) 2391–2403. doi:10.1109/ECOC.2008.4729102.
11. 40-Gigabit-capable passive optical networks (NG-PON2): Transmission convergence layer specification, ITU-T, ITU Stand. G.989.3. (2015).
12. D. Suvakovic, H. Chow, N.P. Anthapadmanabhan, D.T. Van Veen, A.J. Van Wijngaarden, S. Member, T. Ayhan, S. Member, C. Van Praet, G. Torfs, X. Yin, P. Vetter, A low-energy rate-adaptive bit-interleaved passive optical network, *IEEE J. Sel. AREAS Commun.* 32 (2014) 1552–1565.
13. J. Kani, Power saving techniques and mechanisms for optical access network systems, *J. Light. Technol.* 31 (2012) 1–1. doi:10.1109/JLT.2012.2222347.

14. M. Roppelt, F. Pohl, K. Grobe, M. Eiselt, J.-P. Elbers, Tuning methods for uncooled low-cost tunable lasers in WDM-PON, in: *2011 Optical Fiber Communication Conference and Exposition and the National Fiber Optic Engineers Conference*, Los Angeles, CA (2011): pp. 1–3. doi:10.1364/NFOEC.2011. NTuB1.

15. M.J. Wale, Options and trends for PON tunable optical transceivers, in: *37th European Conference and Exhibition on Optical Communication*, Geneva, Switzerland (2011) pp. 1–3. doi:10.1364/ECOC.2011. Mo.2.C.1.

16. N. Cvijetic, D. Qian, J. Hu, 100 Gb/s optical access based on optical orthogonal frequency-division multiplexing, *IEEE Commun. Mag.* 48 (2010) 70–77. doi:10.1109/MCOM.2010.5496880.

17. K. Kanonakis, I. Tomkos, H.G. Krimmel, F. Schaich, C. Lange, E. Weis, J. Leuthold, M. Winter, S. Romero, P. Kourtessis, M. Milosavljevic, I.N. Cano, O. Prat, An OFDMA-based optical access network architecture exhibiting ultra-high capacity and wireline-wireless convergence, *IEEE Commun. Mag.* 50 (2012) 71–78. doi:10.1109/MCOM.2012.6257530.

18. J. Han, Y. Wang, W. Li, R. Hu, Z. He, M. Luo, Q. Yang, S. Yu, Demonstration of a spectrum efficient 8 × 1.25 Gb/s electrical code-division multiplexing passive optical network based on wavelet packet transform coding, *Opt. Eng.* 54 (2015) 56104. doi:10.1117/1.OE.54.5.056104.

19. T.A. Ozgur Can Turna, M. Ali Aydin, A dynamic energy efficient optical line terminal design for optical access network, *Comput. Networks.* 522 (2015) 260–269. doi:10.1016/j.comnet.2008.06.002.

20. J. Zhang, T. Wang, N. Ansari, Designing energy-efficient optical line terminal for TDM passive optical networks, in: *34th IEEE Sarnoff Symposium* (SARNOFF), Princeton, NJ (2011). doi:10.1109/SARNOF.2011.5876436.

21. J.I. Kani, S. Shimazu, N. Yoshimoto, H. Hadama, Energy-efficient optical access networks: Issues and technologies, *IEEE Commun. Mag.* 51 (2013) 22–26. doi:10.1109/MCOM.2013.6461185.

22. J. Kani, Enabling technologies for future scalable and flexible WDM-PON and WDM/TDM-PON Systems," *IEEE J. Select. Topics Quant. Electron.*, 16 (2010), no. 5, pp. 1290–1297.

23. H.H. Noriko Iiyama, H. Kimura, A novel WDM-based optical access network with high energy effeciency using elastic OLT, in: *Proceedings of the 14th Conference Optical Network Design and Modeling (ONDM)*, Kyoto, Japan (2010): pp. 1–6.

24. B. Skubic, D. Hood, Evaluation of ONU power saving modes for gigabit-capable passive optical networks, *IEEE Netw.* 25 (2011) 20–24. doi:10.1109/MNET.2011.5730524.

25. A. Dixit, B. Lannoo, G. Das, D. Colle, M. Pickavet, P. Demeester, Dynamic bandwidth allocation with SLA awareness for QoS in ethernet passive optical networks, *J. Opt. Commun. Netw.* 5 (2013) 240. doi:10.1364/JOCN.5.000240.

26. A. Dixit, B. Lannoo, D. Colle, M. Pickavet, P. Demeester, ONU power saving modes in next generation optical access networks: progress, efficiency and challenges, *Opt. Express.* 20 (2012) B52–B63. doi:10.1364/OE.20.000B52.

27. G.C. Sankaran, K.M. Sivalingam, ONU buffer elimination for power savings in passive optical networks, in: *IEEE International Conference on Communications*, Kyoto, Japan (2011). doi:10.1109/icc.2011.5962483.

28. P. Vetter, D. Suvakovic, H. Chow, P. Anthapadmanabhan, K. Kanonakis, Energy-efficiency improvements for optical access, *IEEE Commun. Mag.* 52 (2014) 136–144.

29. D.R.D. Ren, H.L.H. Li, Y.J.Y. Ji, Power saving mechanism and performance analysis for 10 Gigabit-class passive optical network systems, in: *2nd IEEE International Conference Network Infrastructure and Digital Content*, Beijing, China (2010): pp. 920–924. doi:10.1109/ICNIDC.2010.5657932.

30. P. Sarigiannidis, A. Gkaliouris, V. Kakali, On forecasting the ONU sleep period in XG-PON systems using exponential smoothing techniques, in: *Global Communications Conference (GLOBECOM)*, Austin, TX (2014): pp. 2580–2585. doi: 10.1109/GLOCOM.2014.7037196

31. D.A. Khotimsky, D. Zhang, L. Yuan, R.O.C. Hirafuji, S. Member, D.R. Campelo, Unifying sleep and doze modes for energy-efficient PON systems, *IEEE Commun. Lett.* 18 (2014) 688–691.

32. K. Kumozaki, R. Kubo, J. Kani, Y. Fujimoto, N. Yoshimoto, Proposal and performance analysis of a power-saving mechanism for 10 Gigabit class passive optical network systems, in: *Proceeding of the Conference on Networks and Optical Communications (NOC)*, Valladolid, Spain (2009): pp. 87–94.

33. L. Shi, B. Mukherjee, S.S. Lee, Energy-efficient PON with sleep-mode ONU: Progress, challenges, and solutions, *IEEE Netw.* 26 (2012) 36–41. doi:10.1109/MNET.2012.6172273.

34. S.W. Wong, L. Valcarenghi, S.H. Yen, D.R. Campelo, S. Yamashita, L. Kazovsky, Sleep mode for energy saving PONs: Advantages and drawbacks, in: *IEEE GLOBECOM Work shops*, Honolulu, HI (2009). doi:10.1109/GLOCOMW.2009.5360736.

35. E. Wong, Energy efficient passive optical networks with low power VCSELs, in: *21st Annual Wireless and Optical Communications Conference*, IEEE, Kaohsiung, Taiwan (2012): pp. 48–50.

36. E. Wong, M. Mueller, C.A. Chan, M.P.I. Dias, M.C. Amann, Low-power laser transmitters for green access networks, in: *Photonics Global Conference*, IEEE, Singapore (2012): pp. 6–8.

37. N. Zulkifli, R.A. Butt, S.M. Idrus, Comparative analysis of cyclic and watchful sleep modes for GPON, in: *IEEE 6th International Conference Photonics*, IEEE, Kutching (2016): pp. 6–8.

38. R.O.C. Hirafuji, K.B. Cunha, D.R. Campelo, A.R. Dhaini, D.A. Khotimsky, The watchful sleep mode : A new standard for energy efficiency in future access networks, *IEEE Commun. Mag.* 58 (2015) 150–157.

39. H. Bang, J. Kim, S.S. Lee, C.S. Park, Determination of sleep period for cyclic sleep mode in XG-PON power management, *IEEE Commun. Lett.* 16 (n.d.) 98–100. doi:10.1109/LCOMM.2011.111011.111322.

40. B. Skubic, A. Lindström, E. In De Betou, I. Pappa, Energy saving potential of cyclic sleep in optical access systems, in: *IEEE Online Conference on Green Communications,* IEEE, Piscataway, NJ (2011): pp. 124–127.

41. M. Chincolii, L. Valcarenghe, J. Cheni, P. Montii, Investigating the energy savings of cyclic sleep with service guarantees in long reach PONs, in: *Asia Communications and Photonics Conference*, Guangzhou, China (2012): pp. 3–5.

42. J.LI, N.P. Anthapadmanabhan, C.A. Chan, K. Lee, N. Dinh, P. Vetter, Sleep mode mechanism with improved upstream performance for passive optical networks, in: *IEEE Selected Areas in Communications Symposium*, IEEE, Sydney, NSW (2014): pp. 3871–3876. doi:10.1109/ICC.2014.6883925.

43. G. Kim, S. Kim, D. Lee, H. Yoo, H. Lim, Dual cyclic power saving technique for XG-PON, *Opt. Express.* 22 (2014) A1310. doi:10.1364/OE.22.0A1310.

44. D.N.V. Fernando, M. Milosavljevic, P. Kourtessis, J.M. Senior, Cooperative cyclic sleep and doze mode selection for NG-PONs, in: *16th International Conference on Transparent Optical Networks*, Graz, Austria (2014) 1–4. doi:10.1109/ICTON.2014.6876392.

45. E. Trojer, E. Eriksson, Power Saving Modes for GPON and VDSL2, in: *13th European Conference on Networks Optical Communication. 3rd Conference Optical Cabling Infrastructure,* Austria, 2008.

46. D. Suvakovic, H. Chow, D. Van Veen, J. Galaro, B. Farah, N.P. Anthapadmanabhan, P. Vetter, A. Dupas, R. Boislaigue, Low energy bit-interleaving downstream protocol for passive optical networks, in: *IEEE Online Conference on Green Communications (GreenCom)*, Piscataway, NJ (2012): pp. 26–31. doi:10.1109/GreenCom.2012.6519611.

47. C. Van Praet, G. Torfs, Z. Li, X. Yin, D. Suvakovic, H. Chow, X.-Z. Qiu, P. Vetter, 10 Gbit/s bit interleaving CDR for low-power PON, *Electron. Lett.* 48 (2012) 1361. doi:10.1049/el.2012.3200.

48. C. Gunaratne, S. Member, K. Christensen, S. Member, B. Nordman, S. Suen, Reducing the energy consumption of ethernet with adaptive link rate (ALR), *IEEE Trans. Comput.* 57 (2008) 448–461. doi:10.1109/MCOM.2010.5621967.

49. R. Kubo, J.I. Kani, Y. Fujimoto, N. Yoshimoto, K. Kumozaki, Adaptive power saving mechanism for 10 gigabit class PON systems, *IEICE Trans. Commun.* E93–B (2010) 280–288. doi:10.1587/transcom. E93.B.280.

50. R. Kubo, J.I. Kani, Y. Fujimoto, N. Yoshimoto, K. Kumozaki, Sleep and adaptive link rate control for power saving in 10G-EPON systems, in: *IEEE Global Telecommunications Conference (GLOBECOM),* Honolulu, HI (2009). doi:10.1109/GLOCOM.2009.5425689.

51. L. Valcarenghi, D.P. Van, P. Castoldi, How to save energy in Passive Optical Networks, in: *13th International Conference on Transparent Optical Networks*, Stockholm, Sweden (2011): pp. 1–5. doi:10.1109/ ICTON.2011.5970994.

52. D. Systems, 10-Gigabit-capable passive optical networks (XG-PON): Transmission convergence (TC) layer specification, ITU Stand. G.987.3. (2014).

53. B. Lannoo, A. Dixit, S. Lambert, D. Colle, M. Pickavet, How sleep modes and traffic demands affect the energy efficiency in optical access networks, *Photonic Netw. Commun.* (2015) 85–95. doi:10.1007/ s11107-015-0504-4.

54. H. Bang, J. Kim, Y. Shin, C.S. Park, Analysis of ONT buffer and power management performances for XG-PON cyclic sleep mode, *IEEE Global Telecommunications Conference (GLOBECOM)*, Anaheim, CA, (2012): pp. 3116–3121. doi:10.1109/GLOCOM.2012.6503593.

55. F. Zanini, L. Valcarenghi, D. Pham Van, M. Chincoli, P. Castoldi, Introducing cognition in TDM PONs with cooperative cyclic sleep through runtime sleep time determination, *Opt. Switch. Netw.* 11 (2014) 113–118. doi:10.1016/j.osn.2013.08.007.

56. Y. Maneyama, R. Kubo, QoS-aware cyclic sleep control with proportional-derivative controllers for energy-efficient PON Systems, *IEEE/OSA J. Opt. Commun. Netw.* 6 (2014) 1048–1058.

57. J. Li, K.L. Lee, N. Dinh, C.A. Chan, P. Vetter, A comparison of sleep mode mechanisms for PtP and TDM-PONs, in: *IEEE International Conference Communications Workshops (ICC),* Budapest, Hungary (2013): pp. 543–547. doi:10.1109/ICCW.2013.6649293.

58. N.P. Anthapadmanabhan, N. Dinh, A. Walid, A.J. van Wijngaarden, Analysis of a probing-based cyclic sleep mechanism for passive optical networks, in: *Global Communications Conference (GLOBECOM),* IEEE, Atlanta, GA (2013): pp. 2543–2548. doi:10.1109/GLOCOM.2013.6831457.

59. M.S. Han, H. Yoo, D.S. Lee, Development of efficient dynamic bandwidth allocation algorithm for XGPON, *ETRI J.* 35 (2013) 18–26. doi:10.4218/etrij.13.0112.0061.

60. M.-S. Han, H. Yoo, B.-Y. Yoon, B. Kim, J.-S. Koh, Efficient dynamic bandwidth allocation for FSAN-compliant GPON, *J. Opt. Netw.* 7 (2008) 783. doi:10.1364/JON.7.000783.

16 Energy Efficiency in Wireless Body Sensor Networks

Ali Hassan Sodhro, Giancarlo Fortino, Sandeep Pirbhulal, Mir Muhammad Lodro, and Madad Ali Shah

CONTENTS

16.1   INTRODUCTION

A variety of visualized applications including the medical domain, for example, video image transmission, vital sign signals monitoring and so on, to the non-medical domain such as sports and entertainment and so on produces a set of technical requirements in terms of energy efficiency. Hence, architectures and protocols are needed as shown in Figure 16.1. To make patients feel comfortable and at ease, sensor nodes should have tiny size and long-lasting batteries. Due to the limited node size, the capacity of the battery is also restricted; therefore, WBSNs must have the ability of efficient energy utilization [1].

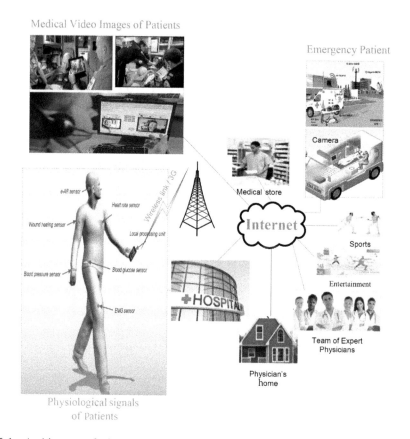

FIGURE 16.1 Architecture of wireless body sensor networks in medical health applications.

The major issue of the deployment of WBSNs is the energy consumption [2]. The recently available battery technologies do not have high energy to produce nodes with adequately long lifetime and satisfactory cost. Furthermore, the reasonably slow progress in battery technology does not promise battery-driven WBSN nodes in the near future, and replacing batteries is simply not a viable option in most cases, such as implanted WBSN nodes. Hence, energy depletion is supposed as the dominant factor in devising and implementation of WBSNs. The key reason of the energy efficiency is to employ energy in such a way that each node remains alive all the time, further to extend battery lifetime of WBSNs. It is very essential for sensor nodes to have efficient energy saving techniques for the confined sources, and the application requirement should be fulfilled according to the presented energy sources.

Additionally, energy saving is examined as a constrained resource for WBSNs, especially when nodes are set up in a remote region; once they utilize the available energy, it is almost impossible to give supplement energy [3–4]. Consequently, balanced energy usage between supply and load is needed to reduce energy drain in WBSNs.

Currently, researchers are focusing on WBSNs for medical healthcare, in terms of bio-signals monitoring and video transmission to deliver clear and insightful pictures of emergency patients [5–9]. Therefore, owing to the first principal application of WBSNs being in long-term healthcare, energy-efficiency is one of the extremely important characteristics. However, the battery lifetime of nodes depends on available energy sources and overall energy depletion.

In today's world, we are witnessing a mind-blowing and significant growth of information and communication techniques for WBSNs. This progress has created a need for their analysis and performance evaluation. Energy efficiency is examined as one of the mature problems of WBSNs and is the cornerstone for medical healthcare applications [10–15]. When one or more sensors have to be implanted, put on a body, or worn by a patient, it is important to reduce the stress on a patient's body caused by battery replacement/recharging. The problem of reducing energy consumption can be tackled by realizing energy-saving techniques at the physical layer.

The development of energy-efficient and battery-friendly WBSNs is a very challenging issue. On one hand, energy-constrained sensors are expected to run autonomously for longer periods. However, it may be very costly to replace exhausted batteries or even impossible in hostile environments. On the other hand, unlike other networks, WBSNs are designed for specific applications.

In recent years, many researchers have proposed several methods of energy saving in WBSNs, and there is a lot of room to continue to optimize energy in battery-limited WBSNs. However, none of the conventional methods is universally applicable. For example, abnormal or emergency medical applications require fast and timely feedback, but this is not the case for other applications, such as in non-critical or normal applications where the latency is not very important. We believe that energy optimization problems should be handled by attentively attending to the application requirements in a more systematic manner.

This research aims to develop the technical solutions for presenting energy-efficient and battery-friendly data transmission to minimize energy consumption and extend battery lifetime of WBSNs. Therefore, battery–driven or data rate control and transmission power control (TPC)-driven or power control algorithms are developed for the validation of theoretical and software-based estimations of energy consumption in WBSNs. In addition, two medical healthcare applications, medical monitoring of the emergency patients and vital sign signals (12-lead ECG) monitoring of patients in emergency situations, have been proposed for simple and cost-effective health monitoring.

16.2 RELATED WORK

Previous research works were focused on protocol design, networking policies, QoS analysis, etc. in wireless body and sensor networks. In Reference [13], WBSNs have been reviewed with an emphasis on physiological parameters, typical network topology, and data rates used for various services. It also discusses core issues that affect physical layer parameters such as channel modeling, antenna design, transmission power, etc. Their research proposed a three-level architecture for WBSNs. Level 1 is known as an intra-body sensor network, in which nodes accumulate vital sign signals and then send them to the intermediate relay node, for example, a server; Level 2 is an inter-body sensor network which shows the communication path between the intermediate relay and some access points; and Level 3 is used as a database for patient record-keeping, but they do not concentrate on energy consumption minimization and battery lifetime extension of WBSNs.

Crosby et al. [16] presented details of wearable devices that classify them into on-body and implantable nodes on the basis of environmental conditions and bio-signals such as electrocardiogram (ECG), glucose, temperature, etc. The authors also figured out that the main source of energy consumption is the inefficient power protocols. In References [17,18], the researchers have compared wireless sensor networks (WSNs) and WBSNs and then examined the major characteristics such as data rate, power level, etc. of the physical layer. The authors further presented their point of view that the combination of energy harvesting and lower energy consumption techniques forms the optimal energy solution for any autonomous system.

Micek et al. [19] discussed the detailed study of energy saving techniques and sources. It was analyzed that different subsystems of a node consume unequal amounts of energy, therefore, necessary steps should be taken for the limited battery budget of WBSNs. Several issues related to network topology, body motion, energy-deficiency, and the overall network lifetime are addressed [20]. In Reference [21], researchers have discussed recent challenges in terms of routing protocols and their energy consumption from the aspect of the IEEE 802.15.6 standard. Besides, most research works have been devoted to the objective of energy optimization and battery lifetime extension in WBSNs.

Multiple experiments were conducted with the help of different technologies for medical emergency detection in WBSNs. The response of the network to the amalgam of the deployed technologies was then analyzed and assessed according to their particular nature and purpose [22].

16.3 PHYSIOLOGICAL SIGNAL MONITORING IN WBSNs

To save the life of emergency patients it is foremost important to deal with human physiological characteristics.

16.3.1 VITAL SIGN SIGNALS

Basic and vital parameters play a major role in evaluating and assessing the health status of the human body and few of them are presented below.

- Electrocardiography (1-lead, 3-lead, and 12-lead): Human heart activity through electrical recordings.
- Electroencephalography (EEG): Representation of the brain activity using electrical signals.
- Respiratory rate: Inspiration and expiration activity within a specific time interval.
- Blood pressure: Level of pushing the blood and its distribution in arteries by the heart.
- Temperature: Ability of the human body to generate and avoid heat.
- Oxygen saturation: Amount of the oxygen level in the blood.

The health condition of patients can be monitored by the frequent analysis of the vital sign signals. Patients with critical ailments such as heart attack must be diagnosed after every single minute. In Reference [23], different requirements of the human body in terms of physiological characteristics are presented. Due to the distinct traits and behavior of the human body, the data rates are changing accordingly.

The time ratio between active and inactive communication intervals is defined as the duty cycle, which is a potential parameter at the MAC layer to determine the status of the radio transceiver. Different researchers adopt distinct parameters to evaluate the performance of WBSNs. In References [21,24], bit error rate (BER), is the focus of attention. It is described that BER plays an important role for wireless monitoring of vital sign signals in WBSNs. To define a desired BER threshold for WBSNs is a very cumbersome job, because the reliability, that is, packet loss ratio (PLR), is a key performance metric which has strong connection with the BER. Therefore, it can be said that PLR will increase as distance between the transmitter node and BS increases. Reference [25] discusses that the BER has linear trade-off with the number of nodes on the patient's body. For example, in a six nodes scenario BER is about 0.001%, which is reasonable to cope-up with performance evaluation in WBSN as shown in Table 16.1. Moreover, in Reference [26] BER values of 0.002% and 0.01%–0.02% are categorized for the line-of-sight and non-line-of-sight scenarios between the transmitter node and BS.

16.4 VIDEO TRANSMISSION IN WBSNs

In this digital world everyone needs a prompt response in a short span of time. Therefore, video is widely deployed as an important asset in WBSNs. Wireless video transmission potentially impacts the starting hours of any emergency incident. The doctors can train nursing and care-giving staff on the basis of a patient's video images to obtain clear and insightful information [27].

TABLE 16.1
Specifications of Sensor Nodes in WBSN

Body Sensors	Endoscopy Image	ECG	Blood Pressure	Respiratory Rate
BER	10^{-4}	10^{-6}	10^{-8}	10^{-7}
Latency	0.5 sec	0.3 sec	0.75 sec	0.6 sec
Traffic generation distribution	Poisson	Constant	Constant	Constant
Message generation rate	1538.46 Byte/sec	500 Byte/sec	512 Byte/sec	1024 Byte/sec
Inter-arrival packet time	0.065 sec	0.20 sec	0.195 sec	0.097 sec

Source: B. Pourmohseni and M. Eshghi, *Journal of Application or Innovation in Engineering and Management*, 2, 393–402, 2013.

Video transmission has got amazing attention due to the magnetic effect of modern digital technologies. MPEG-4 and H.264/advanced video coding (AVC) are the most used standards developed and standardized jointly by the International Organization for Standardization/ International Electrotechnical Commission (ISO/IEC) and the International Telecommunication Union (ITU-T), respectively [28].

16.4.1 Joint Transmission of Video and Physiological Signals

Recently, research in the medical healthcare domain is the topmost priority of developing countries. The concept of pervasive and ubiquitous health has helped everyone from newly born babies to older patients. In addition, telemedicine, which is a well-known platform of wireless video, has broadened its use in medical care, especially in WBSNs, with pleasant and economical output. Moreover, vital sign signals of patients are transmitted in real time with very high desire and requirement. Therefore, the joint integration of video and vital sign features of human body is a leading step in the medical world for physicians and patients to maintain visual inspection of the emergency patient in rural areas [27]. Furthermore, the transmission of video and vital sign signals in WBSNs contains dynamic rate strategy to multiplex the data and also minimize the wasteful resources which were used in the past in conventional telemedicine systems.

16.4.2 Video and Physiological Signals Synchronization

Joint transmission and synchronization of vital sign parameters and video is an incredible choice in medical health applications, because it helps physicians to simultaneously observe the video image and 12-lead ECG of the patient at the emergency location in the hospital. Also, formal and effective practices require the synchronization of both video and physical features for proper medical treatment, diagnosis, and bio-feedback. For example, a patient's safety in anesthesia operating rooms requires the creation of a permanent and accurate record of clinical events with synchronized video and vital sign signals. For diagnosing breathing anomalies in pediatrics and older patients during sleep, the synchronization between video recordings and polysomnographic readings is necessary.

16.5 ENERGY MODEL AND ENERGY OPTIMIZATION IN WBSNs

Energy-efficient and battery-friendly communication is essential for WBSN. In this section, In this section, an energy model [27] is considered for energy optimization in WBSNs. The model parameters can be changed on the basis of application and network requirements. WBSNs with N sensor nodes and BS are assumed, where all nodes wirelessly communicate with each other as well with the BS.

Generally, nodes perform three main tasks in the designed energy model: first, measurement of bio-signals, second, data recording, and third, data communications. As energy deficiency is a key problem, therefore the main purpose is to focus on transmission energy drain in WBSNs. The time interval of sensor S is considered an optimization parameter for energy efficiency during data sensing and transmission. In this regard, we acknowledge that a sensor $i = 1,2,\ldots\ldots,S$, depletes $E_{\mathrm{sen}_i}(b)$ energy while sensing a data packet of size b bits. It also consumes $E_{\mathrm{rs}_i}(b)$ energy in reading and storing b bits, and wastes $E_{\mathrm{tx}_i}(b,d_{ij})$ energy during transmission of a b bit packet with sensor j and distance d_{ij}. E_{st_i} denotes the energy consumption of each node during its active or sleep periods. The energy consumption in the entire WBSN (i.e., of all sensor nodes S) is given in Equation 16.1.

$$E = \sum_{i=1}^{S} \left(E_{\mathrm{sen}_i}(b) + E_{\mathrm{rs}_i}(b) + E_{\mathrm{tx}_i}(b,d_{ij}) + E_{\mathrm{st}_i} \right) \tag{16.1}$$

As the transceiver of every node adopts active, idle, and sleep modes, it is important to discuss the duty cycle (DC) of sensor S, as presented in Equation 16.2.

$$DC_S = \frac{T_{\mathrm{tranON}} + T_{\mathrm{Act}} + T_{\mathrm{tranOFF}}}{T_{\mathrm{tranON}} + T_{\mathrm{Act}} + T_{\mathrm{tranOFF}} + T_{\mathrm{slp}}} \tag{16.2}$$

where

T_{tranON} = time required for sleep to idle transition
T_{tranOFF} = time required for idle to sleep transition
T_{Act} = active/wake-up time of a node
T_{slp} = sleep time of a node
The average current for a sensor S (I_S) is given by Equation 16.3:

$$I_S = DC_S \times I_{\mathrm{Act}} + (1 - DC_S) \times I_{\mathrm{slp}} \tag{16.3}$$

$$\left(I_{\mathrm{Act}} = \frac{I_S - (1 - DC_S) \times I_{\mathrm{slp}}}{DC_S} \right) \tag{16.4}$$

$$I_{\mathrm{slp}} = \frac{I_S \ (DC_S \ I_{\mathrm{Act}})}{(1 \ DC_S)} \tag{16.5}$$

where I_{Act}, I_{slp} show the Current(mA) for active and sleep modes respectively.

$$E_{\mathrm{sen}_i}(b) = bV_{\mathrm{sup}} \times I_{\mathrm{sen}} \times T_{\mathrm{sen}} \tag{16.6}$$

$$E_{\mathrm{rs}_i}(b) = bV_{\mathrm{sup}} \times (I_{\mathrm{str}} \times T_{\mathrm{str}} + I_{\mathrm{rd}} \times T_{\mathrm{rd}}) \tag{16.7}$$

$$E_{\mathrm{tx}_i}(b,d_{ij}) = bE_{\mathrm{tx}} + bd_{ij}{}^{n} \times E_{\mathrm{pamp}} \tag{16.8}$$

$$E_{\mathrm{st}_i} = V_{\mathrm{sup}} \times I_S \times T_{\mathrm{Act}} \tag{16.9}$$

We define V_{sup} as the supply voltage and T_{sen}, T_{str}, and T_{rd} show the time of sensing, storing, and reading only one information bit, respectively, whereas I_{sen}, I_{str}, and I_{rd} show the current (mA) consumed during sensing, storing, and reading processes, respectively. bE_{tx} is the energy drain during transmission of b bits of information, n shows the path loss exponent, and E_{pamp} is the energy consumption of the power amplifier. Finally, T_{Act} is the active time of the sensor, and $DC(<1)$ is the duty cycle. From analysis of the sensor node's operation, we see that more energy is consumed during transmission as shown in Equation 16.10. The term E_{pamp} cannot be optimized because the energy depletion in the *PA* is examined by the hardware designer according to the size and range of the network components. We evaluate the optimization of only bE_{tx}. For further details see Reference [27].

$$bE_{\text{tx}_i} = \frac{Ptx_i \times SF_i \times M_i}{DR} + \frac{Ptx_i \times O_v}{DR \times T_{\text{upd}_i}} \tag{16.10}$$

where Ptx_i, SF_i, and M_i show the transmission power, sampling frequency (in samples per second), and length (in bits per sample) of measurement sample for sensor i, respectively, whereas DR, O_v, and T_{upd_i} represent the data rate, overhead bits, and data updating time interval of sensor i, respectively.

16.5.1 ENERGY OPTIMIZATION CONSTRAINTS

According to the basic health requirements and total emergency reporting time T_{ir}, an expert physician pre-defines the data update interval according to the monitoring level required for sensors attached on his/her body and that specified time acts as the monitoring interval or updating interval for each sensor node. This gives the first constraint on the value of T_{upd_i} as in Equation 16.11.

$$0 < T_{\text{upd}_i} \leq T_{ir}, \text{ for } i = 1, 2, \ldots, S \tag{16.11}$$

There is a limited number of available time slots. Each slot has a *size* T_{slt}, which defines $\dfrac{1}{T_{\text{slt}}}$ slots in a unit time interval. As each sensor is $1/T_{\text{slt}}$ slots this specified interval and that slot can be allotted to one sensor at a time. Equation 16.12 presents the restricted number of time slots.

$$\sum_{i=1}^{S} \frac{1}{T_{\text{upd}_i}} \leq \frac{1}{T_{\text{slt}}} \tag{16.12}$$

Measuring data sample size during two updating intervals, can be restrained to only a fraction X_i of the buffer size B_i, without go beyond $X_i \times B_i$ limit. This defines the second constraint as in Equation 16.13.

$$\left(SF_i \times M_i \times T_{\text{upd}_i}\right) \leq X_i \times B_i \tag{16.13}$$

Total data bits sent during each updating interval are $\left(SF_i \times M_i \times T_{\text{upd}_i} + O_v\right)$;, this expression will be restricted due to the time slot size DT, data rate DR, and the overhead O_v, in the form of constraints, as in Equation 16.14.

$$\left(SF_i \times M_i \times T_{\text{upd}_i} + O_v\right) \leq DT \times DR$$

$$\text{that is, } T_{\text{upd}_i} \leq \min\left(\frac{X_i \times B_i}{SF_i \times M_i}, \frac{DT \times DR - O_v}{SF_i \times M_i}\right); \quad \text{for } i = 1, 2, \ldots S \tag{16.14}$$

16.5.2 COST FUNCTION

The cost function is composed of energy consumption and latency functions. The energy consumption of sensor i in Equation 16.10 can be further expressed in Equation 16.15.

$$bE_{\text{tx}_i} = E_0 + bE_{\text{tx}_i} \times T_{\text{upd}_i} \tag{16.15}$$

where, the first term E_0 is constant at each updating interval T_{upd_i}, and merely changes from node to node counting on the vital sign signal measurement, whereas the second term varies and depends on the overhead data. The other parameter of the cost function is the latency, Lat, which is defined as the time spent in collecting and transmitting data sample from transmitter node to BS, respectively, as shown in Equation 16.16.

$$Lat = \sum_{i=1}^{S} \gamma_{2_i} \times \frac{T_{\text{upd}_i}}{2} \tag{16.16}$$

The final cost function is

$$C_i = \gamma_{1_i} \times bE_{\text{tx}_i} + \gamma_{2_i} \times \lambda \times Lat_i \tag{16.17}$$

where γ_{1_i} is the specific weight given to the power consumption in sensor i, γ_{2_i} is the coefficient exploiting high priority for minimizing latency Lat_i of sensor i, and λ is the term shows importance to latency related to energy drain. The final optimization problem based on the cost function and the constraints is expressed as in Equation 16.18.

Minimize

$$\sum_{i=1}^{N} \left(\gamma_{1_i} \times bE_{\text{tx}_i} + \gamma_{2_i} \times \lambda \times Lat_i \right)$$

constrained to

$$0 \leq T_{\text{upd}_i} \leq \min\left(T_{ir}, \frac{X_i \times B_i}{SF_i \times M_i}, \frac{DT \times DR - O_v}{SF_i \times M_i} \right) \tag{16.18}$$

$$\sum_{i=1}^{N} \frac{1}{T_{\text{upd}_i}} \leq \frac{1}{T_{\text{slt}}}$$

16.6 ENERGY-EFFICIENT TECHNIQUES FOR WBSNs

In this section, we review the conventional energy saving approaches to deal with the energy consumption problem of WBSNs. The classification of energy-efficient techniques is presented in Figure 16.2.

16.6.1 RADIO OPTIMIZATION SCHEME

The radio module is the integral part that causes battery drain of sensor nodes. To minimize energy consumption during data transmission, radio factors, for example, coding and modulation, power level, and the antenna, are considered.

16.6.1.1 Transmission Power Control (TPC)

It has been reviewed that energy-efficiency at the physical layer is obtained by adapting the radio transmission power level. With the help of cooperative topology control with adaptation (CTCA) the power level of every node is adjusted frequently to take into account the random and non-uniform energy depletion profile of the nodes. Therefore, a node with a higher level of unused energy may

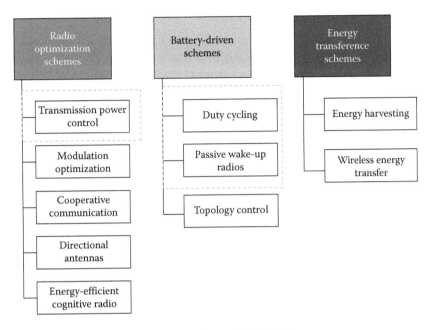

FIGURE 16.2   Classification of energy-efficient schemes for WBSNs.

increase its transmission power, which will probably empower neighboring nodes to decrease their power level to save energy [29]. Nevertheless, the TPC mechanism not only affects energy level but also channel quality, interference, and coverage. As the transmission power decreases, interference decreases, but the delay increases because more hops will be needed to forward a data packet. Finally, transmission power impacts the network topology due to variation in the node's connectivity during communication between the transmitter node and BS.

16.6.1.2   Modulation Optimization

This aims to find the optimum modulation parameters that result in the minimum energy consumption of the radio. For instance, energy is consumed by circuit power and transmission power. For short distances, circuit power consumption is greater than the transmission power, whereas for longer ranges the signal power becomes dominant. Many researchers have tried to establish a relationship between the constellation size, the information rate, the transmission time, and the distance between the nodes. Cui et al. [30] have examined the level of energy consumed to fulfill the BER and delay requirement of the networks, and then minimized energy drain with the optimization of transmission time. Costa et al. [31] have presented the energy saving performance of three modulation schemes with optimal the parameters to obtain the minimum energy consumption between sensor nodes.

16.6.1.3   Cooperative Communications

These schemes improve the quality of the received signal by manipulating several single-antenna devices that cooperate to create a virtual multiple-antenna transmitter. The fact is that the data are usually listened to by neighboring nodes due to the disseminating nature of the wireless on-body channel. Therefore, by retransmitting to these nodes it is possible to create spatial diversity and encounter multi-path fading and shadowing. Jung et al. [32] showed that cooperative transmission enhances the communication range and harmonizes the duty cycling of relay nodes. Cui et al. [33] and Jayaweera [34] compared the energy consumption of both single input single output (SISO) and virtual-multiple input and multiple output (V-MIMO) systems and showed that MIMO saves more energy with smaller end-to-end delay over certain transmission distances, but it requires extra overhead energy for proper training.

16.6.1.4 Directional Antennas

These types of antennas transmit and receive signals in one direction at a time, which improves transmission range and energy efficiency. Directional antennas may require localization techniques to be aligned, but multiple communications can occur in close proximity, resulting in the spatial reuse of bandwidth. In contrast to omni-directional nodes which transmit in unwanted directions, directional antennas limit overhearing and, for a given range, require less power. Thus, they can improve network capacity and lifetime while influencing delay and connectivity [35,36].

16.6.1.5 Energy-Efficient Cognitive Radio

The communication channel can be chosen by a very dynamic and efficient cognitive radio (CR), which further adjusts network parameters between the transmitter and BS. The fundamental software-defined radio (SDR) technology is anticipated to produce entirely reconfigurable wireless transceivers which automatically tunes their transmission parameters as per network needs to improve context-awareness. However, CR requires more energy as compared with traditional devices because of raised complexity for innovative and sophisticated functionalities [37]. Hence, offering an energy-efficient cognitive radio–based sensor network is a main hurdle in increasing battery energy. Along with that, present cognitive radio studies are spotlighted in the transmission power control [38], and residual energy-based channel allocation.

16.7 BATTERY-DRIVEN SCHEMES

The idle states are major sources of energy consumption at the radio transceiver. Battery-driven schemes aim to put the radio in sleep state/mode and adjust its activity to save energy in WBSNs.

16.7.1 Duty Cycling

This arranges the state of a sensor node's radio on the basis of the network activity to support the sleep mode. Duty cycling–based techniques are classified into three categories, for example, on-demand, asynchronous, and scheduled [39]. The protocols related to duty cycling are usually the most energy-efficient but they cause long sleep latency, because the transmitter node waits a long time to wake up the receiver. Furthermore, it is practically impossible for any node to broadcast at the same time to its all neighbors. The performance of the network is affected by key parameters such as, preamble length, listen and sleep periods, and slot time. In recent research little focus is paid toward adjustment of the sensor node active period to optimize energy consumption [40]. Additionally, low duty cycling consumes less energy, however it increases communication delay. For further details see [41–44].

16.7.2 Passive Wake-Up Radios

Duty cycling misuses energy because of less important wake-ups, low-power radios are utilized to awake a node when it sends/receives packets, at the same time energy hungry natured radio is employed for communication. Ba et al. [45] have discussed a network that includes passive radio frequency identification (RFID) wake-up radios called wireless identification sensing platform (WISP)-based nodes and RFID readers. A passive RFID wake-up radio utilizes the energy dissipated by the reader transmitter to activate the slept sensor node. However, every node cannot be operated with RFID readers because they consume a large amount of power, which is one of the biggest problems.

16.7.3 Topology Control

When a large number of nodes is deployed for ensuring good space coverage, it is desirable to deactivate some nodes while maintaining network operation and connectivity. Topology control protocols exploit redundancy to dynamically adjust the network topology and reduce the number of active nodes according to the specific requirements of applications. Those sensor nodes that do not guarantee connectivity or coverage can be turned off to further prolong the lifetime of WBSNs. Misra et al. [46] have proposed an energy optimization technique to maintain network coverage by activating only a subset of nodes with the minimum overlap area.

16.8   ENERGY TRANSFERENCE

Currently energy transference, that is, wireless energy transfer methods and energy harvesting, is receiving a lot of attention. These two approaches can recharge the battery of WBSNs without any human intervention.

16.8.1   ENERGY Harvesting

Emerging energy harvesting (EH) technologies have contributed a lot to enable WBSNs to work longer and effectively [47]. Compared to typical conventional sensors, rechargeable nodes can work continuously for a long time. EH architectures usually need a energy prediction pattern to create an efficient and effective way to manage obtained power. Actually, sensor nodes necessitate estimation of energy transformation to adapt their behavior vigorously and continuously until the next recharge cycle. Hence, they can optimize the value of related entities, for example, the duty cycle, bit rate, transmission power, etc. to minimize energy drain with respect to the regular interval occurring rate and significance of the harvestable source. It is essential to note that sensor nodes are considered energy constrained even during the energy harvesting process, so they still require adopting energy conserving approaches.

16.8.2   WIRELESS Energy Transfer

The rapid advancement in wireless energy transfer (WET) is anticipated to raise the sustainability of WBSNs. Therefore, they can be perpetually operational because these techniques can be utilized to transmit power between wearable devices without any information exchange between the transmitter and the BS. WET in WBSNs can be obtained by, electromagnetic (EM) radiation and magnetic resonant coupling methods. Xie et al. [48] present an omni-directional EM radiation method having ultra-low-power requirements and low sensing activities (such as temperature, light, and moisture). It is owing to EM waves affected from sudden decline in energy-efficiency over distance, and active radiation approach can put security concerns to the human beings. However, magnetic resonant coupling emerges to be the most favorable technique to address the energy requirements of WBSNs, because of its high efficiency at long distance. WET has already been employed in WBSNs to power on-body and implantable sensor nodes. The emerging use of wireless power/energy transfer technology overcomes the energy constraint, and it is possible now to adopt the network elements in a more manageable way. In this regard, some researchers have already developed mobile chargers that directly transfer power to deployed nodes. One of the new challenges raised by WET technologies is energy cooperation, because it is vital to transfer energy to neighboring nodes. So, in future wireless networks, sensor nodes are envisioned to be efficient in harvesting energy from the environment and transferring that energy to their neighbors, to build an autonomous network. To do this, a suitable solution is the multi-hop energy transfer which leads to the design of wireless charging protocols and energy cooperative systems.

16.9   EXPERIMENTAL RESULTS AND DISCUSSION

In this section, we present the experimental results of energy consumption analysis in WBSNs by considering energy models.

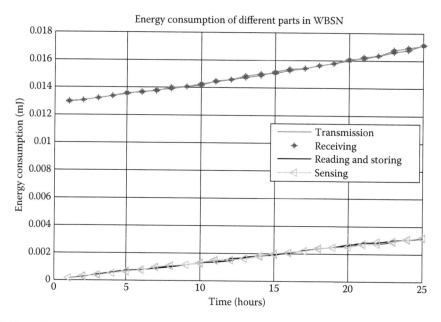

FIGURE 16.3   Energy consumption of different parts in WBSNs.

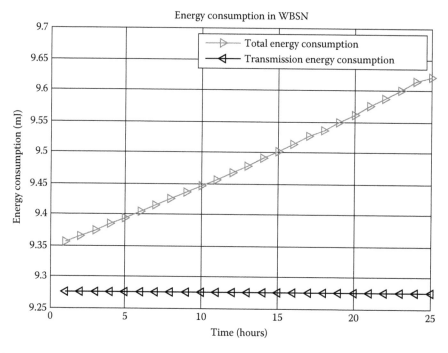

FIGURE 16.4   Total and transmission energy consumption in WBSNs.

Figure 16.3 shows the energy consumption of four main parts of a WSBN: sensing, reading and storing, transmitting, and receiving of the sensor nodes. The transmitting and receiving parts consume high energy while the sensing and reading and storing parts consume much less energy in WBSNs.

Figure 16.4 reveals the relationship between time and energy consumption of transmission parts and the total energy of four parts in WBSNs. We observed that transmission energy is less in comparison to the total energy.

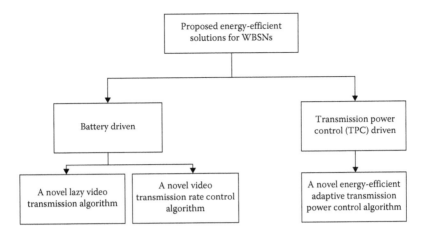

FIGURE 16.5   Proposed energy-efficient solutions in WBSNs.

16.10   PROPOSED ENERGY-EFFICIENT SOLUTIONS FOR WBSNs

To reduce energy dissipation, each sensor node uses battery-model-based rate-control algorithms and power-control algorithms to adjust the data rate and transmission power level based on the RSSI level at the base station (BS). For further details see [49,50] and [51–55] (Figure 16.5).

16.11   CONCLUSIONS AND FUTURE RESEARCH

In this research, an energy-efficient health assessment with energy consumption and optimization model is presented for achieving more energy saving in WBSNs. In addition to fulfilling the necessities of human beings and encountering major hurdles in the medical health domain, this chapter studies and presents typical conventional energy consumption and battery-model-based approaches for vital sign signals and video transmission in WBSNs. Furthermore, physiological signal monitoring and its synchronization with video transmission is discussed.

The fact that the existing energy saving techniques encourages the research on looking for the development of novel energy-efficient and battery-friendly solutions, which guarantee low-power consumption, more energy-saving and longer battery lifetime in WBSNs. For that purpose, the limitations of existing energy saving and battery-model-based techniques are first examined and other potential battery-driven or rate-control and transmission power control (TPC) driven or power control alternatives further explored. It has been evaluated and concluded through theoretical analysis and Monte Carlo simulation in MATLAB® that the constant transmission part of the sensor node consumes more energy than the rest of the parts. Hence, we can say that our research work provides a comprehensive energy-saving technique in WBSNs. In future studies, we will focus on the energy efficiency during video transmission over an integrated network involving the IoT and telemedicine in WBSNs. Furthermore, a novel algorithm will be developed and integrated with an energy optimization model to measure quality of service (QoS) in WBSNs.

REFERENCES

1. T. Liang and D. B. Smith, Energy-efficient, reliable wireless body area networks: Cooperative diversity switched combining with transmit power control, *Electronics Letters*, Vol. 50, No. 22, pp. 1641–1643, 23 October 2014.
2. A. Seyedi and B. Sikdar, Energy efficient transmission strategies for body sensor networks with energy harvesting, *IEEE Transaction on Communications*, Vol. 58, No. 7, July 2010.

3. H. Yoo and N. Cho, Body channel communication for low energy BSNs, In *Proceedings of the IEEE Asia Pacific Conference on Circuits and Systems (APCCAS)*, New York, NY, pp. 7–11, 2008.

4. G. P. Reddy et al., Body area networks, *Journal of Telematics and Informatics*, Vol. 1, No. 1, pp. 36–42, 2013.

5. S. Gonźalez-Valenzuela and X. Liang, *BANs in Autonomous Sensor Networks*. New York, Dordrecht, London: Springer Heidelberg, pp. 17–37, 2013.

6. Y. Hovakeemian et al., A survey on dependability in BANs, *5th International Symposium on Medical Information and Communication Technology (ISMICT)*, Montreux, Switzerland, New York, Dordrecht, London: Springer Heidelberg, pp. 10–14, 2011.

7. M. Chen et al., Body area networks: A survey, *Mobile Networks and Applications*, Vol. 16, No. 2, pp. 171–193, April 2011.

8. A. Bachir et al., MAC essentials for WSNs, *IEEE Communications Surveys & Tutorials*, Vol. 12, No. 2, pp. 222–248, 2010.

9. S. Gopalan and J.-T. Park, Energy-efficient MAC protocols for WBANs: Survey, in *Ultra Modern Telecommunications and Control Systems and Workshops*, Moscow, Russia, pp. 739–744, 2010.

10. R. Carrano et al., Survey and taxonomy of duty cycling mechanisms in wireless sensor networks, *IEEE Communications Surveys & Tutorials*, Vol. 16, No. 1, pp. 1–14, 2013.

11. S. Sudevalayam and P. Kulkarni, Energy harvesting sensor nodes: Survey and implications, *IEEE Communications Surveys & Tutorials*, Vol. 13, No. 3, pp. 443–461, 2011.

12. M. Patel and J. Wang, Applications, challenges, and prospective in emerging body area networking technologies, *IEEE Transaction on Wireless Communication*, Vol. 17, No. 1, pp. 80–88, 2010.

13. H. Cao et al., Enabling technologies for wireless body area networks: A survey and outlook, *IEEE Communications Magazine*, Vol. 47, No. 12, pp. 84–93, 2009.

14. R. Cavallari et al., A survey on WBANS: Technologies and design challenges, *IEEE Communications Surveys & Tutorials*, Vol. 16, No. 3, Third Quarter, pp. 1635–1657, 2014.

15. J. I. Bangas et al., A survey of routing protocols in WBSNs, *MDPI Sensors*, Vol. 14, pp. 1322–1357, 2014.

16. G. V Crosby et al., Wireless body area network for healthcare: A Survey, *International Journal of Ad Hoc, Sensors and Ubiquitous Computing*, Vol. 3, 2012.

17. B. Latré et al., A survey on wireless body area networks, *Journal of Wireless Networks*, Vol. 17, pp. 1–18, 2011.

18. S. A. Gopalan J.-T. Park, Energy-efficient MAC protocols for wireless body area networks: Survey, *International Congress on Ultra Modern Telecommunications and Control Systems and Workshops (ICUMT)*, 2010.

19. J. Micek and M. Kochlan, Energy-efficient communication systems of wireless sensor networks, *Studia Informatica Universalis*, Vol. 11, pp. 69–86, 2014.

20. B. Otal et al., Towards Energy Saving Wireless Body Sensor Networks in Healthcare Systems, *IEEE International Conference on Communications Workshops (ICC)*, Cape Town, South Africa 23–27, May, 2010.

21. A. Boulis et al., Challenges in BANs for healthcare: The MAC communications in ubiquitous healthcare, *IEEE Communications Magazine*, Vol. 50, No. 5, 2012.

22. J. Ko et al., MEDISN: Medical emergency detection in sensors, *ACM Transaction on Embedded Computing System (TECS)*, Vol. 10, pp.1–28, 2010.

23. M. N. Halgamuge et al., An estimation of sensor energy consumption, *Progress in Electromagnetics Research B*, Vol. 12, pp. 259–295, 2009.

24. S. Ullah et al., A review of WBANs for medical applications, *International Journal of Communications, Network and System Sciences*, Vol. 2, No. 8, pp. 797–803, 2010.

25. B. Otal et al., Highly reliable energy-saving MAC for WBSNs in healthcare systems, *IEEE Journal on Selected Areas in Communications*, Vol. 27, No. 4, pp. 535–565, May, 2009.

26. B. Pourmohseni and M. Eshghi, Reliable energy-efficient dynamic-TDMA MAC protocol for WBANs, *Journal of Application or Innovation in Engineering and Management (IJAIEM)*, Vol. 2, No. 7, pp. 393–402, 2013.

27. E. Gonzalez et al., Survey of WBSNs for pre-hospital assistance: Trends to maximize the network lifetime and video transmission techniques, *MDPI Sensors*, Vol. 15, pp. 11993–2021, 2015.

28. I. E. Richardson, *The H.264 Advanced Video Compression Standard*, 2nd ed., West Sussex: Wiley, pp. 81–98, 2010.

29. T. Rault et al., Energy efficiency in WSNs: A top-down survey, *Computer Networks*, Vol. 67, No. 4, pp.104–122, 2014.

30. S. Cui et al., Energy-constrained modulation optimization, *IEEE Transactions on Wireless Communications*, Vol. 4, pp. 2349–2360, 2005.
31. F. M. Costa and H. Ochiai, A comparison of modulations for energy optimization in WSN links, *IEEE GOLOBOCOMM Conference*, Miami, FL, USA, pp. 1–5, 2010.
32. J. W. Jung et al., Cooperative transmission range extension for duty cycle-limited WSNs, *International Conference on Wireless Communication, Vehicular Technology, Information Theory, Aerospace and Electronic Systems. Technology*, Chennai, India, pp. 1–5, 2011.
33. S. Cui et al., Energy-efficiency of MIMO and cooperative MIMO techniques in sensor networks, *IEEE Journal of Selected Areas in Communications*, Vol. 22, pp. 1089–1098, 2004.
34. S. Jayaweera, Virtual MIMO-based cooperative communication for energy-constrained WSNs, *IEEE Transactions on Wireless Communications*, Vol. 5, pp. 984–989, 2006.
35. H.-N. Dai, Throughput and delay in wireless sensor networks using directional antennas, in 5th *International Conference on Intelligent Sensors, Sensor Networks and Information Processing*, Melbourne, pp. 421–426, 2009.
36. A. P. Subramanian and S. R. Das, Addressing deafness and hidden terminal problem in directional antenna based wireless multi-hop networks, *Wireless Networks*, Vol. 16, pp. 1557–1567, 2010.
37. M. Naeem et al., Energy-efficient cognitive radio sensor networks: Parameteric and convex transformations, *MDPI Sensors*, Vol. 13, pp. 11032–11050, 2013.
38. M. Masonta et al., Energy efficiency in future wireless networks: Cognitive radio standardization requirements, *IEEE 17th International Workshop on Computer Aided Modeling and Design of Communication Links and Networks*, Barcelona, pp. 31–35, 2012.
39. G. Anastasi et al., Energy conservation in WSNs: A survey, *Ad-Hoc Networks*, Vol. 7, pp. 537–568, 2009.
40. A. Karagiannis and D. Vouyioukas, A framework for the estimation and validation of energy consumption in WSNs, *Journal of Sensors*, Vol. 2015, pp. 1–13, 2015.
41. A. Hassan and Ye. Li, Medical quality-of-service optimization in wireless telemedicine system using optimal smoothing algorithm, *E-Health Telecommunication System and Networks Journal*, Vol. 2, No. 1, March 2013.
42. A. Hassan and Ye. Li, Novel key storage and management solution for the security of WSNs, published in *TELKOMNIKA Indonesian Journal of Electrical Engineering*, Vol. 11, No. 6, pp. 3383–3390, June 2013.
43. A. H. Sodhro et al., Impact of transmission control protocol window size on routing protocols of WSNs, *Sindh University Research Journal (SURJ)*, Vol. 44, No. 2AB, 2012.
44. A. Hassan and Ye. Li, Battery-friendly packet transmission strategies for wireless capsule endoscopy, *International Federation of Medical and Biological Engineering Conference*, Vol. 42, pp.143–148, 2014.
45. H. Ba et al., Passive wake-up radios: From devices to applications, *Ad Hoc Networks*, Vol. 11, pp. 2605–2621, 2013.
46. S. Misra et al., Connectivity preserving localized coverage algorithm for area monitoring using WSNs, *Computer Communication*, Vol. 34, pp. 1484–1496, 2011.
47. P. Nintanavongsa et al., MAC protocol design for sensors powered by wireless energy transfer, in *Proceeding - IEEE INFOCOM*, pp. 150–154. Turin, 2013.
48. L. Xie et al., Wireless power transfer and applications to sensor networks, *IEEE Wireless Communications*, Vol. 20, pp. 140–145, 2013.
49. A. Hassan et al., Energy-efficient adaptive transmission power control in WBANs, *IET Communications*, Vol. 10, No. 1, pp. 81–90, January 2016.
50. A. H. Sodhro et al., Green and battery-friendly video streaming in wireless body sensor networks, *Journal of Multimedia Tools & Applications* Vol. 76, No. 10, 2017.
51. S. Galzarano et al., A task-oriented framework for networked wearable computing, *IEEE Transactions on Automation Science and Engineering*, Vol. 13, No. 2, pp. 621–638, 2016.
52. G. Smart et al., Decentralized time-synchronized channel swapping for ad hoc wireless networks, *IEEE Transactions on Vehicular Technology*, Vol. 65, No. 10, pp. 8538–8553, 2016.
53. G. Fortino et al., A framework for collaborative computing and multi-sensor data fusion in body sensor networks, *Information Fusion*, Vol. 22, pp. 50–70, 2015.
54. H. Ghasemzadeh et al., Power-aware activity monitoring using distributed wearable sensors, *IEEE Transactions on Human-Machine Systems*, Vol. 44, No. 4, pp. 537–544, 2014.
55. G. Fortino et al., Enabling effective programming and flexible management of efficient BSN applications, *IEEE Transactions on Human-Machine Systems*, Vol. 43, No. 1, pp. 115–133, 2013.

17 Efficient Modulation Schemes for Visible Light Communication Systems

Navera Karim Memon and Fahim A. Umrani

CONTENTS

This chapter is organized as follows. Section 17.1 gives an overview of VLC, followed by a brief discussion on the VLC system architecture in Section 17.2. Section 17.3 highlights the applications of the VLC. The general modulation schemes are given in Section 17.4. Then, Section 17.5 discusses power efficient techniques, while spectral efficient techniques are covered in Section 17.6, and Section 17.7 demonstrates the interference cancellation techniques. This chapter also discusses a comparison of different modulation techniques in terms of power efficiency, spectral efficiency, and interference cancellation and concludes with the best suited technique for power, spectrum, and interference.

17.1   INTRODUCTION TO VISIBLE LIGHT COMMUNICATION (VLC)

Nowadays, transfer of data from one place to another is one of the most essential requirements. Data from one place to another or from one computer to another is carried by local area network (LAN) and wireless LAN (i.e. Wi-Fi). Wireless LAN facilitates fixed and mobile devices (computer, cellular phones, etc.) to communicate via Internet. IEEE has divided Wi-Fi into a few standards known as IEEE 802.11a, 802.11b, 802.11g, and 802.11n. It uses the 2.4–5 GHz radio frequency and has turned into the essential requirement to access the Internet in households, in workplaces, or at open hotspots. At present, 1.5 million radio wave base stations and approximately 5 billion mobile connections are transferring data of around 600 TB every month. As usage of wireless data is increasing exponentially every year, the radio frequency spectrum is getting clogged up day by day and the capacity of wireless data is going down [1]. As the wireless radio frequencies are getting exhausted, therefore, complications are growing. In 2011, to overcome these problems, one of the Professors from the University of Edinburgh named Harald Haas developed a new standard in which data can be transferred through LED by using visible light communication (VLC). This is also known as a light fidelity (Li-Fi) or VLC system [2]. VLC uses the visible light spectrum from 400 THz to 800 THz (780–375 nm) with a data rate up to 1 Gbps as shown in Figure 17.1. VLC is commonly used for indoor communication. It uses LEDs as a transmission source and photodiodes as a receiving component [1].

Due to simultaneous illumination and communication in the VLC, a number of fascinating applications comprising faster communication through illumination structure at homes/offices, networking, communication via headlights in cars, communication in aeroplane/trains, communication via traffic lights, and underwater communication are deployed [3].

As light is restrained to a specific region enclosed in dense walls, the regions of the VLC system ensure higher safety. It provides a higher data rate up to 1 Gbps through a wider unlicensed bandwidth

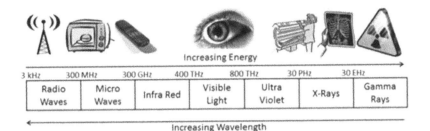

FIGURE 17. 1   Electromagnetic spectrum.

spectrum. It can easily be integrated with the prevailing illumination structure as well as communication system. Apart from the advantages, VLC is also facing some challenges which should be addressed. Generally, the LED would be powered on most of the time for illumination purposes, so the communication is done by fast flickering of the LED (which is not noticeable by the human eye). The flickering rate of the LED is directly proportional to the data rate and LED brightness [4]. In other words, the brighter the LED, the faster the flickering rate, which correspondingly enhances the data rate. The brighter the LED, the more electrical power it consumes. To overcome this issue, dimming techniques can be used. Nevertheless, the dimming mechanism has greater influence over the data rate of communication. It adjusts the mean intensity in accordance with consumer requests, which in turn reduces the attainable data rate of the communication link [5].

Another challenge facing VLC is interference of other lights such as fluorescent light, candescent light, sunlight, reflection of mirrors, etc. The interference can degrade the performance or simply the data rate of the VLC system. Therefore, some modulation and line encoding techniques are analyzed to mitigate or reduce the effects of interference on VLC [6].

17.2   VLC SYSTEM ARCHITECTURE

There are a number of characteristics which should be analyzed briefly to enterprise, execute, and drive an effective VLC system such as impulse response of a channel, which is responsible for analyzing and confronting the influences of distortions. The configurations of the VLC system are basically divided into two categories: directed LOS channel and non-LOS (diffuse) channel. In the directed LOS channel, there is no need to calculate reflections because the path loss can be simply deduced from the beam divergence of the transmitter, distance between the transmitter and receiver, and receiver size. However, non-LOS or diffuse system employ reflections from the surface of room as well as furniture as shown in Figure 17.2. The configurations are taken into account according to some parameters which include inclination degree between transmitter and receiver, presence of LOS route [3].

Intensity modulation with direct detection (IM/DD) is the modulation technique used to implement VLC channels because of its low cost and lower complexity. Figure 17.3 depicts the basic building block of the VLC channel. Initially, the input electrical data (bits) is mapped to the optical source (LED/LD), then the drive current of the LED/LD is combined with modulating signal $m(t)$ by varying the intensity of LED/LD $x(t)$. This intensity modulated $x(t)$ signal is carried over the wireless channel and then received by the photodetector (PD). The PD accepts the signal with a combination of more than thousands of short wavelengths that create the photocurrent $y(t)$. The photocurrent is instantly related to instant transmitted optical power, which in turn is instantly related to power 2 of the electric signal [1,7].

Figure 17.4 demonstrates a schematic representation of the VLC system. Here, a dimming control block is used with LEDs to precisely regulate the brightness of an LED to the appropriate level of dimming. Dimming is used because an LED is on/off for a very short duration of time, that is, for nanoseconds. Thus, to increase the LED response time (on/off time), drive current is

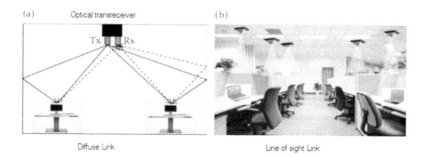

FIGURE 17.2 Optical VLC links: (a) diffuse link; (b) LOS link.

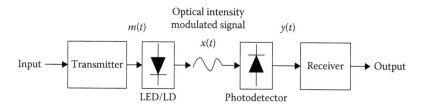

FIGURE 17.3 Basic building block of VLC system with IM/DD.

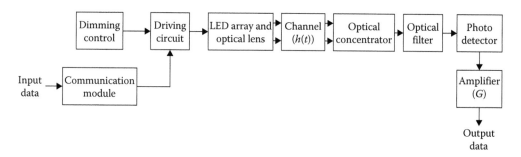

FIGURE 17.4 Block diagram of VLC system.

modulated at moderately higher frequency. This faster LED on/off switching is totally out of sight of human eyes [3].

Consequently, the LED would emit low frequency and high frequency pulses. The brightness (luminous flux) of an LED is instantly related to the thickness of the dimming signal. There are basically two types of optical sources or transmitters: laser diode and LED. The sources can be selected according to user's requirement. Mostly, the LED is more often used instead of a laser due to its dual functionality: communication and illumination. A usual office atmosphere requires the brightness or illuminance around 200 lx to 1000 lx. Basically, two kinds of LEDs are used in VLC which are LEDs that emit one color (i.e. blue, green, and red) and white LEDs. White light in LED can be generated by using one of two ways. The first one generates white light by merging red, green, and blue (RGB) in accurate quantity. The second method uses a blue LED along with a layer of phosphor coating that discharges yellow light. The layer absorbs one segment of light having a short wavelength, then the absorbed light undergoes a shift in wavelength to a wavelength of yellow color. In the next step, the wavelength shifts to red color, and the wavelength is combined with an unabsorbable blue factor, which creates white light [8].

Unlike LED, photodetectors are also of two kinds which are most commonly employed in VLC—PIN photodiode and avalanche photodiode (APD). An APD has higher gain than a PIN but is not excessively used due to the enormous amount of shot noise produced by greater photocurrent. In VLC, the most commonly used photodetectors are PIN photodetectors because they are not expensive, have more tolerance to higher temperature, and can easily be deployed in a flooded area, which has a higher intensity of light [3].

Optical filters are also utilized in VLC systems to overwhelm the deterioration of time-consuming phosphor to squeeze the quicker blue light. Nevertheless, the procedure consumes a significant ratio of received power of the signal. However, almost the entire optical received power is overwhelmed by the multicolored sources for every channel [3].

The effect of limiting the bandwidth of the phosphor layer should be avoided to attain a higher data rate. For this reason, some techniques are used which include blue filtering, pre/post-equalization, or a combo of the three. Blue filtering is used at the receiving side to remove the yellow factors that are very slow in processing. Pre-equalization is used at the source, whereas post-equalization can be used at the receiving side. In addition, some composite modulation techniques can be used to attain a higher data rate. The methodology includes multilevel modulation schemes such as the discrete multi-tone (DMT) modulation technique and quadrature amplitude modulation (QAM) along with optical frequency division multiplexing (OFDM). The data rate could be surged to thousands of Mbps if these techniques are employed along with blue filtering [9–11].

17.3 VLC APPLICATIONS

VLC is a very vast field and may be used in many areas. The areas where VLC may be used are discussed below.

17.3.1 Li-Fi

Initially, VLC was used by Harald Haas in 2011, and he named it light fidelity (Li-Fi). Li-Fi is a VLC technique that offers high speed bi-directional wireless communication. It is equivalent to Wi-Fi. Li-Fi has overcome many problems of Wi-Fi such as interference with other RF signals and low speed. Li-Fi also supports Internet of Things (IoT) by offering 10 Gbps speed [12].

17.3.2 Wireless LANs

VLC can be used to offer wireless communication at very high speeds, up to 10 Gbps. It can be used with a star topology. The LED can be mounted at the roof of the room where users can accommodate it by sitting or standing under the LED light. A VLC star topology architecture can support many users. This type of architecture can be widely used in offices and hospitals due to their high security issues [12].

17.3.3 Hospitals

In ICUs or MRI sections in hospitals, Wi-Fi is not being used because radio signals interfere with MRI signals and other machines and they are harmful for patients too. Comparatively, VLC could be used there because light waves do not create interference. Another reason to use VLC in hospitals is that the light waves are not harmful for patients [13].

17.3.4 Underwater Communication

Wi-Fi employs radio frequency waves for communication purposes. The radio waves are not suitable for underwater communication because of their high conductivity. Unlike Wi-Fi, visible light has low conductivity in water; therefore VLC can easily be employed in underwater communication. The untethered remotely operated vehicle (UTROV) is an application of VLC used mainly for underwater communication. The application is used mainly in maintenance of oceans and opportunity of deployment from the ships [12].

17.3.5 Vehicular Communication

VLC is also suitable for vehicle-to-vehicle communication because all vehicles have visible lights in their vehicular lights. VLC is used to maintain forward collision warnings, warnings of lane change, pre-crash sensors, emergency brake lights, violations of traffic signals, assistance with left/right turns and stop signs, warning of curve speeds, etc. [12].

17.3.6 Visible Light ID System

VLC is also used in ID systems for identification of buildings, rooms, hospitals, subways, aircraft, etc. [12].

17.3.7 Sound Communication System

VLC can also be used in musical signals via RGB (red, green, and blue) LEDs [12].

17.3.8 Information Displaying Signboards

Signboards are used to broadcast data or advertisements at buildings, airports, museums, bus stops, etc. They are mainly made up of bundles of the LEDs and can also be used for the directions of routes [14].

17.4 GENERAL MODULATION TECHNIQUES

The optical channel is totally different from the typical RF channel. The frequency, phase, and amplitude of the carrier signal is modulated in RF systems whereas in optical systems, instead of frequency, phase, and amplitude, the intensity of the optical carrier is being modulated. The modulation techniques are used to regulate the LEDs at required frequencies that hold data to be transferred. IEEE 802.15.7 defined some of the most commonly used modulation techniques [15]; those used in optical wireless communications include OOK, PPM, PAM, DPIM, OFDM, CSK, etc. The modulation techniques integrate run length limited (RLL) series to circumvent prolonged sequences of 0s and 1s (that might generate flicker). RLL also handles data recovery as well as clock synchronization. In case of any error, forward error correction (FEC) techniques such as Reed Solomon (RS) are used [16]. In this section, some of the most commonly used modulation techniques and their features are described [6].

17.4.1 On–Off Keying

On–off keying (OOK) is the most commonly used modulation scheme; it modulates the LED from high to low and low to high. OOK-NRZ uses two distinct voltage levels for 0 and 1. The pulse occupy the whole bit duration of levels (i.e. 1 and 0) whereas OOK-RZ carries a small portion of bit duration. OOK-NRZ consumes more power than OOK-RZ due to its large duty cycle. In VLC, OOK-NRZ is widely employed because it does not have a rest state so it conveys additional data as compared to OOK-RZ [17].

17.4.2 Pulse Position Modulation

Pulse position modulation (PPM) is a modulation scheme in which data is transmitted in short durations at different positions of the pulse train. It is better than OOK in terms of power efficiency due to its short pulses, but requires more bandwidth due to its longer pulse train. PPM also circumvents low frequency and DC components of the spectra. PPM is more complex than OOK because it needs stricter bit as well as synchronization of the symbol at the receiving end. PPM consumes more bandwidth because it has more redundant space among pulses. To improve the spectrum efficiency and complexity, different flavors of the PPM are deployed. Differential PPM (DPPM) is deployed in VLC to attain a greater transmission capacity than PPM due to elimination of all idle time slits in every symbol. Overlapping PPM (OPPM) achieves a higher data rate by overlapping positions of the pulses. Multiple PPM (MPPM) and variable PPM (VPPM) schemes are employed usually to control dimming and transmission of data [6].

17.4.3 PULSE AMPLITUDE MODULATION

Pulse amplitude modulation (PAM) is the simplest modulation technique in which amplitude of the carrier signal is modulated. PAM is widely used in the VLC due to its sufficient bandwidth efficiency. PAM has many intensity levels which can vary the luminous efficacy in the LEDs. It also experiences shifts in color because of discrepancy in the drive current of the LED, as it is dependent on the LED color [18].

17.4.4 PULSE INTERVAL MODULATION

There are many techniques of pulse interval modulation (PIM) which have been proposed and inspected. These techniques have increased data rate, and consume less power by using multiple level amplitude modulation or by involving typical patterns of symbols. PIM encodes the data by injecting blank slits amongst every two symbol pulses. Digital pulse interval modulation (DPIM) is the modest form of PIM, which provides better results in spectrum efficiency than PPM by eliminating empty slots from PPM. DPIM is simpler than PPM because it doesn't need synchronization of the data at the receiver [19].

17.4.5 PWM WITH DISCRETE MULTI-TONE MODULATION

Discrete multi-tone modulation (DMT) is the baseband implementation of OFDM. It is employed for the channels which suffer interference, caused by synthetic atmospheric sources of light such as incandescent light or fluorescent light and so on. Pulse width modulation (PWM) with DMT demonstrates the sturdiness in a diffuse system and decreases the inter-symbol interference by employing the guard interval. DMT with QAM is also employed to offer data rates higher than 513 Mbps and to achieve spectral efficiency as well [20].

17.4.6 PULSE-POSITION-PULSE-WIDTH MODULATION

The pulse-position-pulse-width modulation (PPMPWM) technique is a combination of multilevel PPM and multilevel PWM. The hybrid technique overwhelms the issues of the pulse width and smaller spectral efficiency of PPM by broadening the pulses with the help of PWM. Broadening of the pulse conversely increases the power efficiency as well. It also offers a smaller bit error rate (BER) and stronger power spectrum lines as compared to PPM [21].

17.4.7 PULSE DUAL SLOPE MODULATION

Pulse dual slope modulation (PDSM) is used to avoid inter- and intra-frame flicker. The 0 bit is denoted by the growing edge and the 1 bit is denoted by the dropping edge of the duration of the pulse by keeping the other edge constant (dropping edge is constant for 0 bit and growing edge is constant for 1 bit). Both the bits (0 and 1) are altered by an identical quantity through the redundant bit period. As bits 1 and 0 are counter images to each other, there won't be any inconsistency in brightness during transmission and with redundant slots as well [4].

17.4.8 Orthogonal Frequency Division Multiplexing

Orthogonal frequency division multiplexing (OFDM) is used in VLC due to its spectral efficiency and higher data rate. In OFDM, multiple carrier signals (which are perpendicular to each other) are used to send multiple streams of data at the same time. OFDM also decreases the interference between symbols (namely inter symbol interference (ISI)) and the number of equalizers as well [22].

17.4.9 Color Shift Keying

Color shift keying (CSK) is an intensity modulation technique that is widely used in the optical communication system. In CSK, the data is transmitted through the variation of the color emitted by the LED. The intensity of every bit creates a specific color described in the CIE 1931 color coordinates. Basically, there are seven wavelengths (or colors): violet, blue, cyan, green, yellow, orange, and red. Every symbol has its own intensity which is created by the modulation of RGB chips [15].

QAM on DMT and PAM with multilevels are believed to be the most suitable modulation techniques for VLC as they better offer spectral efficiency and power efficiency as well. OOK and 2-PPM (2 is the number of levels) require equal amount of power. Therefore, some of the techniques offer power efficiency, and others offer spectral efficiency [3].

17.4.10 Selection of Modulation Schemes

The selection of the modulation scheme is based on the specific criteria, which should be described first. The parameters of the criteria are defined below [3].

17.4.10.1 Power Efficiency

VLC has regulations regarding skin and eye safety that require transmission at appropriate optical power. Power efficiency is the most important criteria to be considered for the VLC in order to select the appropriate modulation scheme. The modulation techniques provide average optical power to attain the required BER or SNR [3].

17.4.10.2 Bandwidth Efficiency

Although VLC is supposed to have infinite bandwidth, some elements such as area of photodetector, capacity of channel, and propagation in the multipath confine the available bandwidth. Therefore, the bandwidth or spectrum efficiency is the second-most vital metric to evaluate the modulation scheme for VLC. Power and bandwidth efficiencies are related to each other and their relationship is based on the average duty cycle of the pulse. Accordingly, receivers having one-element practice the photodetectors with larger area. The larger the photodetector area, the higher the capacitance thus causing reduction in the bandwidth of the receiver. Hence, the modulation techniques requesting higher bandwidth are vulnerable to ISI and subsequently suffer from higher power deprivation [23].

17.4.10.3 Transmission Reliability

Another significant factor for selection of the modulation scheme for VLC is transmission reliability. The modulation scheme must be capable of providing at least an adequate rate of error and should also resist ISI occurring due to the multipath induction as well as deviations in DC factor of the data signal. If the transition bits 0 and 1 are absent for a long time, the recovery of the clock could be challenging. Furthermore, consequences of high bits must be circumvented as the output signal will be passed from the high-pass filter at the receiving end. Moreover, the modulation scheme must be unaffected by phase jitter occurring due to deviations in the power of signal, extension of pulses, and distortion of the pulses [3].

17.4.10.4   Other Considerations

It is very necessary to choose the simplest modulation scheme. Although the modulation technique attains power efficiency and/or bandwidth efficiency, it is of no use due to its complexity. One more obligation while assessing the modulation schemes is the capability to discard interference stemming from ambient light sources. The power level of the external light sources is to be filtered by high-pass electrical filters.

17.5   POWER-EFFICIENT TECHNIQUES

VLC has been designed to convey data by using intensity modulation of light radiated from the LED. The LEDs radiate an incoherent thin light spectrum which is focused by current. This phenomenon is known as electroluminescence. As the LEDs release incoherent light, therefore, the information transmitted must be modulated into the LED's optical power and then demodulated into the original data. This modulation and demodulation is termed intensity modulation (IM) and direct detection (DD), respectively. To get the illumination effect, the LED should be powered on all the time, which might challenge the economic cost of the technology. To achieve the dimming effect with smaller cost, some techniques must be used. Restaurants and offices necessitate greater lighting whereas corridors, utility stores, and stairwells can use these dimmable LEDs. However, to attain effective dimming mechanism for a VLC link can reduce the reachable data rate. The dimming effect can be achieved by decreasing the forward current of the LED recognized in the IEEE 802.15.7 task group [15]. Mainly, the dimming is divided into two categories: analog or intensity-based dimming and digital or time-domain-based dimming [6].

17.5.1   ANALOG DIMMING

Analog dimming is provided by altering (decreasing) the magnitude of the forward-driving current of the LED. This phenomenon is known as continuous current reduction (CCR). It is a very modest technique of reducing light to execute [5].

17.5.2   MODULATION TECHNIQUES FOR ANALOG DIMMING

Bai et al. [24] have analyzed overlapping pulse position modulation (OPPM) for communication as well as illumination. As the modulation width of the LED is small, therefore, the narrower pulses of the data should be transmitted. In order to enhance the data rate, the authors have used OPPM. OPPM, as the name suggests, permits the pulses to overlap with one another. It divides interval pulses into NQ sub-intervals, where N is the index of overlap and Q is the non-overlapping positions of the pulse. The OPPM signal offers dimming from 0 to $1/Q$ percentage. It does not reach full brightness because of the ratio $1/Q$. The authors have also analyzed that symbols of OPPM have wide pulse width, which is independent of the levels of dimming. This is useful for optical receivers to detect a constant and narrow bandwidth. Then the authors compared the results to OOK, PPM, and VPM in terms of dimming, power, data rate, and flicker. The authors experienced that OPPM offers smaller flicker in every dimming level, covering a realistically large dimming variety [24].

Stefan et al. [25] have investigated the asymmetrically clipped optical orthogonal frequency division multiplexing (ACO-OFDM) technique. It converts initiated bipolar signals into unipolar terms by cutting the negative values at zero. It needs less power for every data rate. It works by using a CCR technique. Therefore, it is well-matched for dimming schemes, as it works better at lower SNR. The dimming range is distributed into dissimilar zones or regions, which depend upon distinct power levels. The power level can essentially be altered to enhance the beneficial dimming series as well as to preserve decent SNR. The authors have also evaluated the performance for BER versus transmitted power. The transmitted power is employed to prompt the adjustment among dissimilar zones [25].

Rajagopal et al. [26] have presented the standard dimming techniques for VLC specified in IEEE 802.15 Task Group 7 [15]. It describes some techniques of dimming, which are OOK, VPPM, and CSK. Dimming with OOK is attained with on/off stages to ensure lesser intensity. Otherwise, the OOK stages could be similar and only the duty cycle of the signal is altered by using the DC component. This time of DC component has been recognized by using a turned-off LED to provide dimming. VPPM alters the duty cycle for every optical character with respect to the required dimming level. The CSK utilizes amplitude dimming by altering the current of the LED. The amplitude variations in CSK are maintained by converting digital data to analog data to permit higher order modulation. The authors have also presented other dimming methods, for instance, dimming by using an idle pattern or dimming with a visibility pattern [26].

17.5.3 Digital Dimming

In digital dimming, a digital pulse of bits (0s and 1s) is modulated to constrain the LED at a stable current level. By using modulation techniques, signal intensity or average duty cycle is controlled for a certain time interval to dim the LEDs. Digital dimming is a matter of choice than analog dimming [5].

17.5.4 Modulation Techniques for Digital Dimming

Lee and Park [27] have examined the performance of VOOK, VPPM, and MPPM in terms of illumination control of LEDs by calculating normalized power and spectrum efficiency. Results showed that the VOOK offered good spectral efficiency as compared to VPPM. Moreover, MPPM controls the brightness level by increasing the number of chips. Alternatively, with less than 50% brightness, MPPM attained a spectrum efficiency of 1 bits/sec/Hz with additional power as compared to VOOK. VOOK and VPPM require the same power, but VOOK has improved spectral efficiency. Results also concluded that MPPM offered wider spectral efficiency and higher power efficiency as compared to VOOK and VPPM [27].

Dimming control is a crucial task of an illuminating system. Appropriate modulation technique helps to attain appropriate power and spectral efficiencies. Multilevel modulation techniques are influential for the spectral efficiency and pulse position modulations are robust for the power efficiency. To attain the profits of both techniques, Yi and Lee [28] proposed that a novel multilevel transmission technique named variable pulse amplitude and position modulation (VPAPM). They have compared VPAPM (when $M = 2$ and $M = 4$) to other modulation schemes such as OPPM, VOOK, RZ-OOK, and VPPM for power and spectrum efficiency. The results show that when $M = 2$, VPAPM and VOOK have identical bandwidth efficiency; thus, the bandwidth efficiency of VPAPM is enhanced by increasing the value of M as equated to the other modulation techniques. However, OPPM ensures static spectrum efficiency. It requires the highest power to attain certain BER over a certain transmission rate. VPAPM just requires addition of minute power for attaining a certain bandwidth efficiency. The choice for M value is influenced by the necessity for predictable power and spectrum efficiency. The smaller the levels of magnitude, the lesser the efficiency of bandwidth, and the lower would be the power [28].

Kim et al. [29] have presented a VLC model and computer simulations to acquire multipath dispersion characteristics. The authors also defined a VLC model by measuring the spectral reflectance and dissemination of spectral power of the LED. Modulation techniques stream the data as well as control the power of the LED. In Reference [29], the authors demonstrated and analyzed three modulation techniques, RZ-OOK, VOOK, and VPPM, to calculate performance of the VLC model for power and spectral efficiency. Results demonstrate that VOOK and VPPM consume the same amount of power but VOOK is better in terms of spectrum efficiency than VPPM. RZ-OOK consumes 3 dB less power than VPPM and VOOK with higher spectrum efficiency [29].

Ntogari et al. [30] have analyzed the execution of a VLC system by using PWM along with discrete multi-tone. PWM is used for dimming while discrete multi-tone is employed for transmission. In the proposed scheme, the authors multiplied the output to the PWM signal to attain dimming. The PWM signals are required to be sampled at the Nyquist rate, because when frequencies of PWM are very small, aliasing would occur, which in turn creates sub-carrier interference. Likewise, when frequency of the PWM exceeds the typical 3 dB bandwidth of the LED, there would be degradation in the performance of the system. To overcome this degradation, the authors used blue filtering to remove the higher bandwidth factor for dimming. Finally, the results conclude that the system has great performance with dimming [30].

17.5.5 COMPARISON OF DIGITAL DIMMING MODULATION SCHEMES

Table 17.1 lists some modulation schemes taken from the aforementioned papers [27–29]. This table compares different modulation schemes in order to conclude that which modulation scheme can reduce the power consumption. This table also calculates power consumption versus brightness, keeping the brightness factor constant at 80%.

17.5.6 OTHER DIMMING TECHNIQUES IN VLC

There are other methods too which can provide dimming, including inverse source coding, idle pattern dimming, and probability-based dimming.

17.5.6.1 Inverse Source Coding

Kwon [31] presented a method for dimming in VLC, known as inverse source coding. The method employs the proportion of 0s and 1s by means of off and on positions to acquire the desired ratio of dimming. Let's suppose the desired ratio of dimming is 50%; then there must be 50% 1 bits in the output. In other words, to attain the dimming ratio, the transmitted data bits are increased by the encoding scheme. The inverse source coding method generally uses Huffman coding, in which just an estimated calculation is done for the desired dimming ratio. The encoding technique adds additional filler bits to attain the desired bits as soon as the dimming ratio increases from 50%. This addition

TABLE 17.1
Comparison of Digital Dimming Modulation Techniques

#	VOOK	VPPM	MPPM	VPAPM	RZ-OOK	OPPM
			Digital Dimming Modulation Schemes			
[1]	0.625 dB power	0.625 dB power	0.325 dB power when $n = 20$ (chips) 0.275 dB power with bound n	—	—	—
[2]	0.12 dB power	0.12 dB Power	—	0.18 dB power when $M = 2$ levels 0.3 dB power when $M = 4$ levels	0.04 dB power	0.2 dB Power
[8]	0.625 dB power	0.625 dB power	—	—	0 dB power at 80% brightness and 0.625 dB power at 50% brightness	—

of filler bits degrades the data rate, therefore, other encoding techniques are recommended for a better data rate [31].

17.5.6.2 Idle Pattern Dimming

Oh [4] described another method for dimming, named idle pattern dimming. In this method, idle pattern of bits can be injected among the data frames. This causes the duty cycle to be diverged to offer deviation in brightness. The idle pattern has been divided into in-band/same frequency pattern and out-of-band/discrete frequency pattern. The in-band pattern does not need any clock variation so the receiver can easily notice it. An out-of-band pattern needs to be directed at a very inferior clock which is not noticed by the receiver. The method is also standardized in the IEEE 802.15.7 group. IEEE 802.15.7 permits injection of either "on" time or "off" time (named compensation time) of the LED into the data frame or into the idle pattern to decrease/upsurge the LED illumination. Other patterns have also been employed in the payload of the frame namely color visibility dimming (CVD). The patterns are in-band and known as visibility patterns. CVD is used for visibility, dimming, and color support. It delivers eminence of communication and caliber of channel to users by using colors. When a CVD frame is in the transmission process, the LED continues to emit light to accomplish the illumination [4].

17.5.6.3 Probability-Based Dimming

Likelihood or probability of discrete codes could be altered to attain the desired dimming condition. In Reference [32], the OOK modulation scheme has been stretched to the PAM to propose a dimming technique based on probability/likelihood. The scheme combines analog dimming with PAM constellation points to provide dimming requirements, developing a hybrid technique. The authors have also discussed the consequences of merging probability-based dimming and analog dimming. In the proposed hybrid technique, dimming is controlled by mutually improving concentration of the PAM and probability. The authors concluded that the optimum dimming can be attained by altering only the probability of a symbol and shifting of intensity is unnecessary [32].

17.5.7 Recommendations

VLC modulation schemes and dimming mechanisms of LEDs need compatibility and synchronization with each other. Section 17.5 describes some of the modulation schemes and some other methods to offer dimming mechanisms to the LED for acquiring better power efficiency. The dimming techniques are mainly divided into analog and digital dimming. The other methods such as idle pattern dimming, probability-based dimming, and inverse source coding methods provide sufficient dimming but they consume more bandwidth than the modulation techniques. Therefore, the modulation techniques are preferably used for power efficiency. Another advantage of using modulation techniques is to enlarge the spectral width of the LED. Digital dimming methods are convenient and they achieve more power efficiency than the analog dimming methods. Some of the digital modulation techniques are described and compared in Table 17.1. It demonstrates that MPPM, RZ-OOK, and OPPM consume less power than VOOK, VPPM, and VPAPM. The multilevel modulation schemes such as MPPM and VPAPM have flexibility of varying the brightness levels without any additional complexity. Although VOOK and VPPM consume almost equal power VPPM is more complex than VOOK.

17.6 SPECTRAL-EFFICIENT TECHNIQUES

Commercial LEDs are used in VLC as the source of transmission, but they have narrower modulation bandwidth. Therefore, the spectral efficiency of the LED will also be narrower. To broaden the spectral efficiency, many techniques are employed such as blue filtering, modulation beyond 3 dB, transmitter equalizaiton, receiver equalization, adaptive equalizaiton, sub-carrier equalization, etc. [6].

17.6.1 BLUE FILTERING

Blue filtering is a method that improves the bandwidth of the light source up to 3 dB. When the complete range of white light is utilized for detection, the modulation bandwidth will be restricted to approximately 2.5 MHz. The yellow phosphor light has a lower time constant, which does not react to the modulated signal in the LED. Consequently, it produces a stable light that saturates the receiving end and hence raises shot noise, causing degradation in the performance of the system. Blue filtering is utilized to improve the modulation bandwidth up to 20 MHz [33,34].

17.6.2 MODULATION ABOVE 3 DB BANDWIDTH

Initially, pulse shaping was the technique used for enhancement of the LED's frequency response. The technique was employed by the optical fiber to decrease the fall and rise time of the light sources. It uses simply a resistor and a capacitor in parallel with the LED in a series connection. When the LED is powered on, the superfluous current will flow via a capacitor, through which the LED will attain its stable condition faster. When the LED is powered off, the capacitor maintains swing out by making the diode work in reverse bias. Pulse shaping has been observed as a simple method that improves LED response time, consequently enhancing the bandwidth above 3 dB [35,36].

17.6.3 TRANSMITTER EQUALIZATION

Minh et al. [37] proposed a plain pre-equalization method used at the transmitter to upsurge functional bandwidth of the LED. The data is modulated and then equalized with the help of three corresponding driver circuits (one for low frequency, one for medium frequency, and one for high frequency). The equalized arrangement would be able to enhance the bandwidth to 45 MHz with 40 to 80 Mbps growth in the data rate. Various resonant equalization techniques are used to equalize many LEDs to further improve the bandwidth of the LED array. The resonant circuitry adjusts the entire array response therefore it can have greater bandwidth as compared to a single light source. The authors employed a collection having 16 LEDs to improve the bandwidth to 25 MHz [33].

17.6.4 RECEIVER EQUALIZATION

Chow et al. [38] demonstrated a post-equalization technique at the receiving end. It is used as a first-order equalizer following photodetector amplification, yielding 50 MHz bandwidth with a data rate of about 100 Mbps [38, 39], whereas Lubin et al. [40] have merged pre- and post-equalization at the transmitter and receiver sides, respectively. The results reported that this hybrid technique can raise the bandwidth up to 65 MHz.

17.6.5 ADAPTIVE EQUALIZATION

Komine et al. [41] have proposed an adaptive decision feedback equalizer (DFE) method with the help of a least mean squares (LMS) process. The decision is made by using the LMS algorithm over the data. The information is then employed to approximate the number of ISI symbols. The results concluded that a remarkable enhancement in data rate can be achieved [41].

17.6.6 SUB-CARRIER EQUALIZATION

Sub-carrier equalization is based on electrical-optical-electrical (EOE), used to upsurge the data rate in the DMT/OFDM multi carrier systems. Initially all the sub-carriers are adjusted to sustain same quality of transmission quality. Therefore, pre-equalization of the signal is a better choice than using post-equalization, because at the receiver side, when the signal is intensified, the noise will also be amplified along with the signal. Vucic et al. [20] have proposed a loading system for optimum and quick convergence. The loading process calculated the power of all sub-carriers and employed the constellation points of M-QAM/M-PSK. The phenomenon has been recognized by power and loading bit correspondingly. This process must be followed with extreme adequate BER along with total power of signal [20].

17.6.7 MODULATION ABOVE 3 DB BANDWIDTH

Siddique and Tahir [42] have proposed a multilevel pulse amplitude modulation technique that regulates the brightness and increases the data rate. The multilevel PAM (ML-PAM) is used to control the illumination by using the interconnected symbols of PAM. Although ML-PAM achieves higher brightness, it has smaller spectral efficiency. Therefore, in Reference [42], the authors have used multi-level MPPM (ML-MPPM) to attain greater brightness along with spectral efficiency. They also compared the ML-PPM with MPPM, OPPM, and VOOK in terms of spectral efficiency improvement, and concluded that the ML-PAM modulation technique achieves greater spectral efficiency as compared to others. Also, it delivers a prospect for adjustment in the brightness with minimum and maximum brightness levels.

17.6.8 COMPARISON OF MODULATION SCHEMES ABOVE 3 DB BANDWIDTH

Table 17.2 lists some modulation schemes taken from the aforementioned papers [27,28,42,43]. This table also presents comparisons of different modulation schemes to help decide on the most appropriate spectral efficient technique. This table demonstrates compares spectral efficiency against brightness with the brightness factor kept constant at 50%.

17.6.9 RECOMMENDATIONS

A number of modulation techniques are employed in VLC to offer transmission and brightness control simultaneously. These modulation techniques acquire smaller spectral efficiency by using a single LED. Table 17.2 comprises the different spectrum efficient modulation schemes. It concludes that VPAPM, OPPM, and ML-MPPM are more spectrum efficient than the others. VPAPM acquires sufficient spectral efficiency due to implementation of multilevel modulation, whereas ML-MPPM uses more than one LED to provide better brightness control and higher data rate. MPPM offers better spectral and power efficiency due to the use of multiple pulses. VOOK, VPPM, and RZ-OOK have sufficient spectrum efficiency with the consumption of less power. Unlike power efficiency, spectrum efficiency is also achieved with digital modulation techniques rather than analog. Spectrum efficiency is also achieved with the use of other techniques such as transmitter/receiver equalization, sub-carrier or adaptive equalization, blue filtering, etc., but these techniques increase the complexity of the system.

17.7 INTERFERENCE CANCELLATION TECHNIQUES

Additional light sources such as incandescent light, sunlight, and fluorescent light generate interference in transmission. The background noise is another source of interference as it allocates a similar frequency band as the data transmission. Therefore, it is necessary to remove the in-band interference to enhance the quality of the signal at the receiving end. Some techniques are introduced to overcome the interference such as the Manchester and the Miller encoding techniques, zero forcing and minimum mean square error, filters, signal clipping, optical beam forcing, etc. [6].

Saha et al. [44] have examined the performance of the VLC system under reflection and delayed propagation using an OOK scheme. They also reduced the ISI created due to multipath reflections with the help of the OFDM. OFDM uses a guard band to degrade the ISI. The authors also used QPSK

TABLE 17.2
Comparison of Modulation Techniques above 3 dB Bandwidth

#				Modulation Schemes			
	VOOK	VPPM	MPPM	OPPM	VPAPM	RZ-OOK	ML-MPPM
[1]	Full (1 bps/Hz) spectral efficiency	Half (0.5 bps Hz) spectral efficiency	0.89 bps/Hz spectral efficiency with $n = 20$ chips Full (1 bps /Hz) spectral efficiency with bound n	—	—	—	—
[2]	Full (1 bps/Hz) spectral efficiency	Full (1 bps/Hz) spectral efficiency		0.9 bps /Hz spectral efficiency	Full (1 bps /Hz) spectral efficiency when $M = 2$ levels 1.5 bps/Hz spectral efficiency when $M = 4$ levels	Half (0.5 bps/Hz) spectral efficiency	—
[3]	Full (1 bps/Hz) spectral efficiency	Half(0.5 bps/Hz) spectral efficiency	—	—	—	Full (1 bps/Hz) spectral efficiency	—
[4]	Full (1 bps/Hz) spectral efficiency	—	0.8 bps/Hz spectral efficiency	1.4 bps/Hz spectral efficiency	—	—	1.7 bps/Hz spectral efficiency

along with OFDM to reduce the ISI and compared the results to the results obtained by using the OOK scheme. It was concluded that more reflections to the walls occurred in the OOK scenario; therefore, there would be more ISI, which in turn degrades the system performance. In the OFDM technique, the guard band and use of a cyclic prefix has helped to reduce the multipath reflections and in turn also reduced the ISI. Therefore, by comparing both techniques, the authors demonstrated that OFDM provides better performance than OOK in terms of SNR and BER for a high speed communication system [44].

Cailean et al. [45] discussed the effects of noise (other light sources) on the VLC system. The noise mostly imposes its influence on the pulse width of the signal. To diminish the effect of noise, the authors compared two encoding schemes: Manchester coding and Miller coding. To get simulation results, the authors used messages with two coding schemes along with different stages of noise. Manchester coding is used with the OOK technique, therefore, it contains merely two codes (positive and negative) and it only points to two arrangements of pulse width—one_bit pulse width or two-fold bit pulse width. The Miller coding is used in applications having MIMO connections to get better bandwidth efficiency. The Miller code comprise three arrangements—one transient pulse width, one and a half transient pulse width, and double pulse width. Results revealed that with the decrease in SNR, pulse width is more affected by distortion. It is also observed that Miller coding has less distortions in the digital signals and is strictly affected by tolerances. In contrast, Manchester coding has less of a distortion effect with digital filtering but the greater tolerance can reimburse pulse width variations. Manchester and Miller coding have the same performance effects in BER [45].

Park et al. [46] explored the environment concerning aircraft wireless communication using a VLC system. The authors have performed a successive interference cancellation (SIC) technique through optimum collation to select the required data symbols and cancel out all the interfering symbols. The scheme is used to mitigate the interference and its outcome is measured through the BER. They also used zero-forcing (ZF) and minimum mean square error (MMSE) procedures to terminate the interference of the system. The study showed that SIC obtains higher performance as compared to the conventional algorithms. The results also concluded that the BER is increased with the use of a linear equalizer as compared to the results taken without an equalizer. The results of the paper can be used in the strategy for as well as the execution of wireless communication systems for aircrafts [46].

17.7.1 Comparison of Interference Cancellation Techniques

Table 17.3 lists some modulation schemes taken from aforementioned papers [44–47]. This table demonstrates the outcomes of the different modulation schemes. The table computes the BER versus SNR. The table compares three types of different methods to avoid interference such as modulation techniques (OOK, QPSK-OFDM), encoding schemes (Manchester, Miller), and the successive interference cancellation methods (ZF, MMSE).

TABLE 17.3

Comparison of Interference Cancellation Techniques

			Interference Cancellation Techniques			
#	OOK	QPSK-OFDM	Manchester Coding	Miller Coding	ZF	MMSE
[5]	0.2 BER	0.1 BER	—	—	—	—
[6]	—	—	0.000001 BER	0.000001 BER	—	—
[7]	—	—	—	—	0.01 BER	0.01 BER

17.7.2   RECOMMENDATIONS

Interference is a vital problem in VLC and it occurs when the signal from neighboring LED lights and other external light sources strike the transmission. The interference is treated as noise, which degrades the transmission data rate. There are different methods to overcome the disturbances caused by the interference like specific modulation and encoding techniques or other interference cancellation methods. Table 17.3 analyzes and compares the different interference cancellation techniques. It demonstrates that encoding schemes such as Manchester and Miller coding schemes offer better BER than the other considered techniques. The BER of both schemes is the same but they are employed in different criteria. The Manchester encoding scheme is mainly utilized to diminish ambient optical noise in single-channel communications whereas the Miller encoding technique is more suited for MIMO applications.

17.8   CONCLUSIONS

LEDs appear to be an energy efficient as well as auspicious lighting arrangement, with low unit prices. The fast switching ability of LEDs permits them to be employed as an optical source in VLC. An (indoor) VLC system covers the unlicensed bandwidth range from 700 nm to 10,000 nm with data rates beyond 2.5 Gbps at lower cost. VLC systems mostly employ diffuse topology due to its convenience for ad hoc LAN networks as it does not need direct alignment between the transmitter and receiver. Despite all the advantages, diffuse links suffer from high dB path loss, multipath dispersion, and inter-symbol interference causing higher power to be transmitted. Therefore, modulation schemes are deployed to control the power and bandwidth requirements. Dimming techniques are employed in VLC systems to reserve energy and to permit the consumers to have full control over brightness. A number of single and multiple carrier modulation techniques are proposed for VLC communication systems. The modulating signal is comprised of data, which is then superimposed to the DC current to drive the LED, generating intensity modulated light that is received by the receiver. In this chapter, different modulation techniques are discussed and analyzed in terms of power and spectrum efficiency. The power-efficient techniques are categorized into analog and digital dimming whereas there are other techniques as well to reduce illumination and offer power efficiency. Many techniques have been reviewed and compared and analysis shows that digital dimming techniques are widely used and are more appropriate in terms of data rate. There are many other challenges too which degrade the performance of VLC such as ambient light sources, shadowing and blocking, and tracking and alignment. External sources of light such as sunlight, reflections of mirrors, and other fluorescent/candescent lights can create noise. To overcome this noise, modulation schemes are used. Shadowing and blocking cause inter-symbol interference which can be mitigated by deploying equalizers at the transmitter as well as receiver. To mitigate the effect of other light sources, some modulation techniques, encoding techniques, and other interference cancellation methods are explained and demonstrated.

The future of VLC systems looks very bright owing to the increasing reputation of LEDs. As LEDs are anticipated to hastily replace the traditional lighting infrastructure, VLC is predicted to be eagerly executed into wide-ranging lighting technologies which conversely will increase the number of applications. The user would be able to communicate via VLC with a "home network," and be able to handle and monitor all activities as done in RF communication systems. The potential of VLC owing to be most commonly used in indoor localization and intelligent transport networks, the concept of smart grid would be extended. It would further allow the end users to employ VLC for traffic updates, information systems, location estimation, communication via traffic lights or car headlights, and most importantly in underwater communication systems and airplanes.

REFERENCES

1. M. Mutthamma, A survey on transmission of data through illumination - Li-Fi, *International Journal of Research in Computer and Communication Technology*, vol. 2, no. 12, pp. 1427–1430, December 2013.
2. P. Bandela, P. Nimmagadda, and S. Mutchu, Li-Fi (light fidelity): The next generation of wireless network, *International Journal of Advanced Trends in Computer Science and Engineering*, vol. 3, no.1, pp. 132–137, 2014.
3. 31. Z. Ghassemlooy, W. Popoola, and S. Rajbhandari. *Optical Wireless Communications System and Channel Modeling with MATLAB*. New York, NY: CRC Press Taylor & Francis Group, 2013.
4. M. Oh, A flicker mitigation modulation scheme for visible light communications, *2013 15th International Conference on Advanced Communication Technology (ICACT)*, PyeongChang, South Korea, January 27–30, 2013. ISBN 978-89-968650-1-8.
5. F. Zafar, D. Karunatilaka, and R. Parthiban, Dimming schemes for visible light communication: The state of research, *IEEE Wireless Communications*, vol. 22, pp. 29–35, April 2015.
6. D. Karunatilaka, F. Zafar, V. Kalavally, and R. Parthiban, LED based indoor visible light communications: State of the art, *IEEE Communication Surveys & Tutorials*, vol. 17, no. 3, pp. 1649–1678, Third Quarter 2015.
7. S. Hranilovic, On the design of bandwidth efficient signaling for indoor wireless optical channels, *International Journal of Communication Systems*, vol. 18, pp. 205–228, 2005.
8. D. O'Brien et al., Visible light communication. In *Short-Range Wireless Communications: Emerging Technologies and Applications*, R. Kraemer and M. Katz, Eds. Chichester, UK: John Wiley & Sons, Ltd., 2009.
9. S. Randel, F. Breyer, S. C. J. Lee, and J. W. Walewski, Advanced modulation schemes for short-range optical communications, *IEEE Journal of Selected Topics in Quantum Electronics*, vol. 16, pp. 1280–1289, 2010.
10. J. Vucic, C. Kottke, S. Nerreter, A. Buttner, K. D. Langer and J. W. Walewski, White light wireless transmission at 200+ Mb/s net data rate by use of discrete-multitone modulation, *IEEE Photonics Technology Letters*, vol. 21, pp. 1511–1513, 2009.
11. M. Hoa Le et al., High-speed visible light communications using multiple-resonant equalization, *IEEE Photonics Technology Letters*, vol. 20, pp. 1243–1245, 2008.
12. L. U. Khan, Visible light communication: Applications, architecture, standardization and research challenges, digital communications and networks, 2016. Available: http://dx.doi.org/10.1016/j.dcan.2016.07.004
13. X.-W. Ng, and W.-Y. Chung, VLC-based medical healthcare information system, *Biomedical Engineering Applications Basis and Communications*, vol. 24, no. 2, pp. 155–163, 2012.
14. S.-B. Park et al., Information broadcasting system based on visible light signboard, Presented at Wireless and Optical Communications, Montreal, Canada, 2007.
15. IEEE 802.15 WPAN Task group 7 (TG7) visible light communication, 2014. Available: http://www.ieee802.org/15/pub/TG7.html
16. O. Bouchet et al., Visible-light communication system enabling 73 mb/s data streaming, in Proceeding of IEEE GLOBECOM Workshops, Miami, FL, 2010.
17. J. Vucic et al., 125 Mbit/s over 5 m wireless distance by use of OOK modulated phosphorescent white LEDs, in Proceeding of 35th ECOC, Vienna, Austria, 2009.
18. S. H. Lee, K.-I. Ahn, and J. K. Kwon, Multilevel transmission in dimmable visible light communication systems, *Journal of Lightwave Technology*, vol. 31, no. 20, pp. 3267–3276, October 2013.
19. Z. Ghassemlooy and S. Rajbhandari, Convolutional coded dual header pulse interval modulation for line of sight photonic wireless links, *IET—Optoelectronics*, vol. 3, pp. 142–148, 2009.
20. J. Vucic, C. Kottke, S. Nerreter, K. Langer, and J. W. Walewski, 513 Mbit/s visible light communications link based on DMT-modulation of a white LED, *Journal of Lightwave Technology*, vol. 28, no. 24, pp. 3512–3518, December 2010.
21. Y. Fan, B. Bai, and R. Green, PPMPWM: A new modulation format for wireless optical communications, in Proceeding of 7th International Symposium on Communication Systems Networks and Digital Signal Processing (CSNDSP), Newcastle, UK, 2010.
22. H. Elgala, R. Mesleh, H. Haas, and B. Pricope, OFDM visible light wireless communication based on white LEDs, in Proceeding of IEEE 65th Vehicular Technology Conference (VTC), Dublin, Ireland, 2007.
23. N. Hayasaka and T. Ito, Channel modeling of nondirected wireless infrared indoor diffuse link, *Electronics and Communications in Japan*, vol. 90, pp. 9–19, 2007.
24. B. Bai, Z. Xu, and Y. Fan, Joint LED dimming and high capacity visible light communication by overlapping PPM, *2010 19th Annual Wireless and Optical Communications Conference (WOCC)*, Shanghai, China. 978-1-4244-7596-4/10/$26.00 ©2010 IEEE.

25. I. Stefan, H. Elgala, and H. Haas, Study of dimming and LED nonlinearity for ACO-OFDM based VLC systems, *2012 IEEE Wireless Communications and Networking Conference: PHY and Fundamentals*, Paris, France.

26. S. Rajagopal, R. D. Roberts, and S.-K. Lim, IEEE 802.15.7 Visible light communication: Modulation schemes and dimming support, *IEEE Communications Magazine*, vol. 50, pp. 72–82, March 2012.

27. K. Lee and H. Park, Modulations for visible light communications with dimming control, *IEEE Photonics Technology Letters*, vol. 23, no. 16, pp. 1136–1138. August 15, 2011.

28. L. Yi and S. Gol Lee, Performance improvement of dimmable VLC System with variable pulse amplitude and position modulation control scheme, *2014 International Conference on Wireless Communication and Sensor Network*, IEEE Computer Society Washington, DC, USA, pp. 81--85, December 2014.

29. J. Kim, K. Lee, and H. Park, Power efficient visible light communication systems under dimming constraint, 2012 IEEE 23rd International Symposium on Personal, Indoor and Mobile Radio Communications (PIMRC), Sydney, NSW, Australia, pp 1968–1973, September 2012.

30. G. Ntogari, T. Kamalakis, J. W. Walewski, and T. Sphicopoulos, Combining illumination dimming based on pulse-width modulation with visible-light communications based on discrete multitone, *Journal of Optical Communications and Networking*, vol. 3, no. 1, pp. 56–65, January 2011.

31. J. K. Kwon, Inverse source coding for dimming in visible light communications using NRZ-OOK on reliable links, *IEEE Photonics Technology Letters*, vol. 22, no. 19, pp. 1455–1457, October 1, 2010.

32. J.-Y. Wang, J.-B. Wang, M. Chen, and X. Songt, Dimming scheme analysis for pulse amplitude modulated visible light communications, *2013 International Conference on Wireless Communications & Signal Processing (WCSP)*, Shanghai, China, pp 1–6, October 2013.

33. H. L. Minh et al., 80 Mbit/s visible light communications using pre-equalized white LED, in Proceeding of 34th ECOC, Brussels, Belgium, 2008.

34. J. Grubor, S. Randel, K. D. Langer, and J. W. Walewski, Bandwidth-efficient indoor optical wireless communications with white light-emitting diodes, in Proceeding of 6th International Symposium CNSDSP, Graz, Austria, 2008.

35. E. Schubert, *Light-Emitting Diodes*. Cambridge: Cambridge University Press, 2006.

36. E. Schubert, N. Hunt, R. Malik, M. Micovic, and D. Miller, Temperature and modulation characteristics of resonant-cavity light-emitting diodes, *Journal of Lightwave Technology*, vol. 14, no. 7, pp. 1721–1729, July 1996.

37. H. L. Minh et al., High-speed visible light communications using multiple-resonant equalization, *IEEE Photonics Technology Letters*, vol. 20, no. 14, pp. 1243–1245, July 2008.

38. C. W. Chow, C. H. Yeh, Y. F. Liu, and Y. Liu, Improved modulation speed of LED visible light communication system integrated to main electricity network, *Electronics letters*, vol. 47, no. 15, pp. 867–868, July 2011.

39. H. L. Minh et al., 100-Mb/s NRZ visible light communications using a post-equalized white LED, *IEEE Photonics Technology Letters*, vol. 21, no. 15, pp. 1063–1065, August 2009.

40. Z. Lubin et al., Equalization for high-speed visible light communications using white-LEDs, in Proceeding of 6th International Symposium on Communication Systems, Networks and Digital Signal Processing (CNSDSP), Graz, Austria, 2008.

41. T. Komine, J. H. Lee, S. Haruyama, and M. Nakagawa, Adaptive equalization system for visible light wireless communication utilizing multiple white LED lighting equipment, *IEEE Transaction on Wireless Communication*, vol. 8, no. 6, pp. 2892–2900, June 2009.

42. A. B. Siddique and M. Tahir, Bandwidth efficient multi-level MPPM encoding decoding algorithms for joint brightness-rate control in VLC systems, *Globecom 2014 – Optical Networks and Systems Symposium*, Austin, TX USA, pp. 2143–2147, December 2014.

43. K. Lee and H. Pak, Channel model and modulation schemes for visible light communications, *2011 IEEE 54th International Midwest Symposium on Circuits and Systems (MWSCAS)*, Yonsei University Seoul, Korea (South), August 2011.

44. N. Saha, R. K. Mondal, N. Tuan Le and Y. M. Jang, Mitigation of interference using OFDM in visible light communication, *2012 International Conference on ICT Convergence (ICTC)*, Jeju Ramada Plaza Hotel Jeju, Korea (South), October 2012.

45. A.-M. Cailean, B. Cagneau, L. Chassagne, V. Popa, and M. Dimian, Evaluation of the noise effects on visible light communications using Manchester and Miller coding, 12th International Conference on Development and Application Systems, Suceava, Romania, May 15–17, 2014.

46. I. H. Park and Y. H. Kim, IEEE Member, and Jin Young Kim, Interference mitigation scheme of visible light communication systems for aircraft wireless applications, *2012 IEEE International Conference on Consumer Electronics (ICCE)*, Las Vegas, NV, USA, January 2012.

47. Y. F. Liu, C. H. Yeh, Y. C. Wang, and C. W. Chow, Employing NRZI code for reducing background noise in LED visible light communication, *2013 18th OptoElectronics and Communications Conference held jointly with 2013 International Conference on Photonics in Switching (OECC/PS)*, Kyoto, Japan, pp. 1–2, 2013.
48. J. B. Carruthers and J. M. Kahn, Modeling of non-directed wireless Infrared channels, *IEEE Transaction on Communication*, vol. 45, pp. 1260–1268, 1997.

Part IV

Big Data and the Internet of Things

18 A Data Aware Scheme for Scheduling Big Data Applications with SAVANNA Hadoop

K. Hemant Kumar Reddy, Himansu Das,
and Diptendu Sinha Roy

CONTENTS

18.1   INTRODUCTION

The last few years has seen unprecedented advancements in the computing paradigm by means of innovations including numerous virtualization and network technologies, service-oriented architectures, and so forth, leading to computation archetypes such as grid computing [1]. However, evolving applications such as bio-medical science and engineering and IoT have accelerated novel data-intensive computation paradigms. Cloud computing is hyped as the future of information technology paradigms in that it can leverage computing as a utility by providing massive computational and data resources in virtualized environments [2].

Hadoop, built on the map-reduce paradigm, provides enormous data management competence to data-intensive jobs [3]. It permits users to exploit computation and data resources of a Hadoop cluster through its map-reduce implementation, managing a number of low-level issues, such as communication, fault tolerance, and data management among others and providing a simple data management interface. Owing to performance improvement and better reliability, Hadoop breaks up a file into several data blocks and puts them in different nodes within the Hadoop cluster [3].

However, Hadoop does not keep track of placement of data blocks among cluster nodes after splitting a data file while making decisions regarding scheduling of jobs. Every job has its own data file requirement and thus scheduling jobs in a placement-oblivious manner leads to runtime data movements among the Hadoop cluster. On the contrary, if jobs could be scheduled among Hadoop cluster nodes with *a priori* information of the location of data blocks they require, then such unnecessary data movements could be reduced at runtime. Such an intelligent data placement and subsequent job scheduling has been observed to improve data-computation co-location and has been a proven technique to improve performance [4]. This chapter discusses a scheduling mechanism that is aware of the location of data needed for jobs within the Hadoop cluster and has been designated as Data Aware Computation Scheduling (DACS). The proposed DACS includes (1) a data partitioning algorithm that segregates blocks into logical partitions in terms of usage of those blocks by Hadoop jobs. (2) It subsequently puts such partitions into different nodes of a Hadoop cluster through a novel partition placement algorithm, and finally (3) schedules computations such that computation and data co-location is exploited. This chapter deals with the design and deployment of DACS and also presents its performance in comparison with Hadoop's Native Data Placement (HNDP).The Hadoop cluster has been deployed using Savanna-Hadoop on OpenStack to provide the data parallel framework. Savanna-Hadoop on OpenStack [5] supports different Hadoop distributions and solves bare cluster provisioning and analytics as a service. The DACS framework has been implemented on top of this infrastructure. A standard map-reduce job, namely weather forecasting and analysis of temperature data from the weather dataset obtained from the National Climatic Data Center (NCDC) [6,7], has been employed to test the efficacy of the proposed DACS.

Of late, there has been a number of research initiatives that have delved into placement of data in map-reduce parlance. In Table 18.1, a summary of the research contributions have been summarized that will help identify the progressive developments in this area of research.

In perspective of the aforementioned works, the DACS discussed in this chapter has its own niche. It does not consider network topology or balancing cluster memory utilization while making decisions regarding where to place data blocks within a Hadoop cluster. Also, DACS does not delve

TABLE 18.1
Summary of Research Findings in the Area of Data Placement in the Map-Reduce Paradigm

Reference	Author(s)	Contribution Summary
[8]	A. Amer et al.	Presented an intuitive grouping of files that is motivated by the generalized caching principle and has been shown to have better performance in distributed file management. Unfortunately, all big data management frameworks such as Hadoop and its offsprings thrive on dividing a file into multiple blocks and thus this work is rooted at a different level of granularity than the map-reduce paradigm.
[9]	J. Wang et al.	Proposed DRAW, a Data-gRouping-AWare system that dynamically observes access patterns of logs regarding which application needed which data. Thereafter DRAW provides clustering data information and alters data outlines such that parallel access of clustered data can be extracted; this has been demonstrated to provide improved performance.
[10]	D. Yuan et al.	Addressed the domain of data intensive e-science workflows through capturing interdependency between datasets and targeting runtime reduction of inter-cluster data movements. This work has load balancing as one of the goals.
[11]	J. Xie et al.	Contributed a speculative task scheduling on map-reduce frameworks based on location information of data needed by tasks. Additionally load balancing of resources utilization was incorporated.
[12]	S. Sehrish et al.	Contributed a framework, namely, MRAP, MapReduce with Access Patterns, that provides a set of APIs for consolidating future data access patterns founded upon *a priori* knowledge of previous data accesses by other map-reduce tasks.
[13]	J. M. Cope et al.	This work is fundamentally different from the other related works in the way that the main focus here is to provide guaranteed robustness while placing data.
[14]	N. Hardavellas et al.	Proposed a novel dual block placement cum replication management scheme, namely, NUCA, for distributed caches that diminish latency access.

into harnessing performance benefits out of Hadoop's parallel access paradigm; rather it attacks another aspect, namely to improve data-compute colocation for reduced runtime data movements.

The rest of the chapter is organized as follows. Section 18.2 provides details of the HNDP scheme with appropriate examples. In the next section, the DACS scheme has been introduced highlighting the essential design guidelines. Section 18.3 details the background of Savanna-Hadoop, which is deployed for performance evaluation of the proposed DACS scheme. Section 18.5 explains the experiments conducted on the deployed system and provides considerations on results found with their implications. Lastly, the conclusions have been discussed in Section 18.6.

18.2   PLACING DATA AS PER HADOOP'S NATIVE STRATEGY

The fundamental strategy of Hadoop [15,16] is to exploit parallel data accesses from among a cluster of nodes (containing slices of data files needed by running task) where data are stored by means of the map-reduce paradigm. While doing so, the native strategy that Hadoop adopts is to place the split data files (called blocks) among the cluster nodes with redundancy. However, which block of a data file goes to which cluster node is not bound by any rule. It is done randomly. Thus, an application program that needs data from a file stored in the Hadoop distributed file system (HDFS) achieves performance benefits. However, the majority of real-world application programs reveal correlation between the data they require. For example, researchers in the meteorology domain are concerned about precise periods for carrying out their analyses [17]. Furthermore, data gathered over the social networking domain also discloses great extents of correlation in terms of country, linguistics, occupation, and some additional features [18]. For the clarity of Hadoop's default strategy for placing data among cluster nodes, this section summarizes the HNDP.

In HNDP strategy, the driving principle in data distribution among cluster nodes is space balance among clusters' memory resources. This is motivated by the possibility of exploiting data retrievals in parallel by multiple jobs from the HDFS. The data files are split into multiple blocks to facilitate these parallel accesses. However, the time taken to access data differs considerably depending on whether data is available on that node or it has to be fetched from other remote nodes at runtime.

Figure 18.1 portrays the default HNDP strategy where it avoids stand for jobs that constitute running application needing data from HDFS and boxes represent blocks of data that are but split fractions of a data file. Block $F_i B_j$ in the figure depicts that the jth block of a file i is present in HDFS. A job T_i if scheduled might discover performance obstacles in the form of remote data access, or in other words, the data needed by task T_i is not present in the node where it is scheduled.

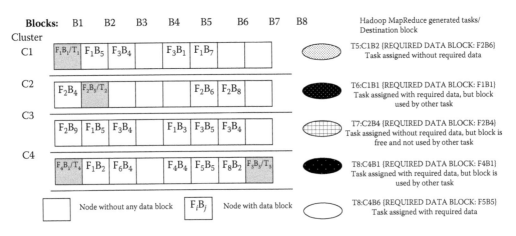

FIGURE 18.1   Illustration of Hadoop's native data placement.

18.2.1 Hadoop's Native Data Placement: An Illustration

To demonstrate HNDP's working principle in this section we take an intuitive example to describe the data placement policy that native Hadoop deployments follow. Table 18.2 portrays the data requirements of eight map tasks among 21 blocks kept in HDFS. The required blocks for every map-reduce job is presented in Table 18.3.

In this subsection, we delineate the working principle of HNDP with an intuitive example where the data blocks required by eight map jobs are shown out of a total of 21 data blocks in the HDFS. Table 18.3 depicts the blocks required by these jobs. Table 18.2 shows the five randomly made partitions for this set of jobs. Hadoop's *namenode* maintains all this information.

As mentioned earlier, Hadoop splits data files into a number of blocks and places these blocks among cluster nodes, usually with a specified degree of redundancy. For the example considered here, we can see that HDFS forms five logical partitions for the 21 data blocks and these five logical partitions are then randomly distributed among four cluster nodes. In HNDP, the goal of placing logical partitions to cluster nodes is governed by the load balancing criterion. For instance, we can see from Table 18.3 that job t2 requires blocks {b2, b3, b6, b11, b13, b15}. Now, this cluster has to be executed on any one of the four cluster nodes. Figure 18.2 illustrates how data blocks of logical paritions are mapped to cluster nodes. For example, it can be seen from Figure 18.2 that task t2 is scheduled to cluster C2. From this information and from Table 18.3, it can be observed that out of the blocks required by job t2, four data blocks are further needed (not present in cluster C2). These are {b11, b13, b6, b9}. These blocks thus have to be accessed from remote clusters at runtime. Table 18.4 shows the number of block movements required for each task for the example considered here.

The proposed DACS scheme attempts to reduce this runtime data movement and in order to do this, we have used two novel algorithms which have been detailed in the following sections.

TABLE 18.2
Random Partitioning of Blocks in HNDP

Partition	Required Blocks				
P1	b7	b2	b5	b15	b11
P2	b1	b3	b6	b9	
P3	b4	b12	b13		
P4	b8	b10	b14	b16	
P5	b17	b18	b19	b20	b21

TABLE 18.3
Example of Data Accesses by Map Tasks

Partition	Required Blocks					
P1	b1	b4	b7	b9	b12	b16	
P2	b3	b6	b11	b13	b15	b2	
P3	b2	b10	b9	b13	b8	b5	
P4	b5	b12	b6	b1	b4	b13	
P5	b6	b14	b4	b10	b12	b18	
P6	b4	b7	b3	b21	b15	b13	
P7	b2	b17	b9	b16	b11	b4	b6
P8	b16	b17	b13	b19	b12	b20	b21

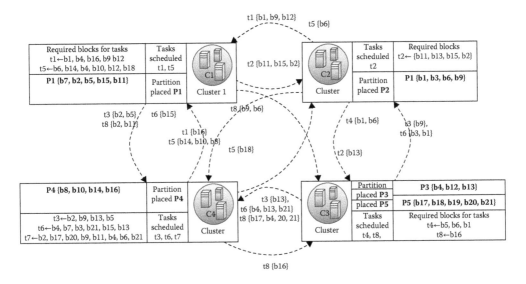

FIGURE 18.2   Mapping the logical data partitions to cluster nodes.

TABLE 18.4

Required Block Movements for HNDP

Task	Data Blocks Required for Executing Task (t$_i$)	No. of Blocks Moved	Assigned Partitions with HNDP	Partition Placed in Cluster ID
t1	{b1, b4,b16, b9, b12}	5	P1={b7, b2, b5, b15, b11}	C1
t2	{b11, b13, b15, b2}	4	P2={b1, b3, b6, b9}	C2
t3	{b2,b9, b13, b5}	4	P3={b4, b12, b13}	C3
t4	{b5, b6,b1}	3	P4={b8, b10, b14, b16}	C4
t5	{b6, b14, b4,b10, b12, b8}	6	P5={b17, b18, b19, b20, b21}	C3
t6	{b4, b7, b3, b1, b15, b13}	6	–	–
t7	{b2, b17, b20, b9, b11, b4, b6, b21}	8	–	–
t8	{b16}	1	–	–

18.3   DATA AWARE COMPUTATIONAL SCHEDULING SCHEME

As mentioned in Section 18.2, the fundamental motivation of the proposed DACS is to reduce access time by efficient placement of data blocks within a map-reduce cluster such that compute and data are collocated. The proposed DACS completes the partition and placement operations as a two-step process, where (1) segregating of data blocks into logical partitions is done in step 1 based on data correlation. This is followed by (2) allocation of these logical partitions to different clusters. These steps have been affected by implementing two algorithms discussed in the two subsequent subsections. For ease of understanding of the logical partitioning and placement of these to clusters in DACS, the same intuitive example has been employed in Section 18.3.3.

18.3.1   Constructing Logical Partitions from Correlation among Data Blocks

In this subsection, we present the logical partitioning scheme based on information of jobs and their data requirements. This information can be obtained from Hadoop's *namenode* logs. From the *namenode* logs, we can extract related information to form a number of tables, the details of which

TABLE 18.5

Cluster Partition Table

Task ID	Partition ID	Cluster ID
t1	P1	C1
t2	P2	C1
t3	P3	C3
t4	P4	C4
t5	P5	C2
t6	P6	C1
t7	P7	C4
t8	P8	C3

TABLE 18.6

Block Requirement Table

Task ID	Blocks							
t1	1	4	7	9	12	16		
t2	3	6	11	13	15	2		
t3	2	10	9	13	5	8		
t4	5	12	6	1	4	13		
t5	6	14	4	10	12	18		
t6	4	7	3	21	15	13		
t7	2	17	20	9	16	11	4	6	21
t8	5	17	13	19	12			

TABLE 18.7

Dependency Table

Partition ID	Blocks						
P1	1	4	7	9	12	16	
P2	3	6	11	13	15	2	
P3	2	10	9	13	5	8	
P4	5	12	6	1	4	13	
P5	6	14	4	10	12	18	
P6	7	3	21	15			
P7	2	17	20	9	16	11	21
P8	17	19					

have been provided later in this section. For example, Table 18.3 contains the information about data blocks that are required for a transaction and Table 18.2 contains information about the logical partitions that are assigned to different clusters depending upon the correlation (dependency) value. Finally, the information pertaining to the obtained logical partitions are assigned to a cluster and such information has been captured in Table 18.5. Tables 18.6 through 18.9 have been prepared as the algorithm for creating logical data partitions executes. Based on correlation of data blocks for execution of jobs, dependency values are computed in Tables 18.8 and 18.9 and the pseudocode for the logical partitioning algorithm is presented in Algorithm 18.1.

TABLE 18.8
Dependency-Based Partition Table

Cluster ID	MaxPvalue (3)	Partition(s)		
C1	0	1	6	2
C2	0	5	7	8
C3	2	3		
C4	1	4	8	

TABLE 18.9
Logical Partition Dependency Table

Partition A	Partition B	Dependency Value
P1	P6	1
P2	P6	1
P5	P7	1
P5	P8	1
P4	P8	1

Algorithm 18.1   Logical Data Partition Algorithm

CPT: Cluster Partition Table {taskId assigned to PartitionId and ClusterId is empty initially}
BRT: Block Requirement Table {tasked, with required blocks for execution}
DPT: Dependency Table {information about assigned blocks to partitions}
BBPT: Dependency-Based Partition Table

For each task ti : BRT
  Set PCT←BRT
  (copy all task id from BR table to PC table)
Endfor
For each task ti : BRT
    For each block Bj :BRT(ti)
        If (No. of replica of Bj in PCT < 3) then
            Move Bj to DPT of same partitioned
            Increment replica value of Bj
      Else
            For each partition Pk : DPT
                For each block Bl : Pk(DPT)
                    If (Bl=Bj) then
                    Map taskId(Bj) and taskId(Bl) with CP table
                    Place the corresponding partitions in DBPT to create a dependency
                    between partitions with clustered
                    Decrement MaxPartition value
                    Endif
                Endfor
            Endfor
      Endif
    Endfor
  Endfor
End.

This section presents the essential features of tables used in the proposed DACS framework. In the case of big data applications, data need to be partitioned based on some criteria and this partition information needs to be stored for further reference. Tables 18.5 and 18.6 present store data partition information, data block information i.e. which block include in which partition and data dependency among these blocks. Table 18.5 shows information about the partitions that are assigned to clusters and the tasks that required the blocks belonging to partitions. With the help of dependency table information (Table 18.5), partitions are efficiently allocated to clusters. Table 18.6 presents the information about the set of blocks that is required for the completion of a particular task. Information stored in Tables 18.5 and 18.6 is static in nature.

In the DACS framework, Tables 18.7 through 18.9 play a major role in creating partitions and efficient allocation of partitions to clusters. Tables 18.7 and 18.8 are occupied in due course of execution of the proposed algorithms. As described in this section, based on the dependencies that exist between blocks and map-reduce tasks, dependency values are calculated in Table 18.8 and the degree of dependence between partitions is computed and stored in Table 18.9. Table 18.8 depicts the cluster-wise dependencies between partitions, where MaxPvalue indicates the maximum dependency between partitions that is allowed. Table 18.7 is populated based on the values of Tables 18.8 and 18.9 that are calculated in the course of execution of the data-aware scheduling algorithms.

18.3.2 Dependency-Based Data Partition Placement

Once logical partitions of data blocks are completed using Algorithm 18.1, these partitions are placed within cluster nodes by dependency value, such that high interdependency between the partitions implies that such data blocks will be placed in same logical partitions. Table 18.8 depicts those logical partitions with a dependency value in excess of zero. Thereafter, such partitions are assigned to a particular cluster node. Table 18.9 forms clusters of partitions with high correlation and arranges them in ascending order to provide high preference to highly interdependent partition groups. Then a data aware co-location scheduling algorithm is used to map the jobs to an appropriate cluster using information from Table 18.5. The goal of this step is to minimize data movements during the map-reduce execution. Detailed procedural steps are presented in Algorithm 18.2. For this scheduling, APIs of Savanna-Hadoop are, employed within the DACS framework.

18.3.3 An Intuitive Illustration

In this chapter, we present an adaptive scheme for forming logical partitions from information of data block requirements by jobs. The *namenode* logs are the primary source for preparing these partitions. Job t_1 requires six data blocks, namely b_1, b_4, b_7, b_9, b_{12}, b_{15}, to complete its run and that has been depicted as $t_1 \rightarrow \{b_1, b_4, b_7, b_9, b_{12}, b_{15}\}$. All other related block requirements have been kept in Table 18.3. All these related blocks are considered to form a single logical partition P1. This is shown in Table 18.7 (DPT). Likewise, job t_5 needs the following data blocks: b_6, b_{14}, b_4, b_{10}, b_{12}, b_{18}. It can be represented as $t_5 \rightarrow \{b_6, b_{14}, b_4, b_{10}, b_{12}, b_{18}\}$. All these blocks are kept in logical partition P_5 as per Table 18.7. Blocks required by both jobs t1 and t5 are b1 and b4 for this example. Although the number of replicas for every data block in Hadoop can be configured, a replication degree of 3 has been widely accepted and can be considered as the de facto standard. In our case, we consider that a data block can be replicated to a maximum degree of 3. So, among blocks b1 and b4, we can see that three replicas of b4 are present in partitions P1, P4, and P5. So a dependency is created between P1 and P5. It is signified by the fact that logical partitions P1 and P5 are placed in cluster C1 as shown in Table 18.8. However, frequently used data blocks with correlation, which have exhausted their 3 replicas; for such cases, dependencies are not required to be preserved. For example, jobs t3 and t5 need b10, which is placed in P3. Thus, Table 18.8 shows no dependency for this. For instance, Table 18.6 comprises information of the block requirement of different tasks from logs of the *namenode*. Logical partitions depicted in Table 18.7 withholds partitions made for the 8 jobs and 21 blocks. Dependency between partitions is preserved in Table 18.8. Column *maxPvalue(n)* in Table 18.8 denotes that a maximum of three replicas can be placed. Table 18.10 preserves statistics pertaining to placement of logical partitions within cluster nodes. The tables are used and updated by the proposed algorithms. Table 18.10 summarizes how DACS' block placement can reduce runtime data movements.

Algorithm 18.2   Data Partition Placement-Allocation

CPT: Cluster Partition Table {taskId assigned to PartitionId and ClusterId is empty initially}
BRT: Block Requirement Table {tasked, with required blocks for execution}
DPT: Dependency Table {information about assigned blocks to partitions}
BBPT: Dependency-Based Partition Table
LPDT: Logical Partition Dependency Table

Set Ci=1
For each partition Pi : CPT
    If (Pi is not exist in DBPT) then
        If (New cluster exist in DBPT) then
            Place partition Pi in new cluster Ci
    Else
            Set Ci = 1
            Place partition Pi in cluster Ci
        Endif
    Endif
Endfor

For each cluster Ci: DBPT
    Set Ci=1
    Set Pi=2
    For each partition Pi : Ci(DBPT)
        If (MaxPvalue(DBPT) < 2) then
            If (Pi and Pi+1 exist in LPD) then
                Increment dependency value
        Else
                Place Pi and Pi+1 with dependency value
            Endif
        Endif
    Endfor
Endfor
End

18.4   OUTLINE OF THE DACS SCHEME

18.4.1   HADOOP DEFAULT FRAMEWORK

The default Hadoop system provides a default file system and map-reduce framework in order to balance the overall data distribution among a cluster of nodes. Hadoop's feature has similar systems to the existing distributed file systems. However, HDFS is a high fault-tolerant system and is developed for low-cost hardware. The use of the map-reduce framework in Hadoop to exploit the potential parallelism and HDFS to distribute the data blocks within the cluster nodes. Hadoop uses a random data placement strategy for data distribution within cluster nodes uniformly in order to achieve performance benefits. However, it has been observed that most practical applications show that some dependency exists among the datasets and map tasks. Therefore, it seems that investigation on data placement strategies among cluster nodes is valuable. In the Hadoop default file system,

TABLE 18.10

Summary of Jobs and Their Data Requirements of Table 18.3 under the DACS Scheme

Task	Data Blocks Assigned to Logical Partitions for Execution	No. of Blocks Moved	Assigned Partition	Partition Placed in Cluster ID
t1	P1={b1, b4, b7, b9, b12, b16}	0	P1	C1
t2	P2={b3, b6, b11, b13,15, b2 }	0	P2	C1
t3	P3={b2, b10, b9, b13, b5, b8}	0	P3	C3
t4	P4={b5, b12, b6, b1, b4, b13}	0	P4	C4
t5	P5={b6, b14, b4,b10, b12,b8}	0	P5	C2
t6	P6={b7, b3, b1, b15}	2 within cluster nodes	P6	C1
t7	P7={b2, b17,b20, b9, b11, b4, b6,b21}	One within cluster nodes	P7	C4
t8	P8={b16}	0	P8	C3

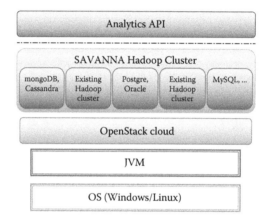

FIGURE 18.3 OpenStack Savanna Hadoop's layered default framework.

stored files are divided into a number of blocks and then these blocks are distributed among cluster nodes for reliability and performance reasons [19]. The block sizes and the replication intensity are configurable and are decided by the system administrator based on its level of granularity.

18.4.2 DACS Framework

In order to assess the efficacy of the proposed DACS, we deploy a virtualized environment using OpenStack [5]. Savanna-Hadoop on OpenStack offers a unified milieu for performing jobs capable of map-reduce capabilities. OpenStack offers a wide range of APIs to support deployment of applications, including services for computing, networking, storing, and other basic services. A multi-node Savanna-Hadoop cluster provides distributed access to all data stored uniformly. Figure 18.3 shows a high level abstraction of the proposed DACS with Savanna Hadoop on OpenStack. The layered framework presents an abstract view of DACS, where analytics API employed on top up SAVANA Hadoop Cluster (SHC). For any user analytic operations, the API interacts with SHC and after processing the query at this level it passes it to the next layer called OpenStack Cloud. The job of the OpenStack Cloud layer includes a check for authorization, VM allocation, migration, QoS measures, and tuning parameters to improve the efficiency of the model.

Savanna Hadoop provides a platform to distribute data needed for Algorithm 18.2 in the Open-Stack environment, where Savanna Hadoop's APIs are used. Figure 18.4 captures a schematic depiction of the DACS framework.

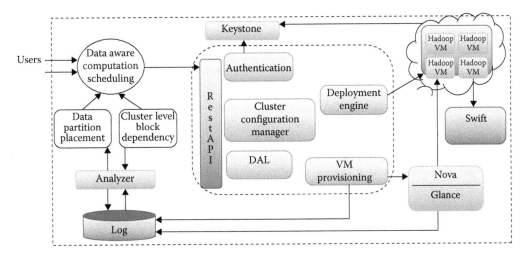

FIGURE 18.4 Schematic depiction of the DACS framework.

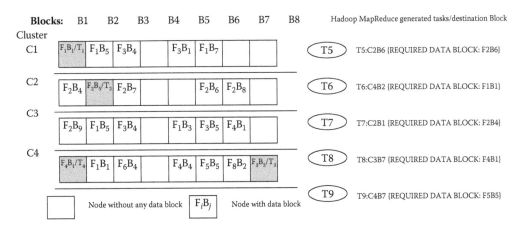

FIGURE 18.5 Illustration of DACS' data placement.

Savanna is integrated with the core OpenStack components such as Keystone, Nova, Glance, Swift, and Horizon. It has a REST API that supports the Hadoop cluster provisioning steps. The Savanna API is implemented as a web server gateway interface (WSGI) server. In addition, Savanna can also be integrated with Horizon, the OpenStack Dashboard, to create a Hadoop cluster from the management console. Savanna also comes with a Vanilla plugin that deploys a Hadoop cluster image. The standard out-of-the-box Vanilla plugin supports Hadoop version 1.1.2.

For the purpose of scheduling applications on Savanna Hadoop's APIs (which in turn uses OpenStack's Nova and REST APIs) are integrated to the proposed DACS. The vital boost in DACS is incorporation of an analyzer block that clusters data blocks into logical partitions depending upon correlation from *namenode* logs that had executed previously. The ensuing section provides details pertaining to experiments carried out to compare performance of the proposed DACS with HNDP.

Figure 18.5 shows a demonstration of the efficacy of the proposed data placement for NHDC with respect to DACS as has been shown in Figure 18.1. The next section presents the experimental work performed and details the performance analysis of the proposed DACS vis-a-vis NHDP.

18.5 EXPERIMENTS AND RESULTS

This section provides details pertaining to experiments carried out to compare performance of the proposed DACS with HNDP. To evaluate the efficacy of DACS over HNDP, this chapter carries out some experiments over Hadoop on top of an OpenStack cluster. To do this, we employed Savanna Hadoop's OpenStack integration.

The hardware platform on which DACS was tested is comprised of four clusters, each with 16 nodes, all having Savanna Hadoop installed on them. Among these 64 nodes, one had been designated as a *Controller Node/Name Node*, another as a *Network Node*, and the remaining 62 nodes were designated as *Compute Nodes/Data Nodes*. The controller node is a Dell Power Edge R610 node of the Dual Intel Xeon Quad-Core E5620 model. Other characteristics include a 2.93 GHz Processor, 12 GB DDR2 memory, and three units of 300 GB SAS HDD/RAID 5. Cent OS 5.0 is the operating system that runs on the *Controller Node*. The *Compute Node* is distinct from the *Controller Node* in that it has a Dell Power Edge R410 with 2.4 Ghz processors. All nodes are enabled by 8-port GB switches.

While carrying out experiments with the aforementioned NCDC dataset, the result of the experimentation may be affected by how datasets are uploaded; since the pattern of data scattering among cluster nodes may depend on whether data is uploaded all at a time, or is uploaded as individual files, one group at a time. To explain this, we can say that for the weather data obtained from NCDC [6,7], the entire dataset may be uploaded at a single go, or it may be uploaded one category at a time, like decade-based or area-based data files at a time. To arrive at an unbiased data distribution pattern, the dataset has been uploaded a total of 20 times, 10 times using the bulk upload strategy; for the rest of the time, data has been uploaded decade-wise. It has been observed that the pattern of overall data distribution remains similar for different runs.

The first experiment was devised to test how DACS can place data effectively in comparison to HNDP. Another experiment was devised to assess the effectiveness of DACS over HNDP as far as runtime movements are concerned. For these two experiments, NCDC weather datasets were used [6,7] in conjunction with a few typical map-reduce applications.

Figure 18.6 depicts the number of data blocks placed in each of the four cluster nodes, namely C1 through C4, by employing HNDP. It shows the same trend in a decade-wise manner between 1901 and 2012. Figure 18.7 depicts the same, but for the proposed DACS. It is readily observed that by employing dependency-based logical clustering and placing correlated partitions in same cluster node, data skewness amongst clusters have been greatly brought down. Figures 18.8 and 18.9 capture the data movements recorded for HNDP and DACS, respectively. Figure 18.10 tracks the progress of

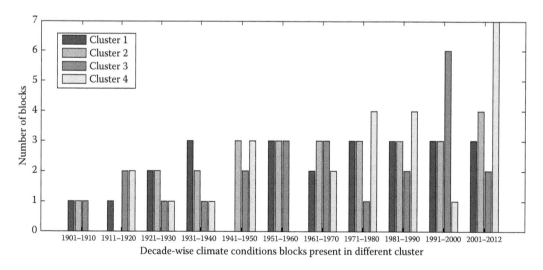

FIGURE 18.6 Block partitioning for the weather dataset using HNDP.

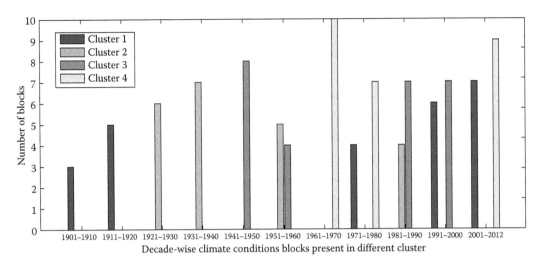

FIGURE 18.7   Block partitioning for the weather dataset using DACS.

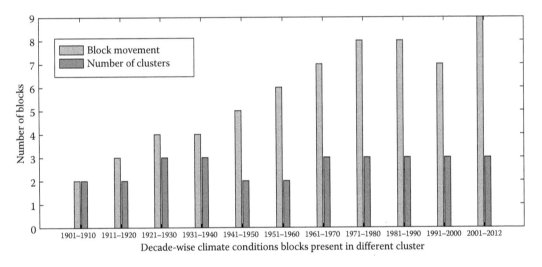

FIGURE 18.8   Data movements recorded for the weather dataset using HNDP.

completion of the jobs, both map and reduce phase, once using HNDP and then for DACS. For this case, the number of replicas for all data blocks have been kept fixed at 3 and this has been represented in the figure as NR = 3.

Relating Figures 18.8 and 18.9, we can see the average block movements have been brought down by about 34% for the weather dataset employed for the study.

Figure 18.10 divulges the percentage of progress of map and reduce phases of the application job for the weather dataset employed for this study, by tracing two executions: first with HNDP and finally with DACS's reorganized data. For all circumstances, data-block replications have been kept fixed at NR = 3. Average improvements in terms of completion times vary approximately around 12%–15%.

In all the experiments with the map-reduce jobs, we have set as many reducers as possible in order to avoid any possible performance bottlenecks, which generally happens in the reduce phase. In an overall sense, DACS provided approximately 34% quicker map phase completions when compared to its HNDP counterpart. Overall execution time of jobs similarly gets reduced by nearly 33% with the DACS strategy.

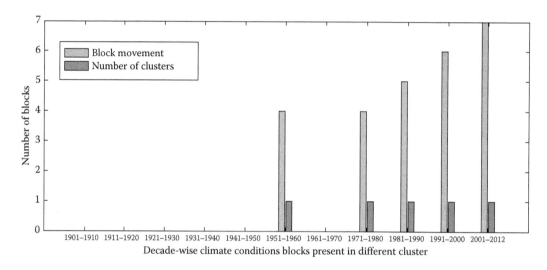

FIGURE 18.9 Data movements recorded for the weather dataset using DACS.

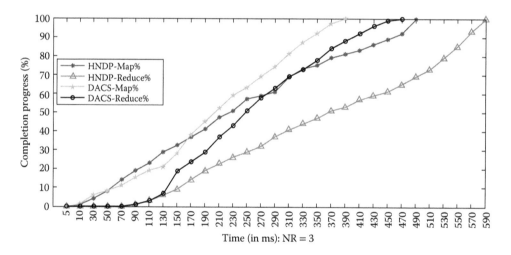

FIGURE 18.10 Comparative % of completion progress with time for the weather dataset.

18.6   CONCLUSIONS

This chapter elaborates a DACS scheme that espouses an approach to improve collocation of data with computational jobs that use the data. This is in conflict to the way that Hadoop distributes data blocks among its nodes by default. Hadoop follows a job-schedule oblivious strategy to place data blocks to cluster nodes. In contrast, the proposed DACS places data blocks intelligently with a prognostic approach that is realized by two novel algorithms presented in this chapter. Experiments carried out divulged that for certain chosen jobs, DACS exhibit substantial improvements in terms of average block movements required at runtime ranging around 34%. Additionally, DACS also shows quicker completion times when compared to HNDP, ranging approximately to 33%–39% for the NCDC dataset. This enhancement may be ascribed to the intelligent logical partitioning and placement algorithms presented in this chapter. In this chapter, data placements have been done by evaluating data dependencies. However, for data that are added over time incrementally, it is much more difficult to evaluate these dependencies. However, entropy can be employed to find indirect dependences between datasets that keep growing and for such cases, we plan to employ entropy-based data dependency calculations and thereafter look to improve upon the data placement problem in Hadoop clusters.

REFERENCES

1. Foster, I., Zhao, Y., Raicu, I., and Lu, S. (2008, November). Cloud computing and grid computing 360-degree compared. In *2008 Grid Computing Environments Workshop* (pp. 1–10). IEEE.
2. Buyya, R., Yeo, C. S., Venugopal, S., Broberg, J., and Brandic, I. (2009). Cloud computing and emerging IT platforms: Vision, hype, and reality for delivering computing as the 5th utility. *Future Generation Computer Systems, 25*(6), 599–616.
3. Dean, J. and Ghemawat, S. (2008). MapReduce: Simplified data processing on large clusters. *Communications of the ACM, 51*(1), 107–113.
4. Ibrahim, S., Jin, H., Lu, L., Qi, L., Wu, S., and Shi, X. (2009, December). Evaluating MapReduce on virtual machines: The Hadoop case. In *IEEE International Conference on Cloud Computing* (pp. 519–528). Berlin Heidelberg: Springer .
5. https://www.openstack.org/software/, last accessed on December 15, 2015
6. http://www.ncdc.noaa.gov/data-access/land-based-station-data/land-based-datasets/integrated-surface-database-isd
7. ftp://ftp3.ncdc.noaa.gov/pub/data/noaa/
8. Amer, A., Long, D. D., and Burns, R. C. (2002). Group-based management of distributed file caches. In *Proceedings of 22nd International Conference on Distributed Computing Systems* (pp. 525–534). IEEE.
9. Wang, J., Shang, P., and Yin, J. (2014). DRAW: A new Data-gRouping-AWare data placement scheme for data intensive applications with interest locality. In *Cloud Computing for Data-Intensive Applications*, L. Xiaolin and Q. Judy, Editors (pp. 149–174). New York, NY: Springer.
10. Yuan, D., Yang, Y., Liu, X., and Chen, J. (2010). A data placement strategy in scientific cloud workflows. *Future Generation Computer Systems, 26*(8), 1200–1214.
11. Xie, J., Yin, S., Ruan, X., Ding, Z., Tian, Y., Majors, J., Manzanares, A., and Qin, X. (2010, April). Improving mapreduce performance through data placement in heterogeneous hadoop clusters. In *IEEE International Symposium on Parallel & Distributed Processing, Workshops and Phd Forum (IPDPSW),* (pp. 1–9). IEEE.
12. Sehrish, S., Mackey, G., Wang, J., and Bent, J. (2010, June). MRAP: A novel MapReduce-based framework to support HPC analytics applications with access patterns. In *Proceedings of the 19th ACM International Symposium on High Performance Distributed Computing* (pp. 107–118). ACM.
13. Cope, J. M., Trebon, N., Tufo, H. M., and Beckman, P. (2009, May). Robust data placement in urgent computing environments. In *IEEE International Symposium on Parallel & Distributed Processing (IPDPS),* (pp. 1–13). IEEE.
14. Hardavellas, N., Ferdman, M., Falsafi, B., and Ailamaki, A. (2009). Reactive NUCA: Near-optimal block placement and replication in distributed caches. *ACM SIGARCH Computer Architecture News, 37*(3), 184–195.
15. http://hadoop.apache.org/ (accessed on 15/11/2014)
16. http://developer.yahoo.com/hadoop/tutorial/module1.html
17. Tripathi, S. and Govindaraju, R. S. (2009, June). Change detection in rainfall and temperature patterns over India. In *Proceedings of the Third International Workshop on Knowledge Discovery from Sensor Data* (pp. 133–141). ACM.
18. Wasserman, S. and Faust, K. (1994). *Social Network Analysis: Methods and Applications* (Vol. 8). Cambridge: Cambridge University Press.
19. Shvachko, K., Kuang, H., Radia, S., and Chansler, R. (2010, May) The Hadoop distributed file system. In *Proceeding of the 2010 IEEE 26th Symposium on Mass Storage Systems and Technologies (MSST)*, pp. 1–10. IEEE, Incline Village, NV.

19 Big Data Computing Using Cloud-Based Technologies
Challenges and Future Perspectives

Samiya Khan, Kashish A. Shakil, and Mansaf Alam

CONTENTS

19.1   INTRODUCTION

The advent of the Internet and rapidly increasing popularity of mobile and sensor technologies have led to an outburst of data in the systems and web world. This data explosion has posed several challenges to systems traditionally used for data storage and processing. In fact, the challenges are so grave that it would not be wrong to state that traditional systems can no longer fulfill the growing needs of data-intensive computing.

The two main requirements of big data analytics solutions are (1) scalable storage that can accommodate the growing data and (2) high processing ability that can run complex analytical tasks in finite and allowable time. Among many others, cloud computing technology is considered an apt solution to the requirements of big data analytics solutions considering the scalable, flexible, and elastic resources that it offers (Philip Chen and Zhang 2014). First, the cloud offers commodity machines that provide scalable yet cost-effective storage solutions. Besides this, the processing ability of the system can be improved by adding more systems dynamically to the cluster. Therefore, the flexibility and elasticity of the cloud are favorable characteristics for big data computing.

Cloud-based big data analytics technology finds a place in future networks owing to the innumerable "traditionally unmanageable abilities and services" that this technology offers. The general definition of future networks describes it as a network that possesses the capabilities to provide services and facilities that existing network technologies are unable to deliver. Therefore, a component network, an enhanced version of an existing network, or a federation of new and existing networks that fulfill the above-mentioned requirements can be referred to as future networks.

The applicability of the big data concept and its relevance to society and businesses alike makes it a potential game changer in the technological world so much so that many people consider this concept just as important to businesses and society as the Internet. This brings us to look at reasons why big data needs to be studied and researched. The answer to this question lies in fundamental principles of statistical science. One of the prerequisites of any statistical analysis is data. Moreover, the higher the number of samples, the better is the computed analysis for the given data. Therefore, more data directly implies better analyses, which in turn means better decision-making.

From an organizational perspective, efficient decision-making can have a significant impact on improving the operational efficiency and productivity of the organization. This notion can be scaled down to the individual level and the availability of better analyses in the form of applications and systems can increase individual productivity and efficiency, manifold. Evidently, the big data concept is capable of bringing about a revolution in the society and business world and can change the way we live our lives, just like the Internet did, years ago.

This technology finds applications in diverse fields and areas. Although the big data problem can model any data-intensive system, there are some established practical applications that have gained popularity amongst the research community and governing authorities. These applications include smart cities (Khan, Anjum, and Kiani 2013), analytics for the healthcare sector (Raghupathi and Raghupathi 2014), asset management systems for railways (Thaduri, Galar, and Kumar 2015), social media analytics (Burnap et al. 2014), geospatial data analytics (Lu et al. 2011), customer analytics for the banking sector (Sun et al. 2014), e-commerce recommender systems (Hammond and Varde 2013), and intelligent systems for transport (Chandio, Tziritas, and Xu 2015), in addition to several others.

This chapter shall illustrate the big data problem, giving useful insights in the tools, techniques, and technologies that are currently being used in this domain, with specific reference to cloud computing as the infrastructural solution for the storage and processing requirements of big data. Section 19.1 provides a comprehensive definition of big data and the two most popular models used for big data characterization, namely the Multi-V model (Section 19.2.1) and the HACE theorem (Section 19.2.2). The lifecycle of big data and the different processes involved have been elaborated upon in Section 19.2. In order to process and analyze big data, several existing mathematical techniques and technologies can be used. Section 19.3 discusses the techniques used for big data processing. This section has been divided into two subsection namely mathematical techniques (Section 19.3.1) and data analytics techniques (Section 19.3.2). The techniques under each of these sub-headings have been described.

Big data has found applications in diverse fields and domains. In order to fulfill the varied requirements of big data applications, six computing models exist. Each of these computing models serve different computing requirements of distinctive big data applications, which include batch processing (Section 19.4.1), stream processing (Section 19.4.2), graph processing (Section 19.4.3), DAG processing (Section 19.4.4), interactive processing (Section 19.4.5), and visual processing (Section 19.4.6). The tools available for each of these computing models have been described in the subsections.

The techniques and computing models need to be efficiently implemented for big data, which require the systems to provide magnified storage and computing abilities. The adaptation of standard techniques for the big data context requires technologies like cloud computing, which has been described in Section 19.5. This section shall examine how big data analytics can use the characteristics of cloud computing for optimal benefit to several real-world applications. The first few subsections introduce cloud computing and describe characteristics (Section 19.5.1), delivery models (Section (19.5.2), deployment models (Section 19.5.3), and how cloud computing is the most appropriate technology for big data (Section 19.5.4). In view of the fact that Hadoop is the most popular computing framework for big data, Hadoop on the cloud (Section 19.5.5) gives the implementation options available for moving Hadoop to the cloud. Section 19.6 summarizes the challenges identified for big data computing in the cloud environment and Section 19.7 discusses directions for future research.

19.2   DEFINING BIG DATA

Several definitions for big data exist owing to the varied perspectives and perceptions with which this concept is viewed and understood. Regardless of the source of a definition, most big data experts possess a unanimous viewpoint on the fact that big data cannot be restricted only to the dimension of volume. Many more dimensions, some of which may be application or data-source dependent, need to be explored before a comprehensive definition for big data can be articulated.

The most accepted definition of big data describes it as huge volumes of exponentially growing, heterogeneous data. Doug Laney of Gartner, in the form of the 3V Model, gave the first definition of big data (Gartner 2016). The fundamental big data characteristics included in this classification are volume, variety, and velocity. However, big data, as a technology, picked up recently after the availability of open source technologies like NoSQL (Salminen 2012) and Hadoop*, which have proven to be effective and efficient solutions for big data storage and processing. Accordingly, the definition of big data was modified to data that cannot be stored, managed, and processed by traditional systems and technologies.

To enhance the technical precision of existing definitions, Shaun Connolly introduced the terms transactions, interactions, and observations, which were added to the big data definition (Samuel 1959). "Transactions" is a term used to describe data that has already been collected and

---

* http://hadoop.apache.org/

analyzed in the past, and "interactions" include data that is collected from things and people. A class of data missed by both these categories is the data that is automatically collected and constitutes "observations." Barry Devlin gave a similar, yet clearer definition of big data, describing it in terms of machine-generated data, human-sourced data, and process-mediated data (Ratner 2003).

From the business-effectiveness perspective, transaction data holds little relevance in view of the fact that it is old data and by the time it is collected and analyzed, the results become obsolete or lesser relevant. On the contrary, new data needs to be analyzed efficiently to provide predictions, which can be used to make timely interventions. Sentiment analysis is an application that works on this perspective and uses big data as signals. This was formalized into a timing and intent-based classification of big data (Buyya 2016).

Besides the above mentioned, there have been several other viewpoints and perspectives on big data. Scholars like Matt Aslett see big data as an opportunity to explore the potential of data that was previously ignored due to limited capabilities of traditional systems whereas some others call it a new term for old applications and technologies like business intelligence (Abu-Mostafa, Magdon-Ismail, and Lin 2012). Regardless of the big data definition one chooses to follow, nothing can take away the fact that big data opens doors to unlimited opportunities and in order to make use of this reserve, we need to develop new or modify the existing tools and technologies.

19.2.1 Multi-V Model

The initial 3V model, given by Doug Laney in the year 2001, includes volume, variety, and velocity as the three fundamental big data characteristics (Gartner 2016). The amount of data included in a dataset indicates the volume of data, which is the most obvious characteristic of big data. The data concerned may come from different sources and can be of diverse types. For instance, data coming from a social media portal includes textual data, audio and video clips, and metadata. In order to accommodate for these different types of data, big data is said to include structured data, semi-structured data, and unstructured data. Lastly, data may be produced in batches, near time or real-time. The speed at which the concerned data is being generated denotes the velocity characteristic of big data. Figure 19.1 depicts the scope of interest for the three Vs mentioned above.

However, this initial model has been expanded in different dimensions by including Vs such as veracity, variability, and value, in addition to several others. One of the most recent adaptations of this model is the 3^2V model (Buyya 2016). This model divides the Vs into three classes, namely, data (volume, variety, and velocity), business intelligence (value, visibility, and verdict) and statistics (veracity, variability, and verdict). Each of these classes include 3Vs specific to their domain and have been described in Table 19.1.

19.2.2 HACE Theorem

HACE stands for heterogeneous, autonomous, complex and evolving (Xindong et al. 2014) and describes big data as a large volume of heterogeneous data that comes from autonomous sources. These sources are distributed in nature and the control is essentially decentralized. This data can be used for exploration of complex and evolving relationships. These characteristics make identification and extraction of useful information from this data excessively challenging. In order to ensure a clear understanding, a detailed description of these characteristics has been given below.

1. *Huge volumes of data from heterogeneous sources*: Volume, as identified by many different modeling frameworks, is one of the fundamental characteristics of big data. In addition to this, with the increasing popularity of the Internet and social networking portals, data is no longer confined to a format or source. Diverse sources of data such as Twitter, Facebook, LinkedIn, organizational repositories, and other external and internal data sources result in data multi-dimensionality. Data representation may also vary as a result

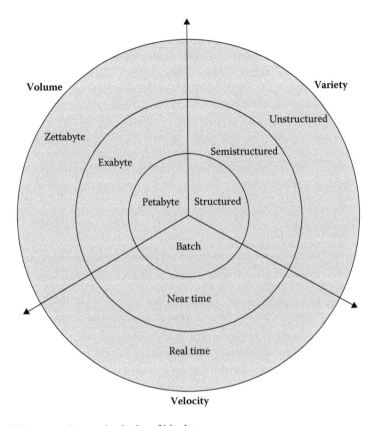

FIGURE 19.1 Volume, variety, and velocity of big data.

TABLE 19.1
Description of 3²V Characteristics

Class	V Characteristic	Description
Data Domain	Volume	Amount of data
–	Variety	Types of data included
–	Velocity	Speed of data generation
Business Intelligence Domain	Value	The business value of the information present in the data
–	Visibility	Insight, hindsight, and foresight of the problem and its adequate solutions
–	Verdict	Possible decision to be made on the basis of the scope of problem, computational capacity, and available resources
Statistics Domain	Veracity	Uncertainty and trustworthiness of data
–	Variability	Variations and complexity of data
–	Validity	Objectivity with which the data is collected

of varying methods of data collection, data collector preferences, and application-specific needs of the system.

2. *Distributed, autonomous sources with decentralized control*: The nature of the sources is usually autonomous and distributed. As a result, the production and collection of data does not require connection to any central control. Many parallels can be drawn between the World Wide Web and big data technology at this level. Just like the web servers that

possess information of their own and can function independently, the sources of big data are also independent entities that do not require any external monitoring or control. More specifically, data sources are guided by local government regulations and market parameters, which result in restructuring of data representations. The distributed nature of sources increases fault tolerance, which is one of the significant advantages of such a system.

3. *Complex and evolving nature of relationships*: Any form of data representation requires exploration of associations and relationships between the different entities involved. Keeping the dynamic nature of this world in view, several spatial, temporal, and other kinds of factors are involved in representing entities and their continuously evolving relationships. Owing to the presence of different data types like audio, video, documents, and time-series data, in addition to several others, this data possesses high complexity.

19.3   BIG DATA LIFECYCLE

The lifecycle of big data includes several phases, which include data generation, acquisition, storage, and processing of data. These four phases have been explained below and illustrated in Figure 19.2.

1. *Generation*: Data is the most rapidly increasing resource in the world. Perhaps the reason for this staggering rise in its generation is the diverse types of devices, entities, and systems involved. With the rapid advancement in technology, devices such as sensors, online portals, social networking websites, and online systems such as online trading and banking, in addition to many others, have come into existence. All these systems, portals, and devices generate data on a periodic basis, contributing to the volume, variety, and velocity of big data.

2. *Acquisition*: Now that we know that big data is being generated by diverse sources, this data needs to be acquired by big data systems for analysis. Therefore, during this stage of the big data lifecycle, the raw data generated in the world is collected and given to the next stage for further processing. Examples of data acquisition systems include log files, sensing systems, web crawlers, and REST APIs provided by portals. Since big data includes different types of data, an efficient pre-processing mechanism is required. Common methods used for this purpose include data cleaning, redundancy reduction, and data integration. This is a crucial step for ensuring data veracity.

3. *Storage*: The sheer volume of big data overwhelms traditional storage solutions. In order to address the challenges posed by big data, as far as data storage is concerned, a distributed

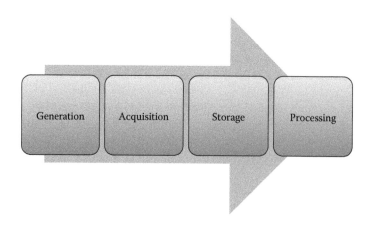

FIGURE 19.2   Big data lifecycle.

file system (DFS) is being put to use. From the first distributed file system, Google file system (Ghemawat, Gobioff, and Leung 2003), to the Hadoop distributed file system or HDFS (Shvachko et al. 2010), there is a range of solutions available in this category. Some of the latest and most popular additions to this category are NoSQL database solutions such as MongoDB* and platforms such as Cassandra†. This phase of the big data lifecycle contributes to the reliability and availability of data.

4. *Processing*: The last stage of the big data lifecycle is processing, during which various analytical approaches and methods are implemented on the available data, for basic and advanced analytics. Similar to traditional data analysis, the objective of big data analysis is extraction of useful information from the available data. Common methods used for this purpose include clustering, classification, and data analysis techniques, besides many others. It is important to mention here that traditional processing techniques have to be adapted to the big data scenario by redesigning them for use with parallel computing techniques such as the MapReduce programming paradigm (Lee et al. 2012). This phase contributes to the value characteristic of big data. The rest of the chapter will discuss this facet of the big data lifecycle in detail.

19.4 TECHNIQUES FOR BIG DATA PROCESSING

Big data processing requires a synergistic approach involving mathematical, statistical, and optimization techniques, which are implemented using established technologies such as data mining, machine learning, and signal processing, in addition to several others, for application-specific processing. This makes big data processing essentially interdisciplinary. An elaborative description of the techniques used for big data processing has been given below.

19.4.1 MATHEMATICAL ANALYSIS TECHNIQUES

19.4.1.1 Mathematical Techniques

Most big data problems can be mathematically modeled and solved using mathematical analysis techniques such as factor analysis and correlation analysis. Factor analysis is mostly used for analysis of relationships between different elements that constitute big data. As a result, it can be used for revealing the most important information. Taking the relationship analysis a step further, correlation analysis can be used for extracting strong and weak dependencies (Ginsberg et al. 2009). Analytics for applications belonging to fields such as biology, healthcare, engineering, and economics require the use of such techniques.

19.4.1.2 Statistical Methods

Statistical methods are mathematical techniques that are used for collection, organization, and interpretation of data. Therefore, they are commonly used for studying causal relationships and co-relationships. It is also the preferred category of techniques used for deriving numerical descriptions. With that said, the standard techniques cannot be directly implemented for big data. In order to adapt the classical techniques for big data usage, parallelization has been attempted. Research fields related to this area are statistical computing (Klemens 2009), statistical learning (Hastie, Tibshirani, and Friedman 2009), and data-driven statistical techniques (Bennett et al. 2009). The economic and healthcare sectors make extensive use of statistical methods for various applications.

---

* https://www.mongodb.com/
† http://cassandra.apache.org/

19.4.1.3 Optimization Methods

Core fields of studies like physics, biology, and economics involves a lot of quantitative problems. In order to solve these problems, optimization methods are used. Some of these methods that have found wide-ranging use, because of the ease with which they can be parallelized, include the genetic algorithm, simulated annealing, quantum annealing, and adaptive simulated annealing (Sahimi and Hamzehpour 2010).

Nature-inspired optimization techniques such as particle swarm optimization and evolutionary programming have also been proven to be useful methods for solving optimization problems. However, these algorithms and techniques are highly storage and computationally intensive. Many research works have attempted scaling of these techniques (Yang, Tang, and Yao 2008; Xiaodong and Xin 2012). An important requirement of big data applications in this regard is real-time optimization, particularly in WSNs (Seenumani, Sun, and Peng 2011).

19.4.2 Data Analytics Techniques

19.4.2.1 Data Mining

Data mining allows extraction of useful information from raw datasets and visualization of the same in a manner that is helpful for making decisions. Commonly used data mining techniques include classification, regression analysis, clustering, machine learning, and outlier detection. In order to analyze different variables and how they are dependent on one another, regression analysis may be used. Companies commonly use this technique for analyzing CRM big data and evaluating varied levels of customer satisfaction and their impact on customer retention.

Taking this analysis further, there may also be a need to cluster similar customers together to analyze their buying patterns or classify them on the basis of certain attributes. Clustering and classification are the techniques used for this purpose. Association rule mining may be used for exploring hidden relationships and patterns in big datasets. Lastly, outlier detection is used for fraud detection or risk reduction by identifying patterns or behaviors that are abnormal.

19.4.2.2 Machine Learning

A sub-field of artificial intelligence, machine learning allows systems to learn and evolve using empirical data. As a result, intelligent decision-making is fundamental to any system that implements machine learning. However, in the big data context, standard machine learning algorithms need to be scaled up to cope with big data requirements. Deep learning is a recent field that is attracting immense research attention lately (Bengio, Courville, and Vincent 2013). Several Hadoop-based frameworks like Mahout are available for scaling up machine learning algorithms. However, several fields of machine learning like natural language processing and recommender systems, apart from several others, face scalability issues that need to be mitigated for development of generic as well as efficient application-specific solutions.

19.4.2.3 Signal Processing

The introduction of Internet and mobile technologies has made the use of social networking portals and devices like mobiles and sensors excessively common. As a result, data is being generated at a never-seen-before rate. The massive scale of data available, presence of anomalies, need for real-time analytics, and relevance of distributed systems gives rise to several signal processing opportunities in big data (Bo-Wei, Ji, and Rho 2016).

19.4.2.4 Neural Networks

Image analysis and pattern recognition are established applications of artificial neural networks (ANN). It is a well-known fact that as the number of nodes increase, the accuracy of the result gets better. However, the increase in node number elevates the complexity of the neural network, both in terms of memory consumption and computing requirements. In order to combat these challenges, the neural network needs to be scaled using distributed and parallel methods (Mikolov et al. 2011). Parallel training implementation techniques can be used with deep learning for processing big data.

19.4.2.5 Visualization Methods

In order to make the analysis usable for the end user, analytics results need to be visualized in an understandable and clear manner. The high volume and excessive rate of generation of data makes visualization of big data a daunting challenge. Evidently, it is not possible to use traditional visualization methods for this purpose. Most systems currently available perform rendering of a reduced dataset to dodge the complications associated with visualization of large datasets (Thompson et al. 2011). However, real-time visualization of big data still presents innumerable challenges.

19.5 BIG DATA COMPUTING MODELS

Different applications make use of distinctive types, volumes, and velocities of data. Therefore, the type of computing models applicable depends on the nature of the application. While some applications require batch processing, others need real-time data processing. This section gives a taxonomy (see Figure 19.3) of big data computing models and discusses the different tools belonging to each of the categories.

19.5.1 BATCH PROCESSING

Theoretically, batch processing is a processing mode in which a series of jobs are performed on a batch of inputs. The MapReduce programming paradigm is the most effective and efficient solution for batch processing of big data. Hadoop, a MapReduce implementation, is identified as the most popular big data processing platform. Therefore, most of the tools described in Table 19.2 are either Hadoop-based or tools that run on top of Hadoop.

19.5.2 STREAM PROCESSING

In some applications, particularly for real-time analytics, data is available in the form of streams or continuous data flows. These data streams need to be processed, record-by-record, in-memory. Stream processing is considered to be the next-generation computing paradigm for big data. The tools available for this purpose have been described in Table 19.3.

19.5.3 GRAPH PROCESSING

Owing to the connected nature of data and the significance of exploring relationships and associations for big data analytics, a graph is perhaps the best-suited mathematical model for a majority of the applications. Big data graph processing techniques work in accordance with the bulk synchronous parallel (BSP) computing paradigm (Cheatham et al. 1996). This computing paradigm is commonly used in cloud computing. Table 19.4 gives a list of graph processing architectures and models for graph processing, also elaborating on their features and supported functionalities.

19.5.4 DAG PROCESSING

In theory, a DAG or directed acyclic graph is described as a finite and directed graph. However, this graph lacks any directed cycles. A DAG processing system represents jobs in the system as vertices of the DAG, which execute in parallel. DAG processing is considered a step up from the MapReduce programming model as it avoids the scheduling overhead prevalent in MapReduce and provides developers with a convenient paradigm for modeling complex applications that require

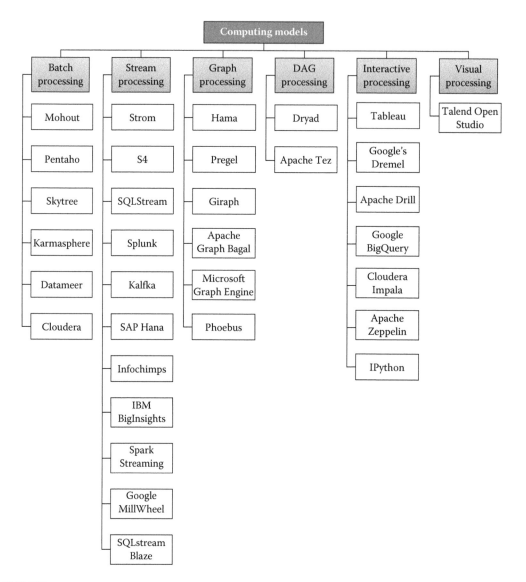

FIGURE 19.3   Big data computing models.

multiple execution steps. Dryad (Isard et al. 2007) is based on a dataflow graph processing-based programming model that supports scalable and distributed programming applications. Another application framework that allows execution of a complex DAG created from tasks is Apache Tez*. This framework is built on top of YARN.

19.5.5 INTERACTIVE PROCESSING

There needs to be a system that sits between big data applications and users to facilitate smooth communication between the two entities, in order to make big data applications usable. For this purpose, large-scale data processing systems that make use of an interactive mechanism are put to use. Some of the tools that are developed for this purpose have been mentioned in Table 19.5.

---

* https://tez.apache.org/

TABLE 19.2

Batch Processing Tools

Tool	Description	Functionality Supported	Link
Mahout	Used for scalable machine learning in the parallel environment	Capable of performing classification, pattern analysis, clustering analysis, regression analysi's, and dimension reduction.	http://mahout.apache.org/
Pentaho	Hadoop-based software platform for business report generation	Allows data acquisition, analytics, and visualization of business data, in addition to plugin-based communication with NoSQL databases such as MongoDB and platforms such as Cassandra.	http://www.pentaho.com/ product/big-data-analytics
SkyTree	A general-purpose server used for machine learning and advanced analytics, which enables optimized machine learning implementation for real-time analytics	Can be used for similarity search, density estimation, dimension reduction, outlier detection, regression, and classification and clustering.	http://www.skytree.net/
Karmasphere	Hadoop-based platform for analysis of business big data	Allows ingestion, iterative analysis, reporting, and visualization of business data.	http://www.karmasphere.com/
Datameer Analytic Solution (DAS)	Hadoop-based platform-as-a-service (PaaS) solution that allows analysis of business data	Includes data source integration, analytics engine, and visualization methods like charts and dashboards. However, data integration is the main focus and it allows data import from structured and unstructured sources. It needs to be deployed with Cloudera-based Hadoop distribution.	http://www.datameer.com/
Cloudera	Apache Hadoop distribution system	Supports Apache tools like Pig and Hive, along with embedded plugins with Oracle.	http://www.cloudera.com/

TABLE 19.3
Stream Processing Tools

Tool	Description	Functionality Supported	Link
Storm	Open-source, scalable, fault-tolerant, distributed system for real-time computation	Storm allows programmers to create a setup for processing of streaming data for real-time analytics. It also supports distributed remote procedure call, an interactive operating system, and online machine learning.	http://storm.apache.org/
S4	A scalable, pluggable, distributed, fault-tolerant computing platform	S4 allows development of applications that require effective cluster management and is used by Yahoo for large-scale search query computation.	http://incubator.apache.org/s4/
SQLStream	A platform that supports intelligent, automatic operations, for processing unbounded large-scale data streams	Pattern discovery in unstructured data.	http://www.sqlstream.com/blaze/s-server/
Splunk	A platform for analyzing machine-generated data streams	Allows analysis of machine-generated log files, structured as well as unstructured.	http://www.splunk.com/
Kafka	Developed for LinkedIn for in-memory management and processing of messaging and stream data	Allows analysis of operational data like service logs and activity data such as user actions for organizational management and usage	http://kafka.apache.org/
SAP Hana	A tool for in-memory processing of data streams	Supports three types of analytics: text analysis and predictive analytics, data warehousing, and operational reporting.	https://hana.sap.com
Infochimps	A cloud suite that provides infrastructure-as-a-service (IaaS) provisioning	Supports cloud streams, cloud-based Hadoop, and cloud queries for private as well as public clouds, in addition to support for Hive, Pig, and Kafka, in addition to several other tools.	http://www.infochimps.com/
BigInsights	A stream-based tool by IBM, used in the Infosphere platform for big data analytics	Used for real-time analysis of data streams and supports Hive, Pig, Lucene, and Oozie, apart from a few others.	http://www-03.ibm.com/software/products/en/ibm-biginsights-for-apache-hadoop
Spark Streaming	Stream processing component of Spark	Framework that allows development of fault-tolerant streaming applications.	http://spark.apache.org/streaming/
Google MillWheel (Akidau et al. 2013)	Framework for fault-tolerant stream processing	Supports development of low-latency data processing applications.	–
SQLstream Blaze	Stream processing suite	Real-time operational intelligence for stream data analytics of machine data.	http://www.sqlstream.com/blaze/

TABLE 19.4

Graph Processing Systems

Tool	Description	Functionality Supported
Hama (Seo et al. 2010)	BSP-inspired computing paradigm that runs on top of Hadoop	Used for performing network computations, graph functions, matrix algorithms, and scientific computations
Pregel (Malewicz et al. 2010)	A graph computation model that approaches problems using the BSP programming model	Allows efficient processing of billions of graph vertices that may be connected to each other via trillions of edges
Giraph (http://giraph.apache.org/)	Scalable, iterative graph processing system	Adapted from Pregel, Giraph allows edge-oriented input, out-of-core computation and master computation
Apache Spark Bagel (Spark 2016)	A Pregel implementation, which is a component of Spark	Supports combiners, aggregators, and other basic graph computation
Microsoft Graph Engine (Microsoft 2016)	In-memory and distributed graph processing engine	Allows high-throughput offline analytics and low-latency online query processing for very large graphs
Phoebus (XSLogic 2016)	Large-scale graph processing framework	

TABLE 19.5

Interactive Processing Tools

Tool	Description	Functionality Supported	Link
Tableau	Hadoop-based visualization environment that makes use of Apache Hive to process queries	Includes three versions: Tableau Desktop (visualization and normalization of data), Tableau Server (business intelligence system for browser-based analytics), and Tableau Public (creating interactive visualizations)	http://www.tableau.com/
Google Dremel (Melnik et al. 2011)	Scalable interactive analysis system	Performs nested data processing and allows querying of very large tables	–
Apache Drill	Interactive analysis system inspired by Dremel, which stores data in HDFS and processes data using MapReduce	Performs nested data processing and supports many data sources and types and different queries	https://www.mapr.com/products/apache-drill
Google BigQuery	An interactive analysis framework that implements Dremel	Uses the Google infrastructure and queries are made on "append-only" tables, allowing super-fast SQL queries.	https://cloud.google.com/bigquery/
Cloudera Impala	A interactive analysis framework that is inspired by Dremel	Capable of supporting high-concurrency workloads and is the true native interactive solution for Hadoop	https://www.cloudera.com/products/apache-hadoop/impala.html
Apache Zeppelin	Interactive data analytics provisioned in the form of a web-based notebook	Integrates with Spark for seamless data ingestion, discovery, analytic's, and visualization	https://zeppelin.apache.org/
IPython	Interactive computing architecture	Interactive shell that works as the kernel for Jupyter (https://jupyter.org/) and supports interactive data visualization	https://ipython.org/

19.5.6   VISUAL PROCESSING

Visualization is the last and most critical phase of any big data application. This is particularly the case for applications where end-users are involved, simply because information not communicated properly is information not communicated at all. One of the most popular tools for visual big data analytics is Talend Open Studio*. It can be integrated with Hadoop and user interfaces can be easily created using its drag-and-drop functionality. Moreover, it also offers RSS feed functionality.

19.6   CLOUD COMPUTING FOR BIG DATA ANALYTICS

John McCarthy gave the world the concept of "utility computing" during the MIT Centennial talk 1961, when he openly spoke about the future of this industry and how computing will share screen with utilities like electricity and water, and be served as such (Qian et al. 2009). Since then, cloud computing has come a long way to become the technology that transforms McCarthy's ideas into reality. However, it was not until 2006 that this technology reached the commercial arena. The introduction of solutions like Amazon's Elastic Compute Cloud (Amazon EC2[†]) and Google App Engine[‡] have been historic milestones in the history of cloud computing.

One of the earliest definitions of cloud computing was given by Gartner, which described this technology as a computing style that delivers scalable and elastic IT-enabled capabilities to customers, as a service, with the support of Internet technologies (Gartner 2016). This definition could not be considered a standard for the sheer simplicity and ambiguity that it entails. The National Institute of Standards and Technology (NIST) gave the industry-standard definition for cloud computing (Mell and Grance 2011).

According to this definition, cloud computing is a technology that allows on-demand, convenient, and ubiquitous network access to computing resources that can be configured with minimal requirement of management and interaction with the service provider. The NIST definition also mentioned the five key characteristics, deployment models, and delivery models for cloud computing. An overview of the NIST definition of cloud computing is illustrated in Figure 19.4. There are three main components of the cloud computing ecosystem, namely, end-user or consumer, distributed server, and data center. The cloud provider provisions the IT resources to the end-user with the help of distributed server and data center.

19.6.1   KEY CHARACTERISTICS

The five key characteristics of cloud computing are as follows:

- *On-demand usage*: The freedom to self-provision the cloud-based IT resources lies in the hands of the end-user.
- *Elasticity*: Some users may require dynamic scaling of IT resources assigned to them. This scaling must be performed on the basis of runtime conditions. Cloud computing's support for this functionality makes it a cost-effective IT provisioning system.
- *Multi-tenancy*: Multi-tenancy is a technology that makes use of virtualization to assign and reassign resources to consumers who are isolated from each other.
- *Ubiquitous access*: The end-users or consumers of the provided cloud services may have different needs, particularly with respect to their ability to use the self-provisioned IT

---

* https://www.talend.com/products/talend-open-studio
† https://aws.amazon.com/ec2/
‡ https://appengine.google.com/

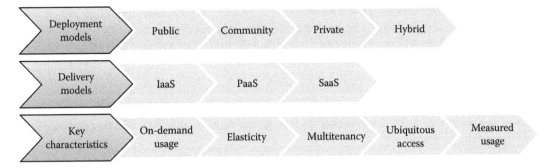

FIGURE 19.4 Cloud computing model defined by NIST.

resources. They may use these resources on any device and using any interface. Therefore, the cloud service must be widely accessible.

- *Measured usage*: In the cloud computing paradigm, the consumer only pays for the services he or she uses. To enable this, the usage data related to each consumer needs to be measured. Billing is done on the basis of this measurement. Besides this, these measurements are also used for monitoring purposes and possess a close relationship with on-demand provisioning.

19.6.2 DELIVERY MODELS

It is evident from the discussion that the cloud provider offers services to the consumers. Obviously, these IT services need to be packaged into combinations to make self-provisioning possible. This pre-packaging of services is called the cloud delivery model. There are three main delivery models used as standards, which have been described below. However, several other delivery models have been created on the basis of applications and functionality that they support. Some examples of these models include database-as-a-service, security-as-a-service, and testing-as-a-service, apart from many others.

19.6.2.1 Infrastructure-as-a-Service (IaaS)

In this delivery model, cloud-based tools and interfaces allow access and management of infrastructure-centric IT resources. Therefore, reserves such as hardware, network, connectivity, and raw IT resources are included in this model. The user is free to configure these resources in the manner that he or she desires. Popular cloud providers of IaaS are Rackspace[*] and Amazon EC[2].

19.6.2.2 Platform-as-a-Service (PaaS)

Although infrastructure forms the base layer and requirement for any kind of development or usage, developers may also require pre-deployed and pre-configured IT resources. This gives them a complete environment to directly work upon, which saves time and effort. Some of the most popular products in this category are the .NET-based environment provided by Microsoft Azure[†] and a Python-based environment by the Google App Engine.

---

[*] https://www.rackspace.com
[†] https://azure.microsoft.com/

19.6.2.3   Software-as-a-Service (SaaS)

The shared cloud service can also host software solutions that can directly be used by consumers on a need basis. Products such as Google Docs*, which is a shared service and provisions documentation software and storage to consumers, are examples of SaaS.

19.6.2.4   Hybrid Models

Apart from the three formal models used for delivery of cloud solutions, the user also has the option to use any combinations of these models. In fact, all the three models may also be combined and provisioned to the end-user.

19.6.3   Deployment Models

The cloud environment has three main characteristics, namely, size, ownership, and access. On the basis of these three characteristics, the deployment model for the cloud environment is determined.

19.6.3.1   Public Cloud

In such a cloud environment, the access is open to all and a third-party cloud provider owns the cloud. Therefore, creation and maintenance of the cloud environment is the sole responsibility of the cloud provider. The size of the cloud is large owing to the "accessible to everyone" nature of it.

19.6.3.2   Community Cloud

The community cloud is an adaptation of the public cloud. The only difference between these two types of deployments is that the community cloud restricts access to the cloud services. Only a small community of cloud consumers can use the services of this cloud. Moreover, such a cloud may involve joint ownership of the community and a third-party cloud provider.

19.6.3.3   Private Cloud

When an individual or organization owns a cloud and limits the access of the cloud to the members of the organization only, the deployment model is known as private cloud. The size of such a cloud is much smaller and staff is employed for performing administration and management of the cloud. It is also important to mention that in such a deployment model, the cloud provider and consumer, both are the owners of the cloud.

19.6.3.4   Hybrid Cloud

A deployment model created using two or more deployment models is called a hybrid cloud.

19.6.3.5   Other Deployment Models

Some other cloud deployment models such as the external cloud, virtual private cloud or dedicated cloud, and inter-cloud also exist.

19.6.4   What Makes Cloud Computing an Ideal Match for Big Data?

Big data solutions have two fundamental requirements. The size of the data is "big." Therefore, in order to store this data, a large and scalable storage space is required. Moreover, the standard analytics algorithms are computing-intensive. Therefore, an infrastructural solution that can support this level of computation is needed. The cloud meets both these requirements well.

First, there are several low-cost storage solutions available with the cloud. Besides this, the user pays for the services he or she uses, which makes the solution all the more cost effective. Second,

---

* https://www.google.co.in/docs/about/

cloud solutions offer commodity hardware, which allows effective and efficient processing of large datasets. It is because of these two reasons that cloud computing is considered an ideal infrastructural solution for big data analytics.

19.6.5   Hadoop on the Cloud

One of the most popular frameworks used for big data computing is Hadoop. It is an implementation of MapReduce that allows distributed processing of large, heterogeneous datasets. There are many solutions that allow moving of Hadoop to the cloud. Some of the popular solutions are the ones provided by Amazon's Elastic MapReduce[*] and Rackspace. There are several reasons why running Hadoop on the cloud is gaining immense popularity, some of which have been listed below.

- *Cost effectiveness*: Running any software on the cloud reduces the cost of the operation considerably. Therefore, this model provides a cost-effective solution for Hadoop-based applications.
- *Scalable resource procurement*: One of the key benefits of the cloud paradigm is that it allows scalable provisioning of resources. This feature is best put to use by the Hadoop framework.
- *Support for efficient batch processing*: Hadoop is a batch-processing framework. In other words, jobs submitted to Hadoop are collected and sent for execution at a fixed and temporal basis. Therefore, the resources are not loaded with work all the time. The cloud allows on-demand resource provisioning. Therefore, resources can be procured and elastically increased and decreased in capability depending on the phase of Hadoop operation, making the solution cost-effective yet efficient.
- *Handles variable requirements of Hadoop ecosystem*: Different jobs performed by Hadoop vary in the quality and quantity of resources they require. For instance, some jobs require I/O bandwidth while others need more memory support. Whenever a physical setup for Hadoop is created, homogenous machines that can support the highest capabilities are used, resulting in wastage of resources. The cloud solution allows formation of a cluster of heterogeneous machines to support the heaviest tasks yet optimize usage.
- *Processing data where it resides*: Most of the organizations keep their data on the cloud. Therefore, processing it on the cloud makes more sense. Migrating the data to a different execution environment will not only be time-consuming, but it will also be inefficient.

The cloud can be used for creation of different types of clusters, with each component of the cluster having a different configuration and characteristics. The available options for running Hadoop in the cloud environment have been described below.

19.6.5.1   Deploying Hadoop in a Public Cloud

Providers such as Hortonworks[†], Cloudera[‡], and BigInsights[§] offer Hadoop distributions, which can be deployed and run on public clouds provided by Rackspace, Microsoft Azure, and Amazon Web Services[¶]. Such a configuration is typically referred to as "Hadoop-as-a-service." The issue with such solutions is that they use IaaS provided by the cloud providers. In other words, the IT resources being used are shared between many customers. This gives the user little control over the configuration of the cluster. As a result, there is no concept of rack awareness that the user

---

[*] https://aws.amazon.com/elasticmapreduce/
[†] http://hortonworks.com/
[‡] http://www.cloudera.com/
[§] http://www-03.ibm.com/software/products/en/ibm-biginsights-for-apache-hadoop
[¶] https://aws.amazon.com/

can configure and access. Besides this, the availability and performance of the cluster are also dependent on the virtual machine (VM) that is being used.

The user is required to install and configure Hadoop on all of the available options, other than Amazon EMR. Amazon EMR provides what can be referred to as 'MapReduce-as-a-Service'. The users can directly run MapReduce jobs on the Amazon EMR-powered cluster, making the whole exercise fairly simple and easy for the developer. If the developer does not wish to use HDFS as the default storage solution, then Hadoop cluster can also be used with S3*. Although, S3 is not as efficient as HDFS, it provides some unparalleled features such as data loss protection, elasticity, and bucket versioning. Some applications may require the implementation of these features, which makes S3 an irreplaceable storage solution. Besides this, Hadoop with S3 may be the storage solution of choice for organizations that already have their data stored on S3. A commercially available solution that uses this configuration is Netflix.

19.6.5.2 Deploying Hadoop in a Private Cloud

The private cloud allows the user to have better control over the configuration of Hadoop in the cloud. Several cloud providers such as IBM's SmartCloud[†] for deployment of InfoSphere BigInsights and IBM's PureData[‡] System offer PaaS solutions, which provide a pre-built setup for convenient deployment of Hadoop. The advantages of using Hadoop on a private cloud are as follows:

- Better control and visibility of the cluster
- Better mitigation of data privacy and security concerns

19.6.5.3 Key Considerations for Deployment

There are obvious advantages of running Hadoop on the cloud. However, it is important to understand that this does not come without problems and potential issues. Some of the things that must be paid heed to before using Hadoop on the cloud are given below.

- The security provided by the Hadoop cluster is very limited. Therefore, the security requirements and criticality of data being shared with the Hadoop cluster need to be carefully examined in advance.
- Typically, Hadoop runs on Linux. However, Hortonworks also provides a Hadoop distribution that works with Windows and is available on Microsoft's Azure Cloud. It is important to identify the operating system requirements and preferences before choosing a cloud-based Hadoop solution.
- Hadoop can never be viewed as a standalone solution. When it comes to designing big data analytics applications, you will need to look beyond the Hadoop cluster and see if the cloud solution supports visualization tools like Tableau and R, to serve your purpose in totality.
- An important consideration that is usually overlooked is data transmission. Is the data already on the cloud or will it have to be loaded from an internal system? If the application needs transferring of data from one public cloud to another, some transmission fees may apply.
- Using the VM-based Hadoop cluster may suffer from performance issues. These arrangements are good solutions only for development and testing purposes or unless performance is not an issue.

---

* https://aws.amazon.com/s3/getting-started/
† http://www-03.ibm.com/software/products/en/ibmsmarnote
‡ https://www-01.ibm.com/software/data/puredata/

19.7   CHALLENGES AND OPPORTUNITIES

Big data computing requires the use of several techniques and technologies. MapReduce and Hadoop are certainly the most popular and useful frameworks for this purpose. Apart from cloud computing, it has also been proposed that bio-inspired computing, quantum computing, and granular computing are potential technologies for big data computing (Philip Chen and Zhang 2014). However, each of these technologies needs to be adapted for this purpose and is not free from potential challenges. Challenges specific to big data computing using these technologies is beyond the scope of this chapter.

Owing to the elasticity and scalability of cloud solutions, this technology is one of the front-runners for big data computing (Talia 2013). With that said, the feasibility and viability of using a synergistic model is yet to be explored. NESSI presented challenges specific to implementation of existing machine learning techniques for big data computing and development of analytics solutions, mentioning the following requirements as fundamental (NESSI 2012):

- There is a need for development of solid scientific foundation, to facilitate selection of the method or technique that needs to be chosen.
- There is a need for development of scalable and efficient algorithms that can be used.
- The developed algorithms cannot be implemented unless appropriate technological platforms have been selected.
- Lastly, the solution's business viability must be explored.

Analytics solutions are expected to be uncomplicated and simple despite the complexity of the problem and the challenges involved. Besides this, distributed systems and parallel computing seems to be an appropriate solution to the big data problem. Therefore, the designed solution has to be inclined towards computational paradigms that work on these foundations. There are several identified challenges related to the use of cloud computing for big data analytics (Hashem et al. 2015; Assunção et al. 2015). Big data in the cloud suffers from several trials, both technical as well as nontechnical. Technical challenges associated with cloud-based big data analytics can further be divided into three categories, namely, big data management, application modeling, and visualization.

Challenges associated with characteristics, storage, and processing of big data are included in big data management. Management of big data is a challenging task considering the fact that data is continuously increasing in volume. Moreover, aggregation and integration of unstructured data collected from diverse sources is also under research consideration. There are two aspects of data acquisition and integration. First, data needs to be collected from different sources. Besides this, the collected data may be of the structured, unstructured, or semi-structured type. Integration of a variety of data types into an aggregation that can be further used for analytics is an even bigger problem.

In order to make predictions, recommendations, or implement any application-specific functionality, the application needs to be modeled. Moreover, this model needs to evolve with the changing needs of the system. The systems of today are generating both static as well as stream data. Therefore, any solution designed for such a system must support integration of different programming models to form a generic analytics engine. It is not possible to create practically viable solutions unless these models make use of resources optimally and exhibit high energy efficiency, which adds to the secondary requirements of the system. The expected future work with respect to the analytics engine can be summarized as follows:

- Development of energy-efficient and cost-effective solutions
- Standardization of solutions

The findings of the analytics engine need to be presented to the user in a manner that the results can be of use. This makes visualization a critical part of any analytics solution. Better and more cost-effective hardware support for large-scale visualization is being investigated (Assunção et al. 2015). In addition to this, there is a pressing need for solutions that can visualize big data efficiently. One of the most recent and popular research topics in visualization is real-time visualization. Since systems are generating data and analytics results in real-time, presenting the same to the user in real-time makes more sense. Visualization in the cloud context suffers from an inherent issue. The network for such application has long been and continues to remain a bottleneck.

Technology needs to be practically viable to support high business value solutions. With that said, cloud-based big data technologies suffer from several non-technical challenges that need to be worked out before widespread adoption can be expected. The lack of generic and standard solutions affects the business value of this technology heavily. Besides this, even though the use of cloud computing has brought down the cost of big data technologies considerably, more cost-effective models that make optimal use of the elasticity property of the cloud need to be devised for the end-users.

Businesses find it difficult to migrate to big data technologies due to lack of skilled staff and debugging-testing solutions available for these technologies. One of the biggest concerns while using big data analytics and cloud computing in an integrated model is security. In view of this, additional challenges like security and privacy risks also exist and need to be mitigated before commercially viable, practical cloud-based big data analytics solutions can be developed and used.

19.8   FUTURE RESEARCH DIRECTIONS

In view of the challenges and issues identified in the previous section, there is a need for a cloud-based framework to facilitate advanced analytics. The analytical workflow is composed of several steps, which include data acquisition, storage, processing, analytics, and visualization. Besides this, each of these steps is composed of many sub-steps. For instance, data acquisition is composed of sub-steps such as data collection, pre-processing, and transformation.

With specific focus on processing and analytics, existing solutions in the field are not generic. There is tight coupling between field-specific analytical solutions and data models used for the same. In addition to this, data model diversity also exists as a fundamental issue for generic framework development. Lastly, the data characteristics used to classify data as big data varies substantially with time and changes from one application to another. Research needs to be directed towards development of a generic analytical framework and cloud-based big data stack that addresses the complexities of the issues mentioned above.

Big data technology applies and appeals to every walk of human life. There is no technology-enabled system that cannot make use of the big data-powered solutions for enhanced decision making and industry-specific applications. However, in order to make this technology commercially viable, research groups need to identify potential "big" datasets and possible analytical applications for the field concerned. With that said, the feasibility and commercial viability of such analytical applications need to be aligned with business and customer requirements.

REFERENCES

Abu-Mostafa, Yaser S., Malik Magdon-Ismail, and Hsuan-Tien Lin. 2012. *Learning from Data.* AMLBook.com.
Akidau, Tyler, Sam Whittle, Alex Balikov, Kaya Bekiroğlu, Slava Chernyak, Josh Haberman, Reuven Lax, Sam McVeety, Daniel Mills, and Paul Nordstrom. 2013. MillWheel. *Proceedings of the VLDB Endowment* 6 (11):1033–1044. doi: 10.14778/2536222.2536229.
Assunção, Marcos D., Rodrigo N. Calheiros, Silvia Bianchi, Marco A. S. Netto, and Rajkumar Buyya. 2015. Big Data computing and clouds: Trends and future directions. *Journal of Parallel and Distributed Computing* 79–80:3–15. doi: 10.1016/j.jpdc.2014.08.003.
Bengio, Yoshua, Aaron Courville, and Pascal Vincent. 2013. Representation learning: A review and new perspectives. *IEEE Trans Pattern Anal Mach Intell* 35 (8):1798–828. doi: 10.1109/TPAMI.2013.50.

Bennett, Janine, Ray Grout, Philippe Pebay, Diana Roe, and David Thompson. 2009. Numerically stable, single-pass, parallel statistics algorithms. IEEE International Conference on Cluster Computing and Workshops. CLUSTER '09, New Orleans, LA, 1–8. doi: 10.1109/clustr.2009.5289161.

Bo-Wei, Chen, Wen Ji, and Seungmin Rho. 2016. Divide-and-conquer signal processing, feature extraction, and machine learning for big data. *Neurocomputing* 174:383. doi: 10.1016/j.neucom.2015.08.052.

Burnap, Peter, Omer Rana, Matthew Williams, William Housley, Adam Edwards, Jeffrey Morgan, Luke Sloan, and Javier Conejero. 2014. COSMOS: Towards an integrated and scalable service for analysing social media on demand. *International Journal of Parallel, Emergent and Distributed Systems* 30 (2):80–100. doi: 10.1080/17445760.2014.902057.

Buyya, Rajkumar. 2016. Big Data Analytics = Machine Learning + Cloud Computing. In *Big Data*, 7–9. Burlington, MA: Morgan Kaufmann Publisher.

Chandio, Aftab Ahmed, Nikos Tziritas, and Cheng-Zhong Xu. 2015. Big-data processing techniques and their challenges in transport domain. *ZTE Communications.*

Cheatham, Thomas, Amr Fahmy, Dan Stefanescu, and Leslie Valiant. 1996. *Bulk Synchronous Parallel Computing—A Paradigm for Transportable Software*, 61–76. doi: 10.1007/978-1-4615-4123-3_4.

Gartner. 2016. Gartner IT Glossary. Gartner.

Ghemawat, Sanjay, Howard Gobioff, and Shun-Tak Leung. 2003. The Google file system. *ACM SIGOPS Operating Systems Review* 37 (5):29. doi: 10.1145/1165389.945450.

Ginsberg, Jeremy., Matthew H. Mohebbi, Rajan S. Patel, Lynnette Brammer, Mark S. Smolinski, and Larry Brilliant. 2009. Detecting influenza epidemics using search engine query data. *Nature* 457 (7232):1012–1014. doi: 10.1038/nature07634.

Hammond, Klavdiya and Aparna S. Varde. 2013. Cloud based predictive analytics: Text classification, recommender systems and decision support. IEEE 13th International Conference on Data Mining Workshops (ICDMW), Dallas, TX, 607–612. doi: 10.1109/icdmw.2013.95.

Hashem, Ibrahim Abaker Targio, Ibrar Yaqoob, Nor Badrul Anuar, Salimah Mokhtar, Abdullah Gani, and Samee Ullah Khan. 2015. The rise of big data on cloud computing: Review and open research issues. *Information Systems* 47:98–115. doi: 10.1016/j.is.2014.07.006.

Hastie, Trevor, Robert Tibshirani, and Jerome Friedman. 2009. *The Elements of Statistical Learning. Springer Series in Statistics*. New York, NY: Springer Science & Business Media. doi: 10.1007/978-0-387-84858-7.

Isard, Michael, Mihai Budiu, Yuan Yu, Andrew Birrell, and Dennis Fetterly. 2007. Dryad. *ACM SIGOPS Operating Systems Review* 41 (3):59. doi: 10.1145/1272998.1273005.

Khan, Zaheer, Ashiq Anjum, and Saad Liaquat Kiani. 2013. Cloud based big data analytics for smart future cities. UCC '13 Proceedings of the 2013 IEEE/ACM 6th International Conference on Utility and Cloud Computing, Washington, DC, 381–386. doi: 10.1109/ucc.2013.77.

Klemens, Ben. 2009. *Modeling with Data*. Princeton, NJ: Princeton University Press.

Lee, Kyong-Ha, Yoon-Joon Lee, Hyunsik Choi, Yon Dohn Chung, and Bongki Moon. 2012. Parallel data processing with MapReduce. *ACM SIGMOD Record* 40 (4):11. doi: 10.1145/2094114.2094118.

Lu, Sifei, Reuben Mingguang Li, William Chandra Tjhi, Kee Khoon Lee, Long Wang, Xiaorong Li, and Di Ma. 2011. A framework for cloud-based large-scale data analytics and visualization: Case study on multiscale climate data. 2011 IEEE Third International Conference on Cloud Computing Technology and Science (CloudCom), Athens, Greece, 618–622. doi: 10.1109/CloudCom.2011.95.

Malewicz, Grzegorz, Matthew H. Austern, Aart J. C. Bik, James C. Dehnert, Ilan Horn, Naty Leiser, and Grzegorz Czajkowski. 2010. Pregel. Proceedings of the 2010 ACM SIGMOD International Conference on Management of data, New York, NY, 135. doi: 10.1145/1807167.1807184.

Mell, Peter M. and Timothy Grance. 2011. doi: 10.6028/nist.sp.800-145.

Melnik, Sergey, Andrey Gubarev, Jing Jing Long, Geoffrey Romer, Shiva Shivakumar, Matt Tolton, and Theo Vassilakis. 2011. Dremel. *Communications of the ACM* 54 (6):114. doi: 10.1145/1953122.1953148.

Microsoft. 2016. Graph Engine. Accessed 2016.

Mikolov, Tomas, Anoop Deoras, Daniel Povey, Lukas Burget, and Jan Cernocky. 2011. Strategies for training large scale neural network language models. IEEE Workshop on Automatic Speech Recognition and Understanding (ASRU), Waikoloa, HI, 196–201. doi: 10.1109/asru.2011.6163930.

NESSI. 2012. Big Data: A New World of Opportunities. Accessed 2016.

Philip Chen, C. L., and Chun-Yang Zhang. 2014. Data-intensive applications, challenges, techniques and technologies: A survey on Big Data. *Information Sciences* 275:314–347. doi: 10.1016/j.ins.2014.01.015.

Qian, Ling, Zhiguo Luo, Yujian Du, and Leitao Guo. 2009. Cloud computing: An overview. 5931:626–631. doi: 10.1007/978-3-642-10665-1_63.

Raghupathi, Wullianallur and Viju Raghupathi. 2014. Big data analytics in healthcare: Promise and potential. *Health Information Science and Systems* 2:3. doi: 10.1186/2047-2501-2-3.

Ratner, Bruce. 2003. *Statistical Modeling and Analysis for Database Marketing.* Boca Raton, FL: Chapman & Hall/CRC.

Sahimi, Muhammad and Hossein Hamzehpour. 2010. Efficient computational strategies for solving global optimization problems. *Computing in Science & Engineering* 12 (4):74–83. doi: 10.1109/mcse.2010.85.

Salminen, Arto. 2012. Introduction to NOSQL. NoSQL Seminar 2012.

Samuel, Arthur L. 1959. Some studies in machine learning using the game of checkers. *IBM Journal of Research and Development* 3 (3):210–229. doi: 10.1147/rd.33.0210.

Seenumani, Gayathri, Jing Sun, and Huei Peng. 2011. Real-time power management of integrated power systems in all electric ships leveraging multi time scale property. *IEEE Transactions on Control Systems Technology* 20:232–240. doi: 10.1109/tcst.2011.2107909.

Seo, Sangwon, Edward J. Yoon, Jaehong Kim, Seongwook Jin, Jin-Soo Kim, and Seungryoul Maeng. 2010. HAMA: An efficient matrix computation with the MapReduce framework. IEEE Second International Conference on Cloud Computing Technology and Science (CloudCom), Indianapolis, IN, 721–726. doi: 10.1109/CloudCom.2010.17.

Shvachko, Konstantin, Hairong Kuang, Sanjay Radia, and Robert Chansler. 2010. The Hadoop Distributed File System. IEEE 26th Symposium on Mass Storage Systems and Technologies (MSST), Incline Village, NV, 1–10. doi: 10.1109/msst.2010.5496972.

Spark. 2016. Spark documentation. spark.apache.org.

Sun, Ninghui, James G. Morris, Jie Xu, Xinen Zhu, and Meihua Xie. 2014. iCARE: A framework for big data-based banking customer analytics. *IBM Journal of Research and Development* 58 (5/6):4:1–4:9. doi: 10.1147/jrd.2014.2337118.

Talia, Domenico. 2013. Clouds for scalable big data analytics. *Computer* 46 (5):98–101. doi: 10.1109/mc.2013.162.

Thaduri, Adithya, Diego Galar, and Uday Kumar. 2015. Railway assets: A potential domain for big data analytics. *Procedia Computer Science* 53:457–467. doi: 10.1016/j.procs.2015.07.323.

Thompson, David, Joshua A. Levine, Janine C. Bennett, Peer-Timo Bremer, Attila Gyulassy, Valerio Pascucci, and Philippe P. Pebay. 2011. Analysis of large-scale scalar data using hixels. IEEE Symposium on Large Data Analysis and Visualization (LDAV), Providence, RL, 23–30. doi: 10.1109/ldav.2011.6092313.

Xiaodong, Li and Yao Xin. 2012. Cooperatively coevolving particle swarms for large scale optimization. *IEEE Transactions on Evolutionary Computation* 16 (2):210–224. doi: 10.1109/tevc.2011.2112662.

Xindong, Wu, Zhu Xingquan, Wu Gong-Qing, and Ding Wei. 2014. Data mining with big data. *IEEE Transactions on Knowledge and Data Engineering* 26 (1):97–107. doi: 10.1109/tkde.2013.109.

XSLogic. 2016. XSLogic/Phoebus. Github.

Yang, Zhenyu, Ke Tang, and Xin Yao. 2008. Large scale evolutionary optimization using cooperative coevolution. *Information Sciences* 178 (15):2985–2999. doi: 10.1016/j.ins.2008.02.017.

20 A Multidimensional Sensitivity-Based Anonymization Method of Big Data

Mohammed Al-Zobbi, Seyed Shahrestani, and Chun Ruan

CONTENTS

20.1 INTRODUCTION

Big Data analytics is where advanced analytic techniques operate on big data sets (Russom, 2011). Hence, analytics is the main concern in big data, and it may be exploited by data miners to breach privacy (Samarati, 2001). In the past few years, several methods that address the data leakage concerns have been proposed for conventional data (Ninghui Li et al., 2007; Qing Zhang et al., 2007). The proposed methods provide remedies for variant types of attacks against data analytics process. Side attack is considered to be one of the most critical attacks (Dwork et al. 2010). This attack is prevalent in medical data, where the attacker owns partial information about the patient. The attacker aims to find the hidden sensitive information by logically linking between his/her data and the targeted data. A side attack can be conducted by either manipulating the query, a state attack, or running malicious code that can transfer the output from other users through the network, a privacy attack. However, a variety of attacks can be triggered by the adversary to interrupt the analytics process by mounting the malicious code, which may cause infinite loop operations or may eavesdrop on other user's operations (Shin et al., 2012).

One powerful optimization algorithm for de-identification procedure known as *k-anonymity* (Bayardo and Rakesh Agrawal, 2005). This approach was proposed for conventional data. *k-anonymity* is a one-dimensional method which highly disturbs the anonymized data and reduces the anonymized information gained (Mohammed et al., 2010). To resolve this matter, various anonymization methods were proposed such as ℓ-diversity (Machanavajjhala et al., 2007), and (*X, Y*) Anonymity (Fung et al., 2007). These extended methods, however, do not resolve the one-dimensional concern, where data are structured in a multidimensional pattern. The multidimensional LKC-Privacy method was proposed to overcome the one-dimensional distortion (Zhang et al., 2014). Later, a more profound method was proposed known as multidimensional top-down anonymization. This method generalizes data to the top-most first and then specializes data based on the best score results (Lublinsky, 2013).

The previously mentioned methods can operate efficiently in conventional data. However, big data specifications and operations concepts are different. Big data operates in a parallel distributed environment, known as MapReduce, where performance and scalability are a major concern (Ren et al. 2013). Researchers amended the previously mentioned anonymization methods so that they can fit the new distributed environment. One of these methods is the two-phase multidimensional top-down specialization method (two-phase MDTDS), which splits the large size of data into small chunks. This technique negatively affects the anonymized information, resulting in increased information loss. Moreover, each chunk of data requires n iterations to find the best score on specialization rounds (Ren et al. 2013). The iterations create n rounds between *map* and *reduce* nodes. Map and reduce may be connected through the network on separate nodes. Therefore, an unknown number of iterations may cause more delay, if map and reduce are on two separate nodes. Hence, this rigid solution may cause a high delay (Ren et al. 2013). Also, the iteration locks both nodes until the end of the process, which disturbs the parallel computing principle.

Another big data privacy method mutates between TDS and bottom-up generalization (BUG) in a hybrid fashion (Zhang et al., 2013, 2014). The method calculates the value of K, where K is defined as workload balancing point if it satisfies the condition that the amount of computation required by MapReduce TDS (MRTDS) is equal to that by MapReduce BUG (MRBUG). The K value is calculated separately for each group of data set, so TDS operation is triggered when the anonymity of $k > K$, whereas BUG operation is triggered when the anonymity of $k < K$. However, the iteration technique of this method is quite similar to MDTDS. Also, finding the value K for each group of record consumes even more time.

In this work, we propose a multidimensional anonymization method that supports data privacy in MapReduce environments. Our multidimensional sensitivity-based anonymization method (MDSBA) is proposed and compared with the MDTDS. MDSBA provides a bottom-up anonymization method with a distributed technique in MapReduce parallel processes. Our method enables data owners to give multiple levels of anonymization for users of multiple access levels. The anonymization is based on grouping and splitting data records, so each group set or domain is anonymized individually. Finally, anonymized domains are merged in one domain. Also, our method embeds a role-based access control (RBAC) model that enforces a security policy on a user's access and ability to run processes. RBAC is chosen over mandatory access control (MAC) for its flexible and scalable user roles and access permissions, and for its popularity over the cloud. However, RBAC and the mapping process will be discussed in a separate paper. Section 20.4 delves into MDSBA grouping and masking processes in detail.

This chapter is structured as follows. The next section introduces some previous *k-anonymity* methods adapted for use in conventional data. Section 20.3 describes our proposed anonymization approach, MDSBA. The experimental evaluation is outlined in Section 20.4. The last section gives our concluding remarks and suggested future work.

20.2 RELATED WORK

The privacy method concept relies on hiding sensitive information from data miners. The hiding principle implies distortion techniques that promote a trade-off between privacy and beneficial data utility. The specialization technique relies on a quasi-identifier (Q-ID), which tends to find a group of attributes that can identify other tuples in the database (Rajeev Motwani, 2007). These identifiers may not gain 100% of data. However, a risk of predicting some data remains high. For instance, knowing the patient's age, gender, and postcode may lead to uniquely identifying that patient with a probability of 87% (Rajeev Motwani, 2007).

Many algorithms were developed to overcome this security breach. *K-anonymity* was proposed by Sweeney (Sweeney, 2002a) who suggested data generalization and suppression for Q-IDs.

K-anonymity guarantees privacy on releasing any record by adhering each record to at least *k* individuals, and this is correct even if the released records are connected to external information. The table is called *k*-anonymous if at least *k*–1 tuples are Q-ID equivalent records. This means the equivalence group size on Q-ID is, at least, *k* (Sweeney, 2002b).

K-anonymity has gained some popularity, but it has some subtle and severe privacy problems, such as homogeneity and background knowledge attacks (Machanavajjhala et al., 2007). The homogeneity leverages if all *k-anonymity* values are identical. ℓ-Diversity was introduced to resolve the privacy breach in *k-anonymity* method (Machanavajjhala, Kifer, Abowd, Gehrke, and Vilhuber, 2008). This algorithm aims to reduce the attribute linkage. It was developed from the fact that some sensitive attributes [S] are more frequent than others in the group. So ℓ-diverse is calculated using the entropy by grouping the Q-ID and then calculating the entropy for the groups using the following formula (Machanavajjhala, Gehrke, Kifer, and Venkitasubramaniam, 2006):

$$\sum_{s \in S} P(qi, s) \log(P(qi, s)) \geq \log(\ell) \tag{20.1}$$

where $P(qi, s)$ is the tuples fractions in qi blocks and S denotes the sensitive attributes.

The previously mentioned methods anonymize data based on the one-dimensional concept. This concept reduces information gained after anonymization and may result in false statistics. Data analytics implements search methods to find a group of records, which can be a part of one-dimensional, two-dimensional, or multidimensional data (Samet, 2006). A popular multidimensional method is proposed to overcome the one-dimensional *k-anonymity* method, known as the LKC-privacy method. LKC-privacy can be applied to multidimensional data, such as a patient's information. The general intuition of LKC-privacy ensures that a Q-ID with a length of *L* and sensitive value of *S* is not greater than Class C. The idea is grouping length of records *L* in the data object *T*, by at least *k* records (Mohammed et al., 2010).

To enhance the LKC-privacy performance, a more profound method was proposed, known as the TDS method (Zhang et al., 2014). This method reverses the LKC-privacy process by generalizing records to the top-most level and then specializing the highest score attributes. TDS, LKC, l-diversity, and all other *k-anonymity* methods tend to have trade-offs. The trade-off between information gained and anonymization loss is presented as $\text{Score}(v) = \dfrac{\text{InfoGain}(v)}{\text{Anony Loss}(v) + 1}$ (Fung et al., 2007).

This equation does not satisfy the form matrix to capture the classification; therefore, Shannon's equation is used for correctness. Shannon's information theory is used to measure information gain (Shannon, 2001). Let $T[v]$ denote the set of records masked to the value v and let $T[c]$ denote the set of records masked to a child value c in child (v) after refining v. Thus, the InfoGain (v) is defined as

$$\text{Info Gain}(v) = E(T[v]) - \sum_c \frac{|T[c]|}{|T[v]|} E(T[c]) \tag{20.2}$$

where E (T[v]) is the entropy of T[v].

In multidimensional TDS (MTDS), Equation 20.2 determines the specialized Q-ID among other suppressed Q-IDs. Initially, all tuples are generalized to the primary root of the taxonomy tree (any) and this value suppresses any Q-ID. Then, the highest score Q-ID is specialized by choosing taxonomy tree child node v, which is known by cut $_j$, where j denotes the taxonomy tree levels. Eventually, the top-most taxonomy tree is the parent that suppresses attributes. The parent level is zero, and the number increases accordingly by moving from the top to the bottom of the tree. MDTDS has recently gained popularity especially in the map/reduce environment (Zhang et al., 2014).

The *k*-anonymization anonymizes Q-IDs by using one of the three different techniques: taxonomy tree, interval, and suppression. The data owner determines the Q-IDs before any anonymization takes a place. Also, data are classified by the miner before starting anonymization. This classification helps the masking process to identify the best technique. The masking techniques deal with data as per tuple or small groups of tuples known as multidimensional process (Fung et al. 2005).

20.3 MULTIDIMENSIONAL SENSITIVITY-BASED ANONYMIZATION

There are three analytics-related parties in Big Data. These are the data owner, the service provider, and the user miner or analyzer. Based on the previously mentioned privacy methods, we introduce a MDSBA framework that implements a bottom-up technique of *k-anonymity*. The framework also adapts a discriminated multiaccess level for users. The framework aims to implement a complete solution for MapReduce operations in big data. The solution basis mimics the parallel distributed processes over MapReduce nodes. This divides the single rigorous anonymization process into multitasks that can be distributed to more than one node. Accessing data for analytics is conducted by many users with multilevel access in the big data environment. This imposes a gradual level of the data access and view. Users with a low-level permission are less trusted by data owners. Therefore, more restrictions are applied to a data view.

All *k-anonymity* extended methods apply similar grouping techniques in anonymizing data. The technique groups records with equivalent values and generalizes nonequivalent values to increase the number of equivalent records. MDSBA promotes mathematical equations to estimate the access levels for users.

20.3.1 THE SENSITIVITY LEVEL DEFINITIONS

The MDSBA method/aims to define the privacy method and masking pattern for each access level. Data owners determine a subset of attributes like Q-IDs and sensitive attributes S. Next, the sensitivity value is determined by the MDSBA equation. The MDSBA process is operated within the RBAC environment. Figure 20.1 presents an algorithm for the MDSBA framework.

Definition 1: Sensitivity level (ψ) implies a scale of data anonymization prominence, so the anonymized data T delegate a multilevel of distorted data.

Definition 2: The *k-anonymity* is the maximum equivalent records number of the ownership level $\bar{k}$. Hence, $\bar{k} = k - i$, where $i \in Z$, and $i = \{k\text{-}1, \ldots, 0\}$ and $\bar{k} \le k$.

Referring to definition 2, the highest number of ownership level $\bar{k}$ describes the lowest user access level, which can be presented by $\bar{k} = k$. The user with a lower value of $\bar{k}$ has a higher access level and privileges. If the sensitivity factor is denoted by ω, then a higher value of $\bar{k}$ leads to a lower value of ω, which is presented by $\bar{k} \, \alpha \, \dfrac{1}{\omega}$. The value of ω can be calculated by finding the probability

---

Algorithm 20.1 Multidimensional sensitivity-based anonymization algorithm

1. Data owner determines the object Q-IDs, and the object sensitive attributes S.
2. Data owner determines the obsolescence object value, aging participation percentage, and object age.
3. The system determines the sensitivity level of S, based on the user access level.
4. System groups the records as per equivalent sensitive value.
5. System applies MDSBA to find the best-anonymized Q-IDs
6. The system applies MDSBA to determine the appropriate masking pattern.
7. The RBAC protects the anonymization processes.

FIGURE 20.1 Multidimensional sensitivity-based anonymization algorithm and relationship with role-based access control.

of the minimum and the maximum values in m number of n Q-IDs. Hence, the maximum sensitivity factor is defined as the highest probable value between Q-IDs, or

$$\omega_{max} = \max(P(qid0), P(qid1)...P(qidm)) \tag{20.3}$$

Moreover, the minimum sensitivity factor is defined as the product of all Q-IDs probabilities, or

$$\omega_{min} = \prod_{i=1}^{m} P[qid_i] \tag{20.4}$$

Based on Equations 20.2 and 20.3, the value of ω can be found between ω_{min} and ω_{max}, as in Equation 20.4:

$$\omega = \omega_{min} + \left(k - \bar{k}\right)\left(\frac{\omega_{max} - \omega_{min}}{k}\right) \tag{20.5}$$

where k denotes the k-anonymity value and $\bar{k}$ denotes the ownership level.

Equation 20.6 collates both terms of ω and aging factor τ to conclude the sensitivity equation ψ. The object's sensitivity level degrades with the data age. The aging factor affects the sensitivity reversely. The older objects are less sensitive compared with the newer ones. Hence, two factors determine the sensitivity level, the sensitivity factor ω and the aging factor τ.

$$\psi = |\omega + \tau| \tag{20.6}$$

where ψ denotes the sensitivity level and τ denotes the aging level

Equation 20.6 is used to anonymize all grouping domains as explained in the next section. The masking process tends to find a close similarity to the sensitivity value. For instance, if $\psi = 0.5$, then any value that falls between 0 and 0.5 is accepted, as described in Table 20.1. However, the closer the probable value is to the sensitive value, the better it is.

20.3.2 THE OBJECT AGING SENSITIVITY (τ)

The aging factor is affected by four different factors. These are the object obsolescence value $\varnothing$, the aging participation ρ, the object age y, and the sensitivity factor ω. The obsolescence value is defined as the critical age before the object sensitivity starts degrading. Thereby, the object aging value is constant, if the age y is smaller than the obsolescence value, or $y < \varnothing$. The object aging value decreases linearly if the age y is greater than the obsolescence value, or $y \geq \varnothing$. Thus, two separate terms are expressed for the age, and $y \geq \varnothing$, as described in Equation 20.7. The aging participation percentage ρ is predetermined by the data owners.

TABLE 20.1
Example of Accepted Probable Values for $k = 5$

Accepted Sensitive Value	Sensitive Value (ψ)	Ownership Level $\left(\bar{k}\right)$
0	0	1
0–0.5	0.5	2
0–0.05	0.05	3
0–0.005	0.005	4
0–0.0005	0.0005	5

TABLE 20.2

The Object Aging Sensitivity τ, When $\varnothing = 10$ and
$\rho = 70\%$ for $\omega = 0.5$

Aging Sensitivity τ (%)	Object Age (Years)
0.35	1
0.35	5
0.315	10
0.158	15
0.095	17
0	20

$$\tau = \begin{cases} y < \varnothing & -\rho \times \omega \\ y \geq \varnothing & -\rho \times \omega \times 0.9 \times (2 - \dfrac{y}{\varnothing}) \end{cases} \tag{20.7}$$

where $\tau \geq 0$, $\varnothing \epsilon \mathbb{Z}$, $\varnothing \geq 0$, $0 \leq \rho \leq 100\%$.

Data owners may set ρ to 0% if their data objects do not mutate with the time factor. Table 20.2 shows an example of aging factor when $\varnothing = 10$ and $\rho = 70\%$ for $\omega = 0.5$.

20.3.3 Masking Operations

Anonymization methods apply data distortion using masking operations. Masking implies a taxonomy tree, suppression, or discretization. The taxonomy tree is the key anatomy for data masking. It implies hiding special information by generalizing it. For example, if the data contains a person's address as "Sydney," then the taxonomy tree contains Australia → NSW → Sydney. The generalization of the first cut is NSW, and the second cut is Australia. Discretization means replacing numerical values with a single interval and is denoted by *Int*. The interval deals with numerical data, where a set of numbers is presented by two numerical values for start and end. Finally, the suppression of value means replacing all its relevant values with the sign * or other characters. This operator is denoted by "Sup" (Benjamin et al., 2011). However, suppression may be presented by other values, such as "any" or "person." Some data can only be suppressed, such as a person's gender {Male, Female}. Other data can only be anonymized by using an interval or taxonomy tree, as shown in Table 20.1. The three masking patterns are presented as <∪Cutj, ∪Supj, ∪Intj> (Zhang et al., 2014). The sensitivity value of the attribute *S* is calculated based on the user access level and other factors. This sensitivity value is the milestone that determines how the masking pattern is applied on Q-IDs. First, the Q-ID data type is chosen, as shown in Table 20.1. Second, the masking pattern is being implemented on the chosen data type. This imposes three masking tools for anonymization, which are taxonomy tree level for ∪Cutj pattern, interval distance for ∪Intj, and the number of suppressed digits for ∪Supj pattern.

The taxonomy tree is constructed based on the hierarchy structure method, which starts from the topmost tree parent node, and moves down to the child of the leaf node v. Each leaf of the tree resides on a separate level, whereas the parent level is denoted by L0, or level 0, and moves downward the tree to Ln, where n denotes the number of levels in each branch of the tree. A probability value is given for each leaf node v depending on the number of subleaves of the children nodes. The probability P measures the rational between the user guess of the anonymized attribute, which is represented by the Ln node, over the total number of subleaves attached to that node. For instance, the student (Certificate) may indicate an anonymized value for any of the following: Primary School,

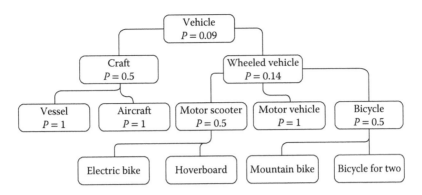

FIGURE 20.2   Taxonomy tree example with probabilities for each node.

High School, Certificate I, Certificate II, Certificate III, and Certificate IV. Hence, the probability of guessing the correct certificate is $P = 1/6 \approx 0.17$. Theoretically, the probability value increases when moving downward the tree. Moreover, the probability value must decrease when the number of subleaves increase. The probability percentage is precalculated by counting the number of leaves or children for each node. Figure 20.2 shows an example of a taxonomy tree, and the probability of each node. For instance, P has 11 child nodes, so the probability of "Vehicle" is $P = 1/11 = 0.09$.

A similar concept is applied in intervals masking patterns. The sensitivity value can be converted to an interval distance, denoted by ϱ. As shown in Table 20.1, determining the masking pattern is subjected to the data type. The value of $\varrho = 1/\psi$ describes the calculated distance. For instance, if $\psi = 0.2$, then $\varrho = 5$. Notice that the value of ϱ is inversely   proportionally with the value of ψ. Technically, creating intervals for digital values implies two essential steps; these are ascending the group of numbers and then deciding the minimum and maximum values in the interval. In our interval masking, the minimum and maximum interval values are given base-five rounding. For instance, if Q-ID represents the patient age as Q-ID(age) = {22,16,19}, then the subgroup is sorted in an ascending order as Q-ID(age) = {16,19,22}. Next, the interval is given a minimum value that is rounded to base-five and a maximum value that is also rounded similarly, so the anonymization is Q-ID(age) = [15–25] and $\varrho = 25-15 = 10$. In the previous example, let us consider $\psi = 0.02$, so $\varrho = 50$. For this reason, the interval is extended to Q-ID (age) = [15–65]. If we suppose that $\psi = 0.4$, then the value of $\varrho = 1/0.4 = 2.5$. In such a case, the interval remains Q-ID (age) = [15–25] because the calculated value $\varrho = 2.5$ is smaller than the original value $\varrho = 10$.

Masking patterns and degree of masking are determined by the sensitivity value, which is in turn influenced by the user access level. The masking degree is calculated based on probability, as shown in taxonomy trees and intervals. In suppression, a similar probability principle is applied to the number of suppressed characters. For instance, a postcode, with one-digit suppression, has a higher probability value than a postcode anonymization with a taxonomy tree. A higher probability value manifests a lower level of security. For instance, if the postcode value 2000 is suppressed with one digit, then the value 200* predicts 10 possible numbers between 0 and 9, so the probability is $P[\text{mask1}] = 0.1$.

The probability method can be applied to the three masking patterns by assigning the minimum and maximum values of probability P. The following two examples show the probability for each masking pattern for year and age. Let the year $y = 2015$, and the used pattern is <sup>. $P[\min] = 0.0001$, for the case $y = ****$, and $P[\max] = 0.1$ for a case such as $y = 201*$.

In the age example, the used pattern is <int>. The $P[\min]$ and $P[\max]$ for the age interval [10–70] imposes the lowest age interval value = 2. Therefore, $P[\max] = P[10–11] = 0.5$. The lowest interval probability imposes the largest contrast between both interval ends. Therefore, $P[\min]$ makes no changes to the original interval. Hence, $P[\min] = P[10–70] \approx 0.0164$.

TABLE 20.3

The Masking Pattern for Some Data Types

Masking Pattern	Data Type
∪Cutj, ∪Intj, ∪Supj	Date
∪Intj, ∪Supj	Integer
∪Cutj	Polynomial
∪Supj	Binomial (ex. yes/no)
∪Intj	Real
∪Supj	Text

The user access level influences the chosen masking pattern for Q-ID. In Table 20.3, some data types can be distorted by any of the three masking patterns, whereas others permit one or two patterns only. For example, postcode can be masked by either one of the three patterns, whereas gender can only be suppressed by ∪Supj. Also, masking methods imply different security levels, as explained in the next section.

20.3.4 GROUPING RECORDS

Three main steps are involved in MDSBA; these are grouping, calculating sensitivity, and applying the proper masking and specialization pattern. The three-procedure algorithm is explained in Figure 20.3. The process starts by creating the G domains. Initial grouping is created to aggregate the records as per equivalent sensitive values. Each group joins a separate domain G, where $G = \{G0, G1, G2 \ldots Gn\}$, and each G_i domain represents one sensitive value only.

The second grouping process, shown in Figure 20.3, is conducted for all Q-IDs. It concludes equivalent and semiequivalent records in one domain, whereas the nonequivalent records are temporarily separated in another domain. The grouping process stratifies the *k-anonymity* conditions, which implies at least some k records or higher are Q-IDs equivalent. However, the equivalent records are exempted from anonymization. The equivalent and semiequivalent domains are denoted by SG domains. SG*i* represents a subgroup of the G domain, where SG = {SG0, SG1,....SG*n*}. The nonequivalent domains are denoted by NG domains. NG*i* represents a subgroup of the G domain, where NG = {NG0, NG1,...NG*n*}.

Before defining the semiequivalent records, we need to know the Q-ID selection criteria. Data owners decide that some of the attributes may cause a privacy violation for some records. The chosen set of attributes is considered risky when they are revealed together as one set. This indicates that choosing Q-IDs must be aggregated in sets. When the owner determines more than one Q-ID set, then anonymization is applied for each set individually. This implies an anonymization iteration based on the number of sets. For instance, let Q-S*i* denotes the Quasi Identifier Set, and suppose the data owner identifies two sets as Q-S1 = {Q-ID1, Q-ID2, Q-ID3} and Q-S2 = {Q-ID4, Q-ID5}. Based on this example, the anonymization process considers three Q-IDs first. The second anonymization process, following the first round, examines two Q-IDs. The two-pair Q-IDs can be divided into two SG and NG groups. However, SG grouping does not contain any semiequivalent records. Instead, all records are equivalent.

So far, we created two separate domains of records, SG and NG. Last, equivalent records are exempted from anonymization, whereas semiequivalent records in the SG and NG domains are anonymized. The difference between SG and NG anonymization is the grouping process that is applied to each domain. The grouping process in the NG domain is applied on one Q-ID only, which is the highest Q-ID probability. This, in turn, causes an extra penalty applied on the rest of the non-grouped Q-IDs. The anonymized number of Q-IDs is higher in the NG domain. In the SG domain,

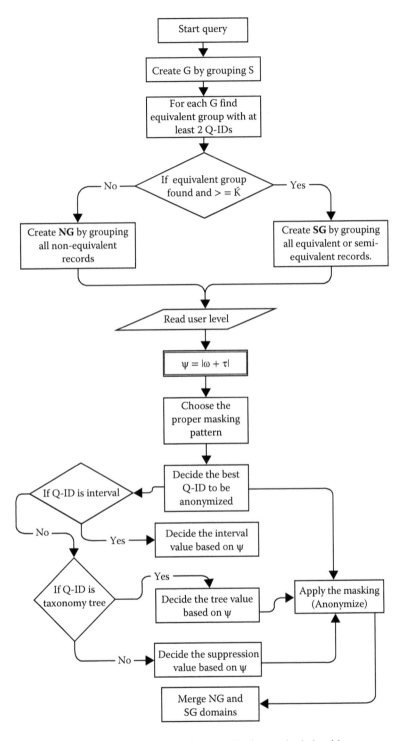

FIGURE 20.3   Multidimensional sensitivity-based anonymization method algorithm.

TABLE 20.4

SG and NG Examples

Q-ID	Max. P(Q-ID)	SG Records	NG Records
Age	0.01	25,25,25,25	25,26,30,32
Sex	0.5	F,F,F,F	M,M,M,M
Postcode	0.001	3001,3008,3027,3095	2074,2057,2055,2051

the grouping process is applied on at least two Q-IDs, whereas the anonymization process is applied to the rest of the nongrouped Q-IDs.

20.3.5 MASK PATTERN

Lemma 20. 1: The masking pattern criterion is chosen based on the probability multiplication. The results are approximate to the maximum sensitivity level as described below:

$$\psi \approx \prod_{i=1} P_i[<\cup \text{Cutj}, \cup \text{Supj}, \cup \text{Intj}] < 1 \qquad (20.8)$$

Let us consider the following examples of how the masking pattern is applied. In this example, let $\psi = 0.125$, and Q-ID = {Age, Sex, Postcode} in SG domain. The maximum probability for each Q-ID is shown in Table 20.4. The created semiequivalent records are grouped based on the highest two Q-IDs, and these are P_{age} and P_{sex}. Suppose that the chosen postcode pattern is the taxonomy tree for the Australian suburbs or <Cutj>. The SG grouped records are grouped with $k = 4$, as shown in Table 20.4. The anonymized postcode should follow the sensitivity rule by finding a probable value $\leq \psi$. The second taxonomy tree probability level is $P_{postcode}$ [Level-2] = 0.125. This probability value was derived from the number of Australian states, or 1/8 states = 0.125. Therefore, the anonymized postcode is Postcode = {VIC, VIC, VIC, VIC}.

Another example is a group of records in an NG domain. Let $\psi = 0.005$ and the NG records are shown in Table 20.2. To find out the best masking pattern for $k = 4$, we need to group a single Q-ID and anonymize the rest of the Q-IDs. The highest probability value is the Sex Q-ID, which can be grouped as sex = {M,M,M,M}. The anonymization process must equalize the four records; to do so, use {Age, Postcode} Q-IDs with a close probability to 0.005. This can be calculated as P_{age} [$\cup$Intj] = X, and $P_{postcode}$ [$\cup$Intj] = Y. Because both Q-IDs are in the <Intj> pattern, then we can assign equal values for each one. $X \times Y \leq 0.005$. Hence, $\sqrt{0.005} \approx 0.07$, so $X = 0.07$ and $Y = 0.07$. This clearly assigns an interval of 1/0.07 $\approx$15 for each Q-ID. The anonymization results for age are Age = {25–40, 25–40, 25–40, and 25–40}. However, the postcode anonymization cannot be replaced by [2050–2065]. This is because one last postcode remains outside the range of the interval. Therefore, the interval range should be extended to include Postcode = {2074}. The final anonymization results for postcode are Postcode = {[2050–2075), [2050–2075), [2050–2075), [2050–2075)}.

The masking pattern is chosen based on the sensitivity value ψ. The algorithm can be used to determine the closest masking pattern. The fact that each Q-ID attribute may accept more than one pattern of anonymization, in this case, the taxonomy tree is chosen first, and then the interval. Because the taxonomy tree is predefined, then it is possible to define the probability value for each v (child) level, this helps the decision of the chosen the masking pattern. The interval pattern is more flexible, and can be generated on demand. Thereby, a base-five mathematical notion is used to determine the best masking pattern, by choosing the closest possible values to the sensitivity value. The algorithm in Figure 20.4 explains the steps of two masking patterns: the interval and the taxonomy tree.

```
Algorithm 20.2 Choosing the masking method
  1.  read the sensitivity value ψ, ownership, and k;
  2.  let ownership is o;
  3.  read masking method m available for A(Q – ID)
  4.    If m = UCut_o then
  5.        Let UCut_o _level is TL;
  6.        'Find the probability of taxonomy tree level close to ψ
  7.        For TL = n to 1 Loop
  8.            n = n – 1
  9.            If Ptaxonomy[n] ≤ ψ
 10.                TL = n                                  End if;
 11.            End if;
 12.        Next For
 13.        Mask A(Q – ID) using PUCut_o[TL];
 14.
 15.        If m = UInti then
 16.            Read smallest value, largest value of the record;
 17.                Interval = largest – smallest;
 18.            If 1 ÷ interval ≤ ψ then
 19.            Change interval to base-five using remainder;
 20.            Smallest_value = smallest – remainder(smallest ÷ 5);
 21.            Largest_value=largest + (5 – remainder(largest ÷ 5));
 22.            Mask A(Q – ID) with smallest_value, and largest_value;
 23.        End if;
       End if;
```

FIGURE 20.4   Masking method algorithm.

20.4   EXPERIMENTAL EVALUATION

Our experiments are conducted on the Adult Database from the University of California/Irvine (UCI) Machine Learning Repository (Becker, 1996), which is a public benchmark for anonymization algorithms experiments. We aim to experiment with our method on conventional data first so that we can compare it with the current MDTDS. Also, we to measure the efficiency of the MDSBA method by monitoring the quality of the anonymized data output for development and improvement. Our future work will focus on big data anonymization in MDSBA. We establish a data comparison according to Mohammed et al. (2010). Data loss and performance approaches are tested within this experiment. The method is comparing this privacy method with public standards such as MDTDS.

Two separate prototypes of Java programs were coded using Java Standard Edition. Each algorithm runs MDSBA and MDTDS, respectively. The comparison concludes testing variable quantity of sensitive values. All experiments on MDSBA were conducted on an Intel Core i5 2.4 GHz with 8 GB RAM.

Two experiments with two rounds of each experiment were conducted. Each experiment was conducted by two different data samples; the first sample contains three values of the sensitive data, whereas the second sample contains only two values of the sensitive data. The sensitive data is related to the user's salary, and the values are either >50k, >50k, or <10k. For each privacy method, the pinch mark promotes the highest score and InfoGain for the anonymized data. The first round suggests only two Q-IDs with two sensitive values, while $k = 5$, $\tau = 0.05$, $\rho = 60\%$, and the sensitive attributes $S = 50\%$, as explained in Equation 20.1. Figures 20.5 and 20.6 illustrate the relationship between the user level and both InfoGain and AnonyLoss, respectively.

The InfoGain measures the entropy or error percentage between the actual data and the anonymized data. Figures 20.5 and 20.6 show expected results for InfoGain and AnonyLoss degrading. Hence, lower ownership level exponentially degrades the data refinement and increases the data suppression. Similar results can be found for $S = 30\%$. Figures 20.6 and 20.7 compare between $S = 50\%$ and $S = 30\%$.

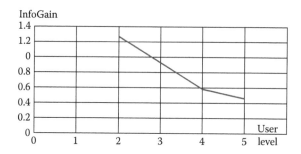

FIGURE 20.5 InfoGain with $S = 50\%$ in multidimensional sensitivity-based anonymization method.

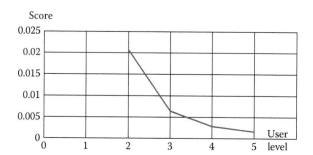

FIGURE 20.6 Score with $S = 50\%$ in multidimensional sensitivity-based anonymization method.

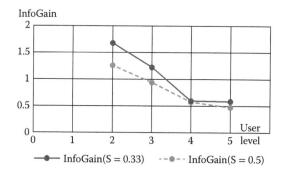

FIGURE 20.7 InfoGain comparison with S.

In the second round of the first trial, both data samples were compared with the InfoGain and score values. Figures 20.7 and 20.8 illustrate the comparison between both data samples, when $S = 50\%$, and a data sample, when $S = 30\%$. The experiment proves an inverse proportionality of the ownership level with the values of Score and InfoGain. This is always true regardless of the sensitive value percentage. For instance, users with ownership level = 2 gain the maximum information. In adverse, users with ownership level = 5 obtain the minimum information gained, and as a result the lowest scores.

In the second experiment of the first round, we compare the MDSBA with MDTDS in relation to the information gained and performance. However, this experiment is a pilot test for the MDSBA method and it has not been examined over big data yet. Therefore, a small number of 15,594 database records are only used in this trial. Because only two Q-IDs are being used, the data anonymization, in MDSBA, is applied to the NG groups only. The SG groups must be equivalent for both

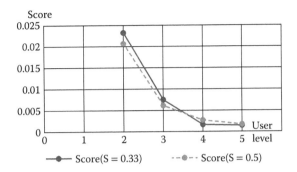

FIGURE 20.8 Score comparison with *S*.

TABLE 20.5
Anonymized Data Sample for $\bar{k}$ = 3, S = 33%

$\bar{k}=3$			
Q-IDs		**Sensitive**	
Age	**Gender**	**Salary**	**No. of Rec.**
60–65	Female	>50k	18
65–70	Female	>50k	4
70–80	Female	>50k	5
15–25	Male	>50k	19
25–30	Male	>50k	115
30–35	Male	>50k	313
35–40	Male	>50k	422
40–45	Male	>50k	407
45–50	Male	>50k	416
50–55	Male	>50k	328
55–60	Male	>50k	196
60–65	Male	>50k	118
65–70	Male	>50k	46
70–75	Male	>50k	14
75–80	Male	>50k	12
80–90	Male	>50k	7

Q-IDs, whereas NG is grouped based on the highest probability value, which is the Gender in this data sample.

After anonymization is completed for verities of ownership levels; an interval distance of Age = 5 is applied for $\bar{k}=3$, Age = 10 is applied for $\bar{k}=4$, and finally Age = 100 is applied for $\bar{k}=5$. In MDTDS, the interval of Age = [0–55] and [55–90] is applied for $k=5$. Notice that the MDTDS method defines an average of interval distance Age = 45, which is a close value to $k=5$. This concludes that the information gained in MDTDS is between $\bar{k}=4$ and $\bar{k}=5$ of MDSBA.

The following Tables 20.5 through 20.7 show a compression of 15,594 records after anonymization. Tables 20.5 and 20.6 illustrate anonymized data samples for $\bar{k}=3$ and $\bar{k}=4$, respectively. The number of compressed records decreases with the increased value of $\bar{k}$. For instance, the number of compressed records in $\bar{k}=3$ is 73 records, and this number decreases in $\bar{k}=4-53$ compressed records, and decreases to six records in $\bar{k}=5$.

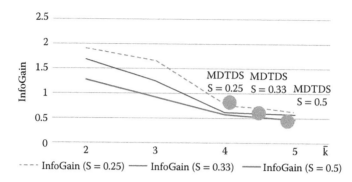

FIGURE 20.9 InfoGain comparison between multidimensional top-down specialization method and multidimensional sensitivity-based anonymization method with different S values.

TABLE 20.6

Anonymized Data Sample for $\bar{k} = 4$, $S = 33\%$

	$k = 4$		
Q-IDs		Sensitive	No. of
Age	Gender	Salary	Rec.
60–70	Male	>50k	118
65–75	Male	>50k	46
70–80	Male	>50k	14
75–85	Male	>50k	12
80–90	Male	>50k	7
15–25	Male	≤50k	1131
25–35	Male	≤50k	1684
35–45	Male	≤50k	1296
45–55	Male	≤50k	795
55–65	Male	≤50k	466
65–75	Male	≤50k	187
75–85	Male	≤50k	49
85–95	Male	≤50k	12
15–25	Female	≤50k	560
20–30	Female	≤50k	1247
25–35	Female	≤50k	1011

TABLE 20.7

Compressed Table for $\bar{k} = 5$, $S = 33\%$

	MDTDS $\bar{k} = 5$		
Q-IDs		Sensitive	No. of
Age	Gender	Salary	Rec.
0–55	Male	≤50k	19
0–55	Female	≤50k	3
55–90	Male	≤50k	2
55–90	Female	≤50k	4
0–55	Male	<10,000	4
0–55	Female	<10,000	3
55–90	Female	<10,000	2
55–90	Male	<10,000	2
0–55	Male	>50k	99
0–55	Female	>50k	101
55–90	Female	>50k	37
55–90	Male	>50k	32

Table 20.7 is compressed by anonymizing the Q-ID of Age by using the MDTDS method. The InfoGain comparison, between Table 20.6 and Tables 20.4 through 20.5 allocates Table 20.6 between $\bar{k} = 4$ and $\bar{k} = 5$ for the three different values of sensitive value S, as shown in Figure 20.9. The figure also shows a slight increase in InfoGain with a decreased percentage of S in both anonymity methods. In conventional data, the MDSBA method is able to provide a gradual level of InfoGain, unlike the rigid anonymization provided by MDTDS.

MDSBA performance is evaluated and compared with MDTDS by experimenting on the computation costs for each method. For this reason, we used 3.8 MB of "adult" data with 32,561 records. The results show a higher calculation cost for a lower value of $\bar{k}$. Moreover, the experiment shows a high cost for MDTDS that is close similar to $\bar{k} = 2$ processing time, as illustrated in Figure 20.10.

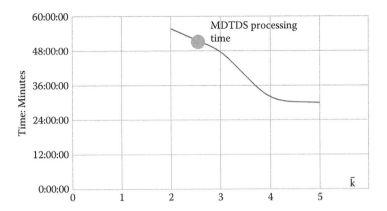

FIGURE 20.10   Processing time comparison.

20.5   CONCLUSIONS

In this paper, the MDSBA method is proposed, studied, and compared with a popular multidimensional privacy method, known as MDTDS. MDSBA is examined for a scalable multiaccess level in conventional data. The multiaccess level discriminates data anonymization by applying extra suppression on the degraded ownership level. Experiments on the MDTDS method showed a low level of gained information with a high cost of computation time. In contrast, MDSBA showed a possibility of raising the level of information gained, while keeping a reasonable computation time in comparison with the MDTDS method. Our experiments pointed out a considerable contrast in anonymized data gained by the MDSBA method. The *k-anonymity* technique is successfully applied to MDSBA; the structure of MDSBA permits a chance of breaking tasks into subtasks, and dividing them among the distributed parallel nodes. Our initial experiment aim was approaching the milestone of the MDSBA anonymization method. Our MDSBA design approach aimed to create gradual user access, by evaluating the degree of Q-ID sensitivity. Also, the aging factor is taken into consideration while calculating the sensitivity level. These procedures reduce the calculation costs by reducing the number of iterations for data anonymization. Our future work will focus on applying MDSBA to a real-world big data environment. A complete framework will include parallel distributed splitting and join processes. Also, our gauge will evaluate the RBAC map procedures in Big Data environment.

REFERENCES

Bayardo, R. J., & Agrawal, R. (2005, April). Data privacy through optimal k-anonymization. In *Proceedings 21st International Conference on Data Engineering, 2005. ICDE 2005*. ICDE 2005 (pp. 217–228). IEEE.
Becker, R. K. (1996). Adults data. Retrieved from ftp://ftp.ics.uci.edu/pub/machine-learning-databases
Benjamin, C. M., Fung, K. W., Ada W. Fu, Philip S. Yu. (2011). *Introduction to Privacy-Preserving Data Publishing: Concepts and Techniques*. Boca Raton, FL: CRC Press.
Dwork, C., Rothblum, G. N., and Vadhan, S. (2010). Boosting and differential privacy. *Paper presented at the Foundations of Computer Science (FOCS), 2010 51st Annual IEEE Symposium on*.
Fung, B. C., Wang, K., and Yu, P. S. (2005, April). Top-down specialization for information and privacy preservation. In *Proceedings 21st International Conference on Data Engineering, 2005. ICDE 2005* (pp. 205–216). IEEE.
Fung, B. C. M., Wang, K., and Yu, P. S. (2007). Anonymizing classification data for privacy preservation. *IEEE Transactions on Knowledge and Data Engineering*, 19(5), 711–725. doi:10.1109/TKDE.2007.1015
Ren, K., Kwon, Y., Balazinska, M., and Howe, B. (2013). Hadoop's adolescence: an analysis of Hadoop usage in scientific workloads. *Proceedings of the VLDB Endowment*, 6(10), 853–864.
Lublinsky, B. (2013). *Professional Hadoop Solutions*. Hoboken, NJ: Wiley.

Machanavajjhala, A., Gehrke, J., Kifer, D., and Venkitasubramaniam, M. (2006, April). l-diversity: Privacy beyond k-anonymity. In *Proceedings of the 22nd International Conference on Data Engineering, 2006. ICDE'06* (pp. 24–24). IEEE.

Machanavajjhala, A., Kifer, D., Abowd, J., Gehrke, J., & Vilhuber, L. (2008, April). Privacy: Theory meets practice on the map. In *IEEE 24th International Conference on Data Engineering, 2008. ICDE 2008* (pp. 277–286). IEEE.

Machanavajjhala, A., Kifer, D., Gehrke, J., and Venkitasubramaniam, M. (2007). L-diversity: Privacy beyond k-anonymity. *ACM Transactions on Knowledge Discovery from Data (TKDD)*, 1(1), 3-es. doi:10.1145/1217299.1217302

Mohammed, N., Fung, B. C. M., Hung, P. C. K., and Lee, C.-k. (2010). Centralized and distributed anonymization for high-dimensional healthcare data. *ACM Transactions on Knowledge Discovery from Data (TKDD)*, 4(4), 1–33. doi:10.1145/1857947.1857950

Li, N., Li, T., and Venkatasubramanian, S. (2007, April). t-closeness: Privacy beyond k-anonymity and l-diversity. In *IEEE 23rd International Conference on Data Engineering, 2007. ICDE 2007* (pp. 106–115). IEEE..

Zhang, Q., Koudas, N., Srivastava, D., and Yu, T. (2007, April). Aggregate query answering on anonymized tables. In *IEEE 23rd International Conference on Data Engineering, 2007. ICDE 2007* (pp. 116–125). IEEE.

Motwani, R., & Xu, Y. (2007, September). Efficient algorithms for masking and finding quasi-identifiers. In *Proceedings of the Conference on Very Large Data Bases (VLDB)* (pp. 83–93).

Russom, P. (2011). Big data analytics. TDWI best practices report, fourth quarter, 19, 40.

Samarati, P. (2001). Protecting respondents identities in microdata release. *IEEE Transactions on Knowledge and Data Engineering*, 13(6), 1010–1027. doi:10.1109/69.971193

Samet, H. (2006). *Multidimensional and Metric Data Structure*: Morgan Kaufmann.

Shannon, C. E. (2001). A mathematical theory of communication. *ACM SIGMOBILE Mobile Computing and Communications Review*, 5(1), 3–55. doi:10.1145/584091.584093

Shin, K. G., Ju, X., Chen, Z., and Hu, X. (2012). Privacy protection for users of location-based services. *IEEE Wireless Communications*, 19(1), 30–39. doi:10.1109/MWC.2012.6155874

Sweeney, L. (2002a). Achieving-anonymity privacy protection using generalization and suppression. *International Journal of Uncertainty, Fuzziness and Knowledge-Based Systems*, 10(05), 571–588. doi:10.1142/S021848850200165X

Sweeney, L. (2002b). Anonymity: A model for protecting privacy. *International Journal of Uncertainty, Fuzziness and Knowledge-Based Systems*, 10(05), 557–570. doi:10.1142/S0218488502001648

Zhang, X., Liu, C., Nepal, S., Yang, C., Dou, W., and Chen, J. (2013, July). Combining top-down and bottom-up: scalable sub-tree anonymization over big data using MapReduce on cloud. In *2013 12th IEEE International Conference on Trust, Security and Privacy in Computing and Communications (TrustCom)* (pp. 501–508). IEEE.

Zhang, X., Liu, C., Nepal, S., Yang, C., Dou, W., and Chen, J. (2014). A hybrid approach for scalable sub-tree anonymization over big data using MapReduce on cloud. *Journal of Computer and System Sciences*, 80(5), 1008–1020.

Zhang, X., Yang, L. T., Liu, C., and Chen, J. (2014). A scalable two-phase top-down specialization approach for data anonymization using mapreduce on cloud. *IEEE Transactions on Parallel and Distributed Systems*, 25(2), 363–373.

21 A Quick Perspective on the Current State of IoT Security
A Survey

*Musa G. Samaila, João B. F. Sequeiros, Acácio F. P. P. Correia,
Mário M. Freire, and Pedro R. M. Inácio*

CONTENTS

21.1 INTRODUCTION

The Internet of Things (IoT) refers to the concept in which everyday objects are embedded with sensors and some communication technology that allow them to sense their environment and *talk* to each other, as well as provide them with unique identifiers in the form of Internet Protocol (IP) addresses that enable them to connect to the Internet. In this concept, computers, smart phones, and tablets are not the only devices that can be connected to the Internet. Consequently, billions of *things* are currently connected to the Internet, and the number of these connected smart devices is increasing at an alarming pace. Gartner predicted that about 6.4 billion connected devices will be in use on the planet in 2016 [1], which portrays the IoT as one of the fastest growing technologies in the world.

Presently, IoT encompasses almost every*thing* that is digitally connected in the world, including, but not limited to, Wireless Sensor Networks (WSNs), smart meters, cameras, pacemakers, pipe flow meters, refrigerators, and connected cars. IoT can be called the next industrial revolution, considering the way it is driving a fundamental shift in business models by creating business opportunities for both individuals and players in different industries. Currently, there are several impressive projections for the impact of IoT on the economy. For example, Gartner [2] estimated that IoT products and services will generate revenue of more than $300 billion. IoT also has the potential to revolutionize the way people live and the way industries, consumers, and businesses interact with the physical world. This is becoming even more apparent as IoT becomes more cost effective and easier to deploy.

Although the IoT is considered one of the fastest growing technologies, and essential for keeping up with fast evolving businesses, it is also identified as one of the fastest growing sources of security threats and vulnerabilities for individuals and enterprises. Today, discussions about the IoT most often start or conclude with the issue of security or rather lack of it. Security experts warn [3] that as the number of web-linked embedded systems, Machine-to-Machine (M2M) communication devices, and Internet-enabled instrumentation systems grows, the potential for security risks and breaches will also grow. Some IoT security compromises that have hit the news recently include a smart light bulbs hack, smart fridges hack and a case of hackers taking control of a car. This chapter will present a quick perspective on the current state of IoT security and its implications for consumer privacy and security by discussing a number of IoT security incidents that made headlines around the world in recent years. The chapter will also consider IoT security requirements and present security versus performance trade-offs and discuss standard IoT security protocols. Additionally, it will review a number of recent IoT security protocols proposed in the literature aimed at addressing some IoT security issues. The objective of this chapter is twofold, first to raise awareness among manufacturers, application developers, and users of IoT devices on the challenges on the horizon with practical examples of the past and the present, and second, to allow researchers to identify open issues and research directions.

The rest of this chapter is structured as follows. Section 21.2 presents the IoT security architecture and requirements. Section 21.3 discusses the scope of IoT cyber security. Section 21.4 focuses on the current IoT threat profiles. Section 21.5 considers how the IoT is under attack. Section 21.6 discusses security versus performance trade-offs in the IoT. Section 21.7 considers the existing protocols and standards, and presents a review of some protocols that have been proposed recently. Finally, Section 21.8 presents the main conclusions.

21.2　IOT SECURITY ARCHITECTURE AND REQUIREMENTS

Considering its present size and estimates produced by many analyst firms along with the numerous components and diverse application domains, it is difficult to specify a general security architecture for the IoT. Defining a uniform set of security requirements for the different application domains is also not an easy task. Determining what information to protect and who should be allowed/denied access; the diverse technologies involved and the dynamic nature of the environment have been identified in [4] as the obstacles to defining and analyzing IoT security and privacy requirements. However, in this section we attempt to present a simplified security architecture and a set of requirements for enabling security operations in a wide range of application areas of the IoT.

21.2.1　IoT Security Architecture

The IoT can generally be divided into four layers: sensing layer, network layer, application support layer and application layer, as shown in the security architecture in Figure 21.1.

21.2.1.1　Sensing Layer

In this security architecture, the sensing layer, which is also called the recognition or perception layer, is further divided into two parts: sensing node and sensing network. Examples of sensing nodes include Radio Frequency Identification (RFID) tags, sensor nodes, cameras, Global Positioning System (GPS), etc. The sensing node is responsible for gathering information and different physical parameters. A sensing node may have basic security features like username and password, or it may

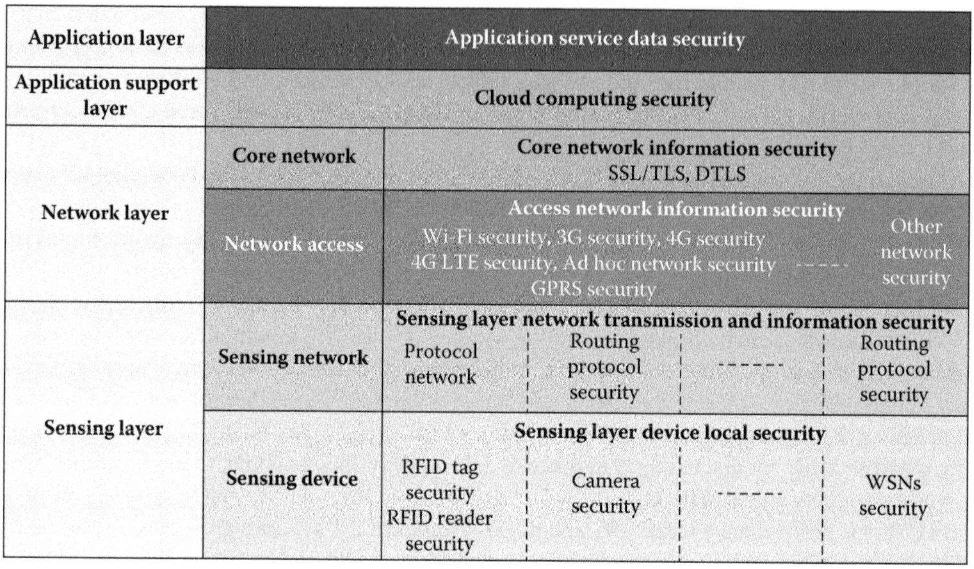

Application layer	Application service data security			
Application support layer	Cloud computing security			
Network layer	Core network	Core network information security SSL/TLS, DTLS		
	Network access	Access network information security Wi-Fi security, 3G security, 4G security 4G LTE security, Ad hoc network security ---- GPRS security		Other network security
Sensing layer	Sensing network	Sensing layer network transmission and information security		
		Protocol network	Routing protocol security	-----	Routing protocol security
	Sensing device	Sensing layer device local security		
		RFID tag security RFID reader security	Camera security	-----	WSNs security

FIGURE 21.1   Internet of Things security architecture.

not have any at all. For some applications like environmental monitoring, physical security of a sensing node cannot be guaranteed. Sensing nodes are usually small in size and constrained in terms of computational and storage capacity and rely on finite energy sources, such as batteries, or sometimes on energy harvesting mechanisms.

The sensing network is the network through which sensing nodes communicate with each other. In a sensing network, a gateway serves as an intermediary device between sensing nodes and the network layer, and links *things* to the Internet. Gateways also provide basic security and manageability. Virtually all communications, such as interconnection between devices, in the sensing network are via radio frequency (RF), which leaves the window open to different kinds of threats. Consequently, it is difficult to set up a reliable security protection mechanism on the sensing network, since most security mechanisms like public key encryption algorithms tend to be resource intensive.

21.2.1.2   Network Layer

The network layer in this security architecture is also known as the transport layer. The operation of this layer is considered key for the IoT network. It is further divided into two parts: access network and core network. The access network is responsible for transmission of data collected in the sensing layer to the core network for onward transmission. It consists of a number of different communication technologies, including, but not limited to, Wi-Fi, 4G Long-Term Evolution (LTE), 3G, and General Packet Radio Service (GPRS). The core network is the Internet. It can be said to have a relatively better protection capability, and the routing protocols used in this layer are similar to those of the standard Internet. However, extremely constrained IoT devices like sensor nodes and RFID devices may be prone to different types of attacks, including counterfeit attacks, unsolicited mails, and Man-in-the-Middle (MitM) attacks. Hence security in this layer is very critical.

21.2.1.3   Application Support Layer

The application support layer sits on top of the network layer and links the network layer to the application layer. This layer supports different cloud computing business service capabilities, such as data processing and data storage. As such, a high level of security is needed in this layer in order to safeguard sensitive data.

21.2.1.4 Application Layer

The application layer is the upper layer in the IoT security architecture and serves as the interface between users and the IoT. Through the application layer, users can connect to the IoT and access a variety of personalized services according to their access rights or subscriptions. Application layer services can cut across different domains, such as environmental monitoring, healthcare, agriculture, logistics, automobile, etc. Since this layer involves integrating numerous business applications in different domains, key issues of concern should include how users will safely access data, user privilege abuse/misuse, privacy issues, and bad password choice, among many others.

21.2.2 IoT Security Requirements

As each connected device could be a potential doorway into IoT systems, the design process of an IoT system should meet stringent security requirements for the particular application. The necessary basic security services required in IoT networks are confidentiality, data integrity, authentication, replay protection, and availability [5]. But there is no universal rule or single security mechanism/solution that can guarantee security in all four layers of the IoT security architecture. Each layer presents different vital security requirements, mainly due to energy, computational, transmission capacity, and storage limitations of many IoT devices. Based on the discussions on the security architecture presented in the previous subsection, we briefly present the security requirements for each layer as shown in Figure 21.2. It should be noted, however, that although some applications may not need all of these requirements, some applications may need other specific security properties in addition to some or all of the ones presented below.

21.2.2.1 Sensing Layer Security Requirements

Sensing layer security requirements vary depending on the applications and the type of devices that are deployed for data collection, and they include:

- In the case of WSNs, *sensor data protection mechanisms* are needed to prevent attackers from accessing data directly from the sensor nodes.
- *Efficient and fast anticollision algorithms* are needed for RFID [6] and hybrid systems where there is integration of sensor nodes and RFID devices [7]. Anticollision algorithms

Application layer	Data security protection Access control protocols Privacy protection Authentication and key management Security education
Application support layer	Secure cloud environment Antivirus software Secure multiparty computation for privacy and confidentiality
Network layer	Node authentication – identity authentication, access control Secure routing, network encryption mechanisms Attack detection, prevention, or mitigation Network access control Cross-domain authentication
Sensing layer	Sensor data protection Efficient and fast anti-collision algorithms Lightweight encryption technology Authentication and key management Physical security

FIGURE 21.2 Security requirements within each layer.

can mitigate interference and prevent collision at the reader when multiple tags transmit simultaneously.

- For most applications, *lightweight encryption schemes* are needed to ensure confidentiality, integrity, and authenticity.
- *Key management mechanisms* are necessary at the data link layer to allow two remote nodes to negotiate security credentials, such as secret keys [8].
- *Physical security* is essential in many applications to protect IoT devices from tampering. For example, some sensor nodes may be deployed in an environment that is open to different adversaries. Such sensors may be left unattended for a very long time; hence they can be easily tampered with.

21.2.2.2   Network Layer Security Requirements

The following are important security requirements in the network layer:

- *Access control* restricts access to a resource to authorized users or devices. It is a vital requirement in this layer and is achieved through encryption, authentication technologies, and for Wi-Fi, through Wi-Fi Protected Access (WPA).
- *Network encryption* (also known as network level encryption), which is implemented above the data link layer, is a requirement that employs cryptographic services for encrypting data in transit using Internet Protocol Security (IPSec).
- An *attack detection, prevention and mitigation* scheme is needed to detect, prevent, or mitigate the effects of network intrusions. Some methods include using anti-DDoS and reliable system updates.
- *Network Access Control (NAC)* is a desirable requirement that ensures that devices comply with a defined security policy before they are allowed to access information in an IoT network.
- *6LoWPAN* is a low-power wireless mesh network that allows IoT devices to use IPv6 to connect directly to the Internet. It also enables heterogeneous integration.

21.2.2.3   Application Support Layer Security Requirements

Application support layer security requirements include:

- A *secure cloud environment* is a crucial requirement considering the amount of data that is being stored and processed in the cloud.
- Strong and reliable *antivirus* software is needed to secure cloud services against attacks.
- *Secure multiparty computation for confidentiality and privacy* is a desirable property because it ensures that input data from users and intermediate results will be protected from snooping by cloud providers as well as other users [9]. Unless there is collusion between cloud providers, multiparty computation between multiple cloud services guarantees the secrecy of input data [10].

21.2.2.4   Application Layer Security Requirements

Application layer security requirements include the following:

- To protect the vast amount of data in the application layer, *data security protection* is an essential security property.
- Different *access control protocols* are needed to effectively manage the way users and various devices access resources in this layer.
- *Privacy protection* of users is an important security requirement that will ensure trust in the IoT. If privacy is taken seriously, users can have confidence that their sensitive private information is well protected.

- In such a heterogeneous network, resolving *authentication and key management* issues will be a very vital security requirement.
- Training and creating awareness among users on the importance of *security education* constitutes a critical security property. This, among other things, will enable users to know how to choose, keep, and manage their passwords. Users should be taught to desist from giving security a secondary consideration until they fall victim to an attack.

21.3 SCOPE AND CHALLENGES OF IOT CYBER SECURITY

There are unintended consequences and ramifications of current trends of connecting a massive amount of devices to the Internet. Perhaps one of the most pressing consequences is that of security. A considerable number of connected devices typically have little or no security measures in place. Vulnerabilities in a single device can lead to the exposure of sensitive information that can eventually expose a whole network to a wide panoply of threats. While IoT network disruption or outage could be caused by a natural disaster, hardware failure, or other things, the most alarming possibility is the significant damage that skilled attackers can cause to a critical IoT infrastructure via cyber attacks. Considering that IoT will continue to be deeply interwoven into the lives of people [11], permeating every aspect of human life like healthcare, power grid, transportation, and business transactions, the repercussions of such attacks can create far-reaching consequences for states, organizations, and individuals, as well as disrupt the aspirations of building a future hyperconnected society. This section presents the scope of IoT cyber security under the following subsections.

21.3.1 Ubiquitous Nature of IoT Devices

In his seminal paper entitled "The Computer for the 21st Century," Mark Weiser [12] predicted 25 years ago that pervasive computing that can learn and adapt to human needs will be so ubiquitous that their presence will hardly be noticed. Through WSNs and RFID technologies, ubiquitous sensing and tracking have today become commonplace. Over the past decades, advances in the semiconductor industry and wireless communication technologies have enabled the miniaturization and reduction in cost of sensors and computing technologies that led to the creation of smaller and more efficient sensor nodes used for different kinds of sensing, ranging from environmental to body sensors [13]. Similarly, advancements in antenna technology as well as wireless communications and microchip fabrication technologies in the past decade have led to significant developments in RFID technology which is used for automatic identification and transmission of the identified data over radio frequency [14]. Essentially, WSNs and RFID have been identified as two key technological enablers of the IoT paradigm [15].

With the aforementioned technological advances, as well as the breakthroughs in wireless connectivity, cloud computing, and widespread adoption of IP-based networking, the growth of IoT devices is getting closer to the envisioned widespread reality. This is particularly obvious, considering the ubiquitous addressing capability of IPv6 [16,17]. Apart from the millions of *smart* devices that are already connected to the Internet, many more devices are still coming online each and every day. The rapid rise in the number of these devices is overwhelming. Although predictions or estimates seem to vary considerably from one analyst firm to another, there are quite a number of predictions on the future size of the IoT in terms of the number of *smart* devices that will be connected to the Internet in 2020. For example, while the prediction of Gartner [1] is that the number will be about 20.8 billion, Business Insider [18] and the Cisco Internet Business Solution Group (IBSG) [19] predicted that about 34 billion and 50 billion devices will be connected by 2020, respectively.

Despite the numerous benefits and advantages of connecting a massive number of *smart* devices to the Internet, including having the greatest potential to transform the future of humanity to a fully connected society, devices in the IoT raise a diverse array of potential security risks.

21.3.2 Perimeter of IoT Networks

As IoT devices are becoming *smarter,* moving from simple sensors to full-fledged computers that are embedded in everything from watches to thermostats to human bodies, they present vulnerabilities that are very similar to those of the traditional computers. Hence, like other computer systems, they become subject to a number of security concerns and can experience different attacks like Denial of Service (DoS), malware, MitM, etc. These concerns are not hypothetical as several real attack incidents have already been reported by the news media, some of which are covered in Section 21.5. Additionally, these devices introduce new security challenges, and as such, some current security solutions may not be effective in safeguarding IoT devices. [20] For instance, the notion of perimeter security, traditionally used to defend computer systems, is becoming obsolete and inadequate as a single measure to defend a network consisting of IoT devices.

Traditionally, the driving principle for enterprise security is to either isolate computer systems from the outside world or shield them through proprietary communication protocols and mechanisms using perimeter security technologies such as the ones shown in Figure 21.3a. These well proven and established security principles involve the use of firewalls, packet filters, encryption, intrusion detection/intrusion prevention systems, security protocols, and authentication. In this era of pervasive connectivity, however, smart devices and other computer systems installed inside secure perimeter of enterprise network that are thought to be safe are still vulnerable. With the proliferation of smart devices, any IoT device with a vulnerability that is connected to the Internet using a public IP address constitutes a hole in the perimeter defenses, thereby making the perimeter porous and easily penetrable [21], as depicted in Figure 21.3b. Today, unfortunately, any security approach to protect a computer network consisting of IoT devices that is based on designing a strong perimeter protection and restricting access to the network or part of it by using perimeter security mechanisms will not be enough to protect the network against cyber attacks. The reason being that IoT systems mostly rely on cloud services for transport and because servers cannot be protected with traditional perimeter defense techniques, there is an increase in cyber attacks against application infrastructures. This brought about the launching of the Software Defined Perimeter (SDP) initiative, in December 2013, with the goal to develop solutions to mitigate such attacks [22].

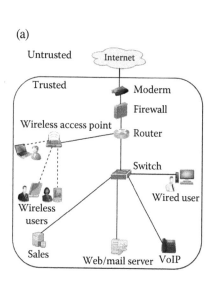 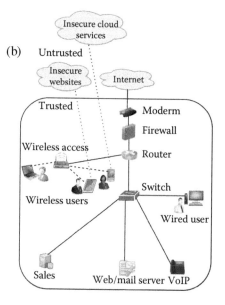

FIGURE 21.3 (a) Architecture based on perimeter security. (b) Architecture based on identity.

Incorporating IoT-enabled embedded systems into organizations and industries as well as adding smart devices to home networks presents a substantive opportunity for attackers to breach the perimeter security of such networks and then target any device in the networks. Although connecting a device to the Internet opens a whole new world of different possibilities for users, it also changes the security boundary. Previously, the *perimeter* was the device. However, after adding networking, the physical *perimeter* no longer exists [23], which implies that perimeter security boundaries are crumbling under the weight of ubiquitous connectivity. Perimeter security has never seen such a time of change. In the face of IoT, the perimeter that used to be defined and defended is fast becoming less effective.

21.3.3 DYNAMIC AND DISTRIBUTED NATURE OF IoT NETWORKS

The IoT has been defined as both a dynamic and distributed global network infrastructure capable of self-configuration based on different interoperable communication protocols, where *things* have physical attributes and virtual identities [24]. In the following subsections, we describe how these two characteristics of IoT networks pose massive security and privacy concerns to the development of the IoT in general.

21.3.3.1 Dynamic Nature of IoT Networks

As smart devices interact in real time with their environment, an adaptive architecture that allows them to dynamically interact with other devices is required. This should include networking technology and networked information processing tailored for dynamic *ad hoc* environments [24]. Thus, the IoT has distinct features that diverge from the traditional Internet. One such feature is the ability to dynamically configure networks of *things* with smart devices entering or leaving a network [25]. This is a very important feature, considering that some IoT devices may be constrained in resources (e.g., battery, memory, computing, and communication capabilities), deployed in harsh and dynamic environments that are open to different adversaries, and left unattended for a long period of time. Another scenario is the deployment of IoT devices in hazardous factory environments to control some infrastructure like factory operations and supply chains. In such scenarios, nodes are expected to organize themselves autonomously in order to perform coordinated tasks through transient ad hoc networking that provides basic means of data sharing and processing. For such systems, it is essential to have a mechanism that can change the network so as to automatically discover active nodes and then map them in. WSNs and web-linked embedded systems, respectively, are good examples for the first and the second scenarios.

Although the IoT promises mobility and dynamic network topology without boundaries with smart *things* joining and leaving the network at all times, establishing trust among devices is crucial. Dynamic configuration is only viable when there is some level of trustworthiness among *things* that can enable them to interact with each other safely and without the risks of negotiating with unfamiliar entities [26]. Unfortunately, dynamic network topology constitutes a possible security vulnerability, making trust negotiation challenging. In addition to this, the dynamic nature of IoT networks poses a challenge to the protection of data confidentiality, data integrity, system integrity, as well as user privacy.

21.3.3.2 Distributed Nature of IoT Networks

The IoT is a distributed technology [26] that is made up of numerous systems, especially the remotely deployed physical *things* that have embedded devices, including sensors, RFID chips, cameras, and miniature GPS receivers, among others, that provide sensing and data collection functionalities. Since IoT systems are more broadly distributed in the physical space than the Internet, analytics in IoT architecture must be highly distributed. Consequently, scalable and distributed means for processing and managing information [27], as well as data use, must become the norm in IoT systems.

The distributed nature of IoT systems, however, presents a number of security challenges. For example, designing security protocols for the distributed systems is a challenging task [28]. Additionally, the distributed nature of ad hoc wireless networks makes them vulnerable to different types of attacks. Furthermore, the remotely deployed devices require mechanisms to ensure secure communication among them and with other trusted entities. Moreover, since IoT devices typically rely on cloud-based services, there is a possibility of malicious attacks such as DoS, identity theft, and service theft.

21.4 IOT CURRENT THREAT PROFILES

A threat profile is a concept that identifies critical cyber assets, threat actors, and threat scenarios in a given application context. It provides organizations with a clear picture of the threats they are facing and enables them to develop stringent measures against such threats. Thus, robust threat profiles are vital in organizations for mounting quick and effective defenses and countermeasures against potential threats. This section presents IoT threat profiles, focusing on their different aspects, including critical cyber assets, threat actors, a brief overview of threat landscape, and the reality of such threats. How the IoT is attracting the attention of attackers and how it is changing cyber threat landscape are other aspects of the threat profiles considered in this section.

21.4.1 IoT Critical Cyber Assets

In the context of IoT, a critical cyber asset can be defined as a piece of equipment, a device, a system, or a facility that, if rendered inaccessible or destroyed, or if its operation is altered or suppressed, would affect the operability, performance, or reliability of the entire IoT network or system. A cyber asset can be hardware, software, or a piece of data. A best practice to strengthen the security of an IoT system is to create an inventory of the different cyber assets [29] in the network, which should include their access points and security requirements. This provides the security administrator with enough information to identify the critical assets.

Identifying vital cyber assets in an IoT system and their security requirements are two key factors in building a robust protection strategy. Typically, IoT critical cyber assets will vary from one application domain to another. Although in a smart grid, for example, the critical assets may include smart meters, smart transformers, smart protective relays and utility data centers, in a smart home, the critical cyber assets may include smart Wi-Fi routers and personal information, such as usernames, passwords, credit card number, social security number, etc.

21.4.2 IoT Threat Actors

In recent years, the IoT has been an industry buzzword with hundreds of both positive and negative press releases [30], and most of the bad news headlines revolve around security incidents caused by different threat actors. In Information Technology (IT) security, a threat actor, also called a malicious threat actor, is a person that uses one or more threat vectors to attack a target, resulting in an incident that has an adverse impact on (or has the potential to affect) the security of an individual or an enterprise. There are several threat actors targeting the IoT with different intentions. Examples of threat vectors that could be used in the IoT threat scenario include fake websites, session hijacking, wireless unsecured hotspots, email links/attachments, mobile devices, malware, Universal Serial Bus (USB) (removable) media, and social engineering. A threat target can be anything of value to the threat actor, such as a critical asset. Threat actors are generally grouped into two categories, namely external (outsider) and internal (insider).

21.4.2.1 IoT External Threat Actors

External threat actors in the IoT are individual hackers, well-funded hackers, organized crime groups, or government entities that seek to gain access to protected information by breaking through and taking over the profile of a trusted entity from outside [31] the IoT network or enterprise. This category of threat actors can launch either active or passive attacks. The attacks are active if the actors participate in the network or generate packets and passive if they only eavesdrop or track the entities using the IoT network. The IoT data breaches that usually make news are typically caused by external threat actors, mainly because they tend to be more severe in terms of negative impact.

21.4.2.2 IoT Internal Threat Actors

In the context of IoT, internal threat actors are people from within an IoT enterprise, or people with legal access rights to an IoT network [32]. They include employees, former employees, business associates, contractors, smart home occupants, and smart city citizens who have privileged information about the security practices of the organization or network. Such people may even know the usernames and passwords of some critical systems. Although malicious insider threats exist, many IoT users and enterprises do not consider these threats as very important ones. Several data breaches resulting from internal threat actors may be unintentional. An external person like a competitor could also provide financial incentives to motivate internal threat actors to sabotage an organization.

21.4.3 BRIEF OVERVIEW OF IoT THREAT LANDSCAPE

In the past, the threat landscape in information security was dominated by malware creators who were seeking attention. But, as cybercrime has now become a business that is driven by a return on investments [33,34], the cyber threat landscape has dramatically changed with numerous capable hackers and cyber criminals everywhere. Additionally, in recent years, IoT is changing the overall cyber threat landscape even further with more devices being connected to the Internet by the day, generating a massive amount of data that must be stored, processed, and analyzed. Considering this massive amount of data being passed back and forth through numerous ill-equipped devices (in terms of security) that may have exploitable vulnerabilities, it is a natural progression for cyber criminals to focus more on the IoT now more than ever before. As a result, the IoT threat landscape is continually evolving, and attacks continue to get more sophisticated, aggressive, and destructive.

The growth of new technologies like the IoT is weakening the defenses of many companies by creating a host of new access points for hackers and cyber criminals to target [35]. There is undeniable evidence that the dependence of people, organizations, and businesses on the IoT is expanding the attack surface for hackers and cyber criminals [36,37]. Cyber criminals are now shifting their focus from desktop computers to mobile devices, especially the IoT devices. Over the last few years, there has been a stream of data breaches in the headlines involving IoT devices [38]. While IoT security threats come from a wide variety of sources, the pattern of attacks can be broadly divided into targeted and opportunistic attacks.

21.4.3.1 IoT Targeted Attacks

In the IoT, a targeted attack is a type of threat in which an adversary persistently pursues a target entity and compromises it while maintaining anonymity. An example of such a threat is the advanced persistent threat (APT) [11], a sophisticated network threat in which an attacker that gains an unauthorized access to an IoT network stays there undetected for quite some time. Usually, the intention of APT is not to cause damage but rather to steal important information. Such attacks normally target specific organizations like financial industries, national defense, and other important government organizations. Therefore, they are often customized and can be modified or improved to suit the target. Attack methods include malicious emails, malware, and malicious sites. The goals of such attacks include business espionage and monetary benefits.

21.4.3.2 IoT Opportunistic Attacks

In opportunistic attacks, the attacker searches for known security flaws to target many different entities by using generic methods in the hope that some of the entities will be vulnerable to the attacks. In such attacks, the adversary usually has a large number of targets in mind and has no a priori knowledge of who the victims will be.

21.4.4 The Reality of IoT Threats

Although IoT has been a key enabler to better and more convenient, comfortable and productive living, as well as representing a technological leap for different people and organizations, it also presents great challenges to network and data security, as increasingly sophisticated cyber threats become a serious concern in the emerging connected world [31]. Hackers and malicious attackers are increasingly using IoT devices as access points to attack other devices in the network that are not adequately secured, and this trend is likely to continue in the foreseeable future. Such threats, which were not known over the past few years, have made their way into headlines in recent years. In recent security news, for example, the Federal Bureau of Investigation (FBI) in the United States (US) warned the general public on the dangers of the IoT and advised users to deal with their IoT devices with caution [39].

With the rapidly evolving phenomena of IoT threats, some individuals and organizations are beginning to come to the reality that the smart *things* around them, like their desktop or laptop computers, are in need of security. However, there are still some users who seem not to pay attention to these threats due to either ignorance of the reality and security implications of the threats, or they assume that because such devices are performing their functions well, it means they are perfectly designed and built. Essentially, some IoT devices are not designed with security in mind. A number of vendors and manufacturers are bypassing many security best practices due to cost, complexity [40], or because adhering to such best practices may reduce their time-to-market. Furthermore, in many IoT devices, communication between system components is not encrypted and data is stored on some devices in plain text. Additionally, many IoT devices do not receive any updates either automatically or manually. Even if updates were available, many users do not know how to update the firmware of their IoT devices. Hence, there is a need for vendors to find a way to securely update smart devices when security issues are found and fixed.

21.4.5 IoT is Attracting the Attention of Attackers

Many IoT and embedded systems lack inherent defenses and there are virtually no firewalls to protect them. Instead, they are left wide open to cyber attacks, relying on simple password authentications and security protocols [21]. This is probably based on the assumption that IoT devices are not attractive targets to cyber criminals and hackers. However, the increase in number and sophistication of targeted attack campaigns against IoT devices that dominated the news headlines lately clearly proves that such an assumption is no longer valid. Today, many industries, including healthcare, banking, power, and hospitality, are embracing the IoT, and more applications (apps) for financial services are being developed and deployed, such developments will definitely attract the interest of financially motivated cyber criminals. Moreover, even if some IoT devices may not contain anything that is of high value to attackers, the fact that such devices are typically embedded deep within networks make them attractive attack targets [41]. Additionally, the easy network access connected *things* provide compared with devices that are adequately protected, and the ignorance of many users with regards to the monitoring of such devices, open an avenue for cyber criminals and hackers to test their skills.

The sheer scale of the IoT is creating an attack surface of unprecedented size, and many more devices will still be connected, which will expand the attack surface even further. As the reliance of humans on the IoT and other Information and Communications Technology (ICT) systems increases, more and more important information, such as sensitive medical records, power grid operational data, water/wastewater treatment plants operational data, banking and nonbanking financial institutions sensitive data, road/rail traffic flow data, and personal sensitive information will be collected by these devices. Any abuse or misuse of such data can be very devastating, and the damages that may arise from such misuse will not only harm individuals and enterprises but could negatively impact several sectors of society, including healthcare, power supply, portable water supply, and transportation systems, just to mention a few. While, on the one hand, the white hat hackers are attracted to IoT devices just to prove the existence of vulnerabilities on IoT networks and devices, on the other hand, IoT devices are attractive targets for many cyber criminals and black hat hackers because one of their primary motives is to launch attack campaigns that can have long-lasting effects on their victims, while hiding behind a shield of digital anonymity [42].

21.4.6 IoT is Changing the Cyber Threat Landscape

There are two key fundamental defense strategies in cyber security: first to examine the network and strive to figure out in advance the possible sources of threats and try to prevent them, and second, to have a mechanism in place that can scan the network traffic so as to detect possible leaks that may occur in compromised devices [35]. The aforementioned defense strategies are possible in the traditional risk management models because organizations own most, if not all, of the cyber assets and the data flowing through them. With the advent of the IoT and other ICT systems, however, most of business today is being done outside the defensive fence of organizations. In addition to this, the mobile and distributed nature of the IoT technology are pushing organizations, assets and data beyond their corporate defense systems, such that potential attackers are presented with a large attack surface. Thus, in this era of IoT, the potential for cyber threat landscape growth is almost unbounded.

IoT is drastically changing the cyber threat landscape, and as a result, both organizations and individuals are exposed to new types of APT scenarios that current risk management schemes cannot defend against. Given the large attack surface created by the number of connected devices, cyber criminals and hackers now search for the weakest link in the cyber security chain [43]. This is because exploiting vulnerability in a single weak link in the security chain could provide attackers with doorways to have unauthorized access to sensitive data, including customer data and personal identifiable information. Currently, when hackers want to attack an organization they do not have to target the well-protected servers, but rather they target ordinary devices that are relatively unmonitored, but are connected to the network, such as smart fridges or smart coffee makers. If they successfully break their way through such smart devices, they can eventually reach their main target.

21.5 IoT UNDER ATTACK

IoT is gaining more acceptance and popularity at a time when threats to data and systems have never been greater. As the IoT becomes more ubiquitous and hackers as well as cyber criminals learn new attack methods, IoT-related attacks are becoming more common. The 2016 Internet security threat report published by Symantec, which covers different facets of cyber security, including IoT security, shows that cyber attacks are increasing in number and sophistication [44]. The recent disaster stories involving IoT security incidents have raised serious security and privacy concerns. This section will discuss how the IoT is under attack, focusing on the following: evolving attack strategies against the IoT, complexity of IoT vulnerabilities, attacks on privacy of users, and emerging IoT attack scenarios.

21.5.1 Evolving Attack Strategies against the IoT

Cyber criminals and malicious hackers, which we will refer to here as attackers, are becoming cleverer in avoiding detection and concealing their paths. As the number of targets grows, attack incentives and strategies also grow. It is estimated that there are about 500,000 security attack attempts every minute [45]. Today, attackers are well-resourced, well-organized, and they are very methodical in approaching their targets for corporate or business espionage, as well as for financial, political, or ideological reasons. Presently, detection and forensic techniques are becoming more challenging such that securing IoT devices against cyber attacks mostly requires more effort, especially as attackers become more skilled in cryptography [46]. According to a recent Spiceworks security survey in the first quarter of 2016, involving about 200 IT professionals in North America and EMEA (Europe, the Middle East, and Africa), 49% of these IT pros pointed to IoT devices as the network-connected endpoints at highest risk in 2016 [47].

Over the years, attackers have discovered a number of ways to attack IoT devices. For instance, attackers can use the Universal Plug and Play (UPnP) protocols that enable devices to automatically find other connected devices on the network to attack IoT devices [48,49]. In addition, exploiting exposed Application Programming Interfaces (APIs) is another option for attackers [50,51]. Exposed APIs are major threats to organizations that use the IoT, and hence it is important for all API calls to be encrypted or authenticated. Moreover, the increasing trend toward *Bring Your Own Device* (BYOD), or rather BYOIoT device to work is another loophole that attackers can exploit to attack an enterprise that is thought to be secured [50]. Furthermore, attackers can also use Domain Name System (DNS) poisoning and hijacking to attack a router, gateway, or hub and take control of an IoT device, respectively [52]. A recent example is the attack incident involving Netgear routers [53], where some cyber criminals were able to bypass the embedded authentication process and change the default DNS to an IP address of a malicious website.

Currently, the IoT can be weaponized and used as a *botnet* [54], or rather *thingbots*. A good example of a botnet for attacking IoT devices is the LizardStresser created by a hacker group called Lizard Squad. This botnet can be used to hijack IoT devices and use them to launch Distributed Denial of Service (DDoS) attack campaigns on banking, gaming, and government websites, among others. The number of cyber criminals that are using the LizardStresser has been rising steadily, and the most disturbing issue is the fact that the botnet source code was released to the public in 2015 [55]. Another issue is that even attackers that do not know how to exploit a technical vulnerability may be able to use the botnet because all that is required is a device default administrative username and password, which many users and some IT professionals do not change. A report showed that 46% of consumers and 30% of IT professionals do not change default administrative passwords [56].

21.5.2 COMPLEXITY OF IoT VULNERABILITIES

Recently, a plethora of vulnerabilities have been discovered in several IoT-enabled consumer products ranging from smart implantable medical devices to connected home appliances. Many more such vulnerabilities are still going to be discovered as new devices are being connected to the Internet daily by the thousands or even millions. For example, a recent study conducted by HP reveals alarming vulnerabilities with 10 top IoT home security systems [57]. The HP study shows that 100% of the security devices studied show significant security deficiencies. The Open Web Application Security Project (OWASP) [58] is designed to provide manufacturers, developers, and consumers of IoT devices with the necessary information about the security issues associated with the IoT, as well as to help them make better security decisions with regards to building, deployment, or using IoT devices. OWASP has identified the following as the most common vulnerabilities in IoT devices:

1. *Username Enumeration*: this allows a malicious person to collect valid usernames by interacting with the authentication mechanism.
2. *Weak Passwords*: allows users to set short and simple account passwords, such as "1234" or "abcd."
3. *Account Lockout*: enables a malicious entity to keep sending authentication attempts after three to five failed login attempts.
4. *Unencrypted Services*: attackers can eavesdrop on the network communication because network services are not properly encrypted.

5. *Two-Factor Authentication*: lack of two-factor authentication mechanisms like a security token or fingerprint scanner.
6. *Poorly Implemented Encryption*: the implemented encryption is poorly configured or not properly updated, e.g., using Secure Sockets Layer (SSL) v2.
7. *Update Sent without Encryption*: Transport Layer Security (TLS) is not used when transmitting updates over the network or encrypting the update file itself.
8. *Update Location Writable*: storage location for update is writable, which allows malicious entities to modify and distribute the modified version to all users.
9. *Denial of Service*: authorized users or devices can be denied access to a service, or to a device.
10. *Removal of Storage Media*: because some devices are exposed in open fields, the storage media can be removed.
11. *No Manual Update Mechanism*: there is no provision that will enable users to manually force an update check for the device.
12. *Missing Update Mechanism*: no provision for device update.
13. *Firmware Version Display and/or Last Update Date*: device does not display current firmware version or last update date.

Essentially, what makes the vulnerability of IoT devices so complex is the fact that almost all communication in the IoT is wireless; therefore it is possible for attackers to eavesdrop on communication between devices in the network. In addition to this, many IoT devices have limited power, memory, and processing capability, and are often battery operated. Such devices cannot handle large amounts of data. Some of them, such as WSNs, are deployed in highly dynamic and harsh environments. Similarly, due to the embedded nature of some IoT devices, some are deployed in factory environments that are noisy with a lot of interference and transmitted messages are often lost due to the lossy nature of such networks [59]. In the aforementioned scenarios, IoT devices are likely to be left on and unattended for a long period of time and thus vulnerable to both logical and physical attacks.

The vulnerability issues of IoT are difficult to deal with because fundamentally most of the devices are not designed with security in mind. There are inherent vulnerabilities in almost every aspect of IoT, including hardware, operating system (OS), communication protocols, APIs, design of the architecture, etc. [60]. Aside from this, when vulnerabilities are found and fixed, the vulnerability patching process in the IoT is another big challenge. For instance, if a firmware update is required in order to patch the vulnerability, it will be a very difficult task, considering the number of devices that may be involved (e.g., it can be thousands in WSNs) [61]. In addition, not every user or IT professional is able to upgrade a custom firmware because it requires extra time and effort.

21.5.3 ATTACKS ON PRIVACY OF USERS

The IoT paradigm enables the cooperation and interoperability between different heterogeneous smart *objects*, including those that have become part of our everyday lives, allowing them to interact autonomously in real time with each other, with the environment, and with humans, giving rise to pervasive collection, processing, and dissemination of different kinds of information. This massive information obviously includes private and personal data that are collected by these devices with or without the consent of the people, which raises serious privacy concerns, thus making privacy a predominant issue in the IoT today [62]. Privacy is considered to be one of the major obstacles for the IoT deployment on a broad scale.

Although privacy can simply be defined as the appropriate use of data [63], in 1968, Westin defined privacy as "the right to select what personal information about me is known to what people [64]" and in [65], privacy is defined as "the faculty and right that a person has to define, preserve and control the boundaries that limit the extent to which the rest of society can interact with or intrude upon. At the same time, he or she retains full control over information generated by, and related to, him or her." Although those definitions may not directly refer to electronic information, they are still valid today. Similarly, in the context of the IoT, in order to ensure privacy, it is expected that users should have the right to know and to control what data are collected about them, who maintains and uses the data, and what purpose the data are used for.

Privacy and security are distinct, but they are complementary properties for IoT services. For instance, one of the purposes of a security policy is to ensure data privacy. This implies that a successful attack on security may eventually result in an attack on privacy. Basically, information security is about preserving the confidentiality, integrity, and availability (CIA) of information. Confidentiality represents an intersection between security and privacy, forming a mutual [66] relationship between the two. Although, in security, confidentiality refers to the property that a trusted party must protect data that have been collected from being released, in privacy, confidentiality is about making sure that such data are not collected, shared, or monitored.

There have been a number of successful attacks on the privacy of IoT users that made headlines in recent years. For example, US researcher Matt Jakubowski [67] exploited vulnerabilities in the Wi-Fi enabled Barbie doll that allowed him to have access to the information system, account information, and the audio files stored on the doll, and direct access to the microphone. Having access to these resources could enable a cyber criminal to spy on children. Another disturbing issue is that tracing the identity of a smart *thing* may lead to automatic identification of the owner. This is because a transaction linked to the same identifier is usually traceable, which in turn shows that the owner is also identifiable. Additional information that IoT devices can reveal about their users include location, interests, habits, and other personal information. In 2015, for example, some researchers at the Synack [68] security firm analyzed 16 smart devices for home automation. They observed that exploiting some vulnerability in those devices could reveal private information, such as user behaviors and activity patterns. It is, therefore, imperative for everyone with a stake in the IoT to take all the necessary measures to prevent malicious entities from having unauthorized access to private information of users.

21.5.4 Emerging IoT Attack Scenarios

Over the past few years, IoT has become both an attack target and attack source, becoming the focus of security threats and the biggest vector of attacks on enterprises and individuals. The recent escalation of cyber attacks stemming from networked devices ranging from home automation devices to smart wearable devices is raising serious security and privacy concerns. These concerns are underscored by recent security breaches that have become more of a common phenomenon, which have the potential to limit the widespread adoption of the IoT technology. One good thing about the hacking incidents is that most of them are being carried out by ethical hackers or white hat hackers, who are only interested in hacking these devices in order to test or evaluate their security, rather than with nefarious or criminal intent. In the following subsections, we present the attack scenarios from 2013 to 2016.

21.5.4.1 IoT Attack Scenarios in 2013

Examples of IoT hacking incidents reported by media outlets in 2013 include research conducted by Nitesh Dhanjani [69], a security researcher that exploited vulnerabilities in the Hue LED lighting system made by Philips. Using a malware, Dhanjani was able to break the weak authentication system and remotely turned on/off the Hue bulbs. In addition, two white hat hackers, Charlie Miller and Chris Valasek [70], have demonstrated taking control of the steering wheel, the braking system, and other functionalities of a Ford car. Although their work assumed physical access to the car onboard computer, a team of researchers at the University of Washington and the University of California earlier [71], in 2010, showed that they could gain wireless access to a car network. The researchers used a sedan car from an unnamed company in their experiment. Moreover, between December 23, 2013, and January 6, 2014, a leading security service provider, Proofpoint [72], found what may be the first *thingbots* attack using more than 100,000 smart household devices such as TVs, home-networking routers, connected multimedia centers, and at least a fridge. Those devices were used to send more than 750,000 malicious emails around the world.

21.5.4.2 IoT Attack Scenarios in 2014

Most of the incidents of IoT attacks reported in 2014 are real attacks that caused immense physical damage and financial losses. One example is the cyber attack that struck a German steel mill [73] in December 2014 that prevented the blast furnace from shutting down properly, resulting in massive damage to the foundry. The attackers used spear phishing and social engineering tactics to infiltrate the corporate network. The attackers demonstrated higher levels of familiarity with conventional IT security, applied industrial control, and production processes. In the same vein, Kaspersky Lab and Interpol [74] reported attacks on automated teller machines (ATMs) in Russia and Eastern Europe. According to Kaspersky, the cyber criminals first gained physical access to the ATMs and inserted a bootable CD that installed malicious software on the systems. After entering a set of numbers, the malware allowed the attackers to make cash withdrawals within four minutes at some specific times on Sundays and Mondays. A similar incident is point-of-sale (POS) attacks using the Tor-camouflaged *chewBacca* payment card-stealing Trojan [75] that targeted dozens of retailers in the United States, Canada, Russia, and Australia. The attacks were performed over the Tor network, which enabled the attackers to mask the actual IP address of their servers as well as encrypt the traffic between their servers and the infected POS machines. The malware captured payment card details and other personal information from about 50,000 customers.

21.5.4.3 IoT Attack Scenarios in 2015

The IoT-related security breaches that made headlines in 2015 include the work of a security researcher Billy Rios [76] who discovered vulnerabilities in smart infusion pumps manufactured by Hospira, an Illinois firm with more than 400,000 intravenous pumps installed in different hospitals around the globe. The vulnerabilities would allow malicious parties to carry out all manner of attacks ranging from raising or reducing the dosage limit to remotely altering the amount of drugs administered to a patient. Furthermore, a team of researchers at the University of Illinois [77] has uncovered some vulnerability in smart watches that attackers can use in order to deduce what character a person wearing a smart watch is typing on the keyboard. The researchers revealed that motion sensors on the Samsung Gear Live smart watch, Apple Watch, and Fitbit could leak information about what a user is typing from the movements of the left wrist. Additionally, researchers from the Rapid7 security firm [78] found critical vulnerabilities in nine Internet-connected baby monitors. According to the researchers, the vulnerabilities could allow an attacker to perform some malicious activities, including monitoring live video feeds, harvesting video clips, changing camera settings, and adding any number of new authorized users who can view and control the device. Lastly, for the year 2015, we consider the VTech toy firm customer data hacking incident [79] that exposed private information of both parents and children, including names, addresses, email addresses, secret questions/answer, and IP addresses. A security analyst Troy Hunt verified that passwords were protected using MD5 hash only, which is not an encryption.

21.5.4.4 IoT Attack Scenarios in 2016

Still continuing with smart toy hacking incidents, in 2016, a security hole was found in the app of the Smart Toy Bear [80] that allowed attackers to gain unauthorized access to the network, which can enabled a malicious entity to launch other attack campaigns. As more smart cars join the IoT, many more vulnerabilities are being uncovered. For example, a research team headed by Stephen Savage, a professor of Computer Science at the University of California, San Diego [81], has discovered a security flaw in the OS of a car that allowed the team to take control of the car within 18 seconds by playing a music file infected with a malware on the CD player of the car. According to Savage, third-party and original equipment manufacturer (OEM) software are responsible for many of the vulnerabilities found in cars recently. In addition, after an in-depth security analysis of a Samsung smart home platform, a group of researchers at the University of Michigan and Microsoft [82] has found security flaws that can enable attackers to take control of a smart home and do all kinds of things ranging from triggering a smoke detector to planting a backdoor PIN code in a digital lock system. Finally, as unmanned aerial vehicles popularly known as drones become commonplace, new research by a computer security team at the Johns Hopkins University [83] has revealed security holes in some commercial drones. The team has discovered three different ways to send rogue commands to the device using a laptop, which includes sending thousands of requests that can overwhelm the processor and crash it. This is a serious cyber threat, since an attacker can take over a drone and use it for a different purpose that could harm other people.

21.6 SECURITY VERSUS PERFORMANCE TRADE-OFFS IN THE IOT

The IoT offers very appealing prospects for almost every area of human endeavor. However, while connecting every device to the Internet comes with added convenience, increased productivity, and a boon to mankind, it is important for users to consider what they may be compromising from a security and privacy point of view. Sometimes the convenience that embracing new technology brings comes at a price. In the case of the IoT, the price is the value of security and privacy that enterprises and individuals lose when their IoT devices connect to the Internet in an insecure manner.

Managing security is among the major challenges facing the IoT, and there is always a tough trade-off between security and performance. Unfortunately, however, many people often prioritize ease and convenience over security and privacy [84]. In the same vein, users of insecure IoT networks knowingly or unknowingly give up varying amounts of privacy and security in order to enjoy the convenience of connected *things*. But data leaks, snooping, etc., are often the result of such sacrifices.

Security decision making requires making wise and informed trade-offs. One fundamental challenge facing IoT designers, manufacturers, and developers is that the more secure an IoT device is, the more inconvenient it is [85]. At first, developing secure IoT hardware, applications, or services may seem expensive. However, failure to observe security best practices in IoT development projects can result in loss of consumer confidence, which can eventually cost the company a lot more in lost sales opportunities than the cost of securing the IoT device or application in the first place [86]. Below we outline some key trade-offs that should be taken into consideration when designing hardware or developing applications for the IoT.

21.6.1 SECURITY VERSUS ENERGY TRADE-OFFS

Energy efficiency and security constitute fundamental challenges for the design and implementation of IoT systems; the two requirements nonetheless are at odds with each other. For example, while low-power consumption is needed for longevity of the energy source, security, which usually drains energy resources, is needed to safeguard data and provide privacy. This issue, therefore, requires making intelligent trade-offs. A security architecture based on configurability and adaptability was developed in [87] to address security versus energy trade-offs in the IoT.

21.6.2 SECURITY VERSUS PROCESSING/MEMORY CAPACITY TRADE-OFFS

When it comes to security implementation, there is always the question of performance and speed. This is because cryptographic schemes are expected to be portable and fast, but with little or no reduction in security performance [88]. On the contrary, most standard security solutions are computationally intensive and require a lot of memory space. Many IoT devices, however, are designed to have a very small footprint with extreme processing and memory constraints. Hence, the requirements of standard cryptographic algorithms are at variance with the requirements of most constrained IoT devices. The key to resolving such trade-offs is to know the specifications of the constrained devices and evaluate the requirements of potential candidate solutions.

21.6.3 SECURITY VERSUS COST TRADE-OFFS

IoT devices, especially those at the sensing layer, such as RFID devices, sensor nodes, and cameras, are expected to be cheap and disposable, because thousands of such devices may be needed for some applications. But there are complex trade-offs between security and hardware/software cost, which require careful business and security/privacy assessments. The fact is that the more resources are added to a device (e.g., memory, processing power, and longer battery life) the more expensive the device will be.

21.6.4 SECURITY VERSUS CONVENIENCE TRADE-OFFS

When people buy a new device, they are usually eager to go home and start using it. From the point of view of many users, it is actually more convenient to buy a device that is already configured with all settings in place. No wonder that in order to satisfy such consumers, devices normally come with default settings that can allow them to perform the basic functions they are intended for, as soon as they are powered on. Most users do not bother to do any security configuration on their devices, even if it is just to enable wireless security settings.

In the case of some IoT devices, the matter is even worse since many of those devices may not have good user interfaces and some must be configured remotely or interfaced with a laptop or desktop computer in order to be properly configured. Here, there is a trade-off between security and convenience, and people would often err on the side of convenience over privacy and security. Although many consumers are unlikely to make the right trade-offs, manufacturers can help to improve security by making some basic security settings on their IoT devices unavoidable. In addition, such IoT devices should not allow the use of weak passwords.

21.7 EXISTING PROTOCOLS AND STANDARDS FOR IOT

In the computing world, the term protocol refers to a set of rules and technical procedures governing communications and interactions between computing devices. Although the traditional Internet supports hundreds of protocols, these protocols cannot be used in the IoT because of the different constraints associated with connected IoT devices. Therefore, the IoT is expected to support hundreds more protocols [89], especially as the number of connected physical heterogeneous objects using a variety of communication technologies continues to grow. The reasons for the increase in the number of protocols include lack of a single universal IoT standard and the fact that there is no harmony in the definition of the IoT itself due to the heterogeneity and interoperability issues [90]. For example, it is difficult to harmonize a standard for bread toasters with a standard for pacemakers.

Accordingly, quite a number of working groups have been created and tasked with the responsibilities of developing and standardizing lightweight protocols for constrained IoT devices [91]. The standards will help application developers and service providers to develop applications and services for the IoT. The working groups include the Internet Engineering Task Force (IETF), the Institute of Electrical and Electronic Engineers (IEEE), the World Wide Web Consortium (W3C), the European Telecommunications Standard Institute (ETSI), and the Electronic Product Code global (EPCglobal) [92].

21.7.1 A BRIEF OVERVIEW OF IoT COMMUNICATION PROTOCOLS

Below we provide brief descriptions of some important IoT communication protocols defined by both IEEE 802.15 working group and IETF. The IEEE 802.15 task group defined the protocols in the two lower layers of the protocol stack shown in Figure 21.4, and the protocols in the upper layers were defined by the IETF.

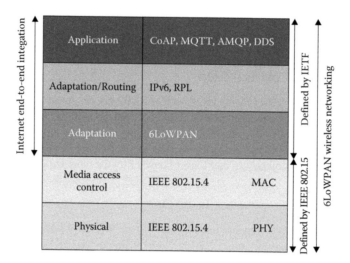

FIGURE 21.4 Internet of Things communication protocols.

21.7.1.1 IEEE 802.15.4

The IEEE 802.15.4 standard, first published in 2003 and revised in 2006, was defined by the IEEE 802.15 task group and is maintained by the same group [93]. This standard enables the constrained sensing devices of the IoT to communicate through the Internet, while coping with low-energy requirements. The IEEE 802.15.4 standard clearly specifies the communication rules for Low-Rate Wireless Personal Area Networks (LR-WPANs) at the lower layers of the stack (Figure 21.4), the Physical (PHY) layer and the Media Access Control (MAC) layer [94]. This lays the foundation for communication protocols at the upper layers, which are not defined in the IEEE 802.15.4 standard. Thus, it serves as the basis for developing other low-power protocol stacks, such as ZigBee, ISA100.11a, WirelessHART, MiWi, and Z-wave specifications.

The IEEE 802.15.4 standard can use 2.4–2.485 GHz frequency band with a typical maximum data rate of 250 kb/s at a maximum output power of 1 mW. It also has a maximum packet size of 127 octets. But the maximum payload size is less than 127 octets (bytes) due to the overhead in a packet (MAC layer headers) [95]. Therefore, the available space for the top layer protocols is usually between 86 and 116 octets [96]. Time Synchronized Channel Hopping (TSCH) capability was integrated into the standard in 2011 to facilitate robust communication by way of hopping and high data rates through synchronization [97]. In 2012, an amendment of IEEE 802.15.4 was proposed to address the issues of inference and fading multipath, which led to the approval of the IEEE 802.15.4e. For more details, we refer the reader to Refs. [98,99].

21.7.1.2 6LOWPAN

IPv6 over Low Power Wireless Personal Area Networks (6LoWPAN) is a low-power adaptation layer protocol [100] that allows wireless transmission of IPv6 packets over the IEEE 802.15.4 networks. In other words, 6LoWPAN can be considered an IPv6 header compression protocol. Instead of using ZigBee or other technologies, 6LoWPAN and IP-based protocols can be used for addressing and routing packets at the network layer and the adaptive application protocols. This provides all IP connectivity from individual sensing nodes to user applications via the Internet with minimum effort on the design of gateway translation protocols [101]. On the contrary, ZigBee usually requires more complex gateways for bridging the different networks. Consequently, two security processes are needed to secure the network, one on the network side and the other on the Internet side [102]. 6LoWPAN also compresses the large IPv6 header (40 bytes) to fit into the

IEEE 802.15.4 frame by removing redundant information like source and destination addresses, since the same information can be inferred from the IEEE 802.15.4 header [91].

21.7.1.3 RPL

The IPv6 Routing Protocol for Low-Power and Lossy Networks (RPL) was standardized in 2011 by the IETF to address the challenging routing issues of 6LoWPAN, such as low-power and lossy radio links and frequent topology changes. RPL supports a variety of IoT applications and uses a Destination-Oriented Directed Acyclic Graph (DODAG) to create a routing topology. That is a directed graph without cycles, which is oriented toward a root node or a border router. The multipoint-to-point communication scheme in RPL supports communication from the devices to the root node using minimal routing state. We refer the reader to [101,103] for more details of RPL.

21.7.1.4 CoAP

The Constrained Application Protocol (CoAP) is an application layer protocol defined within the Constrained RESTful Environment (CoRE) working group of the IETF [104]. The protocol is intended for use in resource constrained devices of the IoT, such as WSN nodes, to enable IP-based HTTP-like interactions [105]. The standard interface between HTTP clients and servers is Representational State Transfer (REST). However, REST cannot be used for lightweight applications like the IoT because it will cause significant overhead. Therefore, CoAP, which is built on top of User Datagram Protocol (UDP), is designed such that it will allow low-power devices to use RESTful services while meeting the energy constraints requirements.

21.7.1.5 MQTT

The Message Queuing Telemetry Transport (MQTT) is a publish/subscribe messaging model that was introduced by IBM and standardized by OASIS in 1999 and 2013 respectively [106,107]. The protocol, designed for communication between IoT devices, serves as an alternative to the conventional client-server model in which a client *talks* directly with an endpoint. This model consists of three main entities: a publisher, a subscriber, and a broker. Resource-constrained IoT devices like sensors are the publishers, applications interested in some topic are the subscribers, and the broker serves as an intermediary between the two [108]. Sensor nodes, for example, connect to the broker and send available data on a particular topic, while subscribers of that topic connect to the same broker to receive available data. Unlike CoAP, MQTT is built over Transmission Control Protocol (TCP). There are various models of Quality of Service (QoS) profiles used in MQTT as discussed in [109].

21.7.1.6 AMQP

The Advanced Message Queuing Protocol (AMQP) is a binary [110] transport layer open standard protocol. It is a high-performance messaging middleware that is designed for enterprise environments, and it can be used in the financial industry. Like MQTT, it runs on top of TCP and uses the publish/subscribe architecture. The protocol supports interoperability and heterogeneity characteristic communications between different IoT devices that support various languages. Unlike MQTT, AMQP uses three different modes of message delivery guarantee models to ensure reliable communication: at-most-once, at-least-once, and exactly-once [110].

- *At-most-once*: In this mode a message is delivered once or never.
- *At-least-once*: In this case there is certainty that a message will be delivered, but it may be delivered multiple times.
- *Exactly once*: In this mode a message will certainly be delivered once.

21.7.1.7 DDS

The Data Distribution Service (DDS) is an open standard middleware designed in the context of naval and aerospace by the Object Management Group (OMG) for M2M real-time communications [111].

While DDS is basically designed to work on top of unreliable transports, such as UDP, it uses multicast UDP within LAN and TCP over WAN [110]. Although it uses the publish/subscribe pattern, DDS does not rely on a central broker for message relay [112]. Instead, it follows a serverless architecture, which allows for high performance information sharing using a Data-Centric Publish-Subscribe (DCPS) scheme [110]. In this model, after establishing a connection between anonymous information publishers and subscribers (participants), the connection between participants bypasses the server and becomes peer-to-peer [113]. This direct message exchange between domain participants enhances urgency, priority, durability, and reliability. DDS offers the scalability, performance, and QoS needed to support IoT applications.

21.7.2 IoT Security Protocols

While quite a number of security protocols have been standardized for the traditional Internet, there are fundamentally two major standard security protocols for IoT: TLS and Datagram Transport Layer Security (DTLS).

21.7.2.1 TLS

TLS is the successor of Secure Sockets Layer (SSL), a cryptographic protocol for providing secure Internet connection between a client and a server using TCP transport. The key difference between the two is that TLS supports newer cipher suites and provides message authentication. TLS is used to provide end-to-end secure communication for applications that require TCP as the underlying transport protocol, such as MQTT. TLS employs a handshake mechanism to negotiate different parameters to establish a secure connection between the client and the server. The protocol is made up of two layers: a *record protocol* for providing a secure connection and a *handshake protocol* for ensuring authentication between devices and negotiating the type of encryption algorithms to be used and cryptographic keys before data exchange.

21.7.2.2 DTLS

The DTLS is basically a variant of TLS for IoT applications that uses UDP as the underlying protocol. DTLS is designed to provide a secure channel between constrained devices of the IoT [114]. In order to be compatible with the unreliable nature of UDP, DTLS adds three additional features to TLS [115]:

* A segmentation/reassembly service for the handshake entity
* A modified record protocol header including a sequence number used both for clear and ciphered operations
* Two new optional flights, Hello-Verify-Request and Client-Hello with cookie.

21.7.3 A Brief Security Analysis of the IoT Protocols

21.7.3.1 IEEE 802.15.4

The IEEE 802.15.4 standard specifies security services at both the physical layer (PHY) and MAC, as well as recommends the implementation of additional security features at the upper layers of the protocol stack due to the resource limitations of IoT devices [116]. The standard supports a number of security modes at the MAC sublayer, including, but not limited to, Advanced Encryption Standard-Cipher Block Chaining-Message Authentication Code (AES-CBC-MAC)-32, AES-CBC-MAC-64, and AES-CBC-MAC-128. It uses 128-bit secret keys known by the communicating entities. The IEEE 802.15.4 standard supports various security functionalities like confidentiality, data authenticity and integrity, replay protection, and access control mechanisms to protect MAC frames [94,116].

While the IEEE 802.15.4 standard recommends the implementation of more security features at the higher layers, nevertheless some attacks can only be addressed at the MAC sublayer [116,117].

Certain optional features specified in the IEEE 802.15.4 standard can reduce security; hence there is a need for designers to exercise care when implementing the standard [118]. Another security limitation of the IEEE 802.15.4 standard is lack of a specific keying model, such as how keys will be created and how the keys will be exchanged, among others [119]. An additional shortcoming is the fact that IEEE 802.15.4-based networks are susceptible to black hole attacks, a type of DoS attack in which an attacker node absorbs or drops every packet passing through it [120].

21.7.3.2 6LoWPAN

Currently, the 6LoWPAN adaptation protocol does not specify any security mechanisms; notwith-standing, discussions in certain relevant documents (Request for Comments (RFC) 4919 [121], RFC 6282 [122] and RFC 6568 [123]) focus on security requirements, threats, and approaches to consider for implementation in the IoT network layer. 6LoWPAN security research proposals and open research challenges have been presented in [94].

21.7.3.3 RPL

The RPL specification has three different modes of security: Unsecured, Preinstalled, and Authenti-cated [124]. It uses Advanced Encryption Standard (AES)/Counter with CBC-MAC (AES/CCM) along with 128-bit keys for MAC generation to support integrity [125]. It also supports confidentiality, data authenticity, semantic security, key management, and protection against replay attacks. The RPL proto-col is designed to employ link layer security mechanisms if they are available and appropriate, or to make use of its own security mechanisms when link layer mechanisms are absent or not suitable [124].

Nevertheless, the protocol is susceptible to different attacks, which include black hole, Sybil, hello flood, selective forwarding, and DoS attacks. Specifically, the authors in Ref. [126] observed that the protocol cannot protect against internal attacks. Using Sybil attacks, they found that an insider can successfully carry out active or passive internal attacks on the RPL network, which could disrupt the network and intercept information on the network respectively. Thus, a malicious node within the network can render cryptographic mechanisms ineffective [126].

21.7.3.4 CoAP

The CoAP protocol in itself does not define any standard for secure architecture; hence it is used in conjunction with DTLS to provide security [124]. Binding DTLS to CoAP means that security is supported at the transport layer and not at the application layer because DTLS is a transport layer protocol. The fundamental AES/CCM cryptographic algorithm is adopted (as in other security schemes in 6LoWPAN environments) to provide confidentiality, integrity, authentication, nonrepu-diation, and protection against replay attacks, since those functionalities are supported in the DTLS [94]. Four security modes are specified for CoAP, namely NoSec, PreshardKey, RawPublicKey, and Certificates.

As mentioned previously, DTLS is essentially TLS with some additional features that enable it to handle the unreliable nature of UDP. Therefore, like TLS, the DTLS handshake process, needed for authentication and key agreement, can significantly impact performance of the resource-constrained devices of the IoT. The impact will be more significant when public-key Elliptic Curve Cryptography (ECC) is employed for authentication and key agreement. In addition, apart from the 25-byte IEEE 802.15.4 overhead, and 10 bytes and 4 bytes for 6LoWPAN and CoAP addressing, respectively, DTLS adds an additional overhead of 13 bytes, leaving only 75 bytes for the application-layer payload [94].

21.7.3.5 MQTT

As a transport protocol designed to provide a lightweight and easy-to-use message transmission, the MQTT specification does not offer anything on top of TCP for secure communication [127]. It does, however, recommend the use of the TLS [RFC5246] protocol for applications that require security. Therefore, it is the responsibility of the implementer to provide appropriate security features [128].

This implies that by default, MQTT connections use unencrypted communication and attackers can eavesdrop on the message or perform MitM attacks on the network. In itself, the protocol relies on a few security mechanisms, which include the use of username and password fields of the MQTT CONNECT packet for authentication and authorization purposes [129]. However, since the communication is not encrypted, the username and password are transmitted in plain text.

While the use of TLS is recommended for applications that require a higher level of security, using MQTT over TLS usually comes at a cost in terms of processing power, memory usage, and transmission overhead. Although the costs of security may not overwhelm the broker, the constrained devices will typically not be able to cope with such demanding requirements because they are not designed for compute-intensive tasks. One alternative to improve the performance of TLS is to use a *session resumption* technique. TLS session resumption allows reusing a previously negotiated TLS session by restoring the secret information of previous sessions when connecting to the server the next time. There are two types of session resumption techniques: Session IDs and Session Tickets [129].

21.7.3.6 AMQP

The use of Simple Authentication Security Layer (SASL) has been specified in the AMQP standard to allow for negotiation of an agreed authenticated mechanism [130], which offers better authentication compared with MQTT. The standard also defines specification for data encryption using TLS. In addition, the design of AMQP allows for separate negotiation of TLS and SASL schemes and upward replacement with alternative mechanisms as they evolve [131].

21.7.3.7 DDS

Apart from urgency, priority, durability, and reliability, one other aspect that has been enhanced by the DDS direct message exchange between domain participants is security. Three extensions have been recently added to the DDS standard. One of them is *DDS-Security v1.0 Beta 1*, which defines the security model and Service Plugin Interface (SPI) architecture for compliant DDS implementations [132,133]. Thus, the current DDS standard includes, among other things, security, high performance, scalability, and data-centric publish/subscribe abstraction [133]. The recent ratification of an advanced security standard for DDS basically provides end-to-end security [134].

21.7.4 REVIEW OF RECENTLY PROPOSED IoT SECURITY PROTOCOLS

Virtually all of the communication protocols considered in the previous sections have some kind of security issues. Most of them must work on top of TLS or DTLS in order to provide a secure communication channel. However, many IoT devices that are highly constrained in terms of available energy, memory capacity, processing power, and communication bandwidth do not meet the resource requirements of these secure communication protocols. This results in insecure connection, poor performance, and latency, among other issues. Consequently, more lightweight secure communication protocols and lightweight security protocols that can cope with the constrained nature of many IoT devices are invaluable and highly needed. To respond to this demand, a number of protocols aimed at providing secure communication in the IoT have been proposed recently.

In [135], for example, an LTE-based Device-to-Device (D2D) secure multihop communication protocol was proposed to address the communication needs of User Equipment (UE) within and outside mobile network radio coverage areas. The proposed protocol enables UEs to become hubs for machine-type traffic using a suitable radio interface for uploading data to the Internet via the mobile network. Using the recent D2D feature of LTE Advanced (LTE-A) that supports Machine-Type Communication (MTC), the authors developed a D2D mesh network for transporting MTC data to the Internet with little impact on the network infrastructure. To achieve this, the UEs aggregate traffic from nearby IoT devices, and the multihop protocol links the LTE-A UEs to a remote node that uploads the data to the Internet. In their proposal, the UE refers to any device equipped with 4G LTE-A communication technology. The proposed protocol uses a lightweight security mechanism, and its security strength lies in sharing encryption keys indexed in the direct mode beacons.

A network security protocol designed to protect resource-constrained devices in the IoT against pattern analysis was presented in [136]. The protocol is also aimed at providing data integrity, identity management, trust management and privacy using minimal processor power. Before any secure message exchange between the server and the constrained client, both parties must share a symmetric key, which the authors referred to as the master key. The master key may be imprinted in the device through physical contact or via software flashing. The key must contain all data types in the message, which includes all alphabets, numbers, special characters, etc. Every element of the data type must have a number of unique IDs that depend on the available memory space. If the client, for example, wants to send a message, containing a single character, to the server, the client randomly selects an ID of that character and sends it instead. The server decodes the message by searching for the corresponding element in the symmetric key. The randomness needed for ID selection is generated by white noise from transistors, operational amplifiers, or resistors. In order to minimize resources consumed by the client, the same protocol can be used for authentication. However, since IoT devices can be tampered with physically, a different key management scheme that has three models is used.

The design and implementation of robust Key Management Protocols (KMPs) is the foundation of secure communication [137]. Hence, in order to create a secure channel between two IoT devices, it is essential to provide a reliable key management scheme that allows two remote devices to negotiate some security credentials like secret keys that can be used to protect data flow [138]. There is a lot of research being conducted currently on authentication and key management in the IoT.

In [139], for instance, a user authentication and key management protocol for heterogeneous WSNs based on the notion of the IoT was proposed. The proposed protocol allows a remote user to securely connect directly to a specific sensor node of interest without first contacting the Gateway Node (GWN). To achieve this, the authors used an authentication model that allows authentication and key agreement by first contacting a particular sensor node. Before deployment, each sensor node is given an identity, and a password key is randomly generated and stored in the memory of each node. The key is also shared with the GWN. The GWN is also given a well-secured randomly generated password that is hidden in its memory and not known by any other node in the network. The protocol specifies two separate registration phases: first, the registration between the GWN and each sensor node, and the second registration is between the GWN and the user, which is initiated by the user on demand. After the registration process, a user can connect to any sensor node of choice by initiating an authentication process; the node then contacts the GWN, etc., thus providing mutual authentication between all parties. The authors employed the use of a rare four-step authentication model, which they believed is most appropriate for resource-constrained devices of the IoT. Although any other security scheme for securely saving user data could be used, the authors used smartcards for user registration and authentication. As a highly lightweight security scheme, the protocol uses only hash and exclusive OR (XOR) computations, which makes it suitable for resource-constrained devices like sensor nodes. The authors asserted that their protocol provides anonymity, and it is resilient to replay attacks, privileged-insider attacks, stolen verifier attacks, stolen smartcard and smartcard breach attacks, impersonation attacks, many logged-in users with the same login ID attacks, GWN bypassing attacks, password-change attacks, and DoS attacks.

The authors in [140], however, discovered some shortcomings in the protocol presented in [139]. Particularly, they found that the protocol is susceptible to a number of cryptographic attacks, such as the stolen smartcard attack, sensor node impersonation attack, MitM attack, and disclosure of session key. Additionally, the protocol does not provide forward/backward secrecy and untraceability (anonymity). To overcome the security weaknesses in [139], the authors proposed an improved version of the protocol. Although the authors used the same four-step sensor node-first authentication model as in [139], there a few vital differences and improvements. For example, unlike the protocol in [139], after successful registration, the shared keys can be deleted on both the GWN and the sensor nodes; this enables the GWN to add more additional nodes to the network. Another improvement is that sensor nodes send authentication messages to the GWN over a secure channel. The authors claimed that their protocol addresses all the vulnerabilities observed in [139], and provides higher security while maintaining the same level of functionality and efficiency.

After analyzing the protocol in [140], the authors in [141] observed that the protocol is vulnerable to some types of attacks, such as a known session-specific temporary information attack and an off-line password-guessing attack. The protocol is also vulnerable to using a stolen-smartcard attack, a new-smartcard-issue attack and a user-impersonation attack. The authors therefore proposed an anonymity-preserving three-factor authenticated key exchange protocol for WSNs. The major improvements over the protocol in [140] include: the use of a different network model that consumes less power, which is more appropriate for IoT–based WSNs; a more efficient four-step remote-user authentication and session key agreement scheme (where the user contacts the GWN first); and the use of a bio-hashing function capable of generating a cancellable fingerprint template. The proposed protocol consists of seven phases: system setup, sensor node registration, user registration, login, authentication and session key agreement, password change, and smartcard revocation. In order to prove its security, efficiency, and functionality, the authors simulated the protocol using the AVISPA software package. They also claimed that the protocol has addressed all the security issues identified in [140].

In the same vein, the authors in [142] have identified some design flaws in the protocol presented in [141], and challenged the claims that the protocol is secure and efficient. The authors demonstrated that the protocol is insecure against replay attacks and DoS attacks. Accordingly, inspired by the protocol in [141], the authors proposed a new lightweight authentication and key agreement protocol using the same assumptions. Nevertheless, they redesigned new sensor node registration, user registration, login, and authentication and session key agreement phases. One of the notable changes and improvements in the proposed protocol is the elimination of what the authors called an *unduly computation* of the *timestamp* (SCT$_i$) in step 2 of the login phase in [141]. The authors argued that the value of SCT$_i$ is sent via an insure channel, hence attackers can eavesdrop the value and compute the corresponding *unique smartcard number* (SCN$_i$). Another improvement is that analysis shows that the proposed protocol has 14 hashes less than the protocol in [141] and 15 hashes less than the protocol in [140]. As a result, the authors claimed that their protocol is very efficient and secure.

DTLS binding is usually recommended for making IoT applications that work on UDP secure. However, there are a few challenges in deploying IoT applications over DTLS in resource-constrained environments. In recent years, researchers have tried to optimize the performance of DTLS for constrained applications in the IoT, and research in this area is still ongoing. The authors in [143], for example, have proposed an enhanced DTLS protocol for the IoT applications. Their protocol is aimed at reducing communication cost of the DTLS as well as improving the weakness of cookie exchange during the handshake process, which could protect against DoS attacks. The authors integrated the proposed protocol inside CoAP in order to minimize the size of the message on the security layer. The process of handshake based on cookies during authentication normally comes with a lot of overhead in terms of energy consumption as well as latency. Hence, the authors used a resource-rich trusted third party (a proxy auto-configuration server) to handle all those operations. This reduces delay and consumption of energy, thereby increasing battery life of resource-constrained devices. The proposed protocol was implemented on the Contiki OS, and when compared with the original DTLS for CoAP, there was a significant decrease in Random Access Memory (RAM) usage, computation time, and packet overhead.

21.8 CONCLUSION

IoT will shape the networks of the future as it represents a revolution already under way, creating a new interconnected smart world through unique interactions between *things*, people, and their environments that are becoming so closely interwoven. With numerous smart *things* already connected to the Internet and vastly more devices on the horizon, in the near future, IoT will arguably be the biggest network the world has ever seen, which promises to improve efficiency, reliability, and productivity in all the fields of human endeavor. This is swiftly becoming a reality as application areas for the IoT impact almost every aspect of our lives. Many executives in industrial sectors that include mining, manufacturing, automotive, and healthcare are realizing that IoT adoption will be vital to the success of their businesses.

While the potential benefits of the nascent IoT are significant, there are numerous potential challenges and barriers to realizing the full vision of the IoT. Security and privacy are among the major impediments and challenges that both enterprises and individuals could face with real-world IoT implementations. Essentially, every new technology usually comes with advantages and prospects, as well as hurdles. The fact that the potential benefits of IoT far outweigh the security risks does not mean, however, that the security risks should be ignored. Ignoring security risks will have significant negative effects on users, which may result in lack of trust. But it is well-established that building consumer trust is crucial for the widespread adoption of the IoT. As adoption of IoT solutions among industries and the incorporation of IoT technology into business processes and services are becoming a global trend, security and privacy should, therefore, be at the forefront of all IoT implementations.

In the near future, billions of devices will work in a more automated fashion than we see today, which may lead to closer interaction with humans; and as it is today, consumers will continue to rely on vendors to provide security, which raises significant concern. Accordingly, IoT devices, applications, networks, and cloud infrastructures must be adequately protected against cyber attacks. One way to achieve this is to intensify research efforts in developing protocols and applications that are more secure, efficient, and energy friendly, and thus appropriate for the constrained devices. In addition, there is a need to build protection into the devices themselves by providing a critical security layer as IoT devices depend less on the corporate perimeter security for protection. In order to ensure data security and privacy in IoT networks, both chip makers and manufacturers of IoT devices must consider security early in the design processes of their chips and devices, respectively. Users also have responsibilities to ensure a safe and effective IoT network, which include changing default usernames, choosing strong passwords, and being careful not to fall prey to phishing and social engineering attacks.

ACKNOWLEDGMENT

The authors wish to thank the Centre for Geodesy and Geodynamics, National Space Research and Development Agency, Toro, Bauchi State, Nigeria, for supporting this work. This work was supported by National Funding from the FCT—Fundação para a Ciência e a Tecnologia, through the UID/EEA/50008/2013 Project.

REFERENCES

1. Gartner. Gartner Says 6.4 Billion Connected 'Things' Will Be in Use in 2016, Up 30 Percent from 2015, November 10, 2015, Accessed June 13, 2016, http://www.gartner.com/newsroom/id/3165317.
2. Gartner. Gartner Says the Internet of Things Installed Base Will Grow to 26 Billion Units by 2020, December 12, 2013, Accessed June 12, 2016, http://www.gartner. com/newsroom/id/2636073.
3. Torsten George. IoT: The Security Risk Iceberg, *SecurityWeek Network*, September 23, 2015, Accessed June 13, 2016, http://www.securityweek.com/iot-security-risk-iceberg.
4. Israa Alqassem and Davor Svetinovic. A Taxonomy of Security and Privacy Requirements for the Internet of Things (IoT), *IEEE International Conference on Industrial Engineering and Engineering Management* (2014): 1244–1248, Accessed December 22, 2016, doi:10.1109/IEEM.2014.7958837.
5. Rinju Ravindran, Jerrin Yomas, and Jubin E. Sebastian. IoT: A review on security issues and measures, *An International Journal of Engineering Science and Technology: (ESTIJ)*, 5 (2015): 348–351, Accessed December 22, 2016.

6. Jianwei Wang, Dong Wang, Yuping Zhao, and Timo Korhonen. Fast Anti-Collision Algorithms in RFID Systems, *IEEE International Conference on Mobile Ubiquitous Computing, Systems, Services and Technologies* (2007): 75–80, Accessed December 22, 2016, doi:10.1109/UBICOMM.2007.34.

7. Ashwini W. Nagpurkar and Siddhant K. Jaiswal. Anti-collision in WSN and RFID network integration, *International Journal of Science and Research (IJSR)*, 4 (2015): 3008–3012, Accessed December 22, 2016.

8. Rodrigo Romana, Cristina Alcaraza, Javier Lopeza, and Nicolas Sklavos. Key management systems for sensor networks in the context of the Internet of Things, *ACM Journal of Computer & Electrical Engineering*, 37 (2011): 147–159, Accessed December 22, 2016, doi:10.1016/j.compeleceng.2011.01.009.

9. Adriana López-Alt, Eran Tromer, and Vinod Vaikuntanathan. On-the-Fly Multiparty Computation on the Cloud via Multikey Fully Homomorphic Encryption, *Proceedings of the Forty-Fourth Annual ACM Symposium on Theory of Computing* (2012): 1219–1234, Accessed December 23, 2016, doi:10.1145/2213977.2214086.

10. Meiko Jensen, Jorg Schwenk, Jens-Matthias Bohli, Nils Gruschka, and Luigi Lo Iacono. Security Prospects through Cloud Computing by Adopting Multiple Clouds, *IEEE 4th International Conference on Cloud Computing* (2011): 565–572, Accessed December 23, 2016, doi:10.1109/CLOUD.2011.85.

11. Mohamed Abomhara, and Geir M. Køien. Cyber security and the Internet of Things: Vulnerabilities, threats, intruders and attacks, *Journal of Cyber Security*, 4 (2015): 65–88, Accessed July 19, 2016, doi:10.13052/jcsm2245-1439.414.

12. Mark Weiser. The computer for the 21st century, *ACM Journal of SIGMOBILE Mobile Computing and Communications Review*, 3 (1999): 3–11, Accessed July 19, 2016, doi:10.1145/329124.329126.

13. Chiara Buratti et al. An overview on wireless sensor networks technology and evolution, *Journal of Sensors*, 9 (2009): 6869–6896, Accessed July 21, 2016, doi:10.3390/s90906869.

14. George Roussos. Enabling RFID in retail, *IEEE Computer Journal*, 39 (2006): 25–30, Accessed July 21, 2016, doi:10.1109/MC.2006.88.

15. Tim Unwin ed., *ICT4D: Information Communication Technology Development*. Cambridge, UK: Cambridge University Press, 2009, pp. 88–89.

16. Sébastien Ziegler and Latif Ladid. Towards a Global IPv6 Addressing Model for the Internet of Things, *IEEE 30th International Conference on Advanced Information Networking and Applications Workshops (WAINA)* (2016): 622–727, Accessed August 4, 2016, doi:10.1109/WAINA.2016.178.

17. Venkata Krishna Kishore Terli, Swetha Prabha Chaganti, Navya Bharathi Alla, and Shobhitha Sarab. Software Implementation of IPv4 to IPv6 Migration, *IEEE Long Island Systems, Applications and Technology Conference (LISAT)* (2016): 1–6, Accessed August 5, 2016, doi:10.1109/LISAT.2016.7494160.

18. Jonathan Camhi. BI Intelligence Projects 34 Billion Devices will be Connected by 2020, *Business Insider*, November 6, 2015, Accessed July 21, 2016, http://www.businessinsider.com/bi-intelligence-34-billion-connected-devices-2020–2015–11.

19. Dave Evans. The Internet of Things: How the Next Evolution of the Internet Is Changing Everything, *Cisco White Paper*, April 2011, Accessed July 21, 2016, http://www.cisco.com/c/dam/en_us/about/ac79/docs/innov/IoT_IBSG_0411FINAL.pdf.

20. Teng Xu, James B. Wendt, and Miodrag Potkonjak. Security of IoT Systems: Design Challenges and Opportunities, *IEEE/ACM International Conference on Computer-Aided Design (ICCAD)* (2014): 417–423, Accessed August 5, 2016, doi:10.1109/ICCAD.2014.7001385.

21. Icon Labs. The Internet of Secure Things—What is Really Needed to Secure the Internet of Things? Accessed August 5, 2016, http://www.iconlabs.com/prod/internet-secure-things-%E2%80%93-what-really-needed-secure-internet-things.

22. Software Defined Perimeter Working Group. Introduction to the Software Defined Perimeter Working Group, Accessed August 5, 2016, https://cloudsecurityalliance.org/group/software-defined-perimeter/#_overview.

23. Jim Reno. Security and the Internet of Things, *CA Technology*, Accessed August 5, 2016, http://www.ca.com/us/lpg/ca-technology-exchange/security-and-the-internet-of-things.aspx.

24. Li Da Xu, Wu He, and Shancang Li. Internet of things in industries: A survey, *IEEE Transactions on Industrial Informatics*, 10 (2014): 2233–2243, Accessed August 7, 2016, doi:10.1109/TII.2014.2300753.

25. Marco Stolpe. The Internet of Things: Opportunities and challenges for distributed data analysis, *ACM SIGKDD Exploration Newsletter*, 18 (2016): 15–34, Accessed August 8, 2016, doi:10.1145/2980765.2980768.

26. Jeroen van den Hoven. Fact sheet—Ethics Subgroup IoT—Version 4.0, *Delft University of Technology*, Accessed August 8, 2016.

27. Gartner. How Does IoT Impact Your Information Management Strategy? March 4, 2015, Accessed August 8, 2016, http://www.gartner.com/newsroom/id/2997917.
28. Redhat. An Intelligent Systems Solution for the Internet of Things, Accessed August 8, 2016, https://www.redhat.com/cms/managed-files/Intelligent%20Systems%20Solution%20for%20the%20IoT.pdf.
29. NERC. Security Guideline for the Electricity Sector: Identifying Critical Cyber Assets, June 17, 2010, Accessed August 12, 2016, http://www.nerc.com/docs/cip/sgwg/Critcal_Cyber_Asset_ID_V1_Final.pdf.
30. Ben Dickson. Why IoT Security Is So Critical, *Crunch Network*, October 24, 2015, Accessed August 11, 2016, https://techcrunch.com/2015/10/24/why-iot-security-is-so-critical/.
31. S. M. Riazul Islam, Daehan Kwak, M. D. Humaun Kabir, Mahmud Hossain, and Kyung-Sup Kwak. The Internet of Things for health care: A comprehensive survey, *IEEE Access Journal*, 3 (2015): 678–708, Accessed August 12, 2016, doi:10.1109/ACCESS.2015.2437951.
32. Jason R. C. Nurse, Arnau Erola, Ioannis Agrafiotis, Michael Goldsmith, and Sadie Creese. Smart Insiders: Exploring the Threat from Insiders Using the Internet-of-Things, *IEEE International Workshop on Secure Internet of Things (SIoT)* (2015): 5–14, Accessed August 12, 2016, doi:10.1109/SIOT.2015.10.
33. India Ashok. BT and KPMG Research Finds Cybercrime has Now Become an Industry, *International Business Times*, July 6, 2016, Accessed August 13, 2016, http://www.ibtimes.co.uk/bt-kpmg-research-finds-cybercrime-has-now-become-industry-1569196.
34. EY. Cybersecurity and the Internet of Things, March 2015, Accessed August 14, 2016, http://www.ey.com/Publication/vwLUAssets/EY-cybersecurity-and-the-internet-of-things/$FILE/EY-cybersecurity-and-the-internet-of-things.pdf.
35. Sam Pudwell. Q & A: Navigating the New Cyber Security Landscape, *ITProPortal*, April 24, 2016, Accessed August 15, 2016, http://www.itproportal.com/2016/04/24/qa-navigating-the-new-cyber-security-landscape/.
36. Michael J. Covington and Rush Carskadden. Threat Implications of the Internet of Things, *IEEE 5th International Conference on Cyber Conflict* (2013): 1–12, Accessed August 13, 2016, ISSN: 2325–5366.
37. David Fletcher. Internet of Things, in evolution of cyber technologies and operations to 2035, Misty Blowers (ed.), *Advances in Information Security*, vol. 63. Switzerland: Springer International Publishing, 2015, p. 29, doi:10.1007/978-3-319-23585-1_2.
38. Alan Grau. How to Build a Safer Internet of Things, *IEEE Spectrum*, February 25, 2015, Accessed August 14, 2016, http://spectrum.ieee.org/telecom/security/how-to-build-a-safer-internet-of-things.
39. Trend Micro. FBI Warns Public on Dangers of the Internet of Things, Accessed August 14, 2016, http://www.trendmicro.com/vinfo/us/security/news/internet-of-things/fbi-warns-public-on-dangers-of-the-internet-of-things.
40. Intelligence. Securing the Internet of Threats, Accessed August 14, 2016, https://www.theintelligenceofthings.com/article/securing-the-internet-of-threats/.
41. Tianlong Yu, Vyas Sekar, Srinivasan Seshan, Yuvraj Agarwal, and Chenren Xu. Handling a Trillion (unfixable) Flaws on a Billion Devices: Rethinking Network Security for the Internet-of-Things, *Proceedings of the 14th ACM Workshop on Hot Topics in Networks*, 5 (2015): 1–7, Accessed August 18, 2016, doi:10.1145/2834050.283409535.
42. Sumanjit Das and Tapaswini Nayak. Impact of Cyber Crime: Issues and Challenges, *International Journal of Engineering Sciences & Emerging Technologies*, 6 (2013): 142–153, Accessed August 14, 2016, ISSN: 22316604.
43. Sheheryar Khan. Four Areas of Security Vulnerabilities Presented by IoT, *PureVPN*, June 20, 2016, Accessed August 15, 2016, https://www.purevpn.com/blog/four-areas-of-security-vulnerabilities/.
44. Symantec. Internet Security Threat Report, April 2016, Accessed August 17, 2016, https://www.symantec.com/content/dam/symantec/docs/reports/istr-21-2016-en.pdf.
45. Advent One. Evolving Protection against Evolving Security Threats, Accessed August 17, 2016, http://www.adventone.com.au/wp-content/files_mf/1470008291Evolvingprotectionagainstevolvingsecuritythreats_Whitepaper.pdf.
46. IDA Tech New. Evolving Security Challenges in the IoT Era, December 19, 2014, Accessed August 17, 2016, https://www.ida.gov.sg/blog/insg/special-reports/evolving-security-challenges-in-the-iot-era/.
47. SpiceWorks. Battling the Big Network Security Hack, Accessed August 17, 2016, https://www.spiceworks.com/it-articles/it-security/.
48. Pierluigi Paganini. Millions Vulnerable UPnP Devices Vulnerable to Attack, *Security Affairs*, October 16, 2014, Accessed August 17, 2016, http://securityaffairs.co/wordpress/29278/cyber-crime/millions-upnp-devices-worldwide.html.

49. Shadi Esnaashari, Ian Welch, and Peter Komisarczuk. Determining Home Users' Vulnerability to Universal Plug and Play (UPnP) Attacks, *IEEE 27th International Conference on Advanced Information Networking and Applications Workshops (WAINA)* (2013): 725–729, Accessed August 17, 2016, doi:10.1109/WAINA.2013.225.

50. Emily Johnson. 6 IoT Security Dangers to the Enterprise, *DARKReading*, April 14, 2016, Accessed August 17, 2016, http://www.darkreading.com/endpoint/6-iot-security-dangers-to-the-enterprise/d/d-id/1325140.

51. Travis Spencer and Jennifer Riggins. APIs Power the Internet of Things, *NORDIC APIS*, January 5, 2015, Accessed August 17, 2016, http://nordicapis.com/apis-power-the-internet-of-things/.

52. Shmulik Regev, Assaf Regev, and Ravid Sagy. Innovative New Solutions for Securing the Internet of Things, *Security Intelligence*, February 22, 2016, Accessed August 17, 2016, https://securityintelligence.com/innovative-new-solutions-for-securing-the-internet-of-things/.

53. Khyati Jain. Critical Netgear Router Exploit Allows Anyone to Hack You Remotely, *The Hacker News*, October 9, 2015, Accessed August 17, 2016, http://thehackernews.com/2015/10/netgear-routerhack.html.

54. Warwick Ashford. LizardStresser IoT Botnet Launches 400Gbps DDoS Attacks, *ComputerWeekly.com*, June 30, 2016, Accessed August 13, 2016, http://www.computerweekly.com/news/450299445/Lizard Stresser-IoT-botnet-launches-400Gbps-DDoS-attack.

55. Botnet Uses IoT Devices to Power Massive DDoS Attacks, *SecurityWeek News*, July 1, 2016, Accessed August 17, 2016, http://www.securityweek.com/botnet-uses-iot-devices-power-massive-ddos-attacks.

56. Tripwire. Tripwire VERT Research: SOHO Wireless Router (In) Security, Accessed August 17, 2016, http://www.tripwire.com/register/soho-wireless-router-insecurity/showMeta/2/.

57. Darlene Storm. Of 10 IoT-Connected Home Security Systems Tested, 100% Are Full of Security Fail, *ComputerWorld*, February 11, 2015, Accessed August 19, 2016, http://www.computerworld.com/article/2881942/cybercrime-hacking/of-10-iot-connected-home-security-systems-tested-100-are-full-of-security-fail.html.

58. OWASP. OWASP Internet of Things Project, Accessed August 19, 2016, https://www.owasp.org/index.php/OWASP_Internet_of_Things_Project?utm_source=datafloq&utm_medium=ref&utm_campaign=datafloq#tab=IoT_Vulnerabilities.

59. CSI-HO-022 - Enabling the IoT: Wireless Low Power and Lossy Networks (LLNs), Accessed August 19, 2016, http://www.cellstream.com/intranet/training/csicourses/265-csi-il-008-wlln.html.

60. Mengmeng Ge and Dong Seong Kim. A Framework for Modeling and Assessing Security of the Internet of Things, *IEEE 21st International Conference on Parallel and Distributed Systems (ICPADS)*, (2015): 776–781, Accessed August 19, 2016, doi:10.1109/ICPADS.2015.102.

61. Rick Blaisdell. The Risks of IoT, *RickCloud*, June 8, 2015, Accessed August 19, 2016, https://www.rickscloud.com/the-risks-of-iot/.

62. Pawani Porambage, Mika Ylianttila, Corinna Schmitt, Pardeep Kumar, Andrei Gurtov and Athanasios V. Vasilakos. The quest for privacy in the Internet of Things, *IEEE Cloud Computing Journal*, 3 (2016): 36–45, Accessed August 22, 2016, doi:10.1109/MCC.2016.28.

63. Security-Marathon. The Difference between Data Privacy and Data Security, May 25, 2016, Accessed August 22, 2016, http://www.security-marathon.be/?p=2007.

64. Alan F. Westin. *Privacy and Freedom*. Washington, DC: Lee Law Review, 1968, p. 166.

65. Karen Renaud and Dora Galvez-Cruz. Privacy: Aspects, definitions and a multi-faceted privacy preservation approach, *IEEE Information Security for South Africa*, (2010): 1–8, Accessed August 21, 2016, doi:10.1109/ISSA.2010.5588297.

66. Wassnaa AL-mawee. Privacy and Security Issues in IoT Healthcare Applications for the Disabled Users: A Survey, Masters Thesis, Western Michigan University, 2015.

67. Samuel Gibbs. Hackers can Hijack Wi-Fi Hello Barbie to Spy on Your Children, *theguardian*, November 26, 2015, Accessed August 21, 2016, https://www.theguardian.com/technology/2015/nov/26/hackers-can-hijack-wi-fi-hello-barbie-to-spy-on-your-children.

68. INFOSEC. How Hackers Violate Privacy and Security of the Smart Home, September 11, 2015, Accessed August 21, 2016, http://resources.infosecinstitute.com/how-hackers-violate-privacy-and-security-of-the-smart-home/.

69. Jamie Condiffe. Philips Hue Light Bulbs Are Highly Hackable, *Gizmodo*, August 14, 2013, Accessed August 23, 2016, http://gizmodo.com/how-philips-hue-light-bulbs-are-highly-hackable-1133092324.

70. Andy Greenberg. Hackers Reveal Nasty New Car Attacks—With Me Behind the Wheel (Video), *Forbes*, August 12, 2013, Accessed August 23, 2016, http://www.forbes.com/sites/andygreenberg/2013/07/24/hackers-reveal-nasty-new-car-attacks-with-me-behind-the-wheel-video/#598181f05bf2.

71. John Markoff. Researchers Show Hoe a Car's Electronics Can Be Taken Over Remotely, *The New York Times*, March 9, 2011, Accessed August 23, 2016, http://www.nytimes.com/2011/03/10/business/10hack. html?_r=1.

72. Proofpoint. Proofpoint Uncovers Internet of Things (IoT) Cyberattack, January 16, 2014, Accessed August 23, 2016, http://investors.proofpoint.com/releasedetail.cfm?releaseid=819799.

73. Pamela Cobb. German Steel Mill Meltdown: Rising Stakes in the Internet of Things, *Security Intelligence*, January 14, 2015, Accessed August 23, 2016, https://securityintelligence.com/german-steel-mill-meltdown-rising-stakes-in-the-internet-of-things/.

74. Amir Mizroch. Hackers Target ATMs in Russia, Eastern Europe, *The Wall Street Journal*, October 8, 2014, Accessed August 24, 2016, http://blogs.wsj.com/digits/2014/10/08/hackers-target-atms-in-russia-eastern-europe/.

75. Kelly Jackson Higgins. Point-of-Sale System Attack Campaign Hits More Than 40 Retailers, *InformationWeek DARKReading*, January 30, 2014, Accessed August 24, 2016, http://www.darkreading.com/attacks-breaches/point-of-sale-system-attack-campaign-hits-more-than-40-retailers/d/d-id/1141247.

76. Kim Zetter. Hacker Can Send Fatal Dose to Hospital Drug Pumps, *Wired*, August 6, 2015, Accessed August 24, 2016, https://www.wired.com/2015/06/hackers-can-send-fatal-doses-hospital-drug-pumps/.

77. Kavita Iyer. Smart Watches Vulnerable to Hacking Says Reserachers, *TechWorm*, September 11, 2015, Accessed August 24, 2016, http://www.techworm.net/2015/09/smartwatches-vulnerable-to-hacking-says-researchers.html.

78. Dan Goodin. 9 Baby Monitors Wide Open to Hacks that Expose Users' Most Private Moments, *arsTECHNICA*, February 9, 2015, Accessed August 24, 2016, http://arstechnica.com/security/2015/09/9-baby-monitors-wide-open-to-hacks-that-expose-users-most-private-moments/.

79. Samuel Gibbs. Toy Firm VTech Hack Exposes Private Data of Parents and Children, *theguardian*, November 30, 2015, Accessed August 24, 2016, https://www.theguardian.com/technology/2015/nov/30/vtech-toys-hack-private-data-parents-children.

80. TREND MICRO. Researcher Discover a Not-So-Smart Flaw in Smart Toy Bear, February 4, 2016, Accessed August 24, 2016, http://www.trendmicro.com/vinfo/us/security/news/internet-of-things/researchers-discover-flaw-in-smart-toy-bear.

81. TREND MICRO. Carjacking by CD? Researcher Shows How a Spiked Song Can Be Used to Hack a Car, February 2, 2016, Accessed August 24, 2016, http://www.trendmicro.com/vinfo/us/security/news/internet-of-things/carjacking-by-cd-research-shows-how-spiked-song-can-hack-a-car.

82. Andy Greenberg. Flaws in Samsung's Smart Home Let Hackers Unlock Doors and Set Off Fire Alarms, *Wired*, February 5, 2016, Accessed August 24, 2016, https://www.wired.com/2016/05/flaws-samsungs-smart-home-let-hackers-unlock-doors-set-off-fire-alarms/.

83. Janet Burns. Johns Hopkins Team Hacks, Crashes Hobby Drones to Expose Security Flaws, *Forbes*, June 13, 2016, Accessed August 24, 2016, http://www.forbes.com/sites/janetwburns/2016/06/13/johns-hopkins-team-hacks-crashes-hobby-drones-to-expose-security-flaws/?ss=Security#373e60f81fa1.

84. Shaun Donaldson. Security Tradeoffs—A Culture of Convenience, *SECURITYWEEK*, April 29, 2013, Accessed December 25, 2016, http://www.securityweek.com/security-tradeoffs-culture-convenience.

85. Maurice Dawson, Mohamed Eltayeb, and Marwan Omar. *Security Solutions for Hyperconnectivity and the Internet of Things*. Hershey, PA: IGI, 2016, p. 93.

86. Susan Snedeker. *Syngress IT Security Project Management Handbook*. Rockland: Syngress, 2006, p. 481.

87. Sandip Ray, Tamzidul Hoque, Abhishek Basak, and Swarup Bhunia. The Power Play: Security-Energy Trade-Offs in the IoT Regime, *IEEE 34th International Conference on Computer Design (ICCD)* (2016): 690–693, Accessed December 25, 2016, doi:10.1109/ICCD.2016.7753360.

88. Rodrigo Roman, Pablo Najera, and Javier Lopez. Securing the Internet of Things, *IEEE Computer Journal*, 44 (2011): 51–58, Accessed December 26, 2016, doi:10.1109/MC.2011.291.

89. Stan Schneider. Understanding the Protocols behind the Internet of Things, *electronic design*, October 9, 2013, Accessed December 27, 2016, http://electronicdesign.com/iot/understanding-protocols-behind-internet-things.

90. Nicolo Zingales. Of Coffee Pods, Videogames, and Missed Interoperability: Reflections for EU Governance of the Internet of Things, *TILEC Discussion Paper No. 2015–026*, December 2015, Accessed December 27, 2016, https://ssrn.com/abstract=2707570.

91. Gustavo A. da Costa and João H. Kleinschmidt. Implementation of a Wireless Sensor Network Using Standardized IoT Protocols, *IEEE International Symposium on Consumer Electronics (ISCE)*, (2016): 17–18, Accessed December 27, 2016, doi:10.1109/ISCE.2016.7797327.

92. Ala Al-Fuqaha, Mohsen Guizani, Mehdi Mohammadi, Mohammed Aledhari, and Moussa Ayyash. Internet of Things: A survey on enabling technologies, protocols, and applications, *IEEE Communication*

Surveys & Tutorials, 17 (2015): 2347–2376, Accessed December 27, 2016, doi:10.1109/COMST. 2015.2444095.

93. Olivier Hersent, David Boswarthick, and Omar Elloumi. *The Internet of Things: Key Applications and Protocols.* West Sussex: John Wiley & Sons, 2011, p. 1.

94. Jorge Granjal, Edmundo Monteiro, and Jorge Sá Silva. Security for the Internet of Things: A survey of existing protocols and open research issues, *IEEE Communication Surveys & Tutorials*, 17 (2015): 1294–1312, Accessed December 28, 2016, doi:10.1109/COMST.2015.2388550.

95. Shahin Farahani. *ZigBee Wireless Networks and Transceivers.* Newton: Newnes, 2008, p. 262.

96. Zhengguo Sheng, Shusen Yang, Yifan Yu, Athanasios V. Vasilakos, Julie A. Mccann, and Kin K. Leung. A survey on the IETF protocol suite for the Internet of Things: Standards, challenges, and opportunities, *IEEE Wireless Communications Journal*, 20 (2013): 91–98, Accessed December 28, 2016, doi:10.1109/ MWC.2013.6704479.

97. Maria Rita Palattella, Nicola Accettura, Xavier Vilajosana, Thomas Watteyne, Luigi Alfredo Grieco, Gennaro Boggia, and Mischa Dohler. Standardized protocol stack for the internet of (important) things, *IEEE Communications Surveys & Tutorials*, 15 (2013): 1389–1406, Accessed December 28, 2016, doi:10.1109/SURV.2012.111412.0015.

98. Sahar Ben Yaala and Ridha Bouallegue. On MAC Layer Protocols towards Internet of Things: From IEEE802.15.4 to IEEE802.15.4e, *IEEE 24th International Conference on Software, Telecommunications and Computer Networks (SoftCOM)* (2016): 1–5, Accessed December 29, 2016, doi:10.1109/SOFT-COM.2016.7772165.

99. IEEE 802.15 WPAN™ Task Group 4e (TG4e), Accessed December 29, 2016, http://www.ieee802. org/15/pub/TG4e.html.

100. Kevin I-Kai Wang, Ashwin Rajamohan, Shivank Dubey, Samuel A. Catapang, and Zoran Salcic. A wearable Internet of Things Mote with Bare Metal 6LoWPAN Protocol for Pervasive Healthcare, *IEEE 11th International Conference on Ubiquitous Intelligence and Computing and 11th International Conference on Autonomic and Trusted Computing and 14th International Conference on Scalable Computing and Communications and Its Associated Workshops* (2014): 750–756, Accessed December 28, 2016, doi:10.1109/UIC-ATC-ScalCom.2014.74.

101. Mai Banh, Nam Nguyen, Kieu-Ha Phung, Long Nguyen, Nguyen Huu Thah, and Steenhaut. Energy Balancing RPL-Based Routing for Internet of Things, *IEEE Sixth International Conference on Communications and Electronics (ICCE)* (2016): 125–130, Accessed December 29, 2016, doi:10.1109/ CCE.2016.7562624.

102. Usama Saqib. Difference between ZigBee and 6LoWPAN Explained with a Simple Arduino + Xbee Setup, *Linkedin*, August 2, 2014, Accessed December 29, 2016, https://www.linkedin.com/pulse /20140802224850-81482458-difference-between-zigbee-and-6lowpan-explained-with-a-simple-arduino-xbee-setup.

103. Oana Iova, Gian Pietro Picco, Timofei Istomin, and Csaba Kiraly. RPL: The routing standard for the intetnet of things… or is it? *IEEE Communications Magazine*, 54 (2016): 16–22, Accessed December 29, 2016, doi:10.1109/MCOM.2016.1600397CM.

104. Angelo P. Castellani, Mattia Gheda, Nicola Bui, Michele Rossi, and Michele Zorzi. Web Services for the Internet of Things through CoAP and EXI, *IEEE International Conference on Communications Workshops (ICC)* (2011): 1–6, Accessed December 29, 2016, doi:10.1109/iccw.2011.5963563.

105. August Betzler, Carles Gomez, Ilker Demirkol, and Josep Paradells. CoAP congestion control for the Internet of Things, *IEEE Communications Magazine*, 54 (2016): 154–160, Accessed December 29, 2016, doi:10.1109/MCOM.2016.7509394.

106. Dave Locke. MQ Telemetry Transport (MQTT) v3. 1 Protocol Specification, *IBM Developer Works Technical Library*, August 19, 2010, Accessed December 29, 2016, http://www.ibm.com/developerworks/ webservices/library/ws-mqtt/index.html.

107. Vasileios Karagiannis, Periklis Chatzimisios, Francisco Vazquez-Gallego, and Jesus Alonso-Zarate. A survey on application layer protocols for the Internet of Things, *Transaction on IoT and Cloud Computing*, 3 (2015): 11–17, Accessed December 29, 2016, https://jesusalonsozarate.files.wordpress.com/2015/01/ 2015-transaction-on-iot-and-cloudcomputing.pdf.

108. Meena Singh, Rajan M. A, Shivraj V. L, and Balamuralidhar P. Secure MQTT for Internet of Things, *IEEE Fifth International Conference on Communication Systems and Network Technologies*, (2015): 746–751, Accessed December 29, 2016, doi:10.1109/CSNT.2015.16.

109. Stefan Mijovic, Erion Shehu, and Chiara Buratti. Comparing Application Layer Protocols for the Internet of Things via Experimentation, *IEEE 2nd International Forum on Research and Technologies for Society*

and *Industry Leveraging a Better Tomorrow (RTSI)* (2016): 1–5, Accessed December 29, 2016, doi:10.1109/RTSI.2016.7740559.

110. Zoran B. Babovic, Jelica Protic, and Veljko Milutinovic. Web performance evaluation for Internet of Things applications, *IEEE Access Journal*, 4 (2016): 6974–6992, Accessed December 29, 2016. doi:10.1109/ACCESS.2016.2615181.

111. Object Management Group. Data Distribution Service V1.4, April 2015, December 30, 2016, http://www.omg.org/spec/DDS/1.4.

112. Julius Pfrommer, Sten Gruner, and Florian Palm. Hybrid OPC UA and DDS: Combining Architectural Styles for the Industrial Internet, *IEEE World Conference on Factory Communication Systems (WFCS)* (2016): 1–7, Accessed December 30, doi:10.1109/WFCS.2016.7496515.

113. Yuang Chen and Thomas Kunz. Performance Evaluation of IoT Protocols under a Constrained Wireless Access Network, *IEEE International Conference on Selected Topics in Mobile & Wireless Networking (MoWNeT)* (2016): 1–7, Accessed December 30, 2016, doi:10.1109/MoWNet.2016.7496622.

114. Jiyong Han, Minkeun Ha, and Daeyoung Kim. Practical Security Analysis for the Constrained Node Networks: Focusing on the DTLS Protocol, *IEEE 5th International Conference on the Internet of Things (IoT)* (2015): 22–29, Accessed December 31, 2016, doi:10.1109/IOT.2015.7356544.

115. Pascal Urien. Towards Secure Elements for the Internet of Things: The eLock Use Case, *IEEE Second International Conference on Mobile and Secure Services (MobiSecServ)*, (2016): 1–5, Accessed December 30, doi:10.1109/MOBISECSERV.2016.7440228.

116. Rakesh Matam and Somanath Tripathy. Denial of Service Attack on Low Rate Wireless Personal Area Networks, *IEEE Twenty Second National Conference on Communication (NCC)* (2016): 1–6, Accessed January 1, 2017, doi:10.1109/NCC.2016.7561148.

117. Roberta Daidone, Gianluca Dini, and Marco Tiloca. Demo Abstract: On Preventing GTS-based Denial of Service in IEEE 802.15.4, *9th European Conference on Wireless Sensor Networks*, (2012): 69–70, Accessed January 1, 2017, https://pdfs.semanticscholar.org/f0af/6cbb0b72b267dcb269db9b4de6c9f280d2bf.pdf.

118. Naveen Sastry and David Wagner. Security Considerations for IEEE 802.15.4 Networks, *ACM 3rd Workshop on Wireless Security* (2004): 32–42, Accessed January 1, 2017, doi:10.1145/1023646.1023654.

119. Syed Muhammad Sajjad and Muhammad Yousaf. Security Analysis of IEEE 802.15.4 MAC in the Context of Internet of Things (IoT), *IEEE Conference on Information Assurance and Cyber Security (CIACS)* (2014): 9–14, Accessed December 31, 2016, doi:10.1109/CIACS.2014.6861324.

120. Krishan Kant Varshney and P. Samundiswary. Performance Analysis of Malicious Nodes in IEEE 802.15.4 Based Wireless Sensor Network, *IEEE International Conference on Information Communication and Embedded Systems (ICICES2014)* (2014): 1–5, Accessed December 31, 2016, doi:10.1109/ICICES.2014.7033873.

121. Nandakishore Kushalnagar, Gabriel Montenegro, and Christian Peter Pii Schumacher. IPv6 over Low-Power Wireless Personal Area Networks (6LoWPANs): Overview, Assumptions, Problem Statement, Goals, *Networking Working Group, RFC 4919* (2007): 1–12, Accessed January 1, 2017, https://tools.ietf.org/html/rfc4919.

122. Jonathan Hui and Pascal Thubert. Compression Format for IPv6 Datagrams over IEEE 802.15.4-Based Networks, *Internet Engineering Task Force (IETF) RFC 6282* (2011): 1–24, Accessed January 1, 2017, https://tools.ietf.org/html/rfc6282.

123. Eunsook Kim, Dominik Kaspar, and J. P. Vasseur. Design and Application Spaces for IPv6 over Low-Power Wireless Personal Area Networks (6LoWPANs), *Internet Engineering Task Force (IETF) RFC 6568* (2012): 1–28, Accessed January 1, 2017, https://tools.ietf.org/html/rfc6568.

124. Reem Abdul Rahman and Babar Shah. Security Analysis of IoT Protocols: A Focus in CoAP, *IEEE 3rd MEC International Conference on Big Data and Smart City* (2016): 1–7, Accessed January 1, 2017, doi:10.1109/ICBDSC.2016.7460363.

125. Jorge Granjal, Edmundo Monteiro, and Jorge Sá Silva. Security for the Internet of Things: A survey of existing protocols and open research issues, *IEEE Communication Surveys & Tutorials*, 17 (2014): 1294–1312, Accessed January 1, 2017, doi:10.1109/COMST.2015.2388550.

126. Faiza Medjek, Djamel Tandjaoui, Mohammed Riyadh Abdmeziem, and Nabil Djedjig. Analytical Evaluation of the Impacts of Sybil Attacks against RPL under Mobility, *IEEE 12th International Symposium on Programming and Systems (ISPS)* (2015): 1–9, Accessed January 1, 2017, doi:10.1109/ISPS.2015.7244960.

127. Todd Ouska. Transport-Level Security Tradeoffs Using MQTT, *IoT Design*, February 22, 2016, Accessed January 2, 2017, http://iotdesign.embedded-computing.com/guest-blogs/transport-level-security-tradeoffs-using-mqtt/#.

128. Oasis Standard. MQTT Vesion 3.1.1, October 29, 2014, Accessed January 2, 2017, http://docs.oasis-open.org/mqtt/mqtt/v3.1.1/os/mqtt-v3.1.1-os.html.
129. HIVEMQ Enterprise MQTT Broker. MQTT Security Fundamentals: TLS/SSL, Accessed January 2, 2017, http://www.hivemq.com/blog/mqtt-security-fundamentals-tls-ssl.
130. Andrew Foster. Messaging Technologies for the Industrial Internet and the Internet of Things: Whitepaper, *PrismTech*, June 2015, Accessed January 2, 2017, http://www.prismtech.com/sites/default/files/documents/Messaging-Whitepaper-040615_1.pdf.
131. Raphael Cohn. A Comparison of AMQP and MQTT, *StormMQ*, Accessed January 2, 2017, https://lists.oasis-open.org/archives/amqp/201202/msg00086/StormMQ_WhitePaper_-_A_Comparison_of_AMQP_and_MQTT.pdf.
132. Object Management Group. OMG and the IioT. *Object Management Group*, December 22, 2016, Accessed January 2, 2017, http://www.omg.org/hot-topics/iot-standards.htm.
133. Object Management Group. DDS: The Proven Data Connectivity Standard for the IoT, Accessed January 2, 2017, http://portals.omg.org/dds/omg-dds-standard/.
134. rti Whitepaper. DDS: The Right Middleware for the Industrial Internet of Things? Accessed January 2, 2017, https://info.rti.com/hubfs/whitepapers/Right_Middleware_for_IIoT.pdf.
135. Gary Steri, Gianmarco Baldini, Igor Nai Fovino, Ricardo Neisse, and Leonardo Goratti. A Novel Multi-hop Secure LTE-D2D Communication Protocol for IoT Scenarios, *IEEE 23rd International Conference on Telecommunications (ICT)* (2016): 1–6, Accessed January 6, 2017, doi:10.1109/ICT.2016.7500356.
136. Sumit Mishra. Network Security Protocol for Constrained Resource Devices in Internet of Things, *Annual IEEE India Conference (INDICON)* (2015): 1–6, Accessed January 6, 2017, doi:10.1109/INDICON.2015.7443737.
137. Savio Sciancalepore, Angelo Capossele, Giuseppe Piro, Gennaro Boggia, and Giuseppe Bianchi. Key Management Protocol with Implicit Certificates for IoT systems, *ACM Proceedings of the 2015 Workshop on IoT Challenges in Mobile and Industrial Systems* (2015): 37–42, Accessed January 6, 2017, doi:10.1145/2753476.2753477.
138. Rodrigo Roman, Cristina Alcaraz, Javier Lopez, and Nicolas Sklavos. Key management systems for sensor networks in the context of the Internet of Things, *Elsevier Journal of Computers and Electrical Engineering*, 37 (2011): 147–159, Accessed January 6, 2017, doi:10.1016/j.compeleceng.2011.01.009.
139. Muhamed Turkanovic, Boštjan Brumen, and Marko Hölbl. A novel user authentication and key agreement scheme for heterogeneous ad hoc wireless sensor networks, based on the Internet of Things notion, *Elsevier Journal of Ad Hoc Networks*, 20 (2014): 96–112, Accessed January 6, 2017, doi:org/10.1016/j.adhoc.2014.03.009.
140. Mohammad Sabzinejad Farash, Muhamed Turkanovic, Saru Kumari, and Marko Hölbl. An efficient user authentication and key agreement scheme for heterogeneous wireless sensor network tailored for the Internet of Things environment, *Elsevier Journal of Ad Hoc Networks*, 36 (2016): 152–176, Accessed January 6, 2017, doi:org/10.1016/j.adhoc.2015.05.014.
141. Ruhul Amin, SK Hafizul Islam, G. P. Biswas, Muhammad Khurram Khan, Lu Leng, and Neeraj Kumar. Design of an anonymity-preserving three-factor authenticated key exchange protocol for wireless sensor networks, *Elsevier Journal of Computers Networks*, 101 (2016): 42–62, Accessed January 6, 2017, doi:org/10.1016/j.comnet.2016.01.006.
142. Sima Arasteh, Seyed Farhad Aghili, and Hamid Mala. A New Lightweight Authentication and Key agreement Protocol for Internet of Things, *IEEE International ISC Conference on Information Security and Cryptology (ISCISC)* (2016): 52–59, Accessed January 6, 2017, doi:10.1109/ISCISC.2016.7736451.
143. Yassine Maleh, Abdellah Ezzati, and Mustapha Belaissaoui. An Enhanced DTLS Protocol for Internet of Things Applications, *IEEE International Conference on Wireless Networks and Mobile Communications (WINCOM)* (2016): 168–173, Accessed January 6, 2017, doi:10.1109/WINCOM.2016.7777209.

22 A Semidistributed Metaheuristic Algorithm for Collaborative Beamforming in the Internet of Things

Suhanya Jayaprakasam, Sharul Kamal Abdul Rahim, and Chee Yen Leow

CONTENTS

22.1 INTRODUCTION

With the arrival of Internet of Things (IoT) in the fifth generation communication systems (5G), it is conceived that all objects will be equipped with wireless communication enabled smart miniature electronic devices in the near future. An inherent feature of such mobile and small devices is the long battery life. Collaborative beamforming (CBF) is a promising solution to address the battery life limitations in a small wireless communication device by pooling the resources of several neighboring devices to establish communication with a distant target.

In CBF, a communication node that intends to transmit a message first broadcasts the message to its neighboring nodes and requests these nodes to cooperate for the transmission. All the cooperating nodes will then collaboratively transmit the message along with the source node to the intended receiver in a directive manner. Thus, CBF effectively increases the communication distance by using less power and less spatial resources. Although CBF was initially proposed specifically for wireless sensor networks (WSNs) (Barriac, Mudumbai, and Madhow, 2004), it is also researched increasingly in other wireless network applications such as cellular networks (Ekbal and Cioffi, 2005; Ng, Evans, and Hanly, 2007), cognitive radio networks (Piltan and Salari, 2012; Zhang et al. 2014), wireless personal area networks (WPANs) (Jeevan et al. 2008), cloud radio access networks (Ha and Le, 2014), wireless body area networks (WBANs) (Ding et al. 2015), and wireless machine-to-machine (M2M) networks (Zhang et al. 2013).

The major challenge in CBF is the synchronization of all collaborating nodes to form a directive beam toward the intended direction due to the distributed nature of the collaborating nodes. When locations of all the collaborating nodes and the target are not explicitly known to each node, the collaborating nodes lack information to synchronize their beamforming coefficients and direct a collaborative beam toward the target in a distributive manner.

This lack of synchronization can be solved if the target node is able to provide feedback in terms of individual channel state information (CSI) to the collaborating nodes. If each collaborating node knows the channel gain of itself as well as every other collaborating nodes to the target, it can then successfully synchronize its phase for CBF. The condition where the collaborating nodes are able to receive feedback from the receiver is called closed-loop beamforming.

However, providing an accurate and rich feedback with complete CSI for perfect beamforming is impractical and consumes a large amount of signaling overhead. Therefore, a more resource friendly, limited feedback scheme is preferred for synchronization in CBF. In a limited feedback closed-loop CBF, the feedback from the target is limited to a few bits that usually consists of coded information to assist the collaborating nodes to perform beamforming.

We propose a new hybrid algorithm called the Combined Hybrid Distributed Joint Activation (CDJA-HYB) to improve upon the legacy synchronization algorithms. The legacy random one-bit feedback (R1BF) method (Mudumbai et al. 2006) and the more recent ADJA method (Thibault et al. 2013) are chosen as the benchmark synchronization schemes. These existing closed-loop beamforming synchronization schemes are combined and further improved using a meta-heuristic inspired controlled randomization. Results show that the proposed CDJA-HYB improves the convergence and the achieved receive signal gain at the receiver in noisy, fading, and time-varying channels with lower feedback compared with the benchmark algorithms.

This chapter is organized as follows: In Section 22.1, an overview of phase synchronization in closed-loop CBF is discussed. Two existing algorithms, namely R1BF and ADJA, are discussed to provide the readers an overview of the bit feedback phase synchronization method. The proposed meta-heuristic inspired semidistributed algorithm called the CDJA-HYB for phase synchronization in closed-loop CBF is presented in Section 22.3. The communication protocol to enable the bit feedback based synchronization is first presented, and this is followed by the description of the proposed CDJA-HYB algorithm. Results of the proposed CDJA-HYB are compared with results of R1BF and ADJA in various cluster sizes as well as in noisy, time-varying, and fading channels in Section 22.4. The chapter ends with conclusions in Section 22.5.

22.2   PHASE SYNCHRONIZATION IN DISTRIBUTED AND COLLABORATIVE BEAMFORMING

Collaborative and distributed beamforming require stringent time, phase, and frequency synchronization. The complex signal at a target when N number of nodes collaboratively transmit at time t, all transmitting with carrier frequency f_c, is (Mudumbai et al., 2006)

$$r(t) = \sum_{k=1}^{N} e^{j2\pi(f_c + \Delta f_k)(t + \Delta t_k)(\Delta \psi_k)} \qquad (22.1)$$

From Equation 22.1, three distinct offsets occur due to the random and distributed nature of the collaborating nodes, which are

Phase offset, $\Delta\Psi_k$: The signal from each node reaches the target with a different phase offset. The phase offset is a combination of the channel rotation, distance variation, and the local oscillator's phase drift. The instantaneous phases of the collaborating nodes differ as the individual oscillators of each nodes drift due to Brownian motion-induced phase noise.

Frequency offset, Δf_k: Ideally, the local oscillators of all collaborating nodes must oscillate with the same carrier frequency. However, because the local oscillators each experience independent drifts, frequency offset is present at each node, where oscillator center frequencies differ slightly from each other. This difference causes a linearly increasing phase offset in time with respect to the reference center frequency.

Time offset, Δt_k: The time reference of each node has to be identical to ensure that all nodes transmit simultaneously. These offsets will cause the combined signal at the target to subsequently become misaligned, causing a loss in the beamforming gain in the target direction.

In an ideal case, there will be no time and frequency offsets, and hence, $\Delta f_k = \Delta t_k = 0$ for $k = 1, \ldots, N$. Furthermore, each node has a perfect knowledge of its phase compensation Ψ_k. The collaborating nodes can produce a coherent signal at the target direction as Equation 22.1 becomes (Mudumbai, Hespanha, Madhow, and Barriac, 2005, 2010)

$$r(t) = e^{j2\pi f_c t} \sum_{k=1}^{N} e^{j(\Delta\psi_k - \psi_k)} \tag{22.2}$$

In reality, the time, frequency, and phase offset all exist, and it is important to perform synchronization to compensate these offsets so that a coherent signal will be received by the target.

A large portion of previous literature on CBF has focused on the synchronization issues. These works focus on either the phase, frequency, or the time synchronizations and sometimes tackle more than one of these synchronizations at once.

The synchronization procedures in the literature can be broadly classified into two categories: closed-loop method and the open-loop method. In the closed-loop method, the synchronization is aided by feedback from the target node. This feedback could either be just one or a few quantized bits or in a form of a rich feedback with semi- or full CSI information. In the open-loop method, the nodes synchronize without any feedback from the target node. The synchronization can be done via intranode communication method where collaborating nodes communicate and exchange information among them to perform synchronization.

We only consider closed-loop CBF. Small mobile sensor nodes in the IoT have the capability of sending simple feedback and thus closed-loop synchronization with a bit feedback method is the focus of this article.

22.2.1 Formulation of Normalized Received Signal Strength

Phase synchronization in closed-loop CBF evaluates the received power at the receiver for synchronization. Under the assumption that the carrier signal frequency f_c is synchronized and unit transmission power is applied at all nodes, the received signal at time slot t, $y(t)$ is (Mudumbai, Barriac, and Madhow, 2007)

$$y(t) = x(t) e^{j2\pi f_c t} \sum_{k=1}^{N} h_k e^{j(\widehat{\phi_k}(t) + \psi_k(t))} + \omega(t) \tag{22.3}$$

where $x(t)$ is the common baseband message signal, $\omega(t)$ is additive noise, and h_k is the propagation channel effects at time slot t. $\widehat{\phi}_k$ is the cumulative phase response at the node n_k, whereby the errors caused by the phase rotation due to the location and the offset of the oscillator are absorbed into the single entity of $\widehat{\phi}_k$. Hence, the beamforming gain, i.e., the normalized baseband received power P_R, is (Mudumbai et al., 2007)

$$P_R = \frac{1}{N}\sum_{k=1}^{N}h_k e^{j\left(\widehat{\phi}_k(t)+\omega_k(t)\right)} + \omega(t)^2 \tag{22.4}$$

The value of this normalized received signal strength (NRSS) is maximized when $\widehat{\phi}_k(t) + \psi_k(t) = 0$, for all k. Hence, the phase rotation $\psi_k(t)$ has to be adjusted to maximize the NRSS at the intended receiver.

The existing phase synchronization methods for closed-loop CBF apply iterative procedures where all nodes apply a phase adjustment to their carrier signal according to the feedback received from the receiver. The initial phase rotation, $\widehat{\phi}_k$ is uniformly and independently distributed across the domain $[-\pi, \pi]$.

In the time-varying channel, the transmitted signals lose their alignment over the time due to time-dependent channel fluctuation; hence, the cumulative phase is modeled as (Thibault et al., 2013)

$$\widehat{\phi}_k(t) = \widehat{\phi}_k[t-1] + D_k[t] \tag{22.5}$$

where $D_k[t]$ is independently and identically distributed uniformly in the range $\left[\dfrac{-\pi}{\alpha}, \dfrac{\pi}{\alpha}\right]$, whereby the constant α determines the severity of the time fluctuation. For all cases, the iterative procedure is repeated until a stopping criterion of maximum iteration, g_{max} is achieved.

22.2.2 RANDOM ONE-BIT FEEDBACK

The R1BF algorithm proposed in (Mudumbai et al., 2005) provides a solution for each collaborating node to iteratively and independently perturb its phase rotation based on only 1 bit of feedback from the receiver at every iteration. If the received signal is improved compared with the previous iteration, the receiver sends bit "1" back to the collaborating nodes as a response, and sends bit "0" otherwise. If a feedback "1" is received, the collaborating nodes retain the phase perturbation performed during the previous iteration and update the phase rotation to

$$\psi_k(t) = \psi_k[t-1] + \Delta_k \tag{22.6}$$

Δ_k is the random phase perturbation that is uniformly distributed in the range $\left[\dfrac{-\pi}{\beta}, \dfrac{\pi}{\beta}\right]$. The variable β is a constant value that determines the convergence of the algorithm. This phase update is done on the basis that there exists constant but unknown phase offsets and channel gain between the transmitting collaborating nodes and the receiver, that is, the channel is static.

The work in Mudumbai et al. (2006) improves upon the R1BF algorithm to consider a time-varying channel. The channel fluctuation in response to time causes misalignment of the phases over time. Quality factor q is introduced to reflect the expected signal deterioration due to the channel effects such that $P_R(t) = q \cdot P_R[t-1]$.

The R1BF algorithm is a simple and a fully distributed closed-loop scheme that allows every node to progressively synchronize its initial phase for successful beamforming, which has been proven through numerous test beds (Mudumbai et al., 2006; Quitin et al. 2013; Seo, Rodwell, and Madhow, 2008). However, the R1BF has a disadvantage of slow convergence and limited upper bound (Thibault et al. 2013).

22.2.3 ADVANCED DETERMINISTIC JOINT ACTIVATION

The deterministic joint activation (DJA) algorithm is a deterministic algorithm that progressively rotates the phase of one collaborating node at every iteration until the maximum NRSS is achieved. The phase rotation is not chosen randomly, rather it systematically tries out all possible solution from a predefined set of solution where

$$S = \left\{ k \times \frac{2\pi}{K} \right\}, \, k = 0, \, \cdots, \, K-1 \tag{22.7}$$

The algorithm starts with all nodes having no phase rotation and set size $K = 2$, whereby each node has an option of rotating its phase to either 0 or π from the available set $S = [0, \pi]$. The value of K is updated to $K = 2K$ every time $(K-1)$ N generation, and hence, the nodes progressively refine their phase rotation. For example, the solution set is refined to $S = \left[0, \frac{\pi}{2}, \pi, \frac{3\pi}{2} \right]$ when $K = 4$.

This deterministic measure ensures that the algorithm could achieve an NRSS very close to the optimal solution in an ideal channel.

As K increases, the time taken to find a solution that improves upon the previous NRSS increases exponentially. As such, the DJA algorithm is switched to the R1BF algorithm after the phase selection with $K = 4$ is completed. This joint DJA and R1BF algorithm is called the ADJA algorithm.

The ADJA algorithm is an elegant deterministic method that seeks all possible solutions from a progressively increasing set of phase rotations to converge the main beam close to the theoretical realization of the distributed beamforming. However, as the authors of the ADJA have noted, the algorithm is not fully distributed as a certain amount of network coordination is needed to index and initiate transmissions (Thibault et al., 2013). Furthermore, convergence to maximum NRSS cannot be guaranteed with the use of the ADJA algorithm due to the discretization of the phase.

22.3 A SEMIDISTRIBUTED PHASE SYNCHRONIZATION IN DISTRIBUTED AND COLLABORATIVE BEAMFORMING: COMBINED HYBRID DISTRIBUTED JOINT ACTIVATION

The drawbacks of both the R1BF and ADJA algorithms motivate the need for a more robust yet fast synchronization algorithm. Hence, the concept of population-based metaheuristics is incorporated into the proposed synchronization method.

In this section, a simple time division–based communication protocol is proposed to allow distributed synchronization in closed-loop CBF. This is followed by a detailed explanation on the proposed CDJA-HYB algorithm.

22.3.1 TIME DIVISION DUPLEX COMMUNICATION PROTOCOL FOR THE PHASE SYNCHRONIZATION

In the optimization of closed-loop beamforming suggested in this article, all collaborating nodes have to work independently to identify the best phase offset that achieves the highest gain at the receivers.

In the closed-loop synchronization algorithms suggested previously, feedback from the receiver is sent after every pilot transmission. In our population-based method, feedback is only sent after a block of pilots are transmitted to the receiver. The nodes decide on their next block of phase choices

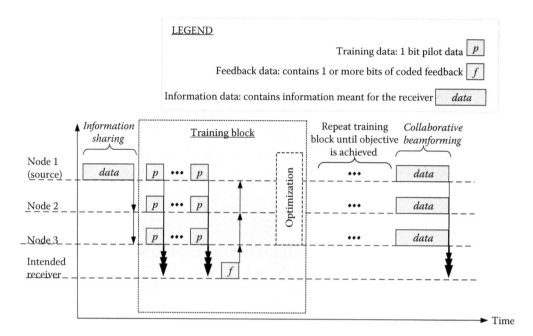

FIGURE 22.1 Timing diagram of the proposed communication protocol for closed-loop optimization with three collaborating nodes.

according to the feedback received from the receiver as well as the information on the phase offsets used in the current block. Therefore, each collaborating node makes a better calculated decision at the next block due the wealth of information available at the node. Figure 22.1 shows the communication protocol for the proposed algorithm.

The overall flow of the communication protocol is depicted in Figure 22.1.

22.3.1.1   Stage 1: Information Sharing

A node in the cluster that wishes to send data to the receiver broadcasts the data to the neighboring nodes within its range. All the other nodes receive and store these data in their cache. No other information sharing occurs until these data are successfully transmitted via CBF or a time-out occurs.

22.3.1.2   Stage 2: Pilot Broadcast

For the next B time slots, each node transmits an identical bit "1" pilot signal with an initial phase perturbation for each slot. Once B slots are over, all nodes listen for a fixed time slot for feedback from the intended receiver.

22.3.1.3   Stage 3: Feedback

The receiver sends b bits of feedback every $B = 2^b - 1$ time slots using the contention method. This feedback is in the form of coded data consisting of b bits. The feedback indicates which of the B pilot transmissions gained the best improvement compared with the previous block.

22.3.1.4   Stage 4: Update and Optimize

The feedback is processed by the individual nodes and the phase offset for the next b transmissions is determined according to the feedback. Stages 2, 3, and 4 are repeated until the maximum number of time slots is reached.

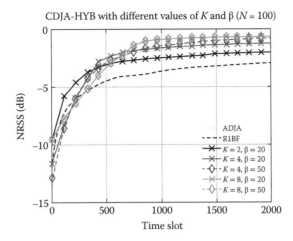

FIGURE 22.2 Combined hybrid distributed joint activation when $b = 3$ in an ideal channel condition $(q = 1)$.

22.3.1.5 Stage 5: CBF
Once the maximum time slot is achieved, each node multiplies the signal with its best beamforming weight and transmits the common data simultaneously. The individual signals add coherently at the receiver.

22.3.2 ALGORITHM DESCRIPTION

The proposed CDJA-HYB synchronization algorithm adopts both ADJA and R1BF to perform the phase synchronization. The feedback from the receiver is reduced by adapting a population-based metaheuristic method where the receivers now send b bits of feedback after every $B = 2^b - 1$ time slots instead of sending feedback after every pilot transmission.

In the proposed CDJA-HYB, each node n_k determines the phase rotation that is applied onto the carrier signal for the next B time slots. All nodes transmit the 1 bit pilot for the next B slots. The receiver compares the NRSS of the B pilot signals and saves the best NRSS as the *global best* solution and sends b bits of feedback to the collaborating nodes. The feedback indicates in binary, on which of the B slots obtained the best NRSS. If the NRSS of all B transmissions are lower than the previous *global best*, then feedback of "0" is sent instead.

The CDJA-HYB algorithm is divided into two stages. For the first stage, the ADJA algorithm is combined with the population-based metaheuristic concepts. Instead of systematically choosing one solution for a single node n_k at every time slot, a few nodes are chosen concurrently for phase rotation for the next K time slots. The number of nodes chosen is $k' = \left(\dfrac{B}{K} \right)$. Although the choice of phase rotation is selected from the fixed set of solutions S_k as with the ADJA algorithm, the CDJA-HYB algorithm randomly chooses a solution from the set, unlike the deterministic approach of the former algorithm. This approach is predicted to have a steeper increase in the NRSS compared with the ADJA due to its random and simultaneous phase rotation nature. The batch transmission is repeated until the last batch of collaborating nodes perform the phase rotation. Once this process is completed, the algorithm switches to the second stage whereby it applies the standard R1BF algorithm. It is predicted that the R1BF would give a continuous improvement at a lower rate, given that the limiting upper boundary of the R1BF is bypassed by applying the proposed population-based phase selection method in the first stage.

From Figure 22.2, it can be seen that using a lower value of K guarantees a faster increase toward better NRSS in the beginning of the synchronization process. However, due to limitations of the available set of phase rotations, the improvement comes to a stall by $N(K - 1)$ time slots, similar to the conventional ADJA in Thibault et al. (2013). The R1BF algorithm then replaces the ADJA in the proposed algorithm, providing continual increment in the NRSS, albeit at a lower rate. The value β makes very little difference in the increment rate of the NRSS. For the cases of $K = 4$ and $K = 8$, almost 93% of the maximum NRSS can be achieved by the 2000th time slot. All the combinations shown in the figure outperform the legacy R1BF algorithm. The existing ADJA algorithm, however, tends to have faster initial convergence and also reaches the best NRSS of about 93% of the maximum value for this case of ideal channel. However, the proposed combined phase rotation and feedback method is able to finally achieve similar NRSS value toward the end. For the 2000 time slots considered in Figure 22.2, 1000 feedback bits were allocated for the feedback mechanism in R1BF and ADJA, whereas the proposed algorithm only utilized 857 feedback bits for the feedback. Furthermore, it will be shown in the Section 22.5 that the proposed algorithm tends to perform better in noisy, time-varying, and fading channel scenarios.

22.4 RESULTS AND DISCUSSIONS

This section analyses the results obtained via the existing R1BF and ADJA algorithms, as well as the proposed CDJA-HYB algorithm. First, the effect of the cluster size on the beamforming ability is analyzed. An analysis on how the cluster size and distribution affects the ability of the algorithm to converge to the maximum NRSS is presented.

All algorithms are run for 2000 time slots. This will in fact not provide the optimal solution possible for each of the algorithms in all cases. If each algorithm is allowed to run for an infinite amount of time, the optimal solution will be obtained. However, because a maximum time window within which phase coherence can be maintained exists in real-time communications, the algorithms are limited to 2000 time slots (Thibault et al., 2013). Monte Carlo simulations of 10^5 trials each are run to obtain the average NRSS.

Practical channel conditions such as a time-varying channel, noisy channel, as well as a fading channel are analyzed in the upcoming sections. The quality factor q is set to 0.9, with reference to previous works (Mudumbai et al. 2006; Thibault, Faridi, Corazza, Coralli, and Lozano, 2013).

22.4.1 Effects of Cluster Dimensions

The average NRSS obtained via the proposed CDJA-HYB phase synchronization algorithms for various cluster dimensions is simulated via 10^5 trials of Monte Carlo simulations.

From Figure 22.3, it can be seen that the NRSS of all algorithms reduce almost linearly as the number of collaborating nodes N increases, whereas the size of the cluster disk and number of time slots are fixed. This does not indicate that the upper bound of the algorithm reduces as the number of nodes increases. Rather, the fixed number of time slots limits the final NRSS of the algorithms. Because the possible combinations of the phase rotations increase with respect to N, the algorithms require more time to achieve better NRSS. The proposed semidistributed CDJA-HYB shows almost a similar pattern to the other two algorithms with NRSS only 0.5 dB lower than ADJA. The R1BF algorithm, however, performs very poorly in comparison with the other two algorithms.

Results in Figure 22.4a compare the average NRSS when the size of the cluster is varied, although maximum time slot and number of nodes are fixed to 2000 and 16, respectively. Figure 22.4b on the other hand presents the average NRSS for various cluster disk sizes for a higher number of collaborating nodes, $N = 128$. From both figures, it can be seen that the average NRSS achieved within the 2000 time slots reduces drastically as the disk size increases for the R1BF algorithm. As the disk size increases, the difference in the initial phase rotation between nodes becomes more defined. Because the R1BF algorithm limits its phase rotation per iteration with the constant β, a longer time is needed for the algorithm to converge, hence the lower value of NRSS as the disk size R increases. The same problem is seen for the CDJA-HYB, but to a lesser extent, because the CDJA-HYB algorithm switches to R1BF once the population-based deterministic phase selection stage is completed.

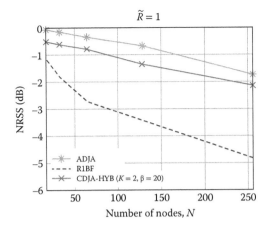

FIGURE 22.3   Comparison of the average normalized received signal strength achieved after 2000 time slots for various numbers of nodes when the disk size is fixed at $R = 1$.

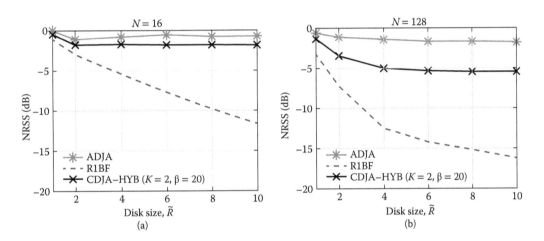

FIGURE 22.4   Comparison of the average normalized received signal strength achieved after 2000 time slots for various disk sizes when the number of nodes is fixed at (a) $N = 16$ and (b) $N = 128$.

The ADJA algorithm also records reduced average NRSS value for the case of $N = 128$, although the effect of disk size on ADJA is not prominent when $N = 16$. Because nodes placed on a larger disk need a more refined solution, and ADJA deterministically attempts all possible solutions in its sample set before refining the sample set, the time taken to achieve this solution becomes longer as the disk size increases. This effect is not seen for the case of $N = 16$ because the lower number of nodes reduces the time slot needed to achieve better NRSS.

22.4.2   EFFECTS OF TIME-VARYING CHANNEL

This section analyzes the effect of a time-varying channel on the R1BF, ADJA, and CDJA-HYB algorithms. In a time-varying channel, the channel fluctuation is time dependent, which affects the alignment of the received signal over time. Therefore, it is important that the algorithms are robust against a time-varying channel to ensure that the NRSS growth achieved over time can be maintained for coherent signal reception. The channel phase response of a collaborating node n_k affected by a time-varying channel is modeled as detailed in Equation 22.5.

Figure 22.5 shows that ADJA and CDJA-HYB can obtain about −3.5 dB by the 2000th time slot when the channel fluctuation constant α is 20. The results of ADJA and R1BF shown here are similar to that obtained in Thibault et al. (2013). The CDJA-HYB is able to obtain similar results to ADJA with lower feedback from the receiver.

Figure 22.6 provides a graphical illustration on how the channel fluctuation constant α affects the NRSS when the time slot is limited to 2000. A lower value of α indicates fast fluctuation, which may also indicate high mobility of a node. The CDJA-HYB achieves similar performance to the

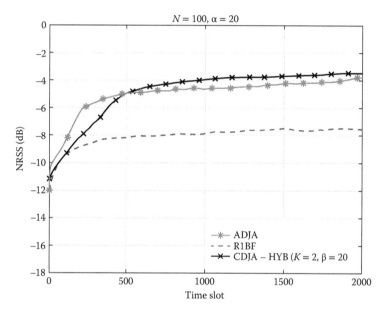

FIGURE 22.5 Comparison of the average normalized received signal strength when the channel drift parameter, α = 20.

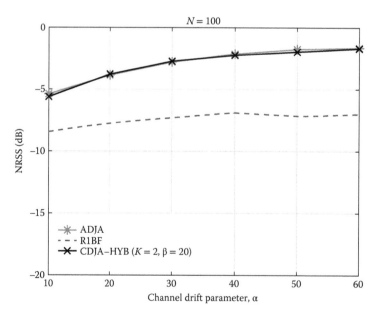

FIGURE 22.6 Comparison of the average normalized received signal strength achieved after 2000 time slots for various channel drift parameters, α.

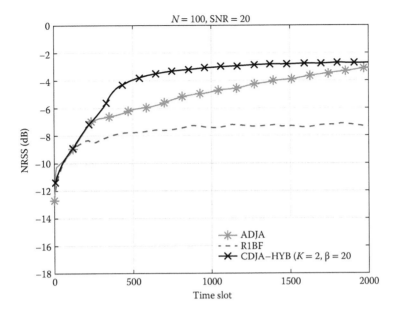

FIGURE 22.7 Comparison of the average normalized received signal strength when the signal-to-noise (SNR) ratio = 20.

legacy ADJA synchronization method. However, it has to be noted that lower feedback amounts are required in the proposed method.

22.4.3 EFFECTS OF NOISY CHANNEL

This section analyzes the effect of AWGN on the R1BF, ADJA, and CDJA-HYB algorithms. The q factor is maintained at 0.9 to represent a practical case.

Figure 22.7 shows an example where the noise at the receiver is 20dB. For the R1BF algorithm, the rate of NRSS increment remains slow and seems to reach a threshold by the 2000th time slot. ADJA and CDJA-HYB achieve at least –3dB by the 2000th time slot. However, the ADJA's NRSS increment is slowed down when compared with its performance in the ideal conditions. Comparatively, the proposed CDJA-HYB has a steeper increase in the 1000th time slot as a result of the randomization in the CDJA-HYB algorithm.

Figure 22.8 shows the relationship between receiver noise, SNR, and the NRSS achieved by the algorithms when the time slot is limited to 2000. It can be seen that under the presence of noise, the proposed CDJA-HYB always performs better than the legacy ADJA.

22.4.4 EFFECTS OF FADING

This section analyzes the effect of fading on the R1BF, ADJA, and CDJA-HYB algorithms. A near-ground WSN application is considered. Therefore, a time-invariant channel with predominantly large-scale fading is assumed (Ahmed and Vorobyov, 2009). Hence, a log-normal distributed random variable is assumed for the channel, with the fading $a \sim \exp\left[N\left(0, \sigma^2\right)\right]$, where $\sigma^2 = 0.2$.

Results in Figure 22.9 show that ADJA and CDJA-HYB settle at the NRSS value of -4.3dB. The CDJA-HYB, however, has a faster rate of increase compared with ADJA due to the more random nature of the CDJA-HYB.

The same behavior is recorded for all three algorithms for the case of the log-normal channel and SNR = 20 with the time-varying parameter $\alpha = 20$ at the receiver, as shown in Figure 22.10. The

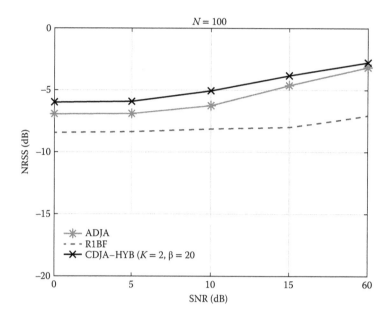

FIGURE 22.8 Comparison of the average normalized received signal strength achieved after 2000 time slots for various signal-to-noise ratio (SNR) values.

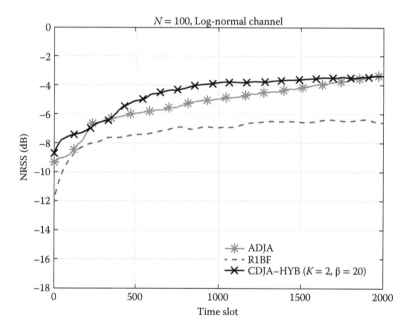

FIGURE 22.9 Comparison of the average normalized received signal strength under log-normal channel conditions.

controlled random nature of CDJA-HYB allows the algorithm to identify the optimal phase rotations and increase the NRSS faster and to a higher value compared with the deterministic ADJA and R1BF.

Next, a moderate time-varying channel is introduced to the channel with $\alpha = 20$. This could indicate outdoor nodes with only a shadowing effect with moderate mobility. Results in Figure 22.11 show that the addition of a time-varying channel on top of the fading and SNR does not change the convergence nature of the ADJA and CDJA-HY. However, the final achieved NRSS is lower due to

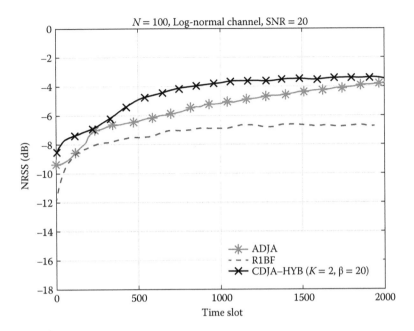

FIGURE 22.10 Comparison of the average normalized received signal strength under log-normal channel conditions when signal-to-noise ratio (SNR) = 20.

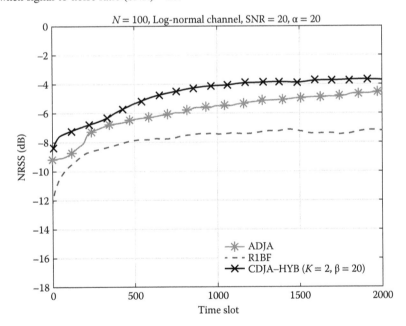

FIGURE 22.11 Comparison of the average normalized received signal strength under log-normal channel conditions when $\alpha = 20$, signal-to-noise ratio (SNR) = 20.

the effect of time variation. The CDJA-HYB continues to perform better than the ADJA synchronization method.

22.5 CONCLUSIONS

In this chapter, a new algorithm called CDJA-HYB is introduced to enable phase synchronization in closed-loop CBF with limited feedback. The CDJA-HYB is a semidistributed phase synchronization method based on the existing semidistributed ADJA and R1BF algorithms. The proposed algorithm adopts the concepts of population-based metaheuristic algorithms to achieve synchronization faster with less feedback. The results show that the proposed algorithm achieves better convergence rate and good maximum achievable NRSS for time-varying, noisy, and fading channels when compared with the legacy ADJA algorithm. Furthermore, results show that the proposed idea works best in less dense and larger clusters, hence it has the potential to be applied in applications such as M2M communications.

It has to be noted that CDJA-HYB still requires some minimal coordination from either a cluster head or the intended receiver (target) to edify the nodes on the number of collaborating nodes. For this reason, the algorithm is still considered a semidistributed synchronization algorithm. There is still scope for improvement in the proposed algorithm for it to be made into a fully distributed phase synchronization algorithm.

The examples provided in this chapter only consider one intended receiver. The work can be further extended to incorporate phase synchronization when more than one unintended receiver needs to be nullified while beamforming toward the intended receiver. Further, cases where there exists more than one intended receiver are also an interesting future direction.

REFERENCES

Ahmed, M. and S. Vorobyov (2009, June). Node selection for sidelobe control in collaborative beamforming for wireless sensor networks. In 10th IEEE Workshop on Signal Processing Advances in Wireless Communications, Perugia, Italy, pp. 519–523.

Barriac, G., R. Mudumbai, and U. Madhow (2004). Distributed beamforming for information transfer in sensor networks. In Proceedings of the 3rd International Symposium on Information Processing in Sensor Networks (IPSN '04), New York, NY, pp. 81–88. ACM.

Ekbal, A. and J. Cioffi (2005, May). Distributed transmit beamforming in cellular networks - a convex optimization perspective. In IEEE International Conference on Communications (ICC), Seoul, South Korea, pp. 2690–2694.

Ha, V. N. and L. B. Le (2014). Joint coordinated beamforming and admission control for front haul constrained cloud-RANs. In IEEE Global Communications Conference (GLOBECOM), Austin, TX, pp. 4054–4059.

Jeevan, P., S. Pollin, A. Bahai, and P. Varaiya (2008, May). Pairwise algorithm for distributed transmit beamforming. In IEEE International Conference on Communications, Beijing, China, pp. 4245–4249.

Mudumbai, R., B. Wild, U. Madhow, and K. Ramch (2006). Distributed beamforming using 1 bit feedback: From concept to realization. In Proceeding of Allerton Conference on Communication, Control, and Computing, Monticello, IL, pp. 1020–1027.

Mudumbai, R., G. Barriac, and U. Madhow (2007, May). On the feasibility of distributed beamforming in wireless networks. IEEE Transactions on Wireless Communications 6(5), 1754–1763.

Mudumbai, R., J. Hespanha, U. Madhow, and G. Barriac (2005, Sept). Scalable feedback control for distributed beamforming in sensor networks. In Proceedings of International Symposium on Information Theory (ISIT), Adelaide, SA, pp. 137–141.

Mudumbai, R., J. Hespanha, U. Madhow, and G. Barriac (2010, Jan). Distributed transmit beamforming using feedback control. IEEE Transactions on Information Theory 56(1), 411–426.

Ng, B., J. Evans, and S. Hanly (2007, June). Distributed downlink beamforming in cellular networks. In IEEE International Symposium on Information Theory, Nice, France, pp. 6–10.

Piltan, A. and S. Salari (2012, June). Distributed beamforming in cognitive relay networks with partial channel state information. IET Communications 6(9), 1011–1018.

Quitin, F., M. Rahman, R. Mudumbai, and U. Madhow (2013, March). A scalable architecture for distributed transmit beamforming with commodity radios: Design and proof of concept. IEEE Transactions on Wireless Communications 12(3), 1418–1428.

Seo, M., M. Rodwell, and U. Madhow (2008, June). A feedback-based distributed phased array technique and its application to 60-GHz wireless sensor network. In IEEE MTT-S International Microwave Symposium Digest, Atlanta, GA, pp. 683–686.

Thibault, I., A. Faridi, G. Corazza, A. Coralli, and A. Lozano (2013, April). Design and analysis of deterministic distributed beamforming algorithms in the presence of noise. IEEE Transactions on Communications 61(4), 1595–1607.

Zhang, C., L. Guo, R. Hu, and J. Lin (2014, April). Opportunistic distributed beamforming in cognitive radio networks with limited feedback. In IEEE Wireless Communications and Networking Conference (WCNC), Istanbul, Turkey, pp. 893–897.

Zhang, X., D. Wang, L. Bai, and C. Chen (2013). Collaborative relay beamforming based on minimum power for M2M devices in multicell systems. *International Journal of Distributed Sensor Networks 2013*, 1–9.

Index